Proceedings

Today, a steadily growing store of information is called for in order to understand the increasingly complex technologies used in modern automobiles. Functions, modes of operation, components and systems are rapidly evolving, while at the same time the latest expertise is disseminated directly from conferences, congresses and symposia to the professional world in ever-faster cycles. This series of proceedings offers rapid access to this information, gathering the specific knowledge needed to keep up with cutting-edge advances in automotive technologies, employing the same systematic approach used at conferences and congresses and presenting it in print (available at Springer.com) and electronic (at SpringerLink and Springer Professional) formats.

The series addresses the needs of automotive engineers, motor design engineers and students looking for the latest expertise in connection with key questions in their field, while professors and instructors working in the areas of automotive and motor design engineering will also find summaries of industry events they weren't able to attend. The proceedings also offer valuable answers to the topical questions that concern assessors, researchers and developmental engineers in the automotive and supplier industry, as well as service providers.

Peter E. Pfeffer
Editor

8th International Munich Chassis Symposium 2017

chassis.tech plus

Springer Vieweg

Editor

Prof. Dr. Peter E. Pfeffer
Munich University of Applied Sciences
Munich, Germany

ISSN 2198-7432 ISSN 2198-7440 (electronic)
Proceedings
ISBN 978-3-658-18458-2 ISBN 978-3-658-18459-9 (eBook)
DOI 10.1007/978-3-658-18459-9
Springer Heidelberg Dordrecht London New York

Springer Vieweg

Printed on acid-free paper

This Springer Vieweg imprint is published by Springer Nature
The registered company is Springer Fachmedien Wiesbaden GmbH
The registered company address is: Abraham-Lincoln-Str. 46, 65189 Wiesbaden, Germany

WELCOME

The development of highly and fully automatic driving and the increasing electrification of the powertrain now face chassis development with new challenges too. Innovative chassis systems have to provide solutions for automated driving. The efficient chassis of the future also has to keep an eye on CO_2 targets, comfort and customer focus at all times. A modern chassis has to provide for this in the form of innovations while taking the physical and mechanical interdependencies into account. Confronting these new developments is a challenge for simulation and testing.

To allow an exchange of experience and a constructive discussion of current topics, the 8th International Munich Chassis Symposium chassis.tech plus to be held on 20 and 21 June 2017 will bring together numerous experts from all over the world to discuss chassis, steering, brakes and tires / wheels.

At the symposium, well-known keynote speakers will offer a comprehensive overview of new approaches for solving these challenges. Speakers from industry and research will deal with current developments in the four parallel strands for chassis, steering, brakes and tires / wheels. The lecture program will be rounded off by interdisciplinary sections at the beginning and end of the event.

We are happy to welcome you to the hotel 'Bayrischer Hof' in the heart of Munich and hope you will enjoy this thought-provoking event.

Prof. Dr. Peter E. Pfeffer
Munich University of Applied Sciences
Scientific Director of the Symposium

INDEX

CHASSIS.TECH PLUS SECTION

KEYNOTE LECTURES I

KEYNOTE LECTURES II

KEYNOTE LECTURES III

PARALLEL STRAND I

NEW CHASSIS

CHASSIS COMPONENTS AND HISTORICAL ASPECTS

PARALLEL STRAND II

VIRTUAL DEVELOPMENT METHODS

OBJECTIVE ASSESSMENTS

CHASSIS.TECH SECTION

DEVELOPMENT METHODS

DRIVING COMFORT

CHASSIS CONTROL SYSTEMS

STEERING.TECH SECTION

STEERING SYSTEMS AND FUNCTIONS

VIRTUAL DEVELOPMENT METHODS

STEERING FEEL

BRAKE.TECH SECTION

HIGH-PERFORMANCE BRAKES AND TEST METHODS

BRAKE WEAR PARTICLE EMISSIONS AND MATERIALS

BRAKE SYSTEMS OF THE FUTURE

TIRE.WHEEL.TECH SECTION

DEVELOPMENTAL AIMS AND OBJECTIFICATION

TIRE TESTING AND SIMULATION

WHEEL TECHNOLOGIES AND TRENDS

SPEAKERS

Dr. Jochen Asbeck
Mubea Fahrwerksfedern GmbH

Urs Bauer
Robert Bosch GmbH

Michael Baumann
Daimler AG

Halil Beglerovic
AVL List GmbH, Austria

Dr. Wolfgang Bernhart
Roland Berger GmbH

Dr. Christoph Böhm
Daimler AG

Emre Boyraz
Dr. Ing. h.c. F. Porsche AG

Stephan Brückner
Hankook Tire Europe GmbH

Alessandro Contini
Hyundai Motor Europe
Technical Center GmbH

Christian Cramer
Continental Reifen Deutschland GmbH

Ronnie Dessort
TESIS DYNAware GmbH

Eric Eckermann
AutoHistorica

Martin Elbs
IPG Automotive GmbH

Rafael Frankenstein
Central Agency for PTI /
FSD Fahrzeugsystemdaten GmbH

Prof. Dr. Stefan Gies
Volkswagen AG

Alberto Girelli Consolaro
Ford Research
and Innovation Center Aachen

Adam Goellner
Honda R&D Americas, Inc., USA

Dr. Sebastian Gramstat
AUDI AG

Theodoros Grigoratos
Joint Research Centre (JRC),
European Commission (EC), Italy

Dr. Stefan Grüner
Robert Bosch Automotive Steering GmbH

Knut Heidsieck
ZF Friedrichshafen AG

Dr. Veit Held
Adam Opel GmbH

Axel Honisch
Hyundai Motor Europe
Technical Center GmbH

Carsten Jablonowski
AUDI AG

Jorge Alberto Jáuregui
Bosch Engineering GmbH

Joshua C. Johnson
Honda R&D Americas, Inc., USA

Dr. Young Deuk Kim
Hyundai Motor Company,
South Korea

Dr. Matthijs Klomp
Volvo Car Group, Sweden

Kentaro Komori
Honda R&D Co., Ltd., Japan

Rob Kraaijeveld
FEV Europe GmbH

Dr. Detlef Kube
UNIWHEELS Automotive
(Germany) GmbH

Dr. Armin Kunz
Robert Bosch GmbH

Dr. Sebastian Labitzke
ITK Engineering GmbH

Bastian Leistner
Faculty of
Mechanical Engineering,
Chemnitz University
of Technology

Dr. Jinhuai Lin
MdynamiX AG

Paul Linhoff
Continental Teves
AG & Co. oHG

Christof Lischka
BMW Group

Christian Ludwig
Hyundai Motor Europe
Technical Center GmbH

Carlo Lugaro
TASS International,
The Netherlands

Prof. Dr. Giampiero Mastinu
Department
of Mechanical Engineering,
Politecnico di Milano, Italy

Paul Milbaier
Robert Bosch
Automotive Steering GmbH

Dr. Saskia Monsma
HAN University
of Applied Sciences,
The Netherlands

Anand Mule
Tata Technologies Limited, India

Minh-Tri Nguyen
Institute of
Internal Combustion Engines
and Automotive Engineering (IVK),
University of Stuttgart

Dr. Anton Obermüller
AUDI AG

Septimiu Popa
Institute for
Manufacturing Technologies
of Ceramic Components
and Composites (IFKB),
University of Stuttgart

Maximilian Georg Reisner
Institute for
Automotive Engineering
Dresden (IAD), TU Dresden

Pavel Sarkisov
Institute for
Automotive Engineering
Dresden (IAD), TU Dresden

Christian Schubert
Institute of
Automotive Engineering (IfF),
TU Braunschweig

Christian Strümpler
TAKATA AG

Felix Tigges
Institute of
Automotive Engineering (IfF),
TU Braunschweig

Christian Wagner
Institute of
Automotive Engineering (IfF),
TU Braunschweig

Daniel Wagner
Institute of
Automotive Technology (FTM),
TU Munich

Dr. Jens Werner
ThyssenKrupp
Carbon Components GmbH

Prof. Dr. Burkhard Wies
Continental Reifen
Deutschland GmbH

Prof. Dr. Gerd Wittler
Faculty of Mechatronics
and Electrical Engineering,
University of Applied Sciences
Esslingen

Tim Wright
Danisi Engineering S.r.l., Italy

Tohru Yoshioka
Mazda Motor Corporation, Japan

Amir Zare
BMW Group

CHASSIS.TECH PLUS SECTION

KEYNOTE LECTURES I

The chassis as game changer for automated and electrified driving

Prof. Dr. Stefan Gies, Dr. L. Liesner, Volkswagen AG

This manuscript is not available according to publishing restriction. Thank you for your understanding.

P.E. Pfeffer (Ed.), *8th International Munich Chassis Symposium 2017*, Proceedings, DOI 10.1007/978-3-658-18459-9_1

The chassis as game changer for automated and electrified driving

Prof. Dr. Stefan Gies, Dr. L. Liesner, Volkswagen AG

This manuscript is not available according to publishing restriction. Thank you for your understanding.

P.E. Pfeffer (Ed.), *8th International Munich Chassis Symposium 2017*, Proceedings, DOI 10.1007/978-3-658-18459-9_1

The chassis of the all-new AUDI A8

Carsten Jablonowski, Dr. C. Schimmel, V. Underberg, AUDI AG

P.E. Pfeffer (Ed.), *8th International Munich Chassis Symposium 2017*,
Proceedings, DOI 10.1007/978-3-658-18459-9_2

The chassis of the all new Audi A8

chassis.tech plus 20.06.2017

C. Jablonowski, Dr. C. Schimmel, V. Underberg

2 The chassis of the all new Audi A8, Jablonowski, Dr. Schimmel, Underberg, AUDI AG

The chassis of the all new AUDI A8

Contents

Introduction

- Target Positioning
- Model Range, Technical Specification

Technical Target Values of Chassis Characteristics

- Ride Comfort
- Driving Dynamics

Main Focus of Chassis Development

Axle Concepts

Central Chassis Control

Damper Technology

Dynamic All Wheel Steering DAS

Electrical Active Body Control eABC

Integration of Active Chassis Control Systems

Chassis Characteristics of the 5th Generation AUDI A8

3 The chassis of the all new Audi A8, Jablonowski, Dr. Schimmel, Underberg, AUDI AG

Introduction
Target Positioning

The all new Audi A8 is designed to be...

› the spearhead of the product portfolio, strongly shaping the brand's image
› A synonym for prestige in the C/D-Segment

Driving Experience:

› „light und dynamic appearance in chassis characteristics"
› „symbiosis of comfort und athleticism"

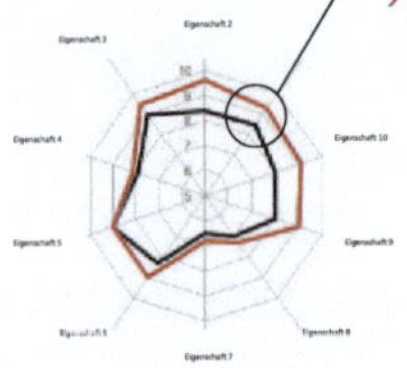

Audi A8 - Chassis Characteristics
— A8 (Gen. 4) — A8 (Gen. 5)

4 The chassis of the all new Audi A8, Jablonowski, Dr. Schimmel, Underberg, AUDI AG

Introduction
Model Range, Technical Specification

Model Range, Vehicle Date (Launch)

› quattro all wheel drive with self locking differential
› Air suspension with damper control as standard
› Different chassis characteristics for the short and long wheel base
› Two Engines: 6 Cyl. TDI (210kW), 6 Cyl. TFSI (250kW), more variants to come
› All variants come as a 48V Mild Hybrid

5 The chassis of the all new Audi A8, Jablonowski, Dr. Schimmel, Underberg, AUDI AG

The chassis of the all new AUDI A8
Contents

Introduction
- Target Positioning
- Model Range, Technical Specification

Technical Target Values of Chassis Characteristics
- Ride Comfort
- Driving Dynamics

Main Focus of Chassis Development

Axle Concepts

Central Chassis Control

Damper Technology | Dynamic All Wheel Steering DAS | Electrical Active Body Control eABC

Integration of Active Chassis Control Systems

Chassis Characteristics of the 5th Generation AUDI A8

6 The chassis of the all new Audi A8, Jablonowski, Dr. Schimmel, Underberg, AUDI AG

Technical Target Values for Chassis Characteristics
Ride Comfort/ Driving Dynamics

The Audi A8 is designed to offer „Best-in-Class“ ride comfort. That means:

› **Highest comfort requirements** in all markets, particularly to meet the demand in China and the USA

› **Simultaneously**, the extensive use of active chassis systems ensures **driving** dynamics at the same level as the key competitors

7 The chassis of the all new Audi A8, Jablonowski, Dr. Schimmel, Underberg, AUDI AG

Technical Target Values for Chassis Characteristics
Ride Comfort

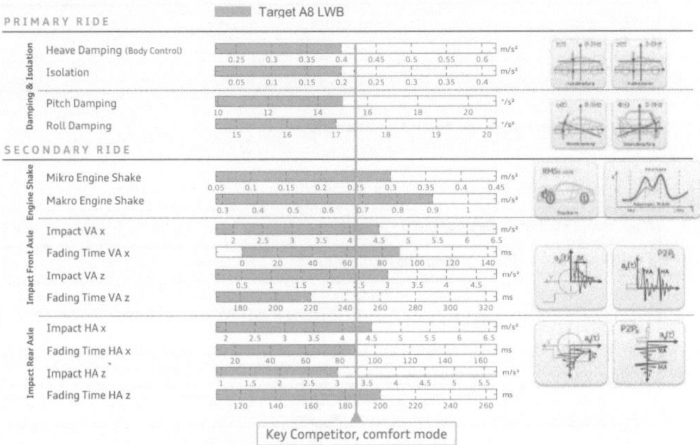

8 The chassis of the all new Audi A8, Jablonowski, Dr. Schimmel, Underberg, AUDI AG

Technical Target Values for Chassis Characteristics
Ride Comfort

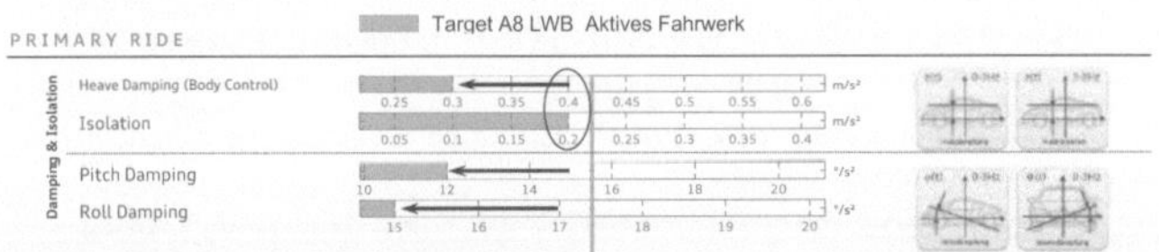

Body Movement

- **Verticale body control to 3 Hz** and **Isolation between 3-8 Hz** are essential for the overall impression of the ride comfort
- **Additional target value for the improvement of the body control with active chassis**

Key Competitor, comfort mode

9 The chassis of the all new Audi A8, Jablonowski, Dr. Schimmel, Underberg, AUDI AG

Technical Target Values for Chassis Characteristics
Driving Dynamics

Target of active chassis systems is an increase in agility and simultaneously a considerable gain of driving stability

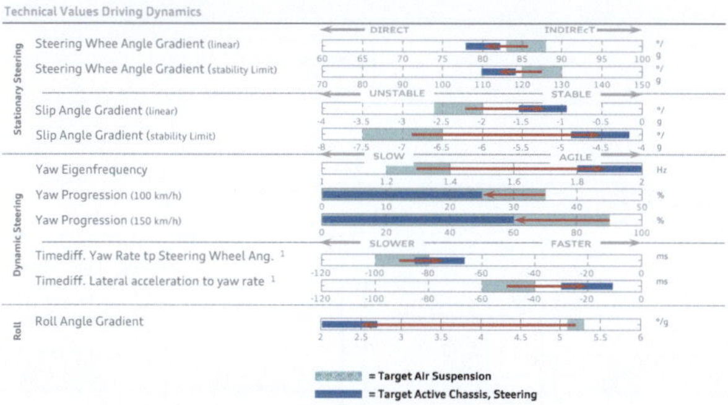

[1]: Ermittelt bei 1,5Hz Lenkfrequenz (entspricht Spurwechsel-Test bei 100km/h)

10 The chassis of the all new Audi A8, Jablonowski, Dr. Schimmel, Underberg, AUDI AG

The chassis of the all new AUDI A8
Contents

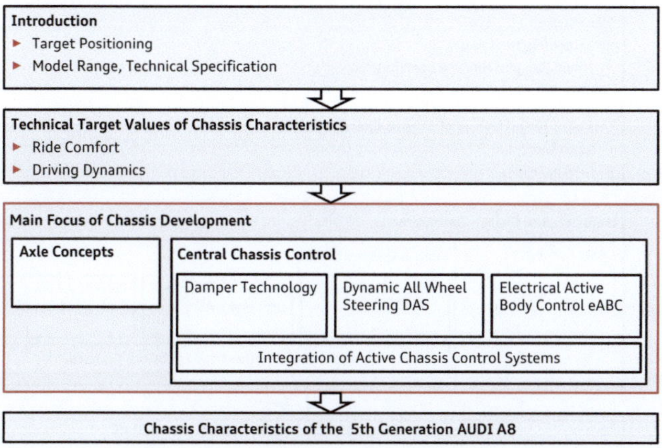

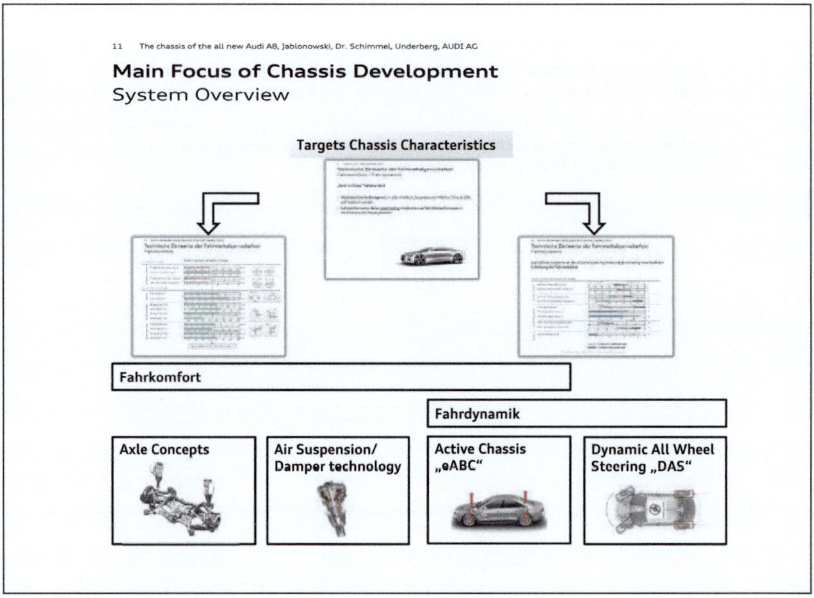
11 The chassis of the all new Audi A8, Jablonowski, Dr. Schimmel, Underberg, AUDI AG
Main Focus of Chassis Development
System Overview
Targets Chassis Characteristics
Fahrkomfort
Fahrdynamik
Axle Concepts
Air Suspension/ Damper technology
Active Chassis „eABC“
Dynamic All Wheel Steering „DAS“

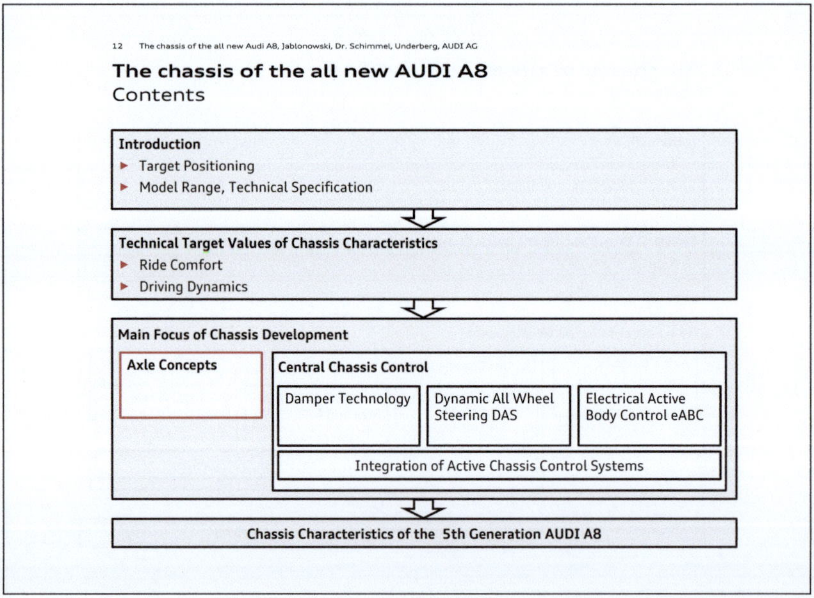
12 The chassis of the all new Audi A8, Jablonowski, Dr. Schimmel, Underberg, AUDI AG
The chassis of the all new AUDI A8
Contents
Introduction
► Target Positioning
► Model Range, Technical Specification
Technical Target Values of Chassis Characteristics
► Ride Comfort
► Driving Dynamics
Main Focus of Chassis Development
Axle Concepts
Central Chassis Control
Damper Technology
Dynamic All Wheel Steering DAS
Electrical Active Body Control eABC
Integration of Active Chassis Control Systems
Chassis Characteristics of the 5th Generation AUDI A8

13 The chassis of the all new Audi A8, Jablonowski, Dr. Schimmel, Underberg, AUDI AG

Main Focus of Chassis Development
Axle Concept

- Multi-link-axeles from the 2nd generation MLB
- Air suspension with damper control as standard equipment, optional active chassis **eABC** (**e**lectrical **A**ctive **B**ody **C**ontrol)
- Electromechanical steering as standard equipment, optional **D**ynamic **A**ll-wheel **S**teering **DAS** with adaptive steering in the front and rear wheel steering
- Wide range of chassis characteristics through Audi Drive Select
- Central Audi Chassis Control

14 The chassis of the all new Audi A8, Jablonowski, Dr. Schimmel, Underberg, AUDI AG

Main Focus of Chassis Development
Increase of rear axle comfort

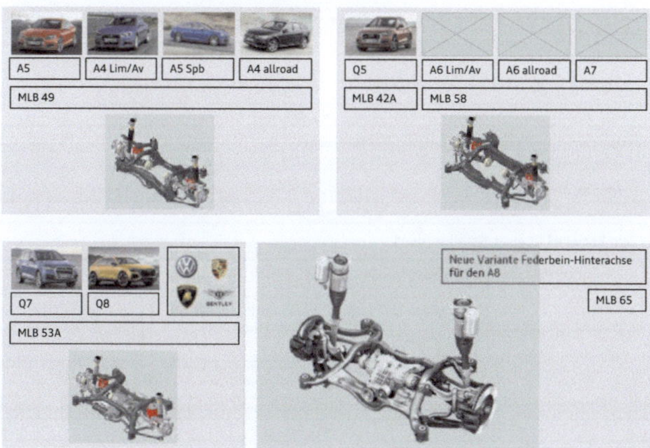

15 The chassis of the all new Audi A8, Jablonowski, Dr. Schimmel, Underberg, AUDI AG

Main Focus of Chassis Development
Increase of rear axle comfort

Descision for a spring strut type axle

- Direct ratio of sping strut reduces initial load of the sub frame mounting
- Reduces secondary spring rate by 15%
- Lower spring rate, lower pressures, easier applikation of the rolling piston at high amplitudes
- Reduced stiffness of damper mountings possible

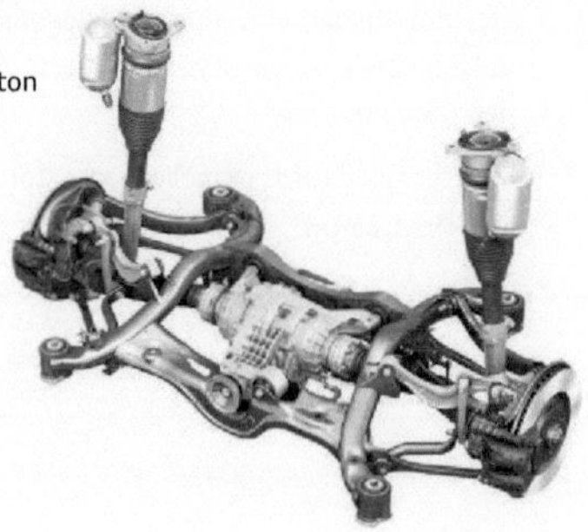

16 The chassis of the all new Audi A8, Jablonowski, Dr. Schimmel, Underberg, AUDI AG

The chassis of the all new AUDI A8
Contents

Introduction
- Target Positioning
- Model Range, Technical Specification

Technical Target Values of Chassis Characteristics
- Ride Comfort
- Driving Dynamics

Main Focus of Chassis Development

Axle Concepts

Central Chassis Control

Damper Technology | Dynamic All Wheel Steering DAS | Electrical Active Body Control eABC

Integration of Active Chassis Control Systems

Chassis Characteristics of the 5th Generation AUDI A8

17 The chassis of the all new Audi A8, Jablonowski, Dr. Schimmel, Underberg, AUDI AG

Main Focus of Chassis Development
Damper Technology

CDCivo and damping control:

- Improved damper hydraulics
- Optimization of the main valve spool
  - Additional part keeps the main valve spool open at low currents and supports during the opening in the pressure stage
  - Reduction of the dynamic force progression at low currents and high freuqent excitations

Audi Damping Control:

- Load specific damper application
- Temperature calculation: constant chassis characteristics at all conditions
- Precise adjustment of damper forces by real-time damper model for the comensation of the damper hysteresis

Cut away view damper CDCivo

18 The chassis of the all new Audi A8, Jablonowski, Dr. Schimmel, Underberg, AUDI AG

The chassis of the all new AUDI A8
Contents

Introduction
- Target Positioning
- Model Range, Technical Specification

Technical Target Values of Chassis Characteristics
- Ride Comfort
- Driving Dynamics

Main Focus of Chassis Development

Axle Concepts

Central Chassis Control

Damper Technology | Dynamic All Wheel Steering DAS | Electrical Active Body Control eABC

Integration of Active Chassis Control Systems

Chassis Characteristics of the 5th Generation AUDI A8

Main Focus of Chassis Development
DAS: Dynamic All-wheel Steering

- **DAS System** consists of adaptive steering, EPS, rear wheel steering and the central control unit (EFP = Elektronische Fahrwerksplattform)
- **The three main focus points** during the steering development of the A8 were:
  - Steering feedback
  - Driving dynamics
  - Drive-Select

Main Focus of Chassis Development
DAS/EPS: Steeringfeedback

- **Intensive Feedback** (at high frequencies) of useful signals while reducing disturbances as a target for the steering systems in the A8

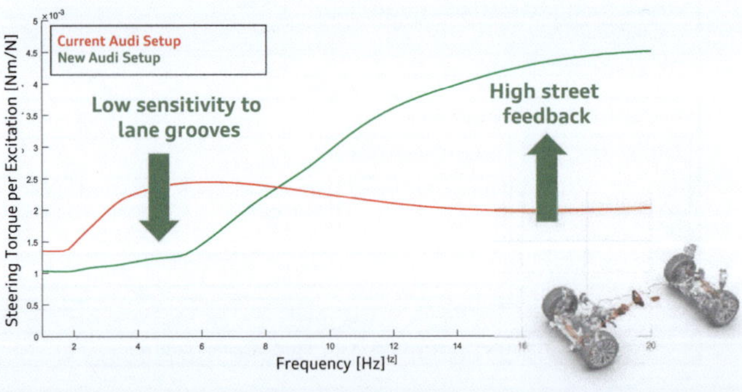

21 The chassis of the all new Audi A8, Jablonowski, Dr. Schimmel, Underberg, AUDI AG

Main Focus of Chassis Development
DAS: Fahrdynamik und Drive Select

- Optimal cooperation of adaptive steering and real wheel steering over the central control enables outstanding **driving dynamics**
  - Small delay between yaw rate and lateral acceleration
  - Small slip angles for linear driving and at the stability limit (improves driving dynamics as well as comfort for passengers in the rear seats)
  - Higher lane change performance
- Increased possibilities for differentiation of Drive Select through free configuration of directness and stability
  - Three levels of directness(sensed steering ratio)
  - Three ride characteristics (slip angle)
  - Three levels of Torque (Steering torque, Feedback)

22 The chassis of the all new Audi A8, Jablonowski, Dr. Schimmel, Underberg, AUDI AG

The chassis of the all new AUDI A8
Contents

Introduction
- Target Positioning
- Model Range, Technical Specification

Technical Target Values of Chassis Characteristics
- Ride Comfort
- Driving Dynamics

Main Focus of Chassis Development

Axle Concepts

Central Chassis Control

Damper Technology | Dynamic All Wheel Steering DAS | Electrical Active Body Control eABC

Integration of Active Chassis Control Systems

Chassis Characteristics of the 5th Generation AUDI A8

23 The chassis of the all new Audi A8, Jablonowski, Dr. Schimmel, Underberg, AUDI AG

Main Focus of Chassis Development
electrical Active Body Control

System targets

- Distinctive comfort experience through maximal body control (1-3 Hz)
- Comfort increase by active control of the roll behaviour
- Better dynamic characteristics via roll moment distribution
- Preview sensors for reduced body movement

24 The chassis of the all new Audi A8, Jablonowski, Dr. Schimmel, Underberg, AUDI AG

Main Focus of Chassis Development
electrical Active Body Control

eABC (electrical Active Body Control):

- **eABC** is an electro-mechanical suspension system, that can apply a force between the body and the wheel independently
- The basis is the air suspension chassis with damping control

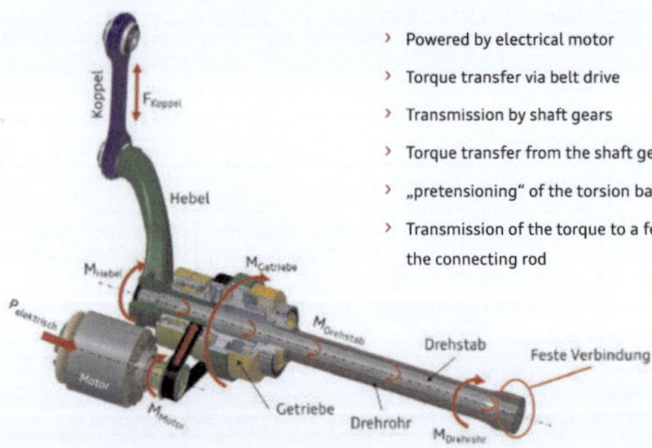

- Powered by electrical motor
- Torque transfer via belt drive
- Transmission by shaft gears
- Torque transfer from the shaft gears to the torque tube
- „pretensioning" of the torsion bar (torsion spring)
- Transmission of the torque to a force on the suspension strut via the connecting rod

25 The chassis of the all new Audi A8, Jablonowski, Dr. Schimmel, Underberg, AUDI AG

Main Focus of Chassis Development
electrical Active Body Control

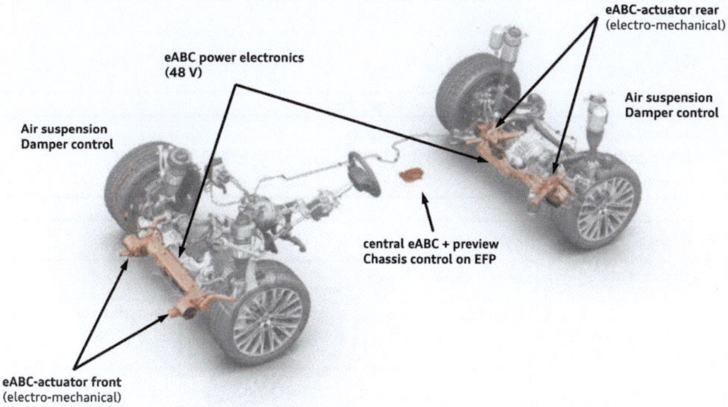

26 The chassis of the all new Audi A8, Jablonowski, Dr. Schimmel, Underberg, AUDI AG

Main Focus of Chassis Development
electrical Active Body Control

System performance:

› Actor torque	**± 1100 Nm** (→ Force transition rod FA/RA: **5,0 / 4,5 kN**)
› Actor dynamics	**3500 Nm/s**
› Max. Power intake	Complete system **5 kW** (Peak per motor 2 kW)
› Control range	bis **5 Hz**
› Power consumption	**10 – 200 W** (je nach Fahrprofil)

System specification:

› Overall transmission ratio	1:189 (Belt drive 1:2,36; Gearbox 1:80)
› ECU	48 V power supply with 12 V logic part, fully integrated in actor units
› System weight	65 kg

27 The chassis of the all new Audi A8, Jablonowski, Dr. Schimmel, Underberg, AUDI AG

The chassis of the all new AUDI A8
Contents

Introduction
- Target Positioning
- Model Range, Technical Specification

Technical Target Values of Chassis Characteristics
- Ride Comfort
- Driving Dynamics

Main Focus of Chassis Development
- Axle Concepts
- Central Chassis Control
  - Damper Technology
  - Dynamic All Wheel Steering DAS
  - Electrical Active Body Control eABC
  - Integration of Active Chassis Control Systems

Chassis Characteristics of the 5th Generation AUDI A8

28 The chassis of the all new Audi A8, Jablonowski, Dr. Schimmel, Underberg, AUDI AG

Main Focus of Chassis Development
Integration of Chassis Control Systems

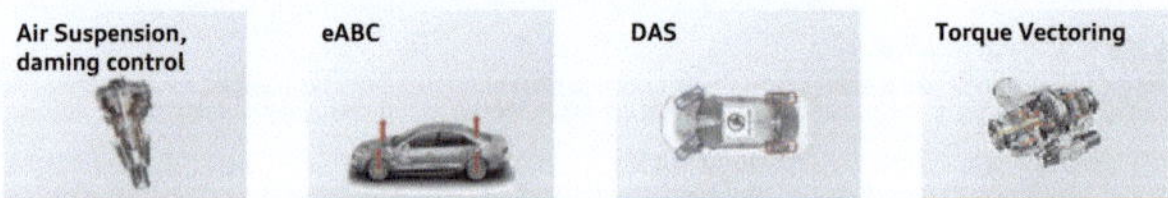

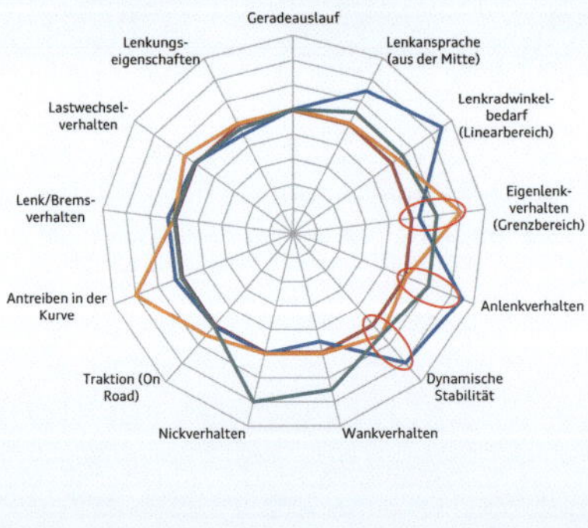

- Control systems overlap in their areas of impact on chassis characteristics in vehicle dynamics
- Efficient application must be ensured
- Inclusion of application strategy in control design

29 The chassis of the all new Audi A8, Jablonowski, Dr. Schimmel, Underberg, AUDI AG

Main Focus of Chassis Development
Integration of Chassis Control Systems

- „Packaging“ of control systems for chassis emphasis on comfort and driving dynamics
- Focusing of the individual systems on their main areas of effect
- Implementation of a suitable system and control architecture
  → central chassis

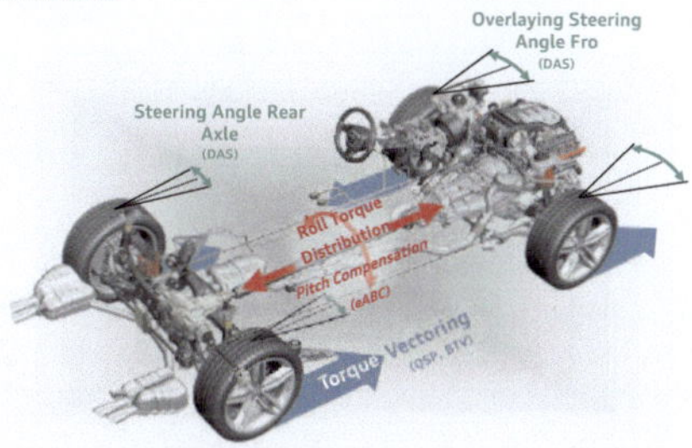

30 The chassis of the all new Audi A8, Jablonowski, Dr. Schimmel, Underberg, AUDI AG

The chassis of the all new AUDI A8
Contents

Introduction
- Target Positioning
- Model Range, Technical Specification

⇩

Technical Target Values of Chassis Characteristics
- Ride Comfort
- Driving Dynamics

⇩

Main Focus of Chassis Development

Axle Concepts

Central Chassis Control

Damper Technology

Dynamic All Wheel Steering DAS

Electrical Active Body Control eABC

Integration of Active Chassis Control Systems

⇩

Chassis Characteristics of the 5th Generation AUDI A8

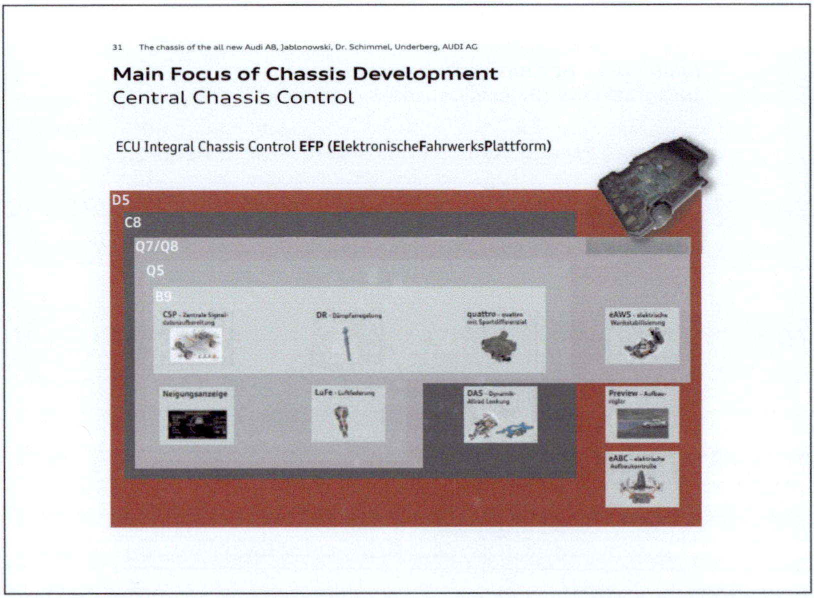
31 The chassis of the all new Audi A8, Jablonowski, Dr. Schimmel, Underberg, AUDI AG
Main Focus of Chassis Development
Central Chassis Control
ECU Integral Chassis Control EFP (ElektronischeFahrwerksPlattform)
D5
C8
Q7/Q8
Q5
B9
CSP - Zentrale Signaldatenaufbereitung
DR - Dämpferregelung
quattro - quattro mit Sportdifferenzial
eAWS - elektrische Wankstabilisierung
Neigungsanzeige
LuFe - Luftfederung
DAS - Dynamik-Allrad Lenkung
Preview - Aufbauregler
eABC - elektrische Aufbaukontrolle

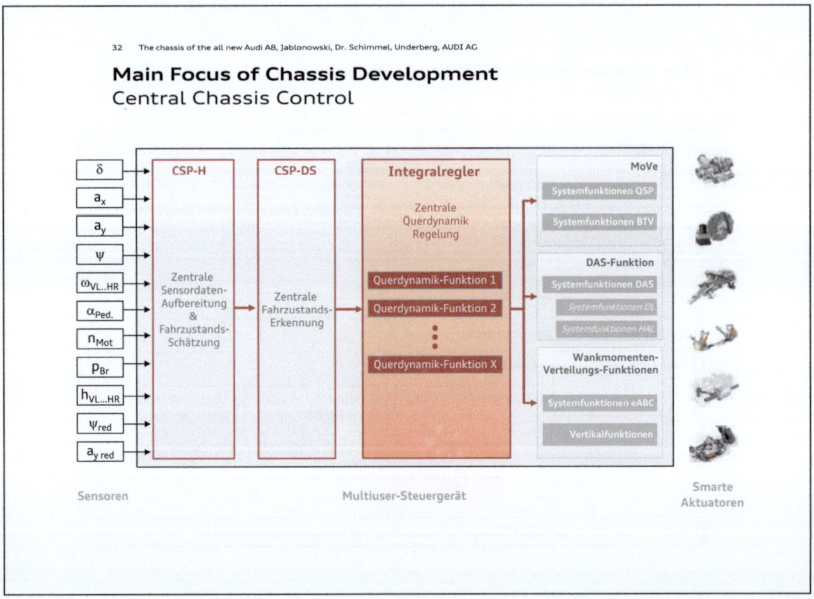
32 The chassis of the all new Audi A8, Jablonowski, Dr. Schimmel, Underberg, AUDI AG
Main Focus of Chassis Development
Central Chassis Control
δ
a_x
a_y
ψ
$\omega_{VL..HR}$
$\alpha_{Ped.}$
n_{Mot}
p_{Br}
$h_{VL...HR}$
ψ_{red}
$a_{y\,red}$
CSP-H
Zentrale Sensordaten-Aufbereitung & Fahrzustands-Schätzung
CSP-DS
Zentrale Fahrzustands-Erkennung
Integralregler
Zentrale Querdynamik Regelung
Querdynamik-Funktion 1
Querdynamik-Funktion 2
Querdynamik-Funktion X
MoVe
Systemfunktionen QSP
Systemfunktionen BTV
DAS-Funktion
Systemfunktionen DAS
Systemfunktionen DL
Systemfunktionen HAL
Wankmomenten-Verteilungs-Funktionen
Systemfunktionen eABC
Vertikalfunktionen
Sensoren
Multiuser-Steuergerät
Smarte Aktuatoren

33 The chassis of the all new Audi A8, Jablonowski, Dr. Schimmel, Underberg, AUDI AG

Main Focus of Chassis Development
Central Chassis Control

- **Improved efficiency during the application of the vehicle dynamics in the integral chassis control**

Virtual Pre-Application

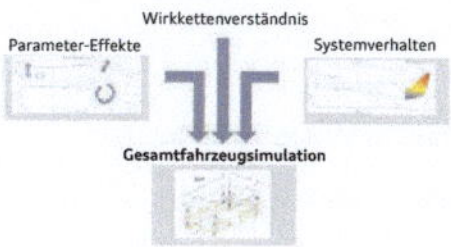

70% of the application of vehicle dynamics accomplished virtually

Final Application in the vehicle

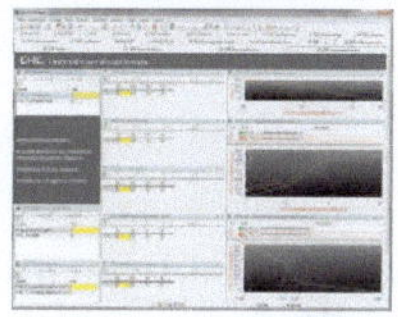

Simultaneous application of all systems based on characteristics possible

34 The chassis of the all new Audi A8, Jablonowski, Dr. Schimmel, Underberg, AUDI AG

The chassis of the all new AUDI A8
Contents

Introduction
- Target Positioning
- Model Range, Technical Specification

Technical Target Values of Chassis Characteristics
- Ride Comfort
- Driving Dynamics

Main Focus of Chassis Development

Axle Concepts

Central Chassis Control

Damper Technology | Dynamic All Wheel Steering DAS | Electrical Active Body Control eABC

Integration of Active Chassis Control Systems

Chassis Characteristics of the 5th Generation AUDI A8

35 The chassis of the all new Audi A8, Jablonowski, Dr. Schimmel, Underberg, AUDI AG

Chassis Characteristics of the 5th Generation AUDI A8
Measurements Primary Ride

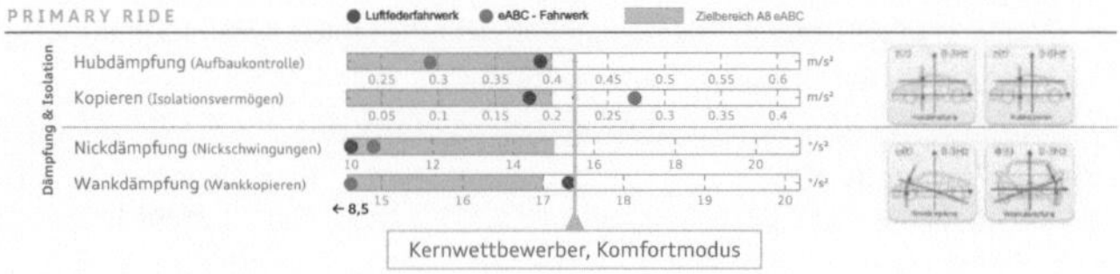

- The **comfort characteristics** of the air suspension chassis of the long wheel base Audi A8 meet the objective target values → Best in Class
- In addition to that, the **eABC** offers an considerable increase of **body control**

36 The chassis of the all new Audi A8, Jablonowski, Dr. Schimmel, Underberg, AUDI AG

Chassis Characteristics of the 5th Generation AUDI A8
Driving Dynamics

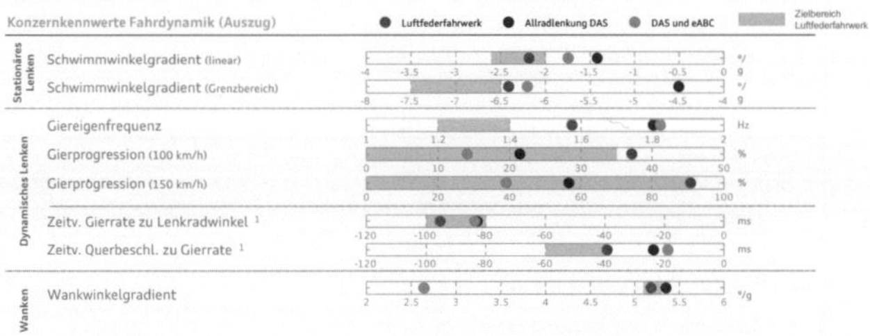

- The combination of the active chassis eABC and the adaptive steering DAS ensure outstanding driving dynamics

[1]: Ermittelt bei 1,5Hz Lenkfrequenz (entspricht Spurwechsel-Test bei 100km/h)

Thank you for your attention

Future brake system architectures for automated driving

Dr. Armin Kunz, Dr. E. Liebemann, Dr. M. Kunz, Dr. S. Strengert,
Robert Bosch GmbH

**This manuscript is not available according to publishing restriction.
Thank you for your understanding.**

P.E. Pfeffer (Ed.), *8th International Munich Chassis Symposium 2017*,
Proceedings, DOI 10.1007/978-3-658-18459-9_3

KEYNOTE LECTURES II

Automotive security is not even a feature – how can we afford that?

Dr.-Ing. Sebastian Labitzke

P.E. Pfeffer (Ed.), *8th International Munich Chassis Symposium 2017*, Proceedings, DOI 10.1007/978-3-658-18459-9_4

Introduction

Many technical specifications of ECU development in the automotive industry send a clear message: adequate cyber security measures are a must. Reasons are manifold because of the variety of potential attackers. However, at the end it mostly comes down to money that the car manufacturers potentially will lose if cyber security is not considered within their products. This financial damage could be induced by reputational damage, confidentiality issues (e.g., lost intellectual property), compliance issues with respect to the law, safety problems because of potentially successful attacks on today's cars – just to name a few. In some cases, the risks are even business-critical, in particular, if manufacturers and suppliers face reputational damage. Reputational damage is induced, for instance, once the customers' expectations on what they get for their money are not met. These expectations may have changed over time. At least, the times are over when horsepower and cubic capacity affected most purchase decision. More and more customers desire connectivity services and connected features as mandatory features their vehicles should be equipped with. Their additional expectation: the car with all its connected features provides adequate safety, as well as cyber security measures against attacks from malicious third parties without an expensive click on "added cyber security" during online configuration of the cars. In turn, if a car gets successfully attacked, the disappointment could result in significant loss of trust in the manufacturer and a reduction of customers' brand loyalty, which results in financial damage.

From a business perspective, the situation gets complicated: on the one hand, manufacturers must deal with the need for cyber security measures that will cost a significant amount of money that should be planned as part of the development costs of any feature. On the other hand, security-by-itself is not even a feature and customers don't want to pay for added cyber security because they expect the cars to be secure. With that, the question arises what the "return on security invest" really is?[1]

In this paper, we show that the investments on cyber security can be justified by the prevention of potential loss, i.e., with the cyber security invest the risks regarding potential financial damage get mitigated. However, a potential loss is fuzzy to estimate and, with that, it will not get clear how much money it is worth to spend for cyber security measures. Therefore, we highlight the key aspects to consider when it comes to the decision on how much money should be spent and what for. We give insights on the motivation of potential attackers, on how management should consider cyber security risk assessment results, and on adequate processes to be implemented. Further-

1 With respect to investments in IT security for server infrastructures, this return is often called "RoSI". Note that calculations for RoSI often don't fit for systems of the vehicle industry.

more, we elaborate on the costs of adequate cyber security processes. Finally, we discuss the need for teams that are actually able to make vehicles more secure, or rather, to prevent manufacturers from losing a significant amount of money due to a holey fence of cyber security.

What is Security if it is not even a feature?

As mentioned, cyber security is not a feature to be added via a car configurator by the end customer. In turn, cyber security is more and more expected as mandatory by customers. This fact is not surprising given that cyber security issues can also induce significant risks with respect to the safety of the driver, his or her passengers, and people around the vehicle. In particular, without appropriately chosen and adequately implemented cyber security controls, safety-related measures could be obsolete.

For more clarification on what security is, we compare it to safety that is also not a feature by itself. Safety measures prevent from faults induced by randomly occurring errors inside the systems, e.g., bit flips due to loosely coupled plug connections or inferences. The security engineers Anderson and Needham state that safety is like programming Murphy's computer (Anderson, 2008). However, cyber security is different from that. Security measures prevent from faults induced by malicious attackers who are trying to, for instance, inject modified messages into the system from outside the system. Such an attack cannot be prevented by most of today's safety measures. For instance, a Cyclic Redundancy Check (CRC) helps to find flipped bits inside a message, but an attacker can generate a message with a perfect CRC although the message is not allowed to be sent through the bus. Anderson and Needham state that cyber security engineering is like programming Satan's computer (Anderson, 2008), i.e., safety protects humans from faulty acting machines, whereas cyber security protects the machines from maliciously acting humans.

So, what is cyber security? Cyber security engineering mitigates the risks regarding scalable attacks on connected vehicle systems. With connectivity, i.e., by opening the vehicles for communication to and from the outside world, the attack surfaces of today's cars get larger. So far, often only non-scalable attacks on vehicles were imaginable, i.e., attacks that costs a certain effort for each of any vehicle to be attacked. For instance, disengaging the wheel nuts of a car, manipulating brake hoses, or more modern attacks like attacking the infotainment system with compromised compact discs are examples for such non-scalable attacks. In short, for performing a non-scalable attack the attacker must be present at the vehicle to be attacked. With connectivity, we are more and more facing scalable attacks, i.e., attacks that can be performed remotely, even by sitting on a sofa at home. Moreover, scalable attacks can potentially be performed on many vehicles at the same time. Note that it is not sufficient to only secure the interfaces implemented to connect the vehicles. If a vehicle system gets "connected", the overall system has to be secured.

Based on this perspective, cyber security engineering brings those measures into today's connected vehicle systems that are necessary to keep vehicles secure and, therefore, if possible, as safe as not connected vehicles. However, cyber security comprises not only a certain amount of additional initial invest. Like in the IT/server business, it is a process starting at the first phase of development of an E/E system and ending at end-of-life of the last vehicle on the road. Such a long running process is necessary to be able to react on the volatile methods of attacks for keeping a car safe and secure during its lifetime.

At least because of the impact on safety, cyber security is a must for today's connected vehicles. Furthermore, cyber security is not only something to be added at the interfaces to the outside world of the vehicle. Security has to be considered by each of any electronic component of today's vehicles. Considering cost of the significant number of requirements with respect to cyber security, it gets obvious that cyber security measures and underlying processes are often very expensive. However, the return on security invest is often not that obvious.

In the following, we try to introduce the perspective that selective cyber security measures can lead to pseudo-security, i.e., only assumed cyber security because of the fact that selective measures have been implemented. Furthermore, an invest of only a small amount of money (below a certain threshold) cannot help making cars more secure at all. In this paper, we provide more insights on the determination of the threshold of the minimal invest of cyber security processes and measures for connected vehicles.

The Security Fence and its Holes

From an abstract point of view, we can state that the more connected a system is to the outside world, the more vulnerable the system is expected to be. In particular, E/E systems with a growing amount and complexity of code are more likely to be vulnerable than systems with less software. Joshua Corman says in one of his talks regarding this context: "if it has software substitute the word hackable, and if it has connection substitute the word exposed" (Corman, 2013). That means, the probability to develop software that is not vulnerable at all is very low. In turn, connecting this software or the systems that run the software significantly increases the attack surface. If the attackers form the threats and the bugs in the software make the systems vulnerable, connecting vehicles brings threats and potential vulnerabilities closer together. Note that this is not a statement against connectivity, since connected vehicles are not only demanded by end customers but also constitute the enabler to make traffic more efficient, less dangerous, and less polluting than today. However, on the one hand, we deal with software that is to be secured and, on the other hand, attackers are in place who might have a high motivation to attack connected vehicles for their malicious business. With that, connectivity makes not only positive transformations of today's traffic possible but also expose connected cars to scalable attacks because of the convergence of threats and vulnerabilities.

To cope with the broad range of potential attacks, cyber security engineers try to build "fences", for instance, around the interfaces of the connected vehicles, the network domains inside a vehicle, around the messages to be sent through the communication channels (e.g., CAN bus, ethernet, etc.), around each of any ECU, and around the cryptographic keys stored on an ECU to be used by the cryptographic algorithms. Since a single fence could only serve as a single hurdle for an attacker, cyber security engineering aims at building up fences on different levels and at different components of the cars. Furthermore, the goal is to build those fences in a manner that is as complete, high and strong as needed for providing an appropriate level of cyber security.

In contrast, attackers just have to find a single hole in those fences to be able to drive their malicious business. In the following, we show an example of such a "hole in the fence" in the area of key management: a common goal of cyber security is the protection of ECUs so that it is very difficult for attackers to clone these ECUs. For this, the cryptographic keys of an ECU are to be stored in a protected manner, i.e., an attacker is not able to just copy the keys. For securing keys on ECUs Hardware Security Modules (HSMs) can be used (Köhler & Förster, 2016). However, these HSMs are not built-in on any of today's ECUs that make use of those cryptographic keys, in particular, if legacy ECUs are in use. With that, although the security engineers did a good job on securing the keys on the modern ECU with an HSM, the attacker would only have to attack the legacy ECU with the accessible keys. Afterwards, he or she could extract the keys and build these keys into his or her malicious ECU that he or she could bring into the car to potentially be able to, for instance, inject malicious messages.[2]

The motivation for performing attacks on vehicles are manifold: some attacks might be performed due to a simple lack of other things to do or for the attacker's own questionable reputation. Some other reasons for attacks on vehicles could be politics, terrorist attacks, or an attack might also be motivated by possible financial gains, for instance, by the use of blackmail. Furthermore, also attacks to cause damage to competitors or espionage motivations are conceivable. The amount of malicious motivations that could results in attacks on connected vehicles is tremendously high.

---

2 Note that this attack might be a non-scalable attack. However, scalable attacks could definitely also be possible if cryptographic keys get lost or rather in hands of an attacker

Moreover, some additional attacks constitute the result of the work of people who try to grow awareness and help to make today's and future cars more secure. In particular, most attacks that we recently saw in the media are actually academically driven. Although carried out by a benign attacker, these attacks might induce a significant loss of money for the affected manufacturer because of the reputational damage that often follows such an attack that went through the media. To cope with this broad variety of threats, cyber security engineering is the only chance manufacturers have for minimizing the probability of vulnerabilities of their innovative vehicles.

In summary, cyber security measures build the fences around "the property". The approach to build a high fence in the front of a property, but only a smaller one or even no fence at all for the backyard, might be less money to invest, but also an invest below the threshold of minimal invest necessary to get an appropriate protection against certain cyber security attacks.

The "Return on Security Invest" Issue

Actually, we often experience that the threshold that has to be crossed to get, at least, some security measures in place is significantly higher than naively estimated at the beginning of many projects in the automotive and related industries. In today's cyber security projects, we often face a valley of tears after the risk assessments or following the first abstract security concept, just because of the fact that a certain number of measures have to be in place, before the fences are stable and are no more holey, i.e., before the cyber security measures are operating effectively.

To prevent such a valley of tears, the following prerequisites are necessary to be considered at the beginning of each of any of today's E/E projects:

- A good understanding on management level of what cyber security engineering is and what kind of goals cyber security engineers try to reach, because this awareness provides the basis on top of which decisions on invests can be made
- A team of cyber security experts of different cyber security domains that can estimate the costs at an early stage of the projects (cf. section "Teams of Experts…" for an answer on why teams of experts are crucial for a success of security projects)
- A risk assessment methodology and adequate risk assessments performed by these teams of experts to get a clear understanding of potential attacks on the vehicle systems to be secured, as well as prioritizations of the measures to be implemented

During the projects, some additional important things are to be considered:

- Working alongside an adequate cyber security engineering process (cf. section "Adequate Processes for Adequate Security") to prevent from selectively implemented cyber security measures (cf. section "The Security Fence and its holes)
- Relying on not only those experts who do the concept, but also on review, validation, and verification steps for each of any stage of the security development process, since also experts might do mistakes that could lead to cyber security risks
- Partner with external companies that are specialized in cyber security in the domain of embedded systems, the automotive industry, AND connected components like backend servers, mobile apps, etc. The reason for this constitutes the need for teams of experts to cope with the complex subtopics of automotive cyber security, such as embedded security (for instance, because of the limited resources on an ECU that need to run the cryptographic algorithms), telematics and communication security (for instance, on a bus-level), mobile app security, and, for sure, IT security as well. Note the difference between backend IT security and embedded security. Thus, to ensure secure and applicable, as well as deployable security concepts, make sure that the security partner has a background in automotive cyber security engineering.

In the following, we elaborate on the question whether there is a return on cyber security invests if the aforementioned threshold of minimum invest is passed. As mentioned in the introduction, the return on security invest is often very fuzzy to estimate. However, bringing the invests on cyber security in an adequate perspective helps to estimate how much money a manufacturer should spend and what for.

First, for most products in the vehicle industry, cyber security might not be a business case that leads to gaining more money than selling the products without any respective measures. At least, this perspective is true if others don't face reputational damage because of weak security of their products. With reputational damage on competitors' side, security measures could serve as a unique selling point in the market. To cite Joshua Corman again: "you don't have to swim faster than the shark, just faster than your buddy" (Corman, 2013). However, for estimating the return on security invest, it might not be the best solution to assume that others will fail in implementing adequate cyber security. Moreover, usually there are more than one shark in the water, as Corman adds in his talk. Therefore, marking the tail end of the field could result in significant financial damage because of the increased probability of successful attacks that could be followed by a bad reputation for the affected manufacturers, as well as their suppliers. This potential loss exactly builds the basis on top of which the return on security invest gets obvious. Now, the question might arise, how the potential loss can actually be estimated, at least, in a fuzzy manner.

"Facts, Urgency, Decision Basis"

Recently, we have seen a lot of talks on cyber security risks. Conspicuously, a significant amount of those talks started with a picture of the Jeep hack by Miller and Valasek in 2015 (Greenberg, 2015). These talks did not only a good job on making the topic popular, but also spread some worries, followed by uncertainty and doubt. Two questions on that: did those talks help at all? Did those talks help to go the actual steps in the direction of an increased cyber security for connected vehicles, or at least in the direction of a decision on how much money should be invested and what for? Regarding the first question, yes, these talks definitely helped to grow awareness, which was a good and absolutely necessary starting point for planning budget for cyber security engineering. However, the problem is not that managers of the vehicle industry do not demand secure products. The fuzziness of the topic cyber security and, therefore, answering the question on how much security is "good-enough" constitutes the actual issue. With respect to the second question, it is time to come up with solutions and guidance in the direction of connected vehicles that are designed appropriately secure. Decision making on invest (not only for cyber security measures) needs facts with an evaluation on the urgency, forming a decision basis on top of which management can decide on the amount of invest for cyber security engineering.

Unfortunately, because of the highly heterogenous systems and their varying requirements and use cases, it is impossible to provide a handbook with tables to read the numbers from. Instead, it is necessary to apply an adequate risk assessment methodology to determine the urgency of specific security mechanisms for each scenario/use case/subsystem and back this up by an appropriate security development process.

Adequate Processes for Adequate Security

An appropriate development process starts with comprehensive analyses of the vehicle system. This applies if a whole vehicle, a telematics unit that provides interfaces to the vehicle, a gateway control unit, an ethernet switch for on-board communication via ethernet, any other ECU inside the vehicle, some component in the IT backend, a mobile app to control functions of the car, or anything else in this context is to be designed. Thereby, it is important to not only focus on the component to be designed since cyber security is always something holistic.

For generating the basis on top of which the process can be started, answers to the following questions have to be determined:

- Is the system architecture decided?
- What are the use cases in scope? How do the users interact with the system, how does the system interact with other components or infrastructure outside the car?

- What are the protection goals? Confidentiality, integrity, and availability form the triad of the main security protection goals. The protection goals to decide on are, for instance, safety, IP protection, compliance, data protection – just to name a few
- Are there security requirements that must be satisfied independently from a risk analysis to be performed, for instance, security requirements dictated by the surrounding system?
- Which security enablers exist on the target platforms or within the architecture, e.g., built-in hardware support for security, existing infrastructure components like PKIs that can integrate with the security concept, planned network architecture and communication channels, available memory on the ECU(s), maximum delays allowed for each communication, number of services hosted on each of the ECU(s), etc.

This incomplete list of prerequisites shows that already the input parameters can get complex when it comes to security. To bring this diversity back to a decision basis as discussed before, a team of experts should perform a risk assessment based on an appropriate methodology. Furthermore, the cyber security engineers have to manage the resulting complexity and have to transform technical decisions to be made into a basis of "easy" decision-making suitable for management.

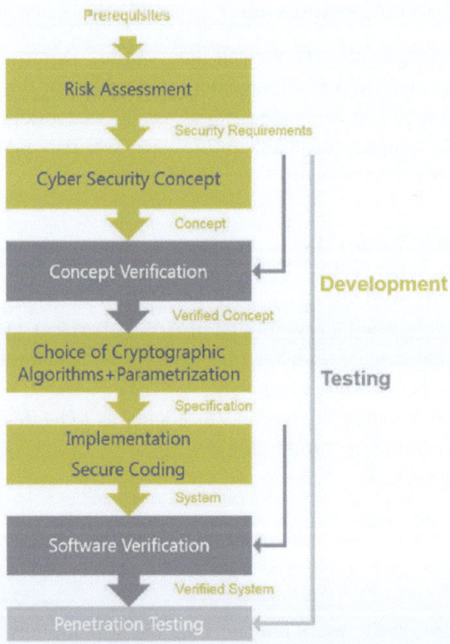

Figure 1: Example of a Cyber Security Engineering Process.

Based on this input information a security process can be started that accompanies the whole development process of the system. Figure 1 shows the subsequent stages of an exemplary cyber security engineering process. Note that the process can also be mapped into the V-model of development. For the sake of readability, we visualized it here as a flowchart from top to bottom. Following a risk assessment that leads to requirements, the security concept is built that includes the integration of security measures into the system architecture, the communication protocols, etc. Afterwards, some manufacturers and suppliers perform the first testing stage to verify the concepts. For instance, the protocols that shall ensure a secure communication can be modelled alongside the protection goals within a verifier. Then, automated verification runs of the protocols are performed to identify potential attacks on the chosen protocols. Within the next stage of the development process, the cryptographic building blocks are chosen for the concept and the parametrization of those algorithms are fixed, before the concept is implemented within the coding phase (see "implementation" considering "secure coding" guidelines; such as (Seacord, 2013)). A phase of verifying the implementation follows, before the general testing of the product takes place (essentially, the right side of the V-Model). Note that code reviews might not only be valuable for the implementation of the cyber security measures. It could also be reasonable to review the whole more or less exposed software based on secure coding guidelines, additionally to the more or less common reviews, for instance, regarding MISRA-C.

Note that at first sight, cyber security processes look similar to the already established safety processes. Safety process nomenclature overlaps with cyber security nomenclature and partly the processes match on the highest level of abstraction (just like other risk management methodologies do, e.g., financial risk assessments). However, the necessary steps to perform a safety risk analysis and a security risk analysis are different once the process steps are performed. Furthermore, security engineers have to have a fundamentally different mindset than safety engineers. Just like safety engineers are good in measures that prevent randomly occurring faults, security engineers build their concepts on top of knowledge on cryptography, which is basically complex math, attacker psychology, and many more things to protect from maliciously acting attackers. Based on this, a collaboration between safety and security engineers is absolutely valuable, whereas a fully joint safety and security engineering process is not expedient.

Teams of Experts for an Adequate Level of Security

We showed that cyber security engineering is a complex topic and we only scratched the surface, i.e., we did not provide a deep dive into the world of cryptography and such. A team of experts is required to perform these processes and the experts must be able to work up the invest on cyber security engineering. In the following, we briefly elaborate on the aforementioned fact that cyber security engineering is nothing to be done by one head for one system. For instance, an embedded security engineer is probably able to work out a concept to securely handle the keys inside the ECU or to implement a certain cryptographic algorithm under restrictive resources. However, another cyber security engineer might be a better choice for handling the weakness of mobile apps. Figure 2 visualizes some very different use cases for which cyber security should be considered.

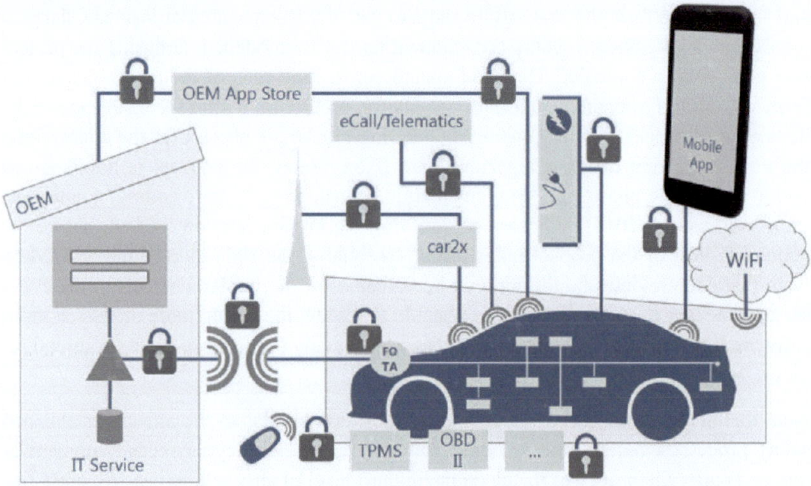

Figure 2: Overview of Different Use Cases for which Cyber Security Measures should be considered.

The Risk of Trust in Rotten Apples

Figure 3: One original image (left image) that was encrypted with the same algorithm (AES) but by use of different modes, i.e., ECB mode (middle image; not secure) and CBC mode (right image).

Finally, it remains the question on what kind of partners are able to provide beneficial support for solving today's cyber security issues in the automotive industry. For answering this, we provide some thoughts on the market of cyber security engineering providers. Since cyber security is a discipline that has been grown already in the 90's, we see a lot of providers in the market that only focused on IT server security and information security for a long time. Some of these companies might face the issue of gaining the domain knowledge and embedded know-how necessary to perform cyber security engineering in the context of connected vehicles. Some other competitors in the market just focus on risk assessments and concepts, as well as penetration testing. However, these are the stages of the process that are difficult to evaluate for the customer, for instance, the car manufacturer. Just a few companies started several years back to bring expertise of different fields and security domains together with domain knowledge out of the vehicle industry. Furthermore, most of these very few companies provide support alongside the whole cyber security process, some with excellent products for embedded security within their portfolio, some who provide product-independent consultancy and the development of individual solutions based on adequate and common cryptography algorithms.

Since we stated that companies risk a "holey fence" when partnering with some of the maybe inappropriate providers on the market, we try to make clear how easy something can go wrong when a security concept has to be developed and implemented. Figure 3 visualizes an image and the result of two types of encryption with the same algorithm, i.e., AES. However, the middle image was encrypted with an old and insecure mode of this algorithm called ECB, whereas the right image was encrypted by the CBC mode of AES and then again visualized. This example shows how easy the fences of security can get holes and should serve as an awareness-growing example with respect to the choice of the right partner for securing today's and future connected vehicles.

Summary

In this paper, we showed that cyber security engineering gets more important due to more and more connectivity of today's vehicles. We elaborated on the perspective that helps to estimate how much money should be spent and what for. For that, we provided insights on attackers' motivation and the return on security invest that is not based on a traditional business case, i.e., gaining more money because of a unique selling point. Adequate cyber security measures could only serve as a unique selling point if competitors fail in this field just because the "sharks will bite the hindmost". However, the potential money a company could lose if it is one of the worst in implementing adequate cyber security measures constitutes a basis on top of which a return on sufficient cyber security invest is very likely.

In the course of this paper, we also showed that risk analyses and adequate cyber security development processes are key for getting an appropriate protection against today's and future attacks. We also showed that not only a well-implemented and well-structured process lead to adequate cyber security, but also cyber security experts from different domains of cyber security who team-up within the respective projects, i.e., expertise in embedded security, telematics and communication security, mobile app security, and IT security for servers in the IT backend. Otherwise, the probability increases that the vehicle is attackable, or rather, that one of the cyber security fences comes with a hole that can be exploited by attackers and their malicious business.

Finally, with the title of the paper in mind, we showed that a "return on security invest" is only determinable if risk assessments are performed in the aforementioned manner by teams of experts. Compared to the following stages of the cyber security development process, this initial step constitutes not very much of the overall effort that is to be spent. However, adequate risk assessments ensure that cyber security gets tangible and builds the basis for decidability.

References

Anderson, R. J. (2008). Security Engineering: A Guide to Building Dependable Distributed Systems. John Wiley & Sons.

Corman, J. (2013). Swimming with sharks – security in the internet of things. TEDxNaperville.

Greenberg, A. (2015, July). Hackers Remotely Kill a Jeep on the Highway — With Me in It. wired.com.

Köhler, J., & Förster, H. (2016, Mai). Trusted Execution Environments in Vehicles. ATZ elektronik.

Seacord, R. C. (2013). Secure Coding in C and C++. Addison Wesley.

Acceptance and market for ADAS and automated driving systems

Dr. Wolfgang Bernhart, M. Yoon, Roland Berger GmbH

This manuscript is not available according to publishing restriction.
Thank you for your understanding.

P.E. Pfeffer (Ed.), *8th International Munich Chassis Symposium 2017*, Proceedings, DOI 10.1007/978-3-658-18459-9_5

KEYNOTE LECTURES III

Modularization of vehicle control systems based on the application of object-oriented design principles

Dr. Veit Held and Dr. Andreas Heitmann

Adam Opel GmbH

P.E. Pfeffer (Ed.), *8th International Munich Chassis Symposium 2017*, Proceedings, DOI 10.1007/978-3-658-18459-9_6

1 Introduction

The art of designing great chassis has outgrown the field of pure mechanical engineering a long time ago. Future vehicles will offer a wide range of active and (semi-)automatic driving modes that require complex control structures which encompass almost all chassis systems. This set of features will stretch from basic chassis functions like electronic stability control, traction control, lane keeping assistance, automatic parking aids all the way up to cover sophisticated autonomous driving functions.

The progress from mechanics to highly complex mechatronics can historically be broken into three phases. In the first phase we saw electric motors and computers being added to single components, which then enabled new customer features. In the brake area we started with introducing ABS and ESC, in the steering area we got electric power steering and the suspensions became semi-active. At this stage the challenges were mostly in the physical domain, the problems appeared isolated and could be solved independently for each component. In the second phase we started to combine systems and immediately hit the first 'mild' integration problems. For example we combined EPS + AFS + ARS or ESC + EPB (all these abbreviations are spelled out in the table at the end of this paper). Several actuators were controlled by one algorithm, which added a whole layer of control engineering problems.

Now we are hitting the third level that we may call strong integration, stretching from LKA all the way up to fully autonomous driving. This means that we have to master the deep integration of many sensors, actuators and the complex administration of several algorithms. As an example Figure 1 shows a simplified model of a vehicle with a complex feature set. This car can be driven manually (the driver operates the throttle and the steering), assisted (ACC or LKA also operate the throttle and the steering) or autonomously (driver is out of the loop). Apart from the various algorithms there may be additional state machines to switch the systems on and off in a coordinated fashion.

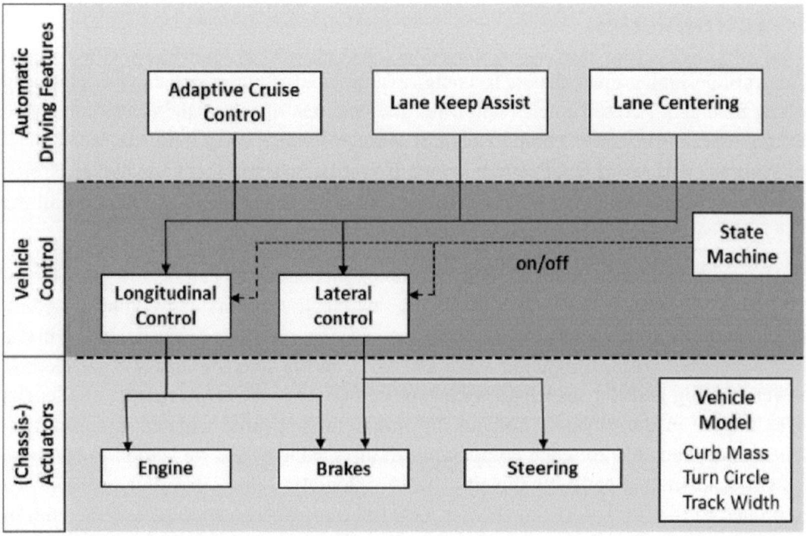

Figure 1 Layer model of advanced driver assist features

The complexity of developing integrated systems on level three is much higher compared to the previous levels of integration – by several orders of magnitude. There are three factors that make the development of highly integrated systems very demanding:

- **Complex interaction of many subsystems.** Individual algorithms rely on signals from other units, which means that these systems have to be developed together.
- **Complex portfolio of functions.** We have to assume that in any particular carline the OEMs will have to offer very different combinations of functions that all have to be developed and validated. There will be cars with only ACC, others with ACC-LKA or automatic high way assist … etc.
- **Complex combinations of vehicle subsystems.** Across the fleet we will have to combine a lot of different sensors, actuators, tires, engines etc. from many different suppliers

As a consequence we cannot keep developing each vehicle case by case. We need to create strictly modular control architectures that will allow us to develop all subsystems and functions independently from each other and still ensure that the combined system will work. We also need to simplify adding or subtracting functions without consequences for the rest. A very important goal is to parallelize the development activities at the suppliers and at the OEM. The requirements for a truly modular architecture can be summarized as follows:

- Provide maximum flexibility to create any desired combination of customer features by having the option to combine any set of technical systems (sensors, actuators, algorithms or vehicle body styles) with minimum effort
- Provide maximum flexibility to exchange, modify or exclude new algorithms or technical subsystems both from suppliers or in-house teams
- Drastically simplify the development work by enabling the development and validation of all subsystems in parallel
- Simplify the usage of simulation tools and benches on component and vehicle level

In this paper we will assume that the technical challenges of making any particular driver assist function work are or will be solved. We want to move on and discuss a few ideas about how we might get to a plug-and-play architecture that enables the parallel and independent development of all subsystems. To get there we will adopt and adapt several principles that were developed in the domain of personal computers many years ago.

2 Modularizing the control systems

In fact, some first and basic concepts towards this goal were introduced with the AUTOSAR-rules (1). AUTOSAR already introduced abstractions on different levels. On the lowest level, AUTOSAR abstracts the microcontroller and provides an API to interact with the microcontroller in a hardware-independent way. The next higher level is about abstracting the ECU which might contain additional internal or external devices (e.g. a CAN controller). On top of that AUTOSAR provides a service layer which unifies all underlying communication facilities, as well as non-volatile memory and diagnostics. Finally, the run-time environment stacked on top of the service layer provides inter-program communication between software components, regardless if the modules are executed on the same hardware or on a different ECU. This network-transparent operating system allows to move OEM-specific applications between ECUs without changing any of the underlying layers.

The AUTOSAR consortium is collaboratively creating a standard operating system available to all automotive OEMs and suppliers and is enabling them to compete on the top level applications without re-inventing the layers below.

In this paper we want to take these ideas further and present an approach how to create a very flexible architecture for chassis control systems on the application layer.

2.1 Alternative one: A central chassis controller

The first idea that comes to mind about how to architect a vehicle with a complex feature set is to introduce a central vehicle or chassis controller, which could come as a dedicated computer or at least as a central software module. The thought is intriguing, because on first sight it seems like one can draw the problem down to the level of plain physics. The basic idea is to study the balance of forces and energies and control all actuators on board with one central computer based on the underlying physical equations. While this will technically work and may even be the best approach for single market entries or high performance vehicles it has huge downsides for high volume applications.

The biggest problem is that it makes the cooperation of OEMs and suppliers very complex, ineffective and inefficient. If such a central computer was developed by the supplier, it would become difficult for the OEMs to change from one supplier to another or to include new subsystems from other sources. If such controllers were developed by the OEMs, then the suppliers would in turn have to adapt their solutions case by case for each OEM. So, all in all nothing is gained to reduce the complexity.

A central computer also makes it difficult to add or change subsystems because every change will need to be incorporated in the central module and it will become increasingly complex to manage and control the different versions across the portfolio. Every newly developed software module creates the need for a big downstream organization to do all the tuning and validation work. Any module that an OEM develops in-house, means that he has to staff up such a team for very specific functions. Solutions from suppliers have the advantage that the OEM can simply use their capacity for a given project and does not have to worry about keeping the engineers trained and their workload constant.

From a business perspective, a central computer will reduce the OEMs options to use the market forces for keeping the cost low and being fast to market. The fact that today we have very competent suppliers and OEMs racing against each other enforces and ensures that we will always see the most recent technology for the cheapest price. With a decentralized design, an OEM can always pick the fastest horse – and get rid of it when he finds a faster one.

All in all one may conclude that a central computer does not do a lot to minimize the pain – it only moves the problems elsewhere.

2.2 Alternative two: Learning from the PC-world

To make some progress, it helps to study other areas. In fact, the problem of designing complex and highly integrated systems showed up in the computer industry many decades ago and got solved over time:

In a typical household or work place, we find very different combinations of computers, printers, input devices, scanners, cameras and operating systems that work together in the most diverse combinations. The miracle is not only that all these combinations work at all, it is also very encouraging that all the individual devices could be developed, validated and brought to market independently. Most Windows applications work without knowing up-front which computer they will be run on (netbooks or workstations) or which devices or printers are attached to the personal computer. Printer manufacturers can develop their devices without knowing what will be printed on them (letters or drawings) or if they will be plugged into a Mac or a PC. In the next chapter, some of the key principles that were invented in the computer industry will be explained and translated to the chassis domain.

3 The object oriented approach to control architectures

A most powerful concept to get to a genuine modular approach is based on the paradigm of an 'object oriented design'. The underlying principles were initially introduced as a concept to break a large software development project into smaller parts, getting them developed by different teams and still making sure that the compiled program ultimately works. Languages were developed that support 'object oriented programming', the most prominent ones being C++ and JAVA. Development of complex office suites and distributed client-server applications became feasible.

If translated to vehicle control, the basic idea is to ignore all traditional signal flow diagrams and take a different view at the hierarchy of vehicle components. The model is to break the space of sensors, control algorithms and actuators into a map of interacting 'objects' that request and grant services from and to each other.

In fact, this is the basic principle that structures the life in a society: A caterer can offer services to any customer by planning the requested event for him, translating the needs into necessary sub-services from food providers, facility brokers or entertainment companies, ordering these services and get the event organized. The caterer does not need to know anything about how to prepare the food or how to play the instruments of the band, he only needs to know what he wants and how to buy the service.

3.1 Case study: Lateral Vehicle Control

The general concept of an object oriented design has been published in many textbooks and papers, see for example (2). For the sake of designing vehicle control systems, the main principles can be summarized in the few points that are listed in the appendix.

The application of these principles will now be explained with an example. Let us assume we want to design a vehicle that may or may not have a lane keeping assist feature on board, an advanced parking aid and a collision mitigation feature. These algorithms will necessarily need a set of sensors to measure the road ahead and around of the car and they must be able to control the speed and the direction of the vehicle. The object oriented design is sketched out in Figure 2 and will now be explained step by step, introducing the ideas that are listed in the appendix.

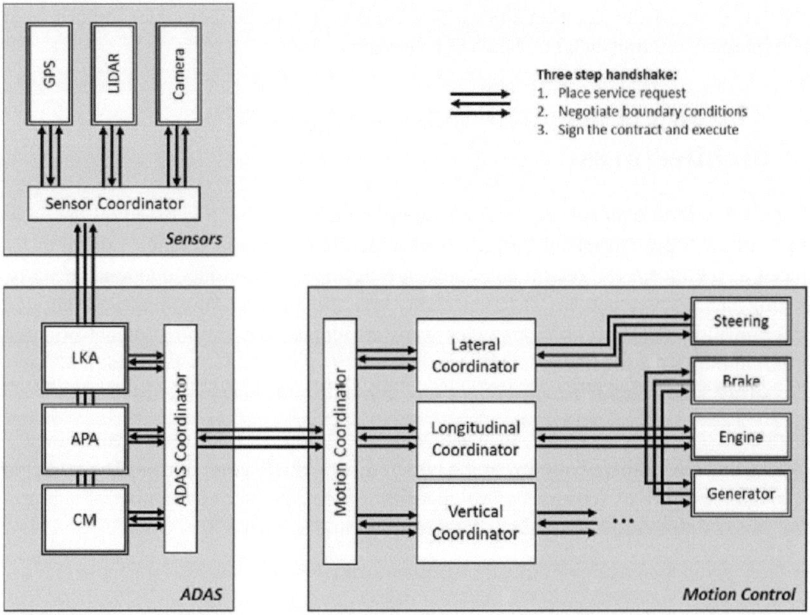

Figure 2 Interacting objects in a vehicle with lane keeping assist, collision mitigation, automatic parking assist

A four step approach to modularizing the system

1. Defining the Objects. The first step is to break the system down into 'objects', which are software modules that interact by requesting or granting services from or to each other.

Objects can be described in standardized forms. A common way is to use the 'unified modelling language', which is a graphical representation of objects and their relation to each other. In Figure 3 three objects are shown which represent steering systems. Objects have attributes which are listed in the top section of each block and 'methods' that are listed below and can be called by other objects. Steering systems have properties like the C-factor or maximum turn rates. These factors may be public so that they can be read by other modules or they may be hidden, like internal tuning factors to compensate for the friction of the shaft. The same goes for the methods, which may be public or not. External objects should be able to ask for a certain turning radius, so there should be a public method like 'set_yaw_rate()'. In Figure 3 it is shown how objects can be grouped in hierarchies, applying ideas like inheritance: Front steering and rear wheel steering systems have a lot in common, so they all get the key properties from a generalized steering object. However, the front wheel steering is directly linked to the steering wheel, so there may be methods like 'set_steering_trq()' that may not apply to rearwheel steering systems.

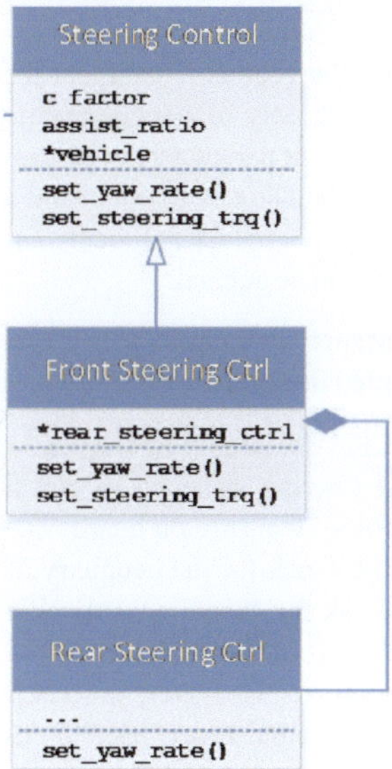

Figure 3 Steering systems modeled as objects

In our example we will need objects in the sensor domain to measure the geometry of the road in front of the vehicle, the LKA-, CM- and APA-modules to calculate trajectories for lane keeping, collision mitigation and automatic parking respectively and objects that represent the actuators on board. However, while LKA, CM and APA all want to command vehicle motion, we cannot allow them to talk to the chassis systems directly for two reasons: (1) multiple systems might want to command actions simultaneously, therefore an arbitrator is needed to prioritize or merge the commands. For example, if collision mitigation is active then parking aid signals must be blocked. (2) the commanded vehicle motion has to be translated to the lower level “language”. Translation is required to avoid a mismatch between the different levels of abstraction.

2. Maximizing Encapsulation. All expertise of an object must be encapsulated within the object and generally not exposed to the outside world (see (3) and (4)). An object should never need to know how a task is done by another object and never ask why it is supposed to do something (like a caterer who does not know how to bake a cake, he only knows, what cake he would like to order).

In our example that means that the whole expertise about how to steer a car must be encapsulated within the chassis systems which immediately implies how the interaction of the modules should be organized. The LKA-module must ask other objects for the geometry of the road ahead of the vehicle and then calculates the geometric path that the vehicle should drive. That means that the LKA-object should only request geometric road parameters from the sensor layer and only communicate path parameters to be executed to the chassis systems. In other words, it should never be allowed to command a steering torque or a steering angle, because that would imply that the LKA module has to worry about the wheelbase or the C-factor of the specific vehicle at hand.

3. Defining the interfaces based on the ‘design by contract’-principle. This is one of the most important ideas and the key to get to a decoupled design (4).

Let us assume that the driver has just switched on the LKA-feature. Now the LKA-module cannot just start feeding path commands to the steering systems. In fact it should not even know which and how many lateral control systems are on board. The only thing that the LKA-object needs is one object that it can ask for the geometry of the road ahead of the vehicle and another one that it can ask for executing a specific path. The knowledge about the physical limits, slip conditions of the road are the expertise of the chassis layer and should be encapsulated there. However, the LKA module has to gather knowledge about the constraints of potential trajectories to ensure that the commanded path is drivable. This knowledge will allow LKA to determine the point in time when an intervention is appropriate and can be supported by the vehicle dynamics.

To get the service we need to introduce a protocol that allows any two units in the vehicle to negotiate services from each other. It is a call-and-response process between a service requester and a service provider that can be broken into three steps (see also the three arrow pattern that links the objects in Figure 2):

(1) **Place a Service Request.** The requesting function informs the provider about the needed service and any boundary conditions that may apply. (An ADAS function may limit acceleration levels to ensure comfort)

(2) **Negotiate boundary conditions.** The service provider processes the request and communicates limiting conditions back that it cannot violate. Those conditions

could arise from vehicle dynamic constraints, limits of the actuators or from adhering to the boundary conditions set in step (1).

(3) **Sign the contract and execute.** The requesting function then orders the service but has to ensure that the request does not violate the pre-conditions communicated in step (2).

The flow is a direct analogy to how human beings organize their interaction or business relationships. If for example a caterer wants to order cakes from a baker which he needs for an event that he is just organizing for a client then the conversation will follow exactly the same process: In the first step he will contact the service provider and pass on all necessary information ("There will be a big event on August 1st with 200 participants. I will need five cakes at 18:00 for the dessert."). The baker will then state his limitations and give the best offer ("I can only do four cakes and cannot deliver before 17:00"). It is important to note that the baker does not need to give the reasons, like a limited capacity of the oven. The customer does not need to know and should not care. If the offer is acceptable for the caterer he will place an order that complies with the limits ("OK, so please deliver four cakes at 17:30").

In our example the LKA module and the lateral control module shall negotiate the service details as shown in Figure 4.

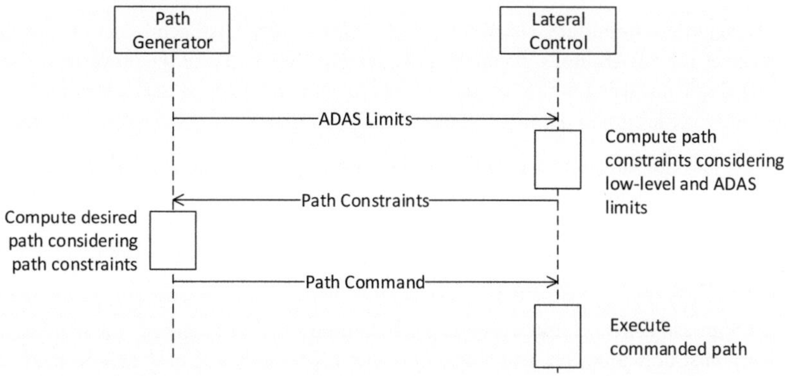

Figure 4 Sequence diagram "negotiation of service" for LKA

During the process the modules will still need to retrieve streams of physical state variables, like wheel speeds, yaw rates etc., but it is crucial to ensure that the choice of signals does not undermine the encapsulation principle. Therefore the LKA module should request the execution of a certain radius or a curvature and not command steering angles.

If the LKA module had to control the wheel angle then it would have to constantly convert the geometry of the desired path into desired steering angles for which it would need to know the wheelbase and the steering properties and this would violate the principle of encapsulation. Commanding steering angles would imply that not only the LKA module, but all ADAS modules that need to influence the lateral motion, would have to be retuned each time when somebody changes anything in the chassis – which we want to avoid.

To elaborate on the LKA-example given above the following paragraphs detail the three-step communication on the signal level:

(1) Place the service request: ADAS Limits. The LKA feature has to comply with certain limits on comfort and safety. For example, functional safety requires to limit the induced lateral acceleration, acceleration rate of the vehicle and the steering wheel torque that the driver might need to override. Comfort requirements might limit the steering wheel torque or the turning rate.

The LKA module does not have detailed knowledge on how to fulfil these limits and delegates them to the lateral control module in the first communication step. The limits might depend on the current driving situation and therefore need to be communicated repeatedly to the lateral control module.

(2) Negotiate boundary conditions: Path constraints. The lateral control module receives the ADAS limits. It can inquire additional limitations of the downstream actuated systems using the same type of protocol. E.g. the steering actuator has a limit on the electric current driving its motor, which in turn is limiting the steering torque.

When several actuators are available, like steering and asymmetric braking, their individual limitations must be combined. Using all imposed limitations, the lateral control module computes the paths which are just within all limits. These limits are communicated back to the LKA module as a set of path constraints like minimal and maximal curvature, and minimal and maximal curvature gradient.

(3) Sign the contract and execute: Path command. Understanding that a drivable path must be within the path constraints, the LKA module is now able to plan its actions. For example it could make an initial decision if an intervention is required by planning an intervention path and checking if the vehicle would stay within a certain area of the lane. If the path of the vehicle becomes critical, the LKA module would activate the lateral control module by transmitting the intervention path data. The path command could consist of an activation flag and a number of coefficients describing the desired path as polynomial in Cartesian, vehicle-centered coordinates. Remember that the path must not be expressed in steering angles. A vector of coefficients that describes the curvature of the track as a polynomial might be a better choice.

4. Minding scalability, extensibility, stability. One of the principles of OOD is the 'open/closed principle' (5). This rule essentially says that systems should be open for extensions but closed for modifications. The LKA module needs to know the curvature of the road ahead of the vehicle, so it has to talk to the sensor layer. However, it should not know about the specific set of GPS-devices or cameras that is on board, because if it had to worry about the actual hardware the design would not be 'closed to modifications'. The way to design the system properly can be learned from Android based smartphones. Any app that needs to know the current location will not have to address the sensors directly. Instead, Android provides a function call 'getLastLocation()', which shields the sensors from the apps. In the case of future chassis systems we will need similar bridges between the ADAS layer and the actuators.

4 Extending the principle to all layers

In the example above the principles were applied to organize the interaction between the ADAS-layer and the chassis layer. However, the approach is universal and can recursively be applied to any layer. In the example above it was critical that the calling unit finds exactly one instance of a service object in the chassis to negotiate the contract and to feed the execution signals to. The specific set of systems on the lower layer must be hidden from the upper unit by these interfacing blocks.

Now let us assume that the ADAS layer wants to command a certain deceleration. The specific vehicle might have a combustion engine or an electric motor, it might have generators for regenerative braking and it will most likely have a traditional friction brake (see Figure 5). Now there must be a block that decomposes the deceleration command and breaks them into specific commands for the subsystems. Again, we want to design a modular system that will keep working even if we remove or change for example the generator. This can be accomplished by applying the same principles as above, except we do it one layer down.

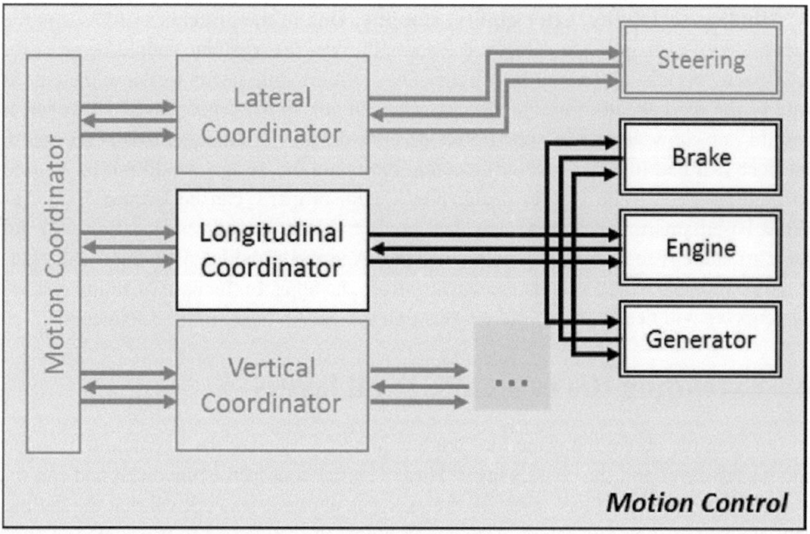

Figure 5 OOD of the longitudinal control system architecture

The design requires that each subsystem includes the same blocks like we saw in the macro structure. Let us assume that the motion coordinator is negotiating the execution of an adaptive cruise request. To serve the upper layers it needs to know about the limitations of the existing systems on board. Applying the principles that were introduced in the previous chapters the sequence of events may work as follows (see Figure 6):

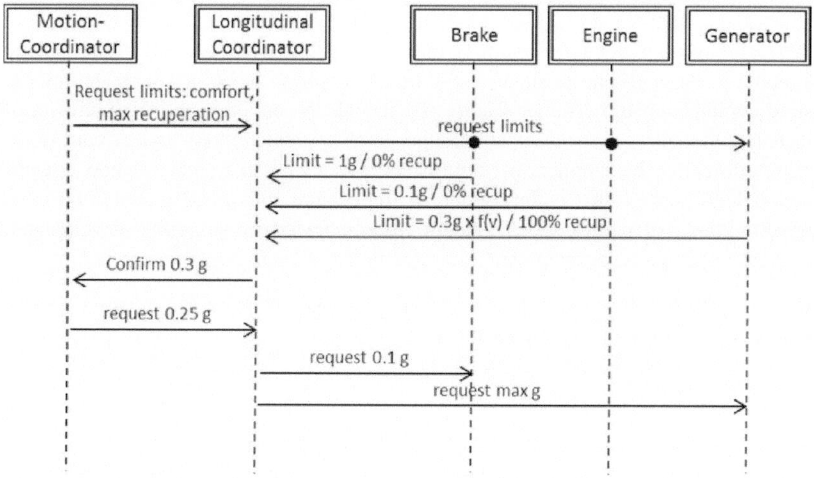

Figure 6 Sequence diagram to execute deceleration requests

The motion generator starts by passing necessary information to the subsystems and requesting the limits for executing a comfort braking manoeuver with the additional boundary condition that we want to maximize the recuperation level from any energy harvesting subsystem. The subsystems respond with their limits. If we have a combustion engine on board then its limit may be 0.1g with zero recuperation potential, if it was an electric motor then both the deceleration and the level of recuperation might be higher. The friction brake has the highest authority while the generator can do less but catch the energy. The longitudinal coordinator will then process these limits and confirm the accumulated boundary condition to the motion coordinator. When the motion coordinator then requests a deceleration level (which must be within the confirmed boundaries) the longitudinal coordinator will distribute the request to the subsystems).

5 Conclusion

Application of the well-known ideas from object-oriented design to traditional automotive controls systems expands their scalability and makes them manageable even when some parts of the system are added or removed or varied.

The encapsulation of control objects hides internal information and minimizes the impact of changes on other systems. Extending the "Design by Contract" methods to allow for dynamic negotiation of the contract enables a control system to react to changes on the fly and to avoid pushing control functions beyond their limits.

When those methods are applied recursively, larger systems can be created without increasing the complexity of object-to-object interaction between parts of the system. Abiding to these principles should give the OEM's high flexibility to create any desired combination of customer features by having the option to combine any set of technical systems (sensors, actuators, algorithms or vehicle body styles) with minimum effort. For the suppliers it would become much easier to add any new component or algorithm to any existing vehicle or architecture, thus creating a lot of business opportunities. All parties should benefit from a drastically simplified development work by enabling the development and validation of all subsystems in parallel.

However, the full advantage of these concepts can only be gained if all partners share their efforts to work on common standards.

Abbreviations

ABS	*Antilock Braking System*
ACC	*Adaptive Cruise Control*
ADAS	*Advanced Driver Assist System*
AFS	*Active Front Steering*
APA	*Automatic Parking Aid*
API	*Application Programming Interface*
ARS	*Active Rear Steering*
CAN	*Controller Area Network*
CM	*Collision Mitigation*
ECU	*Electronic Control Unit*
EPB	Electric Parking Brake
EPS	*Electric Power Steering*
ESC	*Electronic Stability Control*
GPS	*Global Positioning System*
LKA	*Lane Keeping Assist*
OEM	*Original Equipment Manufacturer*
OOD	*Object Oriented Design*

References

1. *AUTOSAR Website.* [Online] www.autosar.org.

2. Martin, Robert C., *"Agile Software Development: Principles, Patterns, and Practices",* Prentice Hall, 2002. p. 149–153

3. Parnas, D. L., *"On the Criteria To Be Used in Decomposing Systems into Modules",* Communications of the ACM 15 (12), 1972, p. 1053–58

4. Dijkstra, Edsger W., *"On the role of scientific thought"..,* Selected writings on Computing: A Personal Perspective, Springer, 1982, p. 60–66

5. Meyer, Bertrand, *"Design by Contract".* Interactive Software Engineering Inc., Technical Report TR-EI-12/CO, 1986

6. Martin, Robert C., *"The Open-Closed Principle"*, The C++ Report, Vol. 8, 1996

Appendix: The principles of an object oriented design approach for chassis control systems

There are four basic principles that should guide the design of a modular architecture for vehicle control systems:

1. Map the systems out as 'Objects': The whole system will be broken down into objects that interact by requesting or granting services from or to each other.

2. Ensure strict Encapsulation: All expertise of an object will be encapsulated within the object and generally not exposed to the outside world. An object should never need to know how a task is done by another object and never ask why it is supposed to do something.

3. Define the interfaces based on the 'design by contract'-principle: The communication between the objects will follow the pattern of

1. Place a service request
2. Negotiate boundary conditions
3. Sign the contract and execute

4. Create scalable and extensible but still stable solutions: Mind the open/closed principle and make sure that the interfaces are stable in the long run and do not depend on specific implementations, such that they should never need to be changed.

Tires and system solutions for the cars of the future

Continental Reifen Deutschland GmbH

Dr.-Ing. Burkhard Wies, Dr. rer. nat. Holger Lange,
Dr.-Ing. Timur Yilkiran

P.E. Pfeffer (Ed.), *8th International Munich Chassis Symposium 2017*,
Proceedings, DOI 10.1007/978-3-658-18459-9_7

Abstract

Already today, far-reaching consequences due to environmental, legal and technological changes and developments are to be foreseen for the tire and automotive industry in the near future. Furthermore, today's customer and market demands lead to an increasing diversification of tire properties and hence of the tire manufacturers product portfolio. The early development of future scenarios and the implementation into business strategies are a key factor for success under conditions in such an extent. As main contributors for this changes and developments the followings aspects were identified:

- Urbanization
- Increasing emissions from road traffic
- Traffic fatalities and injuries
- Connectivity

Deriving from this trends, tires have to fulfill a wide range of functions on a level that hasn't been there yet. In recent years this trend led in a consequence to product specializations into winter, all-season and summer from ECO to U-UHP tire segments as well as to the development of extended mobility devices by providing new TSS (Tire Systems and Solutions) answers for the customers and environmental needs. The capability of retailers to offer this complex product portfolio to the customers can be seen as a key factor to obtain economic advantages in tire selling business. In comparison to replacement customers, OE customers do have significantly higher requirements regarding the target conflicts between vehicle dynamics, rolling resistance and wet performance. Thus, the increasing tendency for marking of OE tires (e.g. * for BMW, NO for Porsche) does make sense to ensure that the customers can remount this perfect performance level on their cars. Enabling, innovative technologies e.g. Monofilament Belt or Thin Capply are meanwhile in serial production, offering the potential to sustainable deliver AA EU-Label values (rolling resistance = A and wet = A).

Based on a worldwide high number of traffic accidents and fatalities Continental follows the strategy "Vision Zero", to decrease the number of traffic injuries dramatically. The development of Advanced Driver Assistant Systems (ADAS) can be seen as the main enabler for this strategy. The intelligent tire will become an integral part of the vehicle and vehicle control system via sensing of inflation pressure, temperature, mileage, life expectancy, load detection and prospectively friction detection.

The interaction between vehicle control systems (ESC) and specific tire designs (rolling resistance optimized tires vs. standard e.g.) have been studied with respect to dry and wet braking performance. The ABS generation 3 with a cycle time of 5 ms ensures shorter braking distances, specifically for low rolling resistant tires due to their changed μ-slip curve.

Besides this safety and information related topics, new materials (e.g. Taraxa Gum) and tire designs offer significant technology & performance improvements but require at the same time deep competence in R&D, manufacturing and processing technology. To bring latest trends and innovations in tire technology into the right market at the right time, the four elements of the future: **Sense**, **Dedicate**, **Perform** and **Sustain** guide the worldwide tire development activities at Continental.

Continental
The Future in Motion
Tire & System Solutions for the Cars of the Future
Chassis.tech plus 2017
8. Internationales Münchner Fahrwerk-Symposium
München, 20. - 21.06.2017
Dr.-Ing. Burkhard Wies - Vice President R&D PLT Replacement ww
Dr. rer. nat. Holger Lange - Vice President R&D PLT Original Equipment ww
Dr.-Ing. Timur Yilkiran - Project Manager Winter Tires

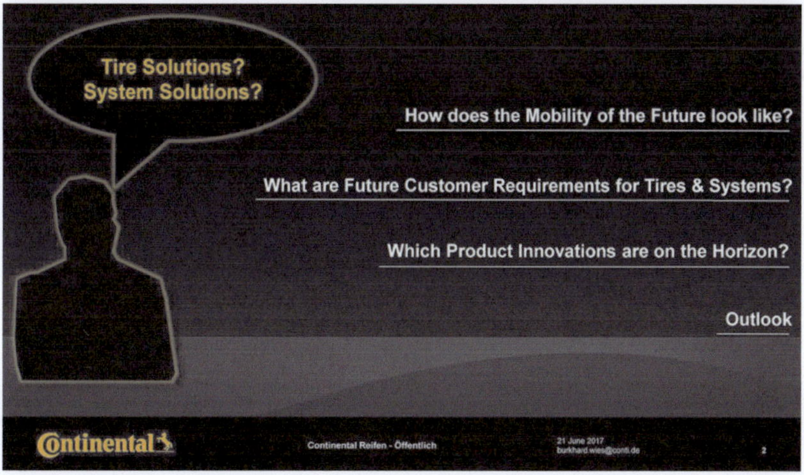
Tire Solutions?
System Solutions?
How does the Mobility of the Future look like?
What are Future Customer Requirements for Tires & Systems?
Which Product Innovations are on the Horizon?
Outlook
Continental
Continental Reifen - Öffentlich
21 June 2017
burkhard.wies@conti.de
2

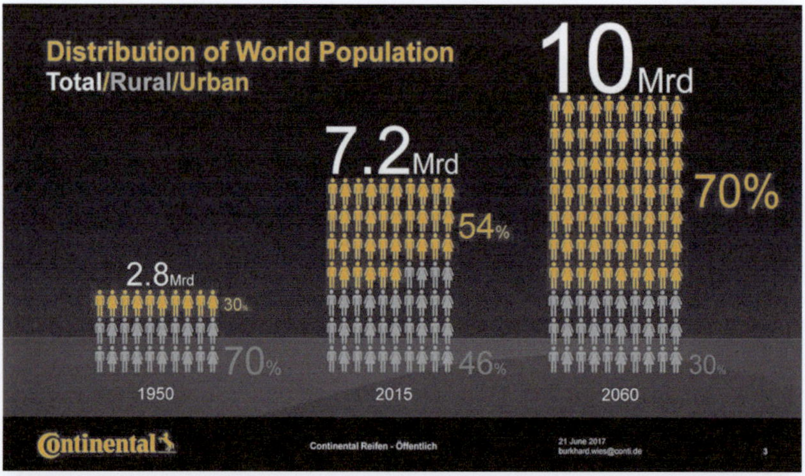
Distribution of World Population
Total/Rural/Urban
2.8Mrd
30%
70%
1950
7.2Mrd
54%
46%
2015
10Mrd
70%
30%
2060
Continental
Continental Reifen - Öffentlich
21 June 2017
burkhard.wies@conti.de
3

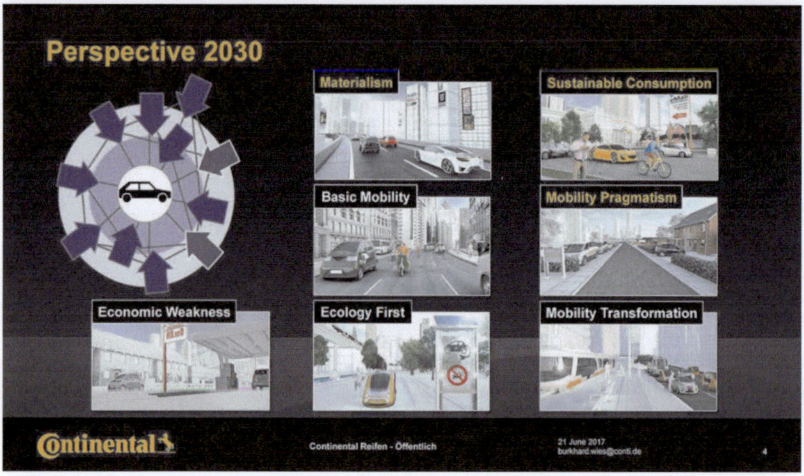
Perspective 2030
Materialism
Sustainable Consumption
Basic Mobility
Mobility Pragmatism
Economic Weakness
Ecology First
Mobility Transformation
Continental
Continental Reifen - Öffentlich
21 June 2017
burkhard.wies@conti.de
4

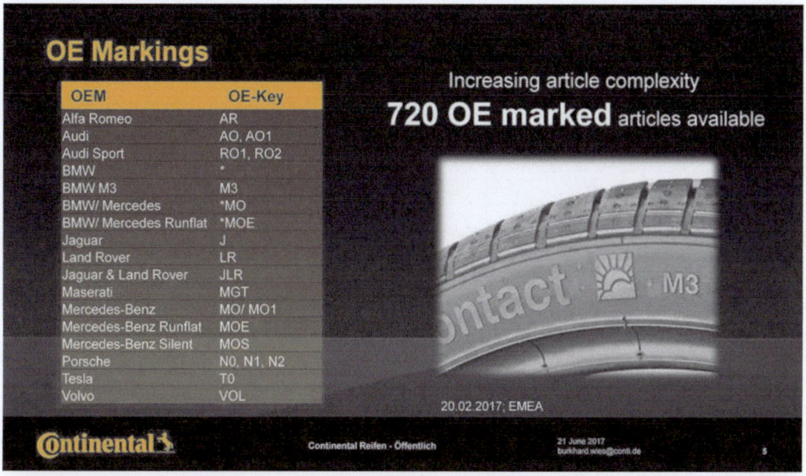
OE Markings
OEM | OE-Key
Alfa Romeo | AR
Audi | AO, AO1
Audi Sport | RO1, RO2
BMW | *
BMW M3 | M3
BMW/ Mercedes | *MO
BMW/ Mercedes Runflat | *MOE
Jaguar | J
Land Rover | LR
Jaguar & Land Rover | JLR
Maserati | MGT
Mercedes-Benz | MO/ MO1
Mercedes-Benz Runflat | MOE
Mercedes-Benz Silent | MOS
Porsche | N0, N1, N2
Tesla | T0
Volvo | VOL
Increasing article complexity
720 OE marked articles available
ntact
M3
20.02.2017; EMEA
Continental
Continental Reifen - Öffentlich
21 June 2017
burkhard.wies@conti.de
5

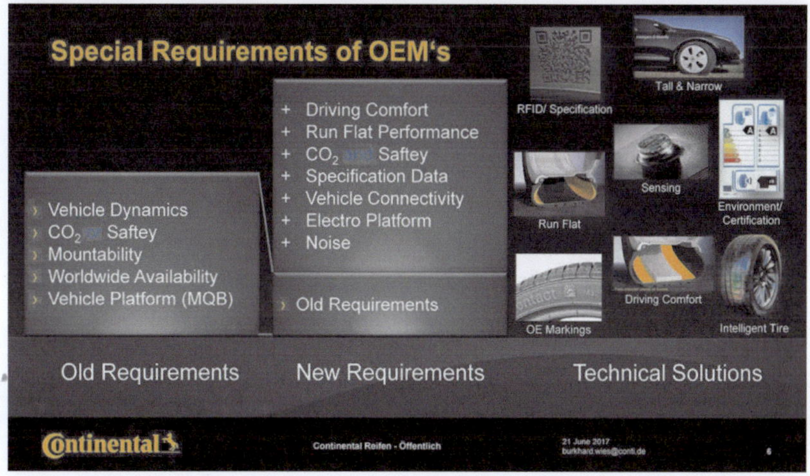
Special Requirements of OEM's
Vehicle Dynamics
CO2 Saftey
Mountability
Worldwide Availability
Vehicle Platform (MQB)
+ Driving Comfort
+ Run Flat Performance
+ CO2 Saftey
+ Specification Data
+ Vehicle Connectivity
+ Electro Platform
+ Noise
Old Requirements
RFID/ Specification
Tall & Narrow
Sensing
Environment/ Certification
Run Flat
OE Markings
Driving Comfort
Intelligent Tire
Old Requirements
New Requirements
Technical Solutions
Continental
Continental Reifen - Öffentlich
21 June 2017
burkhard.wies@conti.de
6

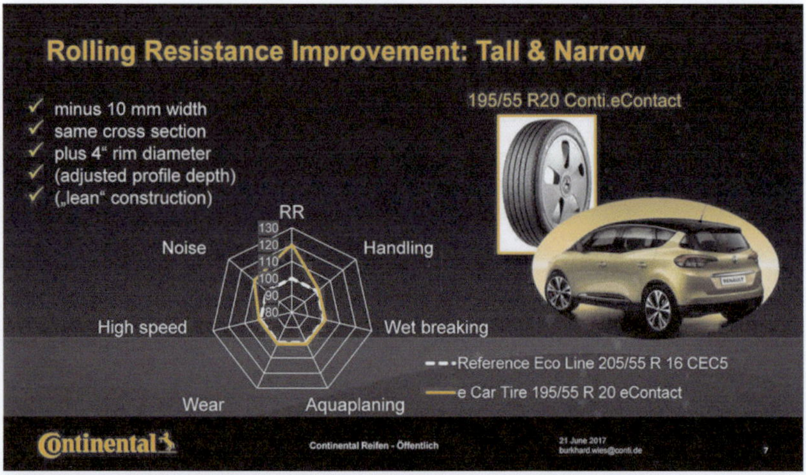
Rolling Resistance Improvement: Tall & Narrow
minus 10 mm width
same cross section
plus 4" rim diameter
(adjusted profile depth)
(„lean" construction)
195/55 R20 Conti.eContact
RR
Noise
Handling
High speed
Wet breaking
Wear
Aquaplaning
130
120
110
100
90
80
Reference Eco Line 205/55 R 16 CEC5
e Car Tire 195/55 R 20 eContact
Continental
Continental Reifen - Öffentlich
21 June 2017
burkhard.wies@conti.de
7

Continental PC-Tires Summer-Portfolio
Ultra High Performance/ Sport
High Performance/ Komfort
Eco
SportContact 6
new!
PremiumContact 6
ContiPremiumC
ContiEcoContact 5
TESTSIEGER 2017
Continental PremiumContact™ 6 225/50 R17 Y
€co-Meister 2017
Continental PremiumContact™ 6 225/50 R17 Y
Auto Bild
225/50 R17 Y – Test Winner „Summer Tire Super Test 2017"
≤ 14"
15"
16"
17"
18"
19"
20"
≥ 21"
Continental
Continental Reifen - Öffentlich
21 June 2017
burkhard.wies@conti.de
8

Continental PC-Tires Winter-Portfolio
WinterContact™ TS 850 P SUV
ContiCrossContact™ Winter
ContiWinterContact™ TS 830 P
SUV
High performance
Mid-sized and compact class
NEW!
WinterContact™ TS 860
Auto Bild
205/55 R16 H – Test Winner
„Winter Tire Super Test 2016"
TESTSIEGER 2016
Continental WinterContact™ TS 860
205/55 R 16 H
€co-Meister 2016
Continental WinterContact™ TS 860
205/55 R 16 H
14" 15" 16" 17" 18" 19" 20" 21" 22"
Continental
Continental Reifen - Öffentlich
21 June 2017
burkhard.wies@conti.de
9

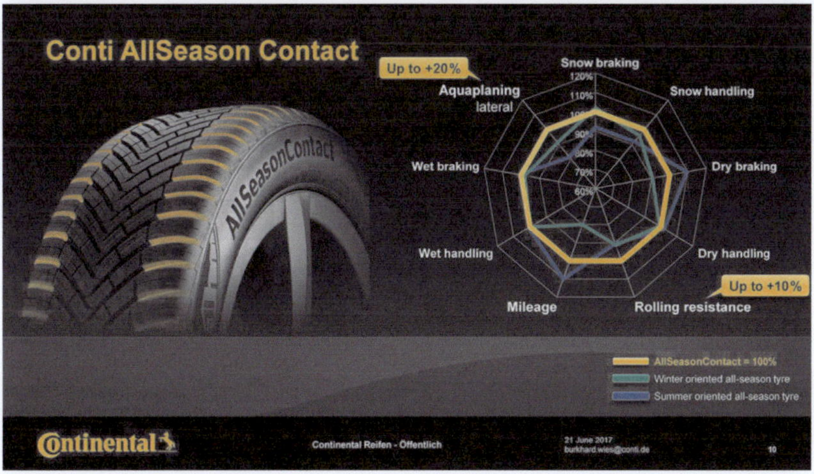
Conti AllSeason Contact
Up to +20%
Aquaplaning lateral
Snow braking
Snow handling
Dry braking
Dry handling
Rolling resistance
Up to +10%
Mileage
Wet handling
Wet braking
120%
110%
90%
80%
70%
60%
AllSeasonContact = 100%
Winter oriented all-season tyre
Summer oriented all-season tyre
Continental
Continental Reifen - Öffentlich
21 June 2017
burkhard.wies@conti.de
10

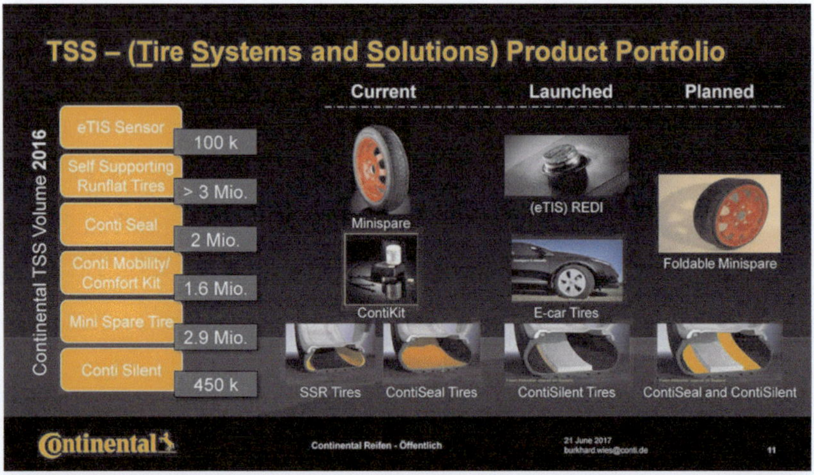

TSS – (Tire Systems and Solutions) Product Portfolio
Current
Launched
Planned
Continental TSS Volume 2016
eTIS Sensor
100 k
Self Supporting Runflat Tires
> 3 Mio.
Conti Seal
2 Mio.
Conti Mobility/ Comfort Kit
1.6 Mio.
Mini Spare Tire
2.9 Mio.
Conti Silent
450 k
Minispare
ContiKit
(eTIS) REDI
E-car Tires
Foldable Minispare
SSR Tires
ContiSeal Tires
ContiSilent Tires
ContiSeal and ContiSilent
Continental
Continental Reifen - Öffentlich
21 June 2017
burkhard.wies@conti.de
11

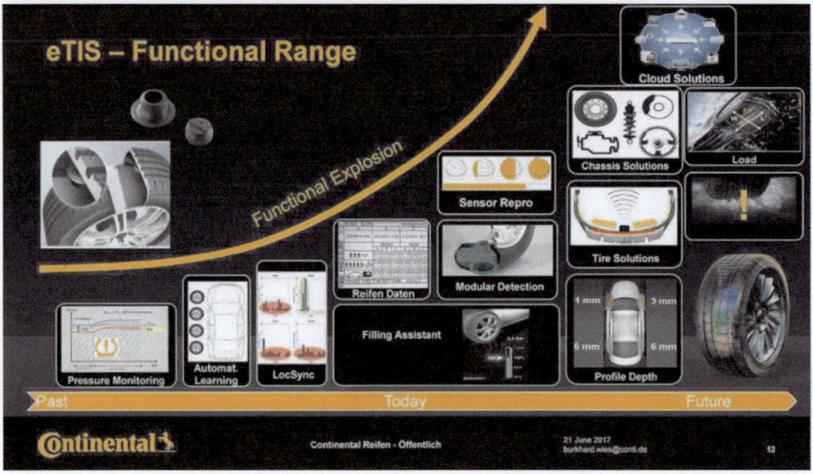

eTIS – Functional Range
Functional Explosion
Cloud Solutions
Chassis Solutions
Load
Sensor Repro
Tire Solutions
Reifen Daten
Modular Detection
Pressure Monitoring
Automat. Learning
LocSync
Filling Assistant
1 mm
3 mm
6 mm
6 mm
Profile Depth
Past
Today
Future
Continental
Continental Reifen - Öffentlich
21 June 2017
burkhard.wies@conti.de
12

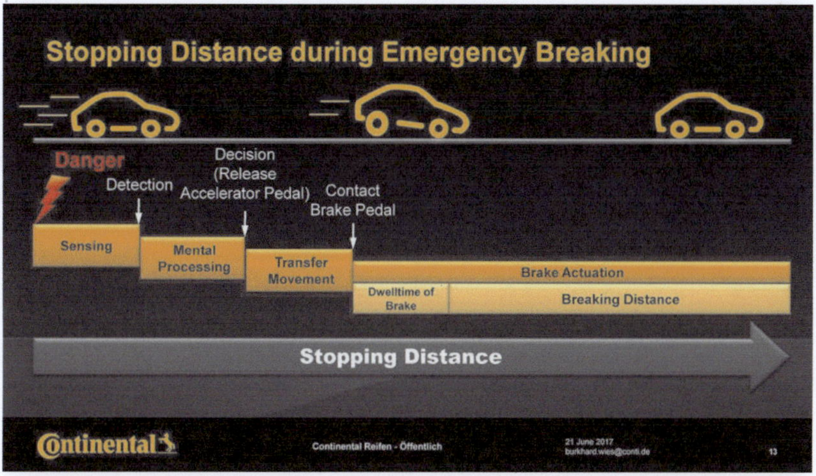
Stopping Distance during Emergency Breaking
Danger
Detection
Decision
(Release
Accelerator Pedal)
Contact
Brake Pedal
Sensing
Mental
Processing
Transfer
Movement
Brake Actuation
Dwelltime of
Brake
Breaking Distance
Stopping Distance
Continental
Continental Reifen - Öffentlich
21 June 2017
burkhard.wies@conti.de
13

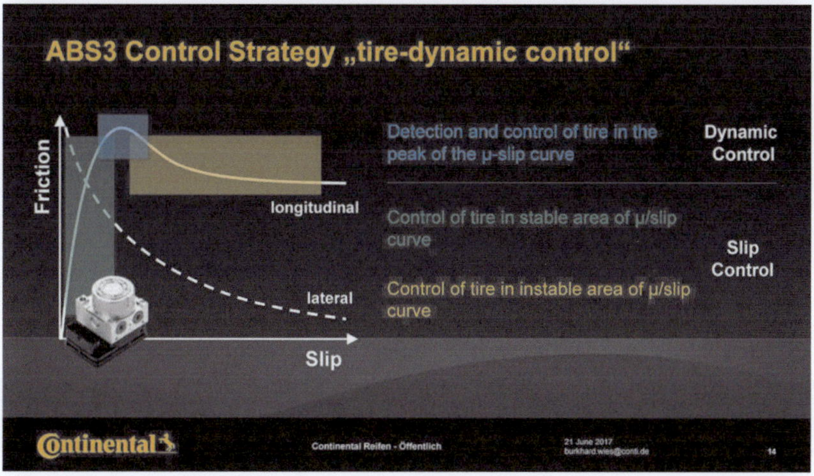
ABS3 Control Strategy „tire-dynamic control“
Friction
longitudinal
lateral
Slip
Detection and control of tire in the peak of the μ-slip curve
Dynamic
Control
Control of tire in stable area of μ/slip curve
Control of tire in instable area of μ/slip curve
Slip
Control
Continental
Continental Reifen - Öffentlich
21 June 2017
burkhard.wies@conti.de
14

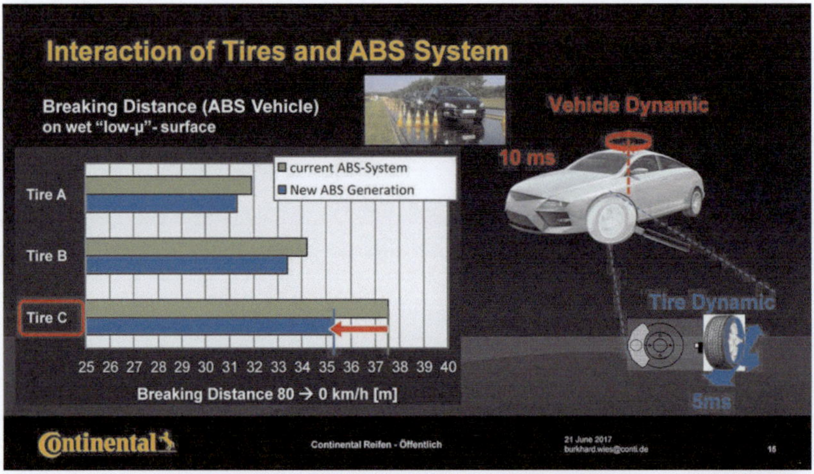
Interaction of Tires and ABS System
Breaking Distance (ABS Vehicle)
on wet "low-µ"- surface
current ABS-System
New ABS Generation
Tire A
Tire B
Tire C
25 26 27 28 29 30 31 32 33 34 35 36 37 38 39 40
Breaking Distance 80 → 0 km/h [m]
Vehicle Dynamic
10 ms
Tire Dynamic
5ms
Continental
Continental Reifen - Öffentlich
21 June 2017
burkhard.wies@conti.de
15

Material Technologies – Dandelion
Tire Ingredients – Future Development: Dandelion Rubber
TARAXA
GUM
Dandelion is unpretentious to climatic conditions
Application in tires & other technical rubber products
Up to 10% replacement of natural rubber fro the rubber tree
Prototype production & testing since 2014
Cooperation with several Institutes (e.g. Fraunhofer IME)
Integration into Serial production expected in 5-10 years
www.taraxagum.de

PARALLEL STRAND I

NEW CHASSIS

Business athlete – exciting driving dynamics for the all-new BMW 5 series

Christof Lischka, BMW Group

P.E. Pfeffer (Ed.), *8th International Munich Chassis Symposium 2017*,
Proceedings, DOI 10.1007/978-3-658-18459-9_8

Summary

The BMW 5 Series has unified sporting agility, dynamic excellence and comfort with each other to the best possible effect for decades.

Fig. 1: The all-new BMW 5 Series

Proudly carrying its heritage, overcoming new challenges and intrepidly facing the increasing pressure of cost-efficiency – these were some of the demands present at the beginning of the development process of the all-new 5 Series. After a lot of hard work during every development step, a decisive chassis setup was produced that could convince both the manufacturer and customers. There was no revolution in terms of technology – just a targeted improving of systems and the architecture. Retrospectively, we can conclusively state that the focused development process, starting with clear targets, continuing with a numerically-optimized chassis design process, and concluding with an integrated chassis set-up, paved the way for fulfilling the goals of the 5 Series as part of BMWs Strategy NUMBER ONE > NEXT. This article will provide a short outline of how BMW developed this successful product using enhanced modern methods for the chassis design and set-up.

1 BMW 5 Series – the story of success has to continue in a 7th generation.

The BMW 5 Series has always ruled its segment as the embodiment of a business sedan. The first BMW 5 Series (E12) was presented at the Frankfurt Motor Show in 1972 as an innovative, midrange four-door sedan with a sporting but elegant design, and which was also powerful and technologically advanced[1]. With the M535i, launched in 1980, BMW opened the next chapter of sheer driving pleasure through its 160 kW/218 hp six-cylinder engine and lively driving dynamics.

Fig. 2: The history of the BMW 5 Series.

The second generation of the 5 Series from BMW, the E28, presented the world with the fastest series-produced diesel vehicle[2] (BMW 524td, introduced in 1983), fortified by a totally re-worked chassis. This was when the number "5" became a synonym for unmistakable driving pleasure in this segment, with innovative new chassis systems such as anti-lock brakes in 1981, electronically controlled dampers, an optional limited slip differential, and speed-dependent Servotronic steering assistance.[3]

---

[1] Ederer, 2017.

[2] Livelifedrive.com , 2016.

[3] Boeriu, 2010.

The E28 received the dual joint axle with wheels suspended by MacPherson struts from the 7 Series and came with a reduced scrub radius to enhance steering comfort. An improved longitudinal stiffness and modified dampers increased the driving and ride comfort. Due to the use of an additional link for the rear suspension, the passenger ride comfort was elevated and the self-steering behaviour became independent from vehicle payload.

The launch of the third generation 5 Series (E34) was a quantum leap over previous generations. An all-wheel drive system was introduced (525iX), and a hydraulically actuated limited slip differential for the rear axle was made available. Improved body stiffness, a lengthened wheelbase and an ideal 50/50 weight distribution improved handling, comfort and safety.

Following the E34, with the E39 the BMW 5 Series became part of a newly created 'premium-class' of vehicles, while still maintaining its sporting and innovative image.[4] In this fourth generation, with the inclusion of Dynamic Stability Control, the BMW 5 Series was acknowledged as an outstanding high-tech representative in its segment. The E39 was also the first vehicle in large-scale production with a nearly all-aluminium chassis, resulting in a weight reduction of 65 kg / 36% and considerably smaller unsprung masses. The touring models were delivered with dampers that were installed nearly horizontally, to ensure a higher storage capacity. For some touring models, height level control combined with air suspension on the rear axle was made standard equipment. Indeed, the magazine "FIRMEN AUTO" nominated the 530d Touring for these and many more reasons as their "business car of the year 2001." [1]

The fifth generation 5 Series (E60) continued the history of unique driving dynamics by introducing Active Front Steering and the Active Roll Stabilization Dynamic Drive. The all-aluminium frontend facilitated a nearly 50/50 axle load distribution and the introduction of the clutch-based all-wheel drive system provided top-level traction and driving dynamics.

Integral Active Steering showed up in the sixth generation of the 5 Series (F10), introducing rear-wheel steering. The Driving Experience Control switch was also launched, with the programs Comfort, Normal, Sport and Sport-Plus included. Adaptive Drive brought with it Active Roll Stabilization and Dynamic Damper Control to maintain refined driving dynamics and ride comfort. The improvements of the chassis further included a double wishbone suspension with dual ball joints and a rear multilink suspension.

[4] Jancke, 2012.

Having proudly acquired a leading position in the business sedan segment over the past several years, the challenge for the 7th generation of the 5 Series became clear: upholding its heritage in the most outstanding fashion.

2 Development

2.1 Development goals

Renewing a successful vehicle involves creating a contemporary interpretation of the major distinguishing elements of the outgoing model. Properties that made the predecessor outstanding have to be adopted to the new boundary conditions laid out by governments and changing customer expectations. Despite the elevated requirements, the vehicle credentials must be forward-thinking – stepping backward is not an option. The essential factor in continuing a story of success is to be simultaneously heritage-conscious, timely and better than the predecessor – independent from legislation.

One of the major goals of the G30 was to continuously reduce CO2 emissions. This entailed global mass reductions from the vehicle itself and an increased PHEV presence in the market. Lightweight construction – often conflicting with crash requirements – and conceiving a vehicle architecture suitable for a large-scale PHEV rollout led the chassis design process. On the other hand, BMW is being pushed towards automated driving technologies, particularly with the 5 Series. The financial strain caused by shifting priorities between conventional development themes and key future technologies implied an extra effort to realize a higher cost efficiency for the development process. To pioneer future technologies, digitalization and the upward compatibility of the E/E systems were two main topics that had to be taken into account.

In general, the customer expectations grew in terms of combining particular chassis options. A result was the extension of the flexibility of configurations, for example by combining Integral Active Steering with xDrive, or by combining xDrive with a lowered suspension setting (sport suspension).

At this point, some attention should be drawn to the competitive situation. The strong competition amongst OEMs has led to changes in the product portfolio. Premium manufactures have conspicuously shifted their image towards a sporty character and profile in the last few years. Competitors have begun occupying the classical turf of BMW, reducing the scope available for a distinct and exclusive driving character for BMW vehicles.

2.2 Development guide: driving dynamic "characters"

A well-defined set of differentiating attributes that make a product unique and individual is critical for continuing a story of success. Chassis systems alone do not definitively lead to a successful product. Integrating a large number of systems – feature counting – is also not the road to sustainable success. Chassis components have to be deployed purposefully to accomplish the mission of creating a distinct and emotionally rich product.

BMW did not introduce new chassis technology in the all-new 5 Series per se. Rather, existing technologies were optimized and an expedient architecture composed. Nevertheless, the new 5 Series objectively performs better than its predecessor, attaining greater perfection in terms of its unique dynamic properties and driving experience.

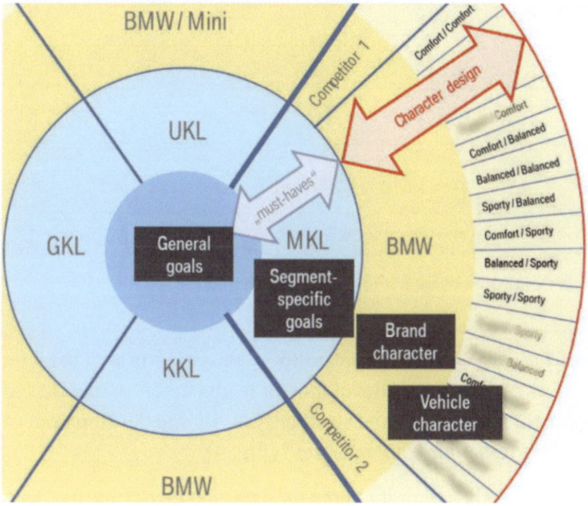

Fig. 3: Schematic deriving of driving dynamic character profiles

To describe the driving character profile of a vehicle, one first has to filter out all elements that facilitate basic mobility without any further aspirations, such as having four wheels and a steering system that is able to turn those wheels in a specific direction. Every car manufacturer will provide these elements. The next stage that has to be filtered out is the class-specific functionality, such as rollover protection for vehicles with high centers of gravity. What remains are brand-specific elements that shape a brand's character. For the BMW brand, this might be precision, ease of handling and a harmonious, linear driving behavior (without overdrawing or exaggerating these elements, such as with special set-ups to improve the noticeability of systems or functionality).

Research at BMW has shown that customers choose our products not only because they have the best performance figures, but also due to the way they feel during driving. This indicates that customers do more than just count features and track top values in objective tests.

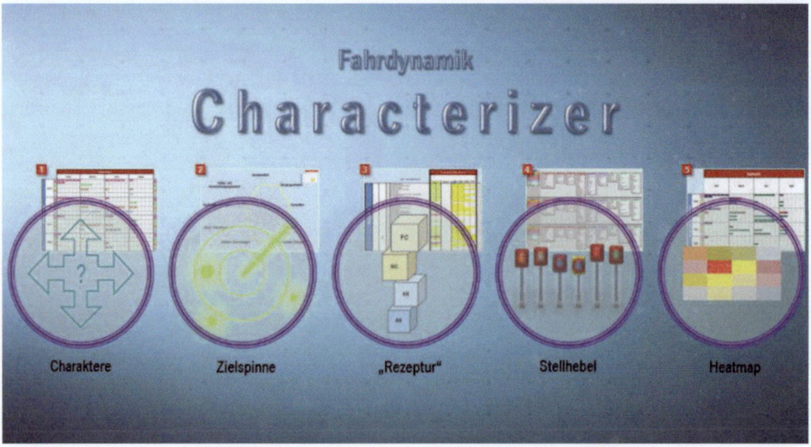

Fig. 4: Screenshot of the chassis develompent tool "Fahrdynamik-Characterizer"

As a result, each vehicle of the BMW product portfolio is classified in a "character matrix." The strict alignment of BMW chassis development to characters is manifested in BMW's software development tools, where the starting point for the development of a new product is the BMW "Fahrdynamik-Characterizer" (Fig. 4).

The characterizing process can be exhausting for some models, and clarity for the driving dynamic ambitions can at times only be achieved in the long run. However, the effort proves itself to be invaluable when the press honors the work we have put into a vehicle during development process.

3 Basic chassis design

3.1 Chassis design process

Every design process starts with the definition of the top-level design goals. In case of the basic chassis design, the subjective driving experience is the top-level goal. At the end of the development process, the product must meet all expectations that customers have when they think about a BMW 5 Series. A consistent set of subjective perceptions leads to a coherent driving dynamics character. The subjective goal set represents all main characteristics of the vehicle, such as steering, straight-line driving or cornering behavior.

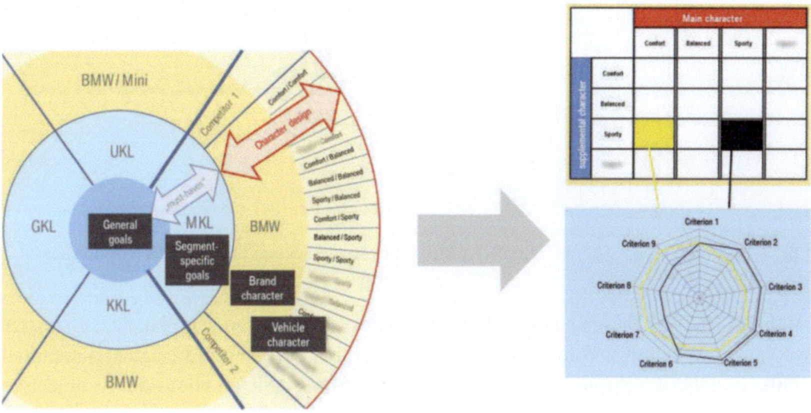

Fig. 5: Different characters result in the final subjective goal sets

For each car segment, class-adequate goals must be met, as well as goals that define the core character properties of each BMW equally. Yet the last fine adjustments of the goals can only be achieved by specifying the individual driving dynamics character for each product separately. Hence, BMW has developed a classification of main and supplemental characters, each a the range from comfort to sporty, resulting in a "character matrix," as illustrated in chapter 2.2. For each combination of main and supplemental character, a generic set of subjective driving dynamics goals are developed that from experience lead to a harmonic character. Fig. 5 displays the different goal levels as well as the subjective goal sets, which result from the character definitions.

As a next step in the process, the objective goals are derived from the subjective goal set, requiring the use of quantification methods such as simulation models. With high levels of experience from previous vehicle generations and numerous results from objectification activities, a relationship can be developed between the subjective and objective vehicle behavior.

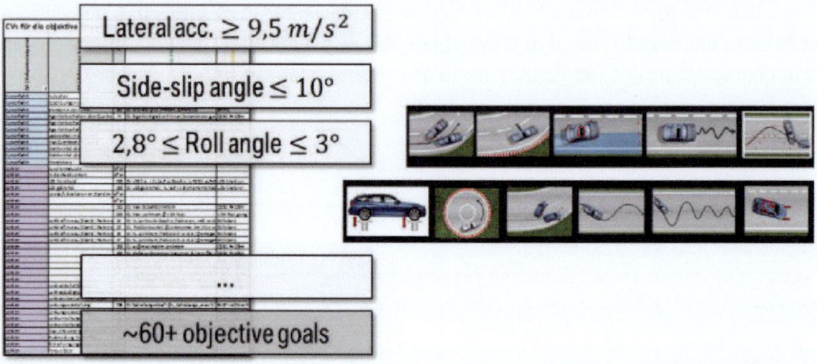

Fig. 6: A comprehensive set of objective goals is key to the successful design of a basic chassis

With the BMW 5 Series having a balanced main and balanced supplemental character we can now define a comprehensive objective goal set from the original subjective goals. For the development of the BMW 5 Series we utilized about 60-70 objective goals for the basic chassis design process (Fig. 6).

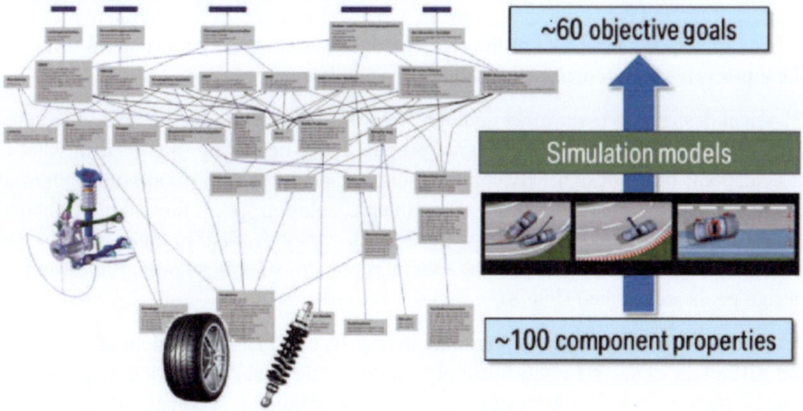

Fig. 7: Simulation models are derived based on dependency graphs

For the design of chassis components, we require the relationship between component and/or system properties as well as the vehicle behavior. To achieve this, dependency graphs help us to manage the complexity of interacting component, system and vehicle properties (Fig. 7).

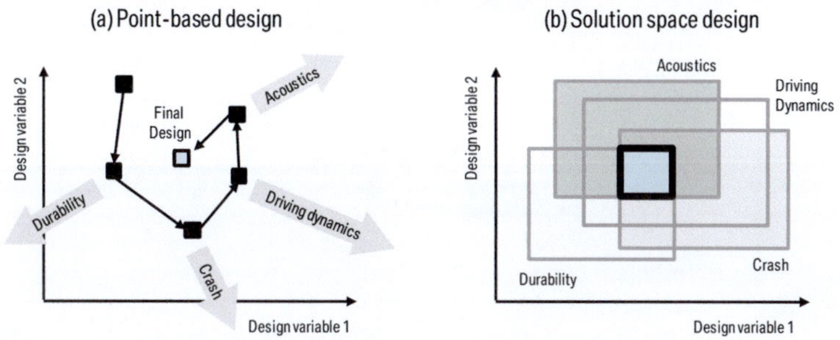

Fig. 8: Design process based on: iterations (a) / permissible design ranges (b) [2].

Starting from that knowledge, we were able to develop complex simulation models that help us to map the influence of the component or system properties (bottom level) on the top-level, which represents the objective vehicle goals.

Classical design processes often do not rely on the systematic development of requirements. Instead, candidate solutions are proposed and iteratively improved in what is called "point-based design." By contrast, solution spaces specify permissible ranges of design variables for each discipline. For example, solution spaces for two different disciplines that seem to be in conflict with each other may overlap. In this case, they can be combined successfully, and define a target region for subsystem properties where all design goals are reached (Fig. 8).

As example for designing based on solution spaces, consider the problem of specifying the stiffnesses of sway bars to fulfill objective vehicle goals such as the magnitude of the roll angle vs. lateral acceleration, as well as the maximum lateral acceleration while cornering (Fig. 9). Solution spaces help to derive the permissible ranges of stiffnesses, which are represented with green dots in the parameter space. Rectangular ranges of the parameter ranges can be defined to decouple the design parameters, such as the green and blue rectangles in Fig. 5. The overlapping regions of both solution spaces offers the permissible range for communal component settings. Such a solution space was employed in the design process of the new BMW 5 Series, in order to create a robust basic chassis design.

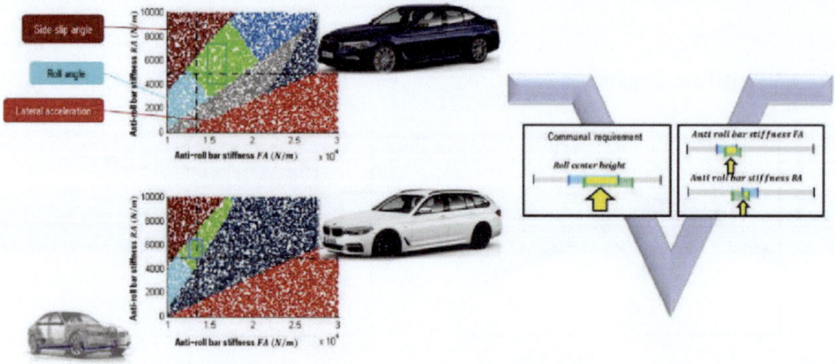

Fig. 9: Deriving requirements for communal components using solution spaces.

3.2 The 7th generation of the BMW 5 Series

The chassis of the 5 Series was completely engineered for its 7th generation (the G30) according to described development process. As a result, the 2017 BMW 5 Series features a redesigned chassis, intelligent lightweight engineering, BMW's renowned balanced weight distribution and a strongly torsion-resistant body. The entirely new development allowed for weight losses of up to 100 kg through the use of magnesium, aluminum and high-strength steel. BMW engineers have succeeded in making the car's handling more agile and precise without compromising comfort; a hallmark of this sport sedan[5].

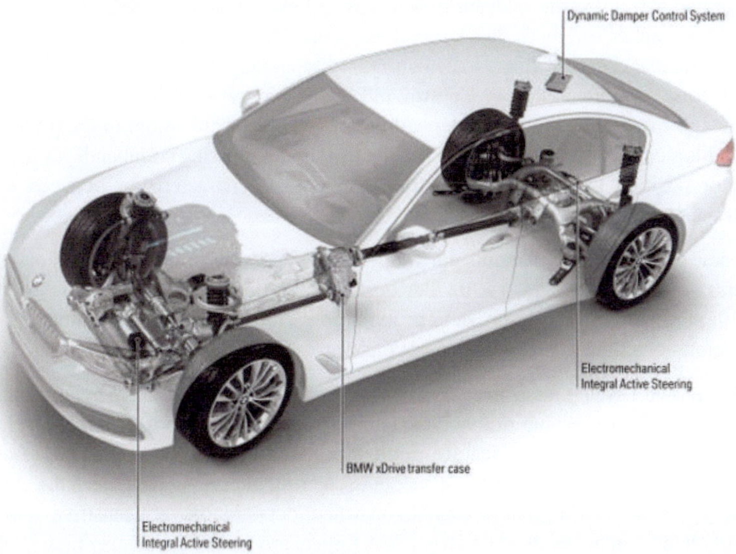

Fig. 10: Chassis control systems of the new 5 Series.

Light-metal components are used extensively in the new five-link rear suspension of the G30 5 series, which is even lighter and stiffer than in the previous model, and provides precise wheel location with excellent tracking stability in all driving situations. The specially calibrated elastokinematics of the suspension, the large axle sub frame

[5] Dexter, 2017.

mounting system, the compression strut connecting the suspension to the body, the high structural stiffness and low unsprung mass all combine to create an excellent overall balance between agility and comfort. The large mounting and supporting system plays a further important role in ensuring acoustic insulation from the powertrain and suspension[6].

The BMW 5 Series can be equipped with a revised electromechanical Integral Active Steering system, while the electromechanical Active Roll Stabilization sees its own series debut. The Driving Experience Control switch features slightly changed programs, allowing the driver to choose between Comfort, Sport, ECO PRO and, optionally, Adaptive modes (the latter if the vehicle is equipped with Dynamic Damper Control), depending on the driving situation and personal preferences. Each mode activates a predefined set-up for the relevant powertrain and chassis components. The Sport and ECO PRO modes can be further tailored using the iDrive Controller, which allows the powertrain and chassis settings to be configured separately. In the new Adaptive Mode, the steering, Dynamic Damper Control and Steptronic transmission are automatically adapted to the driving style and, depending on vehicle specification, to the route. The control logic reacts to inputs such as accelerator inputs, steering commands, or the gear selector position to vary the powertrain and chassis characteristics between being sporty- and comfort-oriented.

The new 5 Series model variants portfolio has been complemented by an EfficientDynamics Edition of the 520d, a 530e iPerformance plug-in hybrid variant, and the BMW M550i xDrive.[7] All engines prove to be more powerful and fuel-efficient, but with less CO2 emissions than in previous generations.[8]

There are two Driving Assistance Systems providing semi-automated driving capabilities in the G30 5 series: the Active Lane Keeping and Traffic Jam Assistants. The latest generation of these systems feature improved image recognition and software algorithms, providing smooth, comfort-enhancing steering interventions to keep the vehicle reliably on course. The system's hands-on-wheel detection is very sensitive, and can identify hand contact even when only a few fingers are touching the steering wheel.

[6] Dexter, 2017.

[7] DeMattia, 2016.

[8] Huma, 2016.

The new 5 Series also features Side-Collision Avoidance. Furthermore, Dynamic Cruise Control (DCC) now simultaneously maintains any selected speed between 30 and 250 km/h, while supporting energy-saving coasting by decoupling the engine from the powertrain. The new and improved Active Cruise Control (ACC) with Stop & Go uses radar to adjust the pre-selected speed to the current traffic situation.[9] Another new function contained in Active Cruise Control with Stop & Go is the evasion aid. If a quick lane-change is required, the evasion aid supports the necessary evasive steering action. However, faster steering and counter steering during an evasive maneuver reduces vehicle instability, which in turn is countered by the vehicle's Dynamic Stability Control (DSC).

Fig. 11: Driving assistance systems of the new 5 Series.

[9] Santer, 2016.

4 Integration

Integrated application plays a central role in reaching the targets for the driving dynamics of a vehicle and realizing its desired character. The application approach combines two central themes: 1) the basic development of the chassis hardware, including the influences of the body structure and the aerodynamics on the vehicle dynamics; and 2)

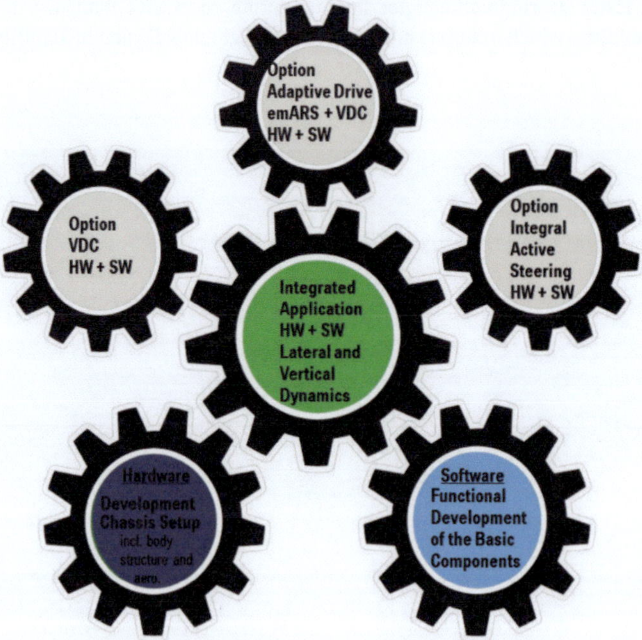

Fig. 12: Central role of integrated application.

the functional development of the software components for the driving dynamics systems. Integral application entails designing and tuning all vertical- and lateral-dynamic systems centrally, minimizing the number of disparate specialist teams and maximizing the efficacy of the integrated functionality.

The implementation of chassis control systems has become more complex, requiring increased consideration for the interaction and interdependencies between the individual systems. A comprehensive, integrated application facilitates the scrutiny of these interdependencies, ensuring a harmonious result to the overall development of the chassis. Accompanying this approach is a distinct ordering of each control system with respect to their dynamic potential and possible contribution to a vehicle's personality:

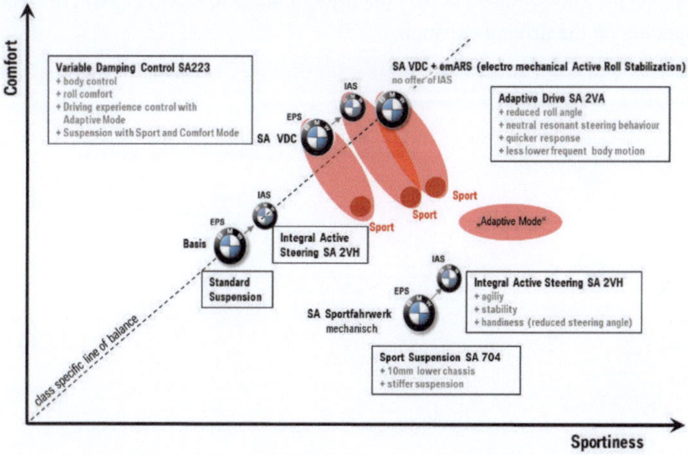

Fig. 13: Influence of chassis systems on driving dynamics character.

A vehicle's targets can only be achieved with a synergetic fusion of hardware- and software-based functions. An illustrative example of this will be given next for the "Adaptive Drive" system (a combination of Dynamic Damper Control, VDC, and Active Roll Stabilization, emARS), to show how its functional characteristics could be developed using block functions over two different control modules.

5 Chassis set-up

Optionally available, variable dampers provide a more comfortable vehicle characteristic in comparison to the base-suspension setup. However, the dampers can also assume a sportier character if SPORT is selected as the driving mode. In addition, the ADAPTIVE mode uses the dampers to vary the driving mode between COMFORT and SPORT, depending on the driving situation.

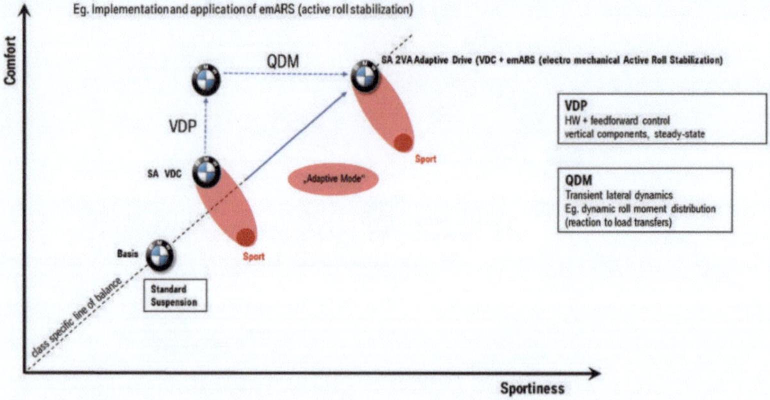

Fig. 14: Influence of application on driving dynamics character.

The integration of Active Roll Stabilization leads initially to a more comfortable driving characteristic, through the decoupling of the roll stabilization from vertical body motions. As such, the application of the underlying function is seeded in the algorithms and logic of the VDP control unit. The personality of the vehicle is then pulled back towards a sportier nature through parameterization of the transient lateral dynamics within the QDM control unit (lateral dynamics management), while simultaneously managing interactions with other programmable functions in the QDM.

6 From iPerformance to M Performace: a wide spectrum for the chassis set-up

The BMW M Performance and iPerformance models are filling the gaps between the core brand BMW and the bookends BMW M und BMW i.

The vibrantly dynamic M550i xDrive captivates with a sporty chassis setup and up to 462bhp engine power, which can propel the car to 100km/h in 4.0 seconds. The sportiness is further accompanied by a high level of suitability for daily use.

Fig. 15: The 5 Series product portfolio.

The M Performance model receives the Adaptive M suspension, which is lowered by 10mm in comparison to the base suspension. Some specific modifications, such as the tuned camber and optimized steering, were essential to accomplish the targets for the handling of the M Performance vehicle.

The M550i xDrive with the M Sport brake package and the optional Adaptive M Suspension Professional enjoys the exclusive combination of the Adaptive M suspension, Integral Active Steering and Active Roll Stabilization. Combined with the rear-biased tuning of the all-wheel drive system (xDrive), the 20” high-performance wheels promote a precise and agile handling while assuring a superb level of traction.

The chassis of the M Performance Model is direct and to the point, providing the most dynamic chassis to the most dynamic motor. In this way, the M550i xDrive closes the gap between BMW's top models and the core models of BMW M, which themselves are designed for maximum performance and racetrack capability.

The highly efficient 530e iPerformance, on the other hand, is a plug-in hybrid variant of the G30 with carbon emissions of just 44g/km. It represents a well-balanced car that can be driven electrically with zero local emissions, while maintaining the dynamic driving experience for which the brand is renowned, providing excellent long-distance driving credentials.

Fig. 16: The grading of the 5 Series M Performance offer.

The iPerformance tag attached to the BMW 530e is synonymous with the transfer of expertise from BMW i to the core brand's cars. Consequently, the 530e comes with a specific chassis set-up to compensate for conceptual challenges with electric vehicles, like an increased vehicle mass and higher moments of inertia. The result is the preservation of the driving dynamics level from the agile base sedan.

The all-new 5 Series has arrived with an unprecedented spectrum of driving dynamic profiles for one product subline. From a fully-fledged PHEV up to a unique expression of sporting excellence in the M Performance model, the 5 Series delivers maximum driving dynamics without sacrificing comfort, thus intensifying driving pleasure.

7 Conclusions

The all-new 5 Series from BMW continues a long story of success. Despite several new challenges, the driving experience of the 5 Series could be brought to a new level, exemplifying precision, ease of handling and a harmonious driving behavior within its cultivated character.

The formula for success entailed focusing on the driving dynamics profile through deriving the driving dynamics properties from a guiding targeted character: the beginning of the development process defines *where* the voyage is headed, by defining the character profile, whereas the integrated application of the chassis systems defines *how* the development process will get there – independent form driveline topologies.

The range of possible character profiles, from iPerformance through the base model to an M Performance variant, and the process of diligently fulfilling the development goals brings the new 5 Series closer to the customer than ever, revealing an even stronger

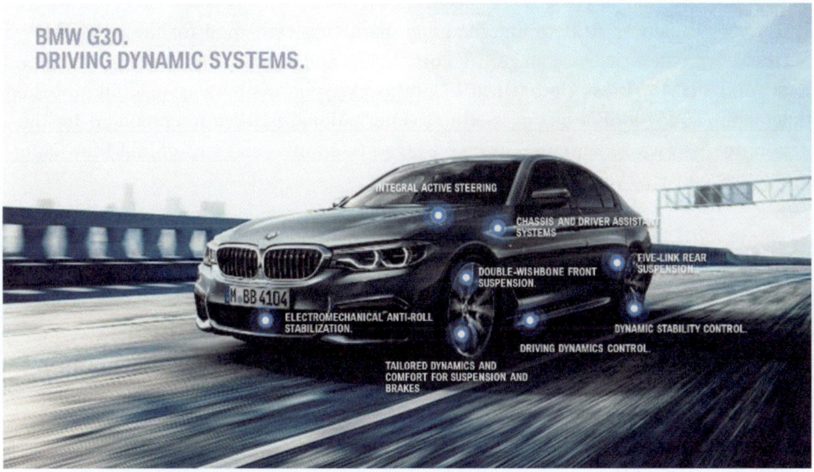

Fig. 17: The 5 Series driving dynamic systems.

emotional appeal. The all-new BMW 5 Series once more defines the athletic and executive interpretation of a true sport sedan, and provides unsurpassed driving dynamics and comfort. In other words, “It remains the epitome of the business sedan.”

7 References

[1] Der neue BMW 5er Touring. BMW Group, June, 11th 2010, P. 63.

[2] M. Zimmermann, J. Edler von Hoessle. "Computing Solution Spaces for Robust Design". International Journal for Numerical Methods in Engineering 94 (2013), pages 290–307.

H. Boeriu, "History of the BMW 5 Series". BMW Blog (07/2010), http://www.bmwblog.com/2010/04/13/history-of-the-bmw-5-series/

N. DeMattia, "World Premiere: The 2017 BMW G30 5 Series". BMW Blog (10/2016), http://www.bmwblog.com/2016/10/12/world-premiere-the-2017-bmw-g30-5-series/.

R. Dexter, "The All-New 2017 BMW 5 Series Sedan: Performance, Redefined". BMW Pressclub Canada (01/2017), https://www.press.bmwgroup.com/canada/article/detail/T0264877EN/the-all-new-2017-bmw-5-series-sedan:-performance-redefined?language=en

B. Ederer, "Business Athlete: international marketing campaign for the new BMW 5 Series Sedan enters decisive stage. TV commercial and online film featuring actor Scott Eastwood". BMW Press Club (01/2017), https://www.press.bmwgroup.com/global/article/detail/T0267406EN/business-athlete:-international-marketing-campaign-for-the-new-bmw-5-series-sedan-enters-decisive-stage-tv-commercial-and-online-film-featuring-actor-scott-eastwood?language=en

E. Huma, "2017 G30 BMW 5 Series Sedan: Driving Dynamics and Efficiency". BMWCOOP.com (10/16), http://www.bmwcoop.com/2016/10/17/2017-g30-bmw-5-series-sedan-driving-dynamics-efficiency/.

C. C. Jancke, "40 Jahre BMW 5er – Als München leuchtete". Handelsblatt (02/2012) http://www.handelsblatt.com/auto/test-technik/40-jahre-bmw-5er-als-muenchen-leuchtete/6211456.html

livelifedrive.com, "History Of The BMW 5 Series – Seven Generations And Counting From 1972". (02/17), http://www.livelifedrive.com/news/history-bmw-5-series-seven-generations-and-counting-1972/44079.

B. Santer, "The new BMW 5 Series Sedan". BMW Press Club (12/16), https://www.press.bmwgroup.com/global/article/detail/T0264349EN/the-new-bmw-5-series-sedan?language=en (02.05.2017).

BMW Group, http://www.bmw.de/de/neufahrzeuge/5er/limousine/2016/business-assistenzsysteme/fahrerassistenz.html (02.05.2017).

A compact rear chassis for the 2018 Honda Odyssey – merits and challenges, and innovations

Joshua C. Johnson, Principal Engineer, Chassis Design
Adam Goellner, Senior Engineer, Vehicle Performance Engineering

Co-authors and collaborators:
Kirk Rhoades, Senior Engineer, Chassis Design
Eric Payne, Engineer, Chassis Design
Derek Horton, Engineer, Vehicle Performance Engineering
Keith Malarik, Senior Engineer, Ride/Handling Performance
Bryan Johnson, Senior Engineer, Ride/Handling Performance
Sara Bradner, Engineer, Ride/Handling Performance
Matthew Hoersten, Principal Engineer, Chassis Design
David Wolf, Senior Engineer, Reliability Engineering

P.E. Pfeffer (Ed.), *8th International Munich Chassis Symposium 2017*,
Proceedings, DOI 10.1007/978-3-658-18459-9_9

1 Abstract

The product objectives of the 2018 US-type Odyssey required a new rear body and chassis architecture. Following principles of man-maximum, machine-minimum, a rear suspension and subframe with an innovative and compact link arrangement were developed. The new chassis not only improved frame packaging with lower mass, but also increased ride and handling performance levels. This paper discloses unique design features of the suspension system, performance improvement levels, and tools and techniques utilized in the development. This paper also describes some unique chassis packaging integration concepts which enabled mass reduction of surrounding parts. New features of the rear chassis are:

- New link arrangement, which enables a simplified frame layout and mass reduction.
- Kinematic and structural improvements, which improve performance.
- A compact subframe, which reduces mass and enables integration of fuel system for mass reduction.
- A patented subframe mount packaging which improves isolation and rigidity.

2 Introduction to US-Type Honda Odyssey

The US-type Odyssey (not to be confused with the Japan-type Odyssey, which is a completely different vehicle built on a different platform) is an eight-passenger minivan with traditional US market sizing. The gross vehicle weight rating (GVWR) is 2,730 kg, with a 3.0-m wheelbase. The Odyssey has a 3.5-l, normally aspirated V6 engine as standard equipment. In addition to the US, the US-type Odyssey is also available in the following markets: Canada, Mexico and Central America, the Middle East, and Korea.

Figure 1. 2018 Honda Odyssey (US-Type)

The Odyssey is a family mover with a focus on utility, operating ease, and efficiency. At the same time, it has been infused with Honda dynamics DNA, which makes the vehicle enjoyable to drive. Safety, utility, and entertainment features, such as on-board internet connectivity, coolers, and even vacuums, are ever-increasing in this vehicle type. To offset the increased mass of such features, engineers strove to optimize the platform structure without compromising performance. As the focus of these vehicles is on the passengers, cargo, and their respective space requirements, engineers also strove to reduce the intrusion of mechanical parts into those spaces through the philosophy of *"Man-Max, Machine-Min."*

3 Honda US Type Light Truck Platform History

The 2018 Odyssey is the fourth derivative model built on the PFP[1] platform, after the 2014 Acura MDX, 2016 Honda Pilot, and 2017 Honda Ridgeline. Honda R&D Americas in Ohio, United States, developed the PFP platform and all of its derivative models. These vehicles are the largest road vehicles in the Honda line up, and are currently produced in two plants in the United States.

Figure 2. Acura MDX, Honda Pilot, Honda Ridgeline

The first three vehicles are uni-body SUV/SUTs with available all-wheel-drive and moderately high ground clearance. However, the Odyssey is a minivan with a low flat rear floor and passenger car level ground clearance. The Odyssey is available in front-wheel drive configuration only. Figure 3 shows the package overlay of the Odyssey. The front seating position is similar to the Pilot SUV, but the ground clearance has been reduced by adjusting the suspension. The third row seating position benefits from a low flat floor, which is also convenient for cargo loading when the seats are stowed. The low, flat floor requires a vertically compact rear chassis and suspension compared to the SUV/SUT derivatives. The SUV/SUTs apply a 4-link suspension with coil-over springs and dampers, which cannot work with the Odyssey floor arrangement. Figure 4 shows the interference of the SUV suspension with the Odyssey floor frame. In past Odysseys, this challenge was solved by deploying a trailing arm suspension type with a single upper arm and separated spring and damper. Figure 5 shows the previous Odyssey rear chassis and frame (note the frame escapement around the upper arm). This design provides sufficient performance for today's minivan market, but with recent innovations, both the package efficiency and performance can be improved to meet future customers' needs and wants.

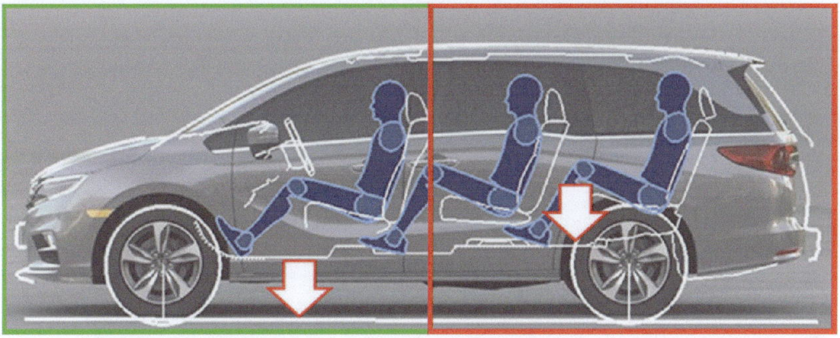

Figure 3. Odyssey Package Overlay

Figure 4. Pilot 4-Link Rear Suspension Overlaid with Odyssey Rear Frame

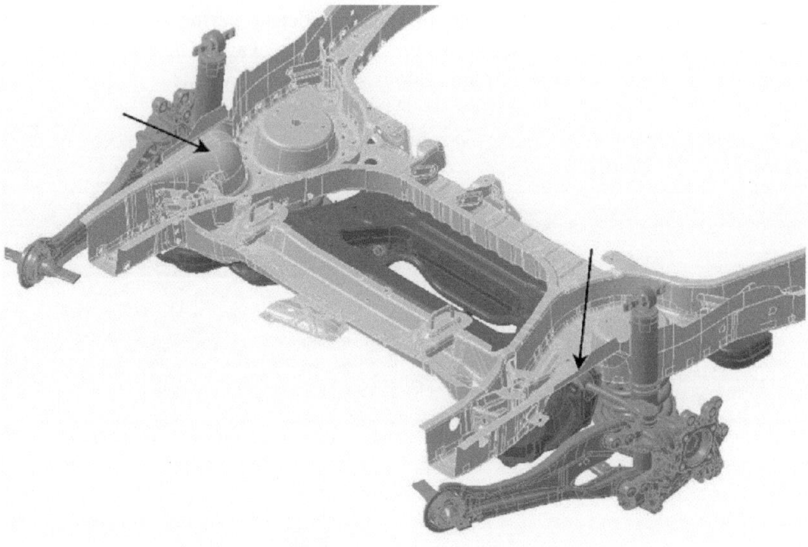

Figure 5. Previous Generation Odyssey Rear Suspension and Frame

4 Vehicle Dynamics and Chassis Development Process

Honda R&D Americas, like many OEM development centers, uses an adaptation of the V-model or V-flow process for vehicle dynamics and chassis development. This process has many gates or steps.

First the vehicle-level dynamic targets are set with strategic positioning against the latest competitors. Then the vehicle level targets are broken down into sub-system targets. For example, a vehicle level yaw gain target would be broken down into targets for steering ratios and front and rear chassis parameters. For a single axle, dozens of subsystem kinematics and compliance targets are developed to meet all of the vehicle level targets. For example, camber rigidity, stroke toe, roll toe, stroke camber, longitudinal toe, etc… are all defined within acceptable limit ranges.

From the sub-system targets and packaging space requirements, a suspension design concept is studied and iterated in order to best-fit the system level targets in the most efficient way. The calculations are done with multibody dynamics solvers. This process typically starts with an existing design concept and applies the design engineers' experience plus a DOE method to arrive at a best-fit solution. At this point, it may become apparent that not all sub-system targets and packaging requirements can be achieved simultaneously, or efficiently. In this case, some targets may be rebalanced and checked against overall vehicle level performance predictions before proceeding. Once an acceptable vehicle performance target level and design concept are realized, the component design work begins.

Individual component specifications are developed, such as knuckle and link arm stiffnesses, and compatible bushing stiffnesses. Then manufacturing-friendly component designs are generated and re-checked in the sub-system calculations. After confirmation, the components are built using prototype tools, tested for agreement with the specifications, and assembled into a vehicle. The vehicle is then loaded onto a kinematics and compliance measurement machine to test the sub-system characteristics against the subsystem targets. Often, some light tuning or adjustment of the components is necessary to achieve all of the sub-system targets. For example, a bushing with 10% higher stiffness may be substituted in one of the links arms to compensate for a deficiency in aligning rigidity achievement. When the sub-system targets are achieved on the measurement machine, the vehicle is released for dynamic tuning by test driver engineers. After tuning specifications are set, the design is updated for release to production tooling.

In the case of the compact rear chassis, a research project for the concept was conducted ahead of the vehicle development phase, including a mule build and test, to ensure the concept was viable. This research phase especially yielded valuable learning on the specific geometry of the link arrangement and balancing of component stiffnesses.

5 Vehicle Objectives

5.1 Ride, handling, and noise, vibration, and harshness (NVH) improvement targets

To improve Honda minivan positioning, the dynamic performance of the vehicle needed to be improved in a few key areas. The subjective improvement chart is shown in figure 6.

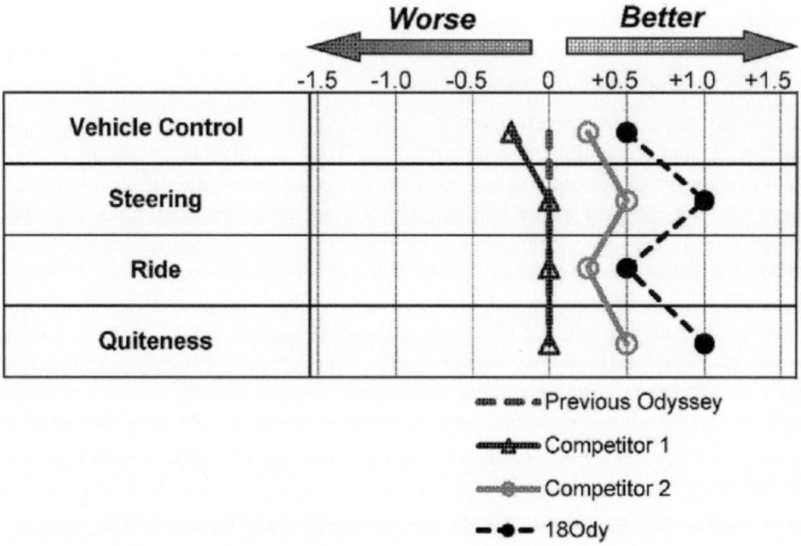

Figure 6. Vehicle Level Subjective Peformance Targets

These high level improvement areas were broken down into more specific targets shown in the table below.

Improvement Regime	Target
Maneuverability and Agility	Improve mid-speed (60 kph) yaw rate gain by 20% while keeping higher-speed yaw rate gain at an acceptable level for higher speed (140 kph) stability
Laden character vs. un-laden character	0.2 Hz yaw rate gain change (laden vs. unladen) reduction of 50% at 100 kph
Flat ride	Maintain the roll character/float feel while improving 3-8 Hz flat ride 0.5 pt subjectively
Impact harshness	0.5 pt subjective improvement
Road noise	Improve mid frequency (80~630 Hz) road noise by 2dB

In the final tuning, the targets were all achieved with some contributions from the compact rear chassis. A few of the target achievement details are shown in figures below. Figure 7 shows a significant increase in yaw rate gain between 60 and 100 kph to class-top levels. These levels improve the maneuverability of the vehicle in city driving and merging onto a highway.

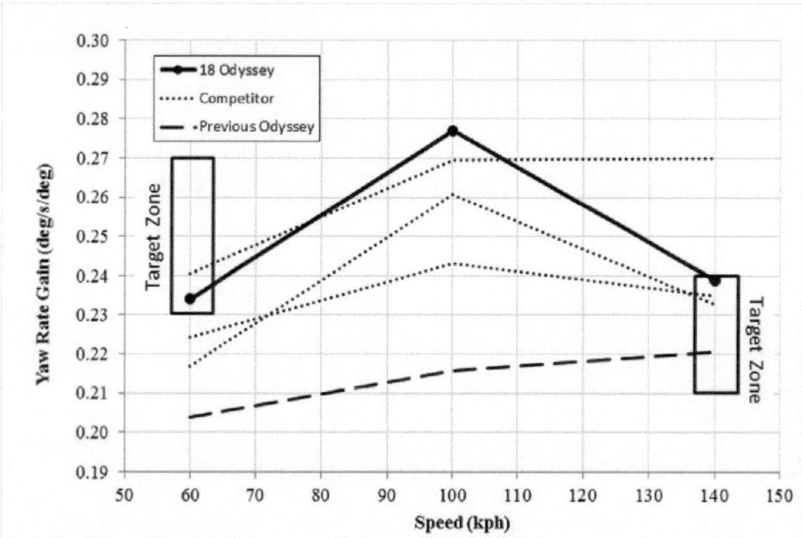

Figure 7. Yaw Rate Gain vs. Speed

This metric is primarily achieved by an increased steering system ratio, but also complemented by the front and rear chassis responsiveness. The yaw rate gain at 140 kph is held to a similar level as the 60 kph setting primarily due to improvements in the rear chassis stability. At higher speeds a lower yaw rate gain is desired for stability. The boxes in Figure 7 show the target zones.

A large difference in character between a lightly loaded vehicle and a fully loaded vehicle is generally not desirable. Especially, if the fully loaded condition creates excessive gain and feels unstable. Figure 8 shows the difference in yaw rate gain between fully loaded and 2 passenger (lightly loaded condition) at 100 kph. This character consistency is enabled primarily by the performance improvements of the rear chassis.

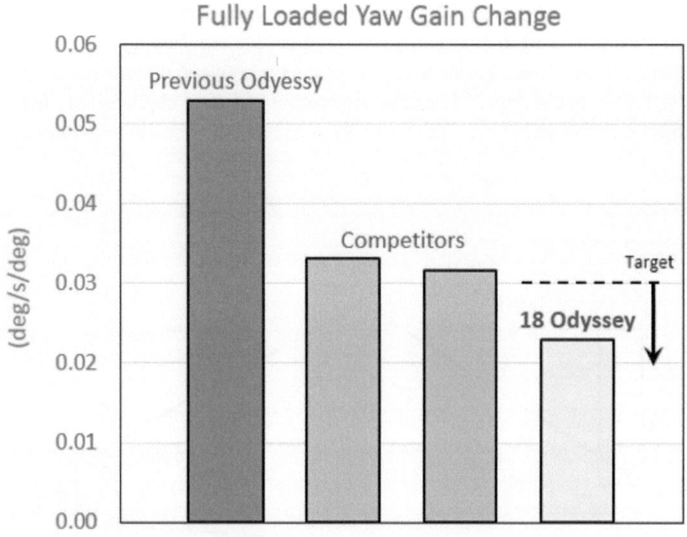

Figure 8. Yaw Rate Gain Change at 0.2 Hz 100 kph [Fully Loaded – Lightly Loaded]

5.2 Hardware package drivers and mass reduction

To reduce overall fuel consumption, mass reduction was targeted in all areas of the vehicle. A more compact rear chassis is not only lighter, but it enables mass reduction in the body and other areas. The combined subframe and suspension system is 4 kg, or approximately 6% lighter than the previous system. The net weight reduction includes the addition of a stabilizer bar sub-system, whose components are 3 kg. So a part-for-part comparison would show 7 kg of mass savings. The majority of the mass savings comes from the compact subframe sub-system, while the suspension parts are weight-neutral, but provide higher performance.

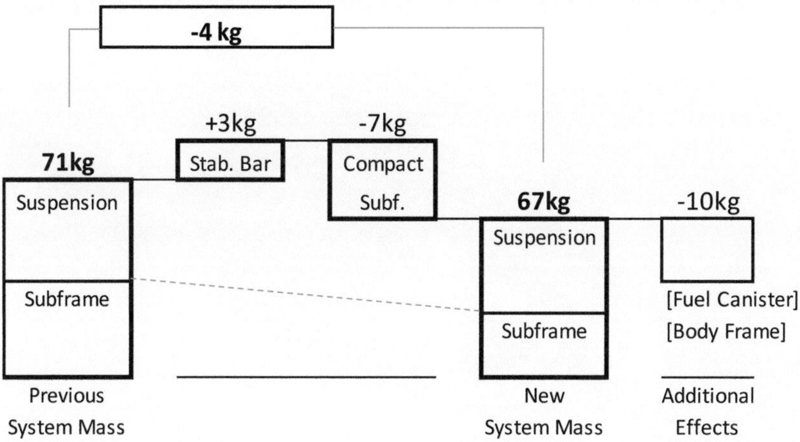

Figure 9. System Mass Chart

For body mass reduction and simultaneous rigidity improvement, a straightened and lowered rear frame was identified with a potential body mass reduction of approximately 8 kg. However, it required compacting the vertical packaging space of the suspension links by approximately 100 mm. This was a key challenge of the compact rear chassis.

Additionally, the fuel canister, Figure 10, was nested between the subframe and the fuel tank, eliminating redundant ground contact protection deployed in the previous Odyssey. The realized mass reduction of the fuel canister system was 2 kg.

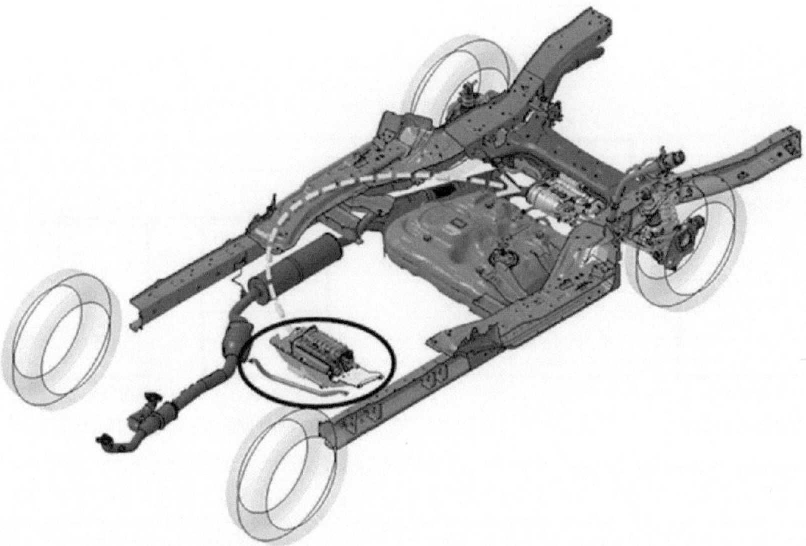

Figure 10. Relocated Fuel Canister Nested between Subframe and Fuel Tank

6 Suspension and subframe design

The suspension configuration is an independent trailing arm arrangement with three lateral links attached to a floating subframe with four mounts, plus a trailing arm connected directly to the body. Of the lateral links, there is a lower arm, an upper arm, and a toe control arm. The spring, damper, and bump stop (auxiliary spring) are independent elements and attached to separate points of the suspension. The spring and damper are attached between the knuckle and the body. The bump stop is attached to the upper arm, and it strikes against the body frame. A stabilizer bar added to the front side of the subframe connects via links to the lower arms.

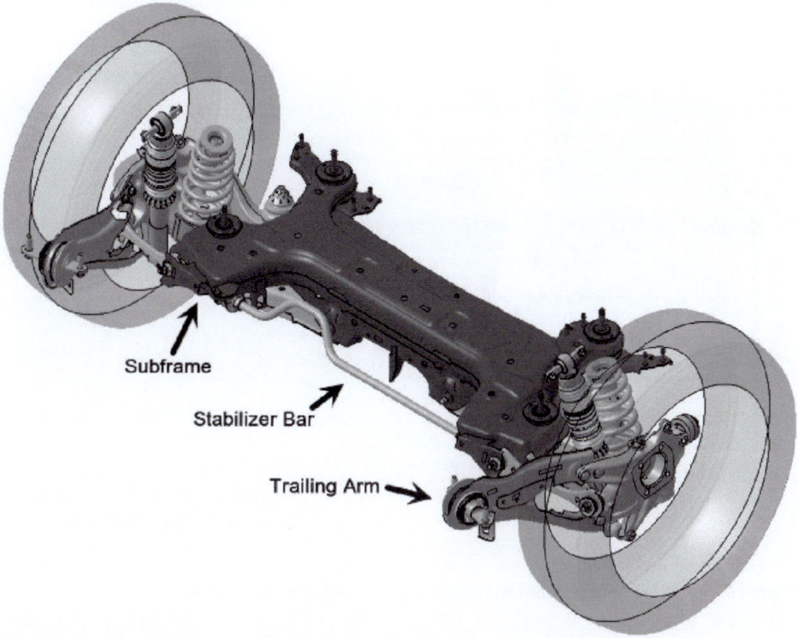

Figure 11. Suspension and Subframe Arrangement

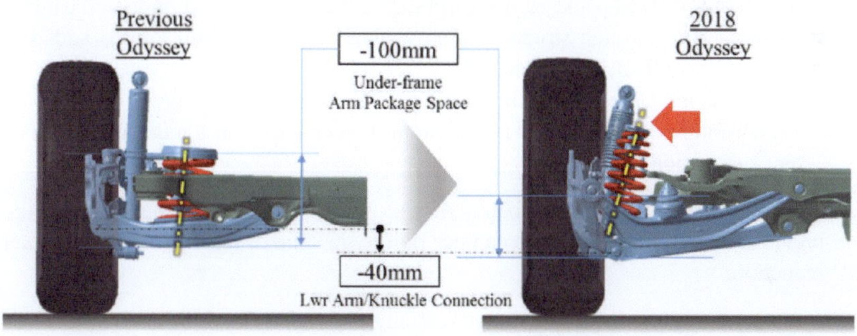

Figure 12. Suspension Link Vertical Packaging

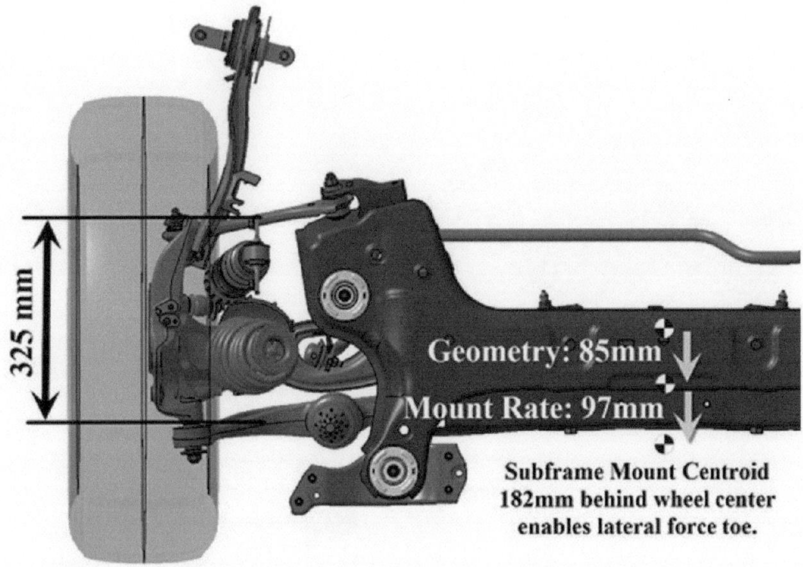

Figure 13. Subframe Mount Centroid (Elastic Center) and Link Arm Span

The following features were incorporated in the design:

1. Reduced link vertical package space 100mm, Figure 12, but also improved camber rigidity by 20% through structural improvements. A pillow-ball joint is added to the lower arm and knuckle connection.
   - New Feature: Link centroid on knuckle below wheel center, Figure 14.
   - New Feature: Smallest known vertical packaging space for a minivan independent suspension.
2. Moved the lower link approximately 40 mm lower from the wheel center than the previous model. This improved camber rigidity in spite of the overall vertical span reduction.
3. Arranged the lateral links to improve rear aligning rigidity by 71% and lateral force toe by 50%, increasing rear cornering power and reducing the rear axle delay feeling.
   - New Feature: Control link and rear upper links X-span and positioning. Control link was placed ahead of the forward subframe mount. The link X-span increased to 325 mm, 55% over the previous model, Figure 13.
   - New Feature: Subframe mount elastic center was set behind the wheel center 182 mm to induce side force toe.

4. Arranged the trailing arm and lateral links to improve mechanical compliance from 1.1 to 3.3 degrees to reduce impact harshness.
5. Applied damper rebound spring and stabilizer bar to balance the ride rate with improved body control. Roll stiffness increased approximately 50% while maintaining ride rate.
6. Improved spring action (lever) ratio from 1.8:1 to 1.1:1 by attaching it to the knuckle.
7. Subframe Compact Package. The subframe was attached to the body by four floating point bushings. Reduced longitudinal packaging dimension by 190 mm. Box architecture changed to beam architecture.
   - New Feature: Rear subframe mount was nested in the frame rail (US patent[2]), as shown in Figure 17, with double shear stay, improved body lateral stiffness of 132%. This also enabled improved isolation (10dB to 24dB), Figure 16, and contact patch stiffness. Isolation is defined as the ratio of the body attachment static stiffness to the bushing dynamic stiffness at 100 Hz. The isolation values shown in Figure 16 are the averages of X, Y, and Z directions.
8. Subframe single lateral beam in line with the lower link for camber rigidity and packaging of fuel canister. The subframe stiffness at the lower arm connection is increased 23.1 kN/mm -> 31.9 kN/mm, 38%.

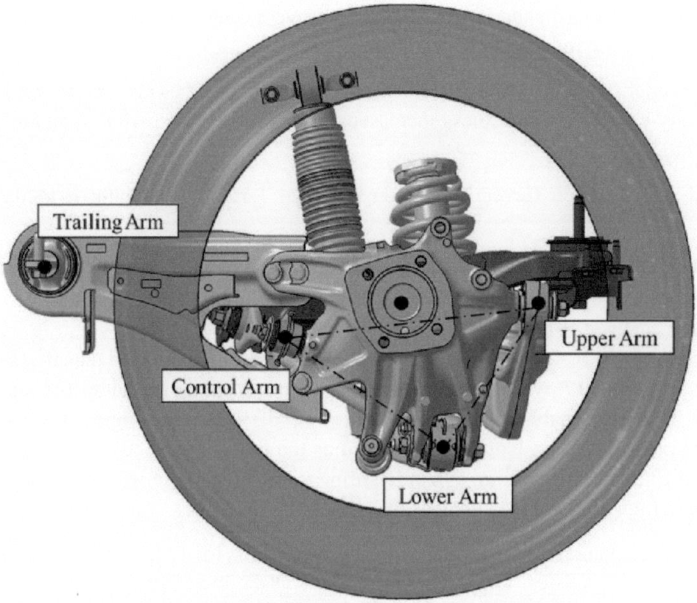

Figure 14. Side View of Link Arrangement

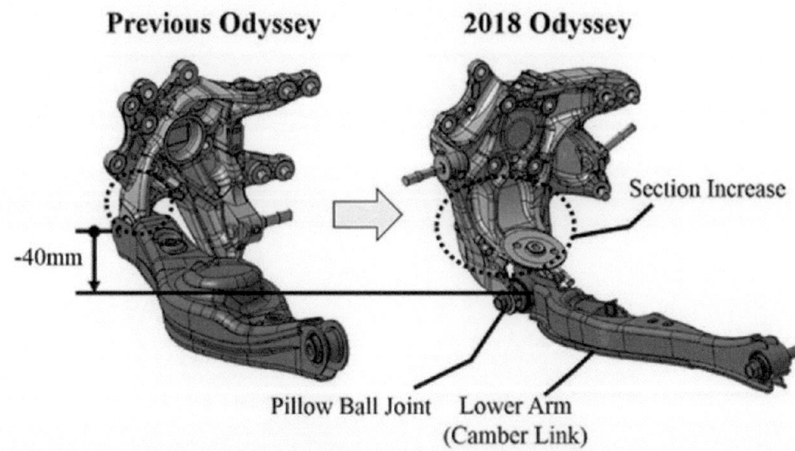

Figure 15. Lower Arm and Knuckle Connectivity for Camber Rigidity

Figure 16. Attachment Isolation Improvements
[Body/Bushing Stiffness, XYZ average]

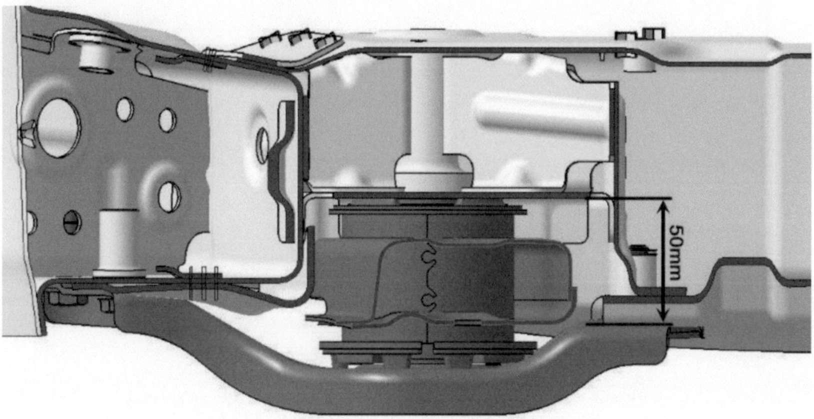

Figure 17. Subframe Mount Nested in Body Frame

7 Technical Challenges and Solutions

7.1 Aligning rigidity vs. lateral force toe

Aligning rigidity is important for initial responsiveness of a rear suspension, but it needs to be balanced with appropriate phasing of lateral force steer to promote progressive understeer for stability. Typical suspension arrangements tend to sacrifice one parameter for the other. This is due to the inherent difficulty in making the suspension apparatus resistant to toe compliance induced by contact patch slip angle moments, but allowing toe compliance due to lateral forces building at the contact patch.

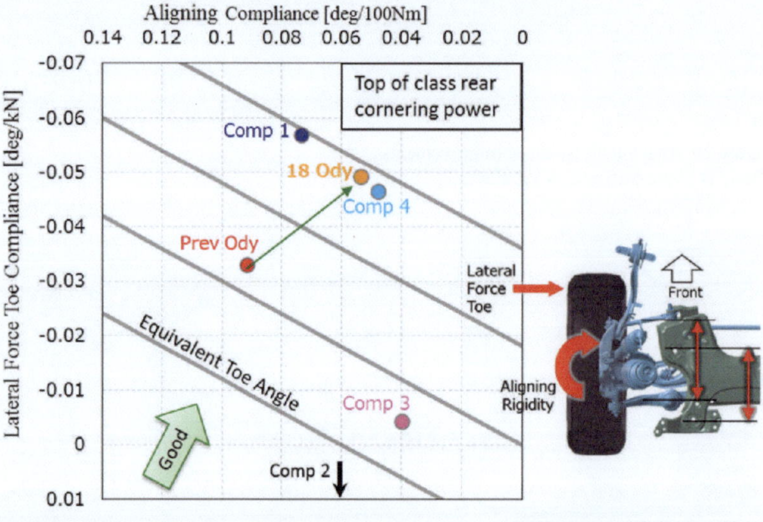

Figure 18. Lateral Force Toe vs. Aligning Compliance

However, the compact chassis of the Odyssey enables high levels of both to be achieved simultaneously. This is due primarily to the relationship of the wheel center, lateral links, and the subframe mount elastic center, Figure 13. The wheel center is well supported against aligning torque due to the longitudinal spacing of the lateral links well ahead and behind the wheel center. Lateral force toe is induced by the placement of the elastic center of the subframe significantly behind the wheel center. The new Odyssey improves self-aligning torque by 71 % over the previous model with increasing lateral force toe by 50%.

7.2 Camber rigidity vs. vertical arm span

An immediate concern with reducing arm vertical span is the reduction in moment arm length supporting camber rigidity. Camber rigidity is important for rear axle responsiveness and stability. Several steps were taken to achieve the camber rigidity target (increased 20%). The lower arm path is the primary contributor, so it was packaged as low in the chassis as possible, while achieving ground clearance requirements. Second, aggressive stiffness targets for the parts in the lower arm load path were set. Especially, the subframe stiffness was increased over the current model by 38%. Additionally, the outboard joint is a pillow ball with a measured stiffness of 50kN/mm. Key values in the path are shown below.

Component	Y-Stiffness (kN/mm)
Knuckle Arm (From Hub Mount Face to Bush Pt)	21
Outer Pillow Ball Joint	50
Lower Arm	37
Inner Bushing	29
Subframe (at Lower Arm Inner Bushing)	32 (+38%)

7.3 Ride comfort vs. handling

The compact rear chassis achieves high levels of handling and ride comfort simultaneously, while the competitors tend to excel in only one of the two areas.

For improving impact harshness, mechanical compliance was increased from 1.1 to 3.3 degrees. In addition, the rear anti-pitch (braking) angle was increased, Figure 19. This effect was created primarily by adjusting the trailing arm bushing point, and thus the side-view instant center of the suspension upward and rearward compared to the previous suspension. Also for impact harshness, the longitudinal stiffness of the chassis was reduced, Figure 20, by applying a more progressive trailing arm bushing. However, as described previously, the camber rigidity (compliance) level is also improved to top class level which is typically only achieved by twist beam suspensions.

A stabilizer bar and damper rebound spring were added in order to lower the suspension vertical rate for improved flat ride. The vertical (bounce) rate, as shown is Figure 21, is reduced by increasing the bump stop gap in the primary stroke range. However, this would result in a loss of vehicle stability in a turning maneuvers, without measures to recover roll stiffness. The stabilizer bar and rebound spring work together to increase the roll stiffness over the base model, without influencing the primary vertical ride rate.

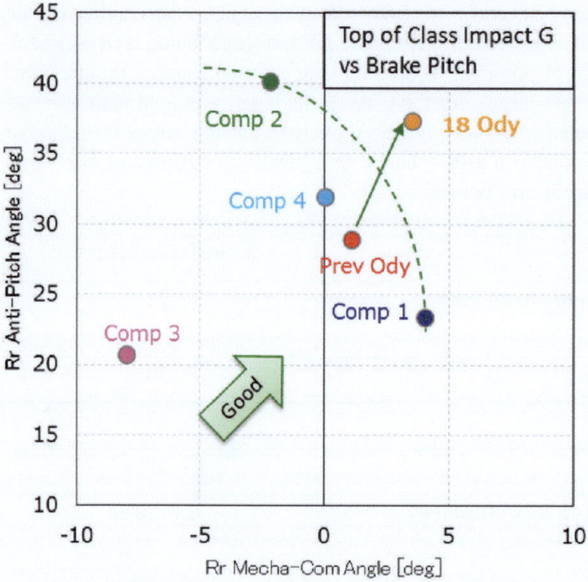

Figure 19. Anti-Pitch Angle vs. Mechanical Compliance Angle

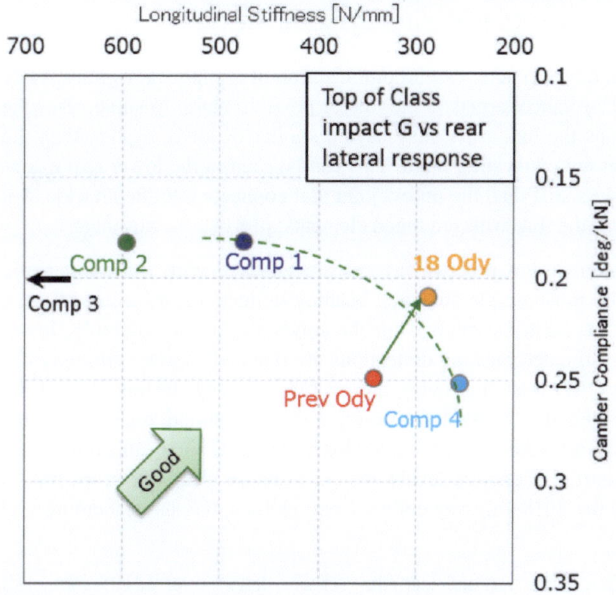

Figure 20. Longitudinal Stiffness vs. Camber Compliance

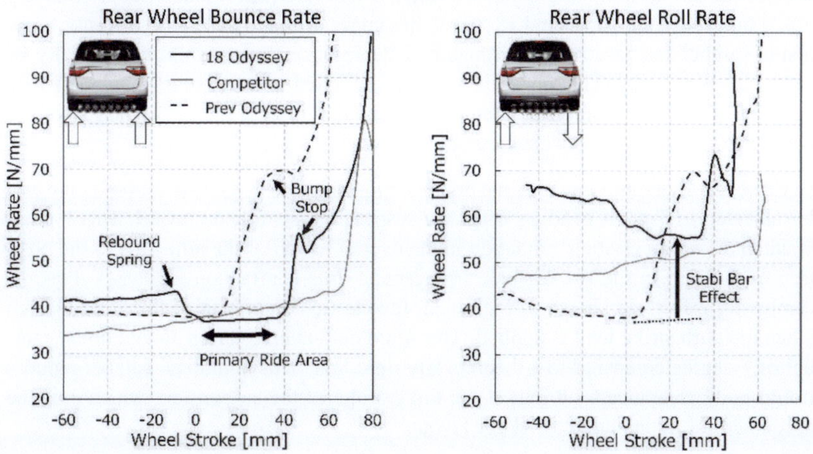

Figure 21. Bounce vs. Roll Rate

7.4 Layout clearances considering dynamic flexure

In modern suspension development, careful optimization of component rigidity is key to achieving high performance targets at the lowest possible mass. In some cases the structural parts, such as the link arms, may have a similar or lower rigidity than the bushings or joints that are connecting them. This can be seen in the lower arm example where the arm is less stiff than the pillow joint that connects it to the knuckle. So, in modern suspensions, the link arms are tuned elements, just like the bushings.

In the usual suspension design process, layouts are checked with CAD kinematic modelling tools. In the most simple analyses, bushing deflections are assumed to be near zero, or allowed to float just enough for the solver to operate smoothly. In sophisticated analyses, estimated bushing deflections are mapped from a multibody dynamics model to the CAD tool, improving the overall accuracy. In this method the metal parts are still assumed to be rigid. However, in rapidly applied, high load events such as driving through a pothole or over a curbstone, the metal parts can exhibit elastic deformation of significant enough levels enough to cause clearance problems. In tight layouts, such as the 2018 Odyssey compact rear chassis, this understanding was extremely important.

In the Honda suspension interference test, the vehicle drives over a 105 mm curb block at speeds up to 50 kph. The requirements are that the tire should not damage the body work or fender liners (light contact is allowed), and there should be no interference of the suspension parts with each other. This event puts a load of approximately 30kN into the bump stop, and exercises the upper arm and subframe to cause a dynamic camber increase of approximately 1 degree over what would theoretically be seen with rigid links. This phenomenon was modelled in a dynamic simulation with metal parts flexure to understand the total potential motion of the suspension links and tire. From there appropriate clearances were set.

Because the upper arm contains the bump stop which strikes against the body, the arm is subjected to 3-point bending when the suspension is stroked upward. This 3-point bending drives an elastic contraction in the overall length of the arm. Due to the position of the arm in the overall link arrangement, this contraction induces a dynamic cambering effect. As shown in Figure 22, the elastic contraction of the arm is 4 mm when the curb strike load is applied. This 4mm deflection induces an additional camber of 1 degree compared to a theoretically rigid link. This additional camber reduces tire-to-body clearance by 8 mm at the limit stroke of the suspension, which must be accounted for in the overall package layout.

This dynamic flexure phenomenon can clearly be seen using high-speed video recorders attached to the underbody. In early testing, the videos, along with strain-gaged arms for load measurement, were used to correlate the suspension motion to CAE predictions. From there, arms of various stiffnesses from the handling tuning phase were simulated for effects to dynamic clearances. After the final link arm specifications were set, a prototype test confirmed the dynamic clearances to be sufficient.

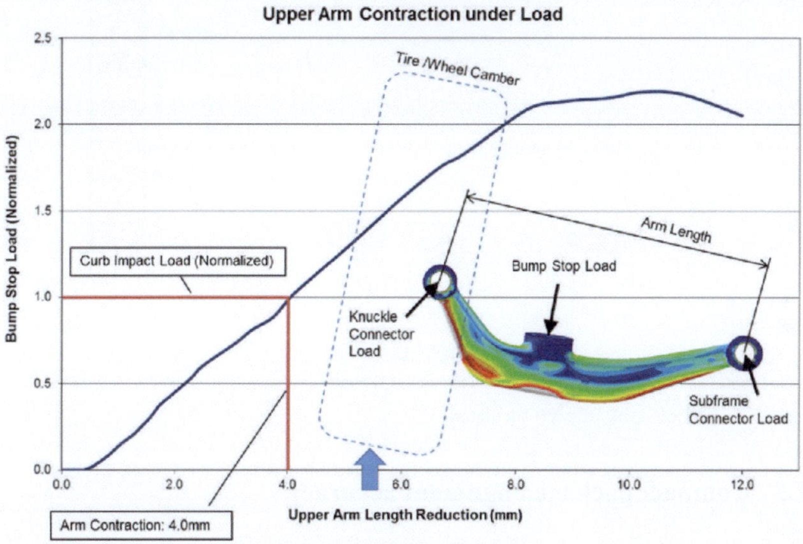

Figure 22. Upper Arm Contraction with Bump Stop Load

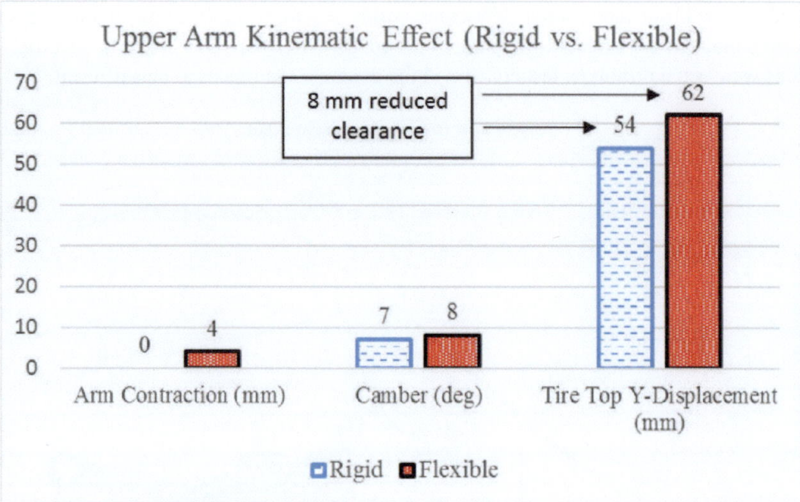

Figure 23. Flexible Link Effect (Calculated)

7.5 Compact package alignment accuracy

Another concern of a compact link arm package is alignment accuracy. Applying standard manufacturing tolerances to a more compact link arrangement could result in a larger population variation for camber and toe. In Honda factories, toe is adjustable at the end of the production line, but typically, camber is not. In most link arrangements, the toe and camber are also interdependent due to kingpin inclination, so toe cannot be adjusted without affecting camber. It is important to understand this interdependency when setting the kinematic points and production tolerances of the components.

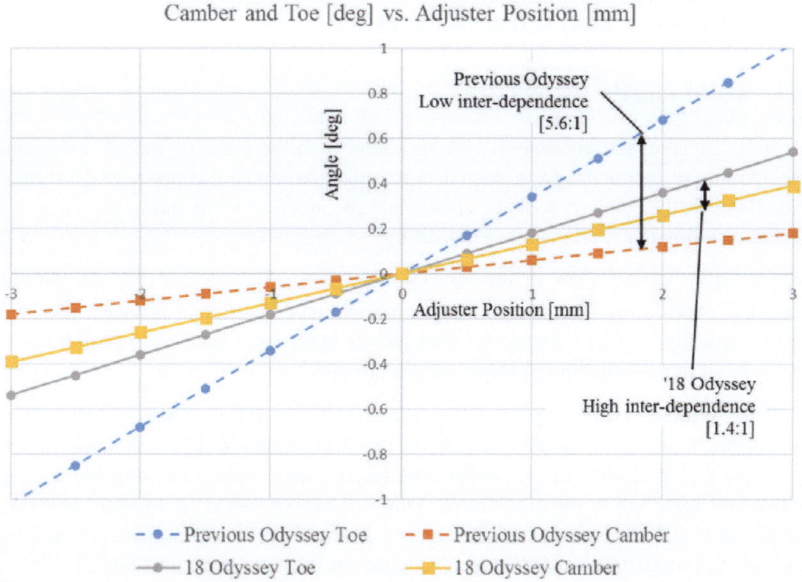

Figure 24. Camber vs. Toe Sensitivity

In the 2018 Odyssey compact rear chassis, the toe and camber interdependency were mapped using a kinematic solver. Then the toe adjustment range was set. Based on the overall sensitivity, tight tolerances were applied to the subframe control points to ensure repeatable camber alignment. The toe is adjustable by a cam bolt applied to the control arm. In the Honda factory, this is adjusted by an automated dynamic alignment machine. Figure 24 shows the toe and camber interdependency. The y-axis shows the change in toe and camber vs. the position of the toe adjusting cam bolt for each suspension. The compact chassis has more interdependency than the previous Odyssey suspension.

8 Summary and Conclusions

8.1 Total merits

Overall, the compact rear chassis is 4 kg lighter than the previous model, even with the addition of functional parts, such as a stabilizer bar and a damper rebound spring. Additionally, the compact package allow mass reduction of surrounding parts mass of 10 kg, resulting in a total mass merit of 14 kg.

The dynamic performance of the compact rear chassis greatly exceeds the current model in terms of agility, ride comfort, and quietness. These performance improvements typically conflict with one another, but the compact achieves a breakthrough to improve all three with reduced mass and package size.

A focus on joints and metal parts stiffnesses of the lower arm load path yielded a 20% improvement in camber rigidity. Further, the link arrangement and the positioning of the subframe mount elastic center enabled simultaneous improvements in aligning rigidity and lateral force toe to top-class levels. This was accomplished while also improving longitudinal and mechanical compliance for ride comfort. Finally, a focus on bushing isolation improvements yielded 2dB improvement in mid frequency road noise.

8.2 Lessons learned

A very tight layout presented some technical issues, which needed to be managed in development of the compact rear chassis.

In particular, a dynamic clearance simulation including metal parts flexure was studied and correlated to on-car testing to determine final layout clearances. Additionally, as component shapes evolved during the chassis development process, this flexure and resulting camber effect became component specifications and sub-system targets.

A compact link arrangement requires high component accuracy to achieve both initial alignment accuracy and performance consistency. After studying kinematic interdependency, special measures were taken to ensure key parts accuracy.

Compacting the chassis was a great challenge. The initial expectation was that performance would be compromised to achieve the desired layout, but through clever link and subframe mount arrangement, plus structural improvements, an overall higher performance and more efficient system was realized. The results of this compact chassis demonstrate that there are still potential improvements worth exploring in purely passive, mechanical suspensions, without resorting to overly complex systems.

9 References

1. Johnson, Joshua C., author; New Light Truck Platform Chassis. SAE Technical Paper 2013-01-0370. April 8, 2013.

2. Goellner, Adam Corley, inventor; Vehicle Suspension System, Suspension Mount Assembly And Method Of Mounting A Subframe To A Frame. US patent US20160288834A1. October 6, 2016.

From i30 to i30: an evolutionary journey for Hyundai C-seg bestseller

Axel Honisch,
Head of Department, HMETC, Vehicle Test & Development

Michael Lugert,
Manager, HMETC, Vehicle Test & Development,
Steering & Suspension

Antonino Pizzuto,
Manager, HMETC, Engineering Design, Chassis Development

Timo Schöning,
Manager, HMETC, Vehicle Test & Development,
Steering & Suspension

Hyundai Motor Europe Technical Center GmbH

P.E. Pfeffer (Ed.), *8th International Munich Chassis Symposium 2017*,
Proceedings, DOI 10.1007/978-3-658-18459-9_10

Abstract

5 years ago the second generation i30 was launched in Frankfurt motorshow. This was the very first Hyundai C-segment vehicle designed for European market.
To be a successful newcomer in this very competitive segment, Hyundai focused on quality, reliability and style, and was well received from customers, press and even competitors.

5 years later we "move on" and a new generation i30 is on the market, with the ambition to be an even stronger competitor for the market leaders.
For pursuing this goal, Hyundai worked hard to improve each performance, with a special focus on R&H. In close co-work with Korea HQ, a 'simultaneous engineering team' has worked in HMETC (Hyundai European Technical Center), integrating state-of-the-art CAE simulation with H-i-L development (i. e. for steering tuning) and road testing.

The results are a fine-tuned front suspension and steering, a redesigned rear suspension, all supported by a stiffer body, which contribute to a more direct and quicker handling, remaining stable and neutral and always comfortable with a precise steering giving the driver a natural feel and a good connection to the road.

Initial drive tests with selected journalist during vehicle development have shown a broad acceptance of the steering feel confirming it as 'natural with clear feedback' (Auto Zeitung, September 7, 2016) and 'good precision' (Auto Motor und Sport, September 1, 2016). (Updated magazine review will be available in Q1 2017, after January test drive event)

HMETC is still working hard to further improve i30 performances for upcoming variants: wagon, fastback, N-brand.

CHASSIS COMPONENTS AND HISTORICAL ASPECTS

mSTARS – modular Semi-Trailing Arm Rear Suspension

Dipl.-Ing. Knut Heidsieck

Dipl.-Ing. Klaus Wallgren

P.E. Pfeffer (Ed.), *8th International Munich Chassis Symposium 2017*,
Proceedings, DOI 10.1007/978-3-658-18459-9_11

ABSTRACT

The new ZF modular Semi-Trailing Arm Rear Suspension – mSTARS – for passenger cars combines exclusive features of the chassis such as an active steering with an electric drive, and easy adaptation to different types of axle configurations. The new ZF mSTARS is also highly flexible: it can be connected easily with integrated electric or conventional traction modules, interchangeably used as both driven and non-driven rear axle in all conventional vehicles with independent rear suspension. Moreover, all variants of the new ZF mSTARS can be implemented in conjunction with the AKC (rear wheel steering) system. This new ZF mSTARS concept constitutes the basis for greater modularity, flexibility and space efficiency while ensuring superior performance in vehicle dynamics and safety.

1 INTRODUCTION

ZF's new mSTARS is to rival traditional multi-link and trailing blade rear suspension designs in chassis performance and system flexibility. It is a rear axle system that combines high modularity and flexibility of the chassis with active steering and other active or passive control systems. Thus it ensures easy adaptation to different types of axle configurations. The capabilities and competitiveness of ZF's new mSTARS is demonstrated within ZF's scope of test vehicles, namely in a passenger car fitted with the new modular rear axle.

2 CURRENT TRENDS, OBJECTIVE & BACKGROUND

In general, the requirements of modern automobile manufacturing and vehicle platforms also apply specifically to chassis systems. To minimize the development, application and investment scope required for individual vehicle types, bodies and drive types, implementation should be as simple as possible based – on a flexible modular system! After all, OEMs must meet the aims of continuous improvement of cost effectiveness while at the same time satisfying key technologies and market trends with their vehicle designs.

In current market trends, customers demand a vast selection of vehicle variants within one model range such as hatchback, fastback, station wagon, coupé, convertible, van, SUV, and SUV coupé. This trend also applies to drive systems: Due to the need of meeting future CO2 limits and emission standards, demand for hybrid systems is one the rise (internal combustion engines with electric motor). In addition, fully electrically-powered vehicles have become the development focus of many car manufacturers.

For premium manufacturers in particular, standing out from competition with regard to vehicle dynamics, safety and ultimately autonomous driving capabilities will be equally important. This aspect pertains to their entire portfolio, from the compact car up to the luxury car sector. Among the technologies that contribute to these objectives are, for example, active stabilizers and steerable rear axles that support the steering angle of the front wheels in a variable and situation-dependent manner. However, for many multi-link like trailing blade rear suspensions this feature would be difficult to implement in combination with a central actuator and central drive. To this end, adequate solutions must to be designed specifically for rear steering assistance – at considerable additional cost as a given, otherwise, this feature would be dispensed with.

Approximately two years ago, ZF has started to develop a rear axle system to meet these objectives: It will not only rival – at least – the more complex multi-link and trailing blade rear suspension designs in chassis performance but also allow a very high degree of modularity and flexibility. This system will especially excel in terms of future-proof driveline and chassis technology: The target was to develop a system that could be combined easily with the Active Kinematics Control (AKC) or with integrated electric or conventional traction modules. Another aim in this context: the new development can also be used as both, driven or non-driven rear axle in all conventional vehicles with a combustion engine (be it a front, rear, standard or all-wheel drive system). Its engineering background: always easy to upgrade and to optimize with regard to installation space. At the very same time, it has to remain cost-effective in comparison with existing designs and similar in terms of performance, i.e. driving dynamics and safety.

3 mSTARS by ZF

3.1 The new basic axle design

To allow modular system architecture and an effective, flexible adaptation to the platform, body and model types, the basic axle, FIGURE 1, must provide ample space for the different upgrade modules. The new system therefore now has a separated spring and damper arrangement as a standard. However, conventional suspension struts can continue to be used – provided some advantages the axle design offers are forgone. The installation space required for a separate arrangement of springs and dampers is provided by specific design details: Simply put, it is an evolutionary development of a semi-trailing link suspension with integral link.

On conventional semi-trailing link suspensions, the lower control arm is designed as a parallelogram and is connected to the wheel carrier and to the body with two kinematic points each. By contrast, the rear of the two outer pivot points has been eliminated on the modular rear axle and has been replaced by an integral link, FIGURE 1. This makes it possible to move the wheel-side suspension parts further outwards that are closer to the rim, ensuring the important space gain. For the first time, a coil/ air spring and the damper can not only be installed separately from each other but also behind a potential output shaft when viewed from the top. This is important for configurations such as rear-wheel drive, all-wheel drive or hybrid systems.

In conjunction with the lower control arm's inclined axis of rotation, the new design with its integral links permits an outstanding spring ratio of approximately 0.7 and a damper ratio of more than 0.9 (values may vary depending on the required application). These high values are crucial for the use of the required springs which have been optimized with regard to installation space (including optimization of diameter) together with semi active dampers even in applications with extraordinary wheel loads. Since larger values for spring and damper ratio create lower suspension forces, other components can also be designed correspondingly lighter. In addition, the possibility of dispensing with suspension struts can increase the trunk width by approximately 10 cm on each side. This even applies to applications with axle upgrade modules such as an electric motor and the AKC rear axle kinematics.

Another special feature of the modular basic axle is the standard toe link on the left and right-hand side where both are located behind the corresponding spring and damper. In the basic configuration, the toe links are attached to the axle carrier by a joint. The requested toe values can thus already be set accurately during production.

In addition, these links can be taken over unchanged for the optional AKC actuator. This fact already indicated that the modularity of the axle does not require complex adaptations of the wheel corner module. In fact, the wheel corner module can remain virtually unchanged in all types of one specific vehicle application. Instead, the new development takes its flexibility from different axle carrier welding assemblies, which have been designed for efficient to simply integrate diverse upgrade options, FIGURE 2.

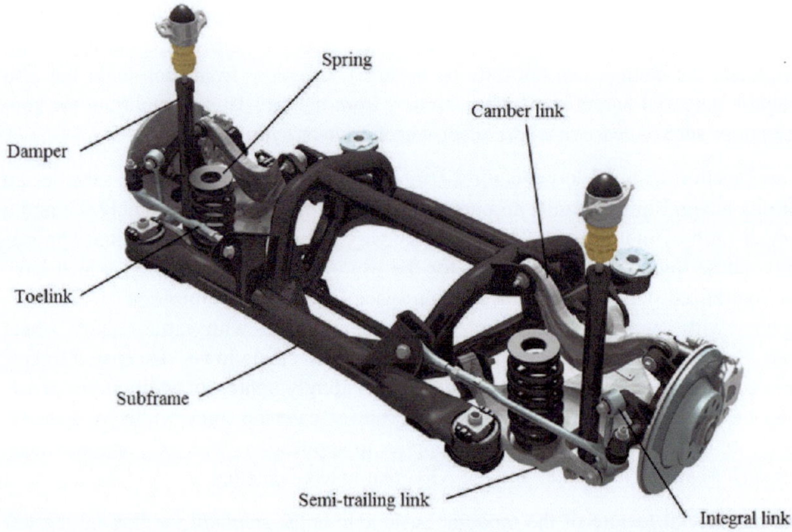

FIGURE 1: Creates the basis for greater modularity, flexibility and space gain while at the same time ensuring high performance: the basic axle of the modular ZF mSTARS with numerous new design features and separated spring and damper arrangement

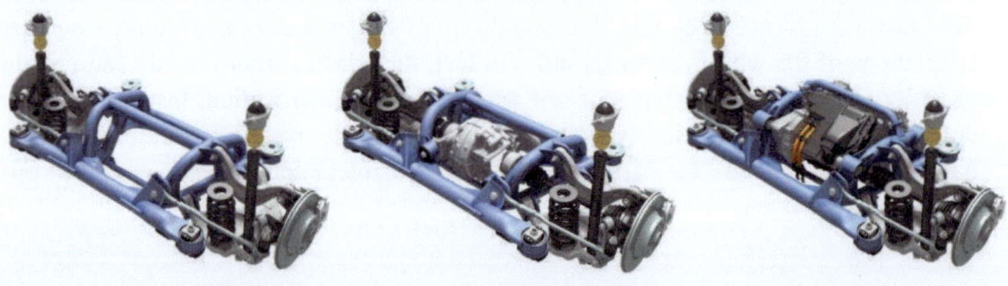

FIGURE 2: Core elements providing flexibility; the different axle carrier designs (blue) including their specific stabilizer arrangements; conventional non-driven rear axles [1], rear axles with conventional axle drive [2] or rear axles with electric traction module [3] are therefore possible with virtually no difference regarding the installation space required; the AKC system can be implemented in conjunction with all three variants.

3.2 The AKC module

As mentioned before, one of the biggest advantages of the mSTARS is that it literally embraces AKC technology and thus, easily and without additional efforts, produces an actively steering rear axle "by wire". Here an electromechanical actuator with integrated electric motor is needed, together with a toothed belt drive, a spindle mechanism, a sensor system and a control unit which communicates with other vehicle systems. In its so-called central actuator version, the AKC system is positioned in the middle of the toe links and it is directly connected to the rear axle carrier. The rack travel of ±30mm can turn the wheels – either in the same direction as the front wheels or in the opposite direction, depending on speed and driving situation.

In a ZF test vehicle based on a compact passenger car, the modular rear axle equipped with AKC replaced the original production axle. In combination with the Dual Pinion EPS (Electrically Powered Steering) at the front, the rear axle demonstrates its advantages, FIGURE 3: The vehicle always remains safely on the track selected by the driver, even during lane changes and overtaking maneuvers on snow and ice while at the same time exceling with improved dynamics. As automation of driving functions increases, it is in particular this AKC and EPS combination that will continue to gain in importance, with regard to safety, comfort and system redundancies. In addition, the AKC system can also carry out various steering tasks without involvement of the front axle.

3.3 Integrated in the axle: the electric motor

Now, against the backdrop of ZF's innovative electric drive combined with power electronics, the customer can select a complete, modular axle concept with fine-tuned kinematics and elastokinematics as well as adapted NVH characteristics. As a result, with minimum efforts, powerful (plug-in) hybrid vehicles or in particular purely electric vehicles can be easily integrated in the modular rear axle concept, including the electric motor

When using the electric axle drive, this can be installed at the center of the axle carrier which can be modified for different traction modules, FIGURE 4. The unit houses the following components: an electric machine (in different types and power spectrums), a gear drive, a differential or a second machine and the power electronics with control software. On vehicles with hybrid configuration, the electric motor at the rear axle also makes it possible to implement advanced driver assistance and intelligent all-wheel drive features.

With the modular rear axle system this all-wheel drive option is also possible in conventional applications: To this end, a suitable axle drive is installed with the passenger car's (front-wheel drive) axle carrier instead of the electric traction module (this design is based on an OEM-integrated cardan shaft). In turn, this approach automatically entails the option to use the rear axle for standard or rear-wheel drive passenger cars, with or without AKC, FIGURE 5.

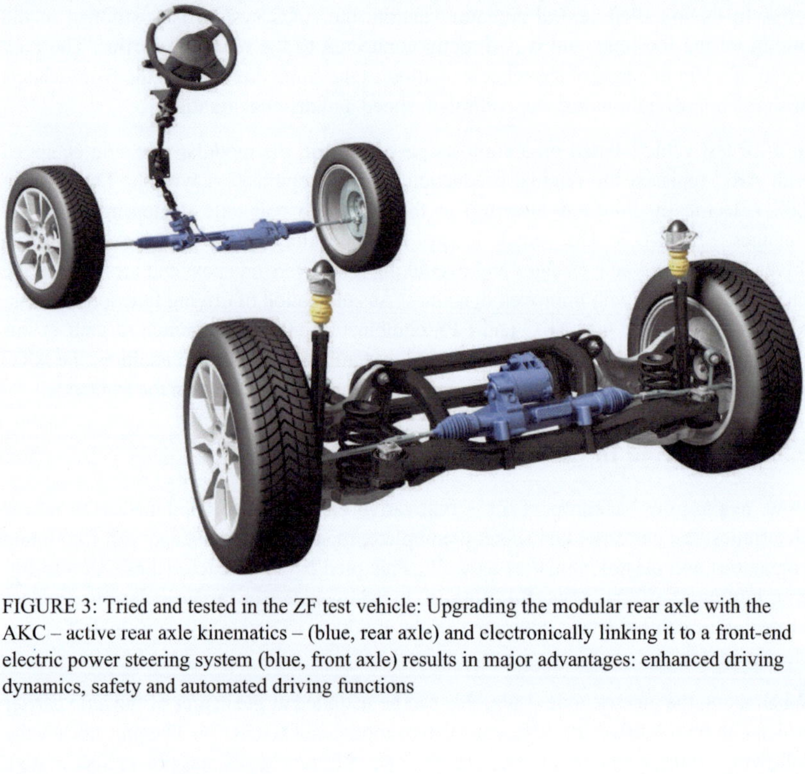

FIGURE 3: Tried and tested in the ZF test vehicle: Upgrading the modular rear axle with the AKC – active rear axle kinematics – (blue, rear axle) and electronically linking it to a front-end electric power steering system (blue, front axle) results in major advantages: enhanced driving dynamics, safety and automated driving functions

FIGURE 4: Compact unit consisting of drive and axle: Upon request, a ZF motor (with an output of up to 150 kW), a transmission, a differential and power electronics are integrated in the axle carrier – more than just a sneak preview on full hybrid vehicles and purely electric vehicles

FIGURE 5: If a ZF axle drive is positioned centrally below the axle carrier – shown here in an example – the modular rear axle is predestined for conventional rear-wheel drive or all-wheel drive layouts

4 CONCLUSION AND PROSPECTS

The new ZF mSTARS as a semi-trailing link suspension combines innovations that promote safety and vehicle dynamics with drive systems that reduce fuel consumption and emissions. In this regard, the mSTARS is the response to the key trends and questions of the automotive industry. At its heart is the basic axle which stands out thanks to its: general ease of application to different vehicle types & axle configurations, significant gain in installation space and superior performance in driving dynamics and safety. In addition, it makes it particularly simple for OEMs to implement active steering on the rear wheels using the AKC system. This is clearly one of the key advantages when compared to previous conventional multi-link or trailing blade suspension constructions and simultaneously a relevant factor on the path to autonomous driving. The modular rear axle is also unique: its new integral link geometry makes it possible to place coil/ air springs and dampers separately from each other behind the output shaft.

Furthermore, the ZF mSTARS offers the possibility to integrate electric control systems. Thus, it enables simple implementation of (plug-in) hybrids and all-wheel drive functions derived thereof or purely electric drives. Using conventional final drive modules, rear-wheel drive and all-wheel drive systems can also be realized for vehicle concepts that only have a front engine as the power source. For the time being, the ZF mSTARS has successfully completed its predevelopment stage and the transition to specific volume production development is imminent.

REFERENCES

(1) Berger, F.; Krimmel, H.: Active Rear Axle Kinematics for Cars.
In: ATZworldwide 116(2014), No. 6, pp. 4-7

Composite chassis components

M.Sc. Alberto Girelli Consolaro
Dipl.-Ing. (FH) Nicole Zandbergen
Dr.-Ing. Paul Zandbergen
Dr.-Ing. Friedrich Wolf-Monheim
M.Sc. Francesco Italiano

Ford Research and Innovation Center Aachen

P.E. Pfeffer (Ed.), *8th International Munich Chassis Symposium 2017*, Proceedings, DOI 10.1007/978-3-658-18459-9_12

1 Introduction

To reduce the fuel consumption of future road vehicles, weight reduction is a key area of focus within the automotive industry. Therefore, chassis components made from fiber-reinforced plastics (composites) are becoming more and more popular relative to state of the art components made from steel or aluminum.

Aeronautics and sports goods industries have been following the path of composite technology for several years now, but today composites are increasingly used in the automotive industry, too. Interior parts, body panels and structures are currently most common applications. In the chassis area, the use of composites is still limited, at least in relation to structural parts. Components such as suspension links, wheel carriers or subframes have very challenging requirements in terms of strength, stiffness and durability. This limits the design freedom and makes a material substitution very difficult. In addition, the challenging requirements on structural chassis components often require carbon fiber reinforced plastics to be used, which increases part costs due to high material prices.

However, components where the implementation of composite materials can generate much better business cases are springs. The Ford Research and Innovation Center Aachen has been working on composite springs (longitudinal and lateral leaf springs as well as coil springs) for the last 15 years, focusing on design principles, materials and manufacturing processes, testing as well as failure mode avoidance strategies.

For springs, glass fiber reinforced plastics are suitable to meet all functional requirements such as strain capabilities, stiffness and durability. The lower material price of glass fiber compared to carbon fiber supports the business case for various applications.

The paper focuses on lateral composite leaf springs, longitudinal composite leaf springs and composite coil springs. Next to the component and system designs, package constraints as well as interface designs to adjacent components are shown.

2 Transverse Leaf Springs

A transverse leaf spring is placed in a vehicle suspension perpendicular to the vehicle driving direction. Different configurations are possible. Typically, they work as four point bending beams. This means that they are connected to the vehicle in four different points: Two inner connectors position the spring relative to the vehicle body or the suspension subframe, two outer connectors position the spring relative to the wheel carrier or to suspension links or arms.

If the leaf spring is not influencing the suspension kinematics, it is referred to as a 'non wheel-guiding leaf spring'. Some examples of this set up can be found in Mercedes Sprinter (front), Chevrolet Corvette (front and rear) and Volvo XC90 (rear) suspensions.

In this configuration, when both wheels move up and down in parallel motion, the leaf spring has a 'C'-shape, offering a certain resistance to the wheel motion. In this case, the leaf spring is working similar to conventional coil springs.

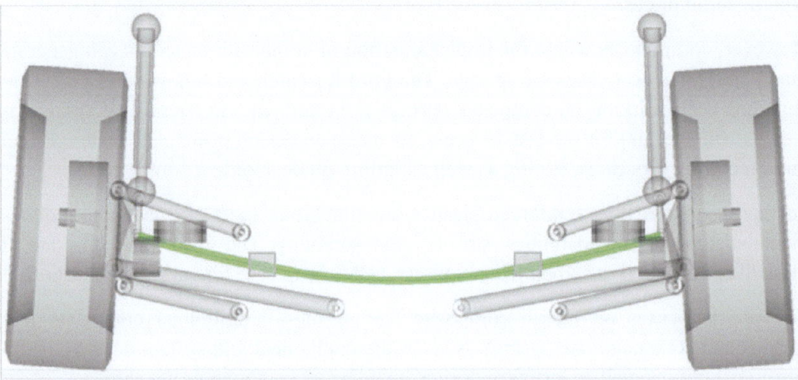

Figure 1: Transverse leaf spring working principle: When the wheels move in-phase (parallel motion), the transverse leaf spring assumes a "C" shape

If both wheels move in out-of-phase motion, the leaf spring forms an 'S' which generates a different resistance to the two wheels. In this case, the leaf spring acts as a conventional anti-roll bar. In a best case, the transverse leaf spring can completely substitute the anti-roll bar. This results in a very efficient solution in terms of cost and weight.

Typically, different requirements are used as input to the development process. Some examples are axle weight, wheel bounce and roll stiffness, durability and strength load cases as well as tuning range. In addition, boundary conditions such as package might limit the space for possible design solutions.

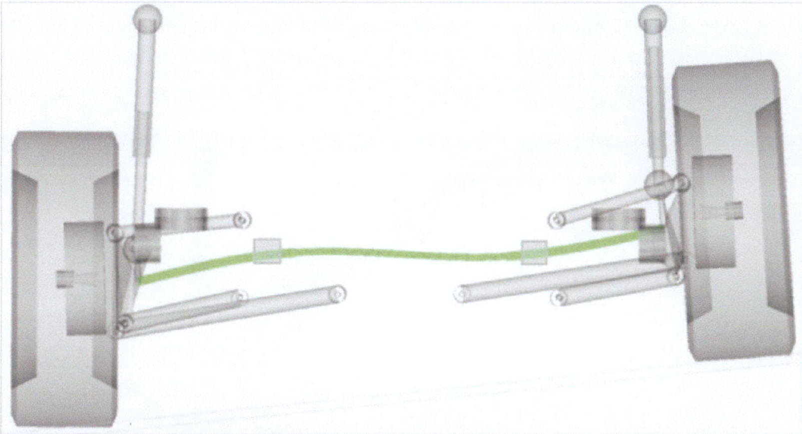

Figure 2: Transverse leaf spring working principle: When the wheels move out-of-phase (roll motion), the transverse leaf spring assumes an ‘S’-shape

There are different manufacturing processes used to produce a leaf spring. Each of them might impose different rules for an optimum design or even have hard constraints (e.g. constant cross section or constant width). The material plays an important role, too. The type of fibers and matrices and the percentage of fibers in the composite, for example, determine the basic mechanical properties of the final part.

A virtual development phase is conducted before building physical prototypes. Analytical and numerical development tools are applied to generate optimal design solutions on system and component level.

In a first step a basic layout is developed using the linear beam theory. The advantage of this approach is to analyze and optimize many designs in a very short time. The input parameters are material properties, connector positions, wheel loads and stiffness targets. For the optimization process different target metrics can be applied.

The leaf spring suspension system is modeled in a multibody simulation environment to analyze the suspension motion in detail. The desired stiffness and loads at suspension or vehicle level can be iterated and optimized. Additionally, other important kinematics and compliance characteristics can be verified.

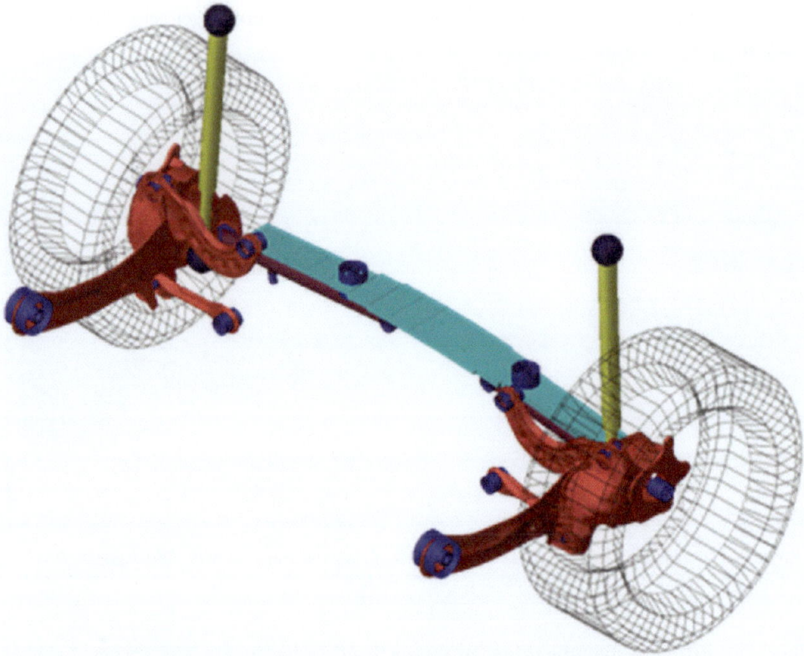

Figure 3: Multi body model of a transverse leaf spring rear suspension

The next step is based on a more refined finite element model of the transverse leaf spring and the connectors. The stresses and strains in the leaf spring can be analyzed in detail, not only the ones in fiber direction, but also in all the other directions. In particular, shear stresses and compression stresses can be critical. Other important results like for example the deformation of the spring can be generated by the FE model. The deformation can be important to check against the package constraints.

The remaining suspension components can also be modelled as flexible components. The interfaces are often the most critical items of a composite leaf spring system. Especially the articulation of the connectors is important to define their requirements.

If the FE model confirms that the requirements are met and the boundary conditions are respected, enough confidence is reached to build physical prototypes. Physical tests are important to validate the developed designs and to correlate the simulation models. Tests are planned on component, system and vehicle level.

When no anti-roll bar is used, the leaf spring needs to generate enough roll support. This can be sometimes challenging. One of the most powerful items to tune is the spring section (width and height). It is desired to have different sections for the center part of the spring (approximately between the inner connectors) which influences mainly the vertical motion and for the outer part which influences mainly the roll motion. For example it might be desired to have a thin center section for a low vertical stiffness with thick outer sections for a high roll stiffness. With regards to the section height it should be noted that the bending stress is a linear function of the section height. Increasing the thickness of the spring will result in higher bending stresses for a given motion. Finally, the freedom in changing sections might be limited by the manufacturing process, e.g. a constraint of a constant section area or a constant width.

Another important tuning parameter is the lateral position of the inner connectors. For a given geometry and length of the leaf spring, a larger spread between the two inner connectors increases the roll support. On the other hand, a big distance between the inner connectors means a bigger movement of the center part of the spring during vertical and parallel wheel motions.

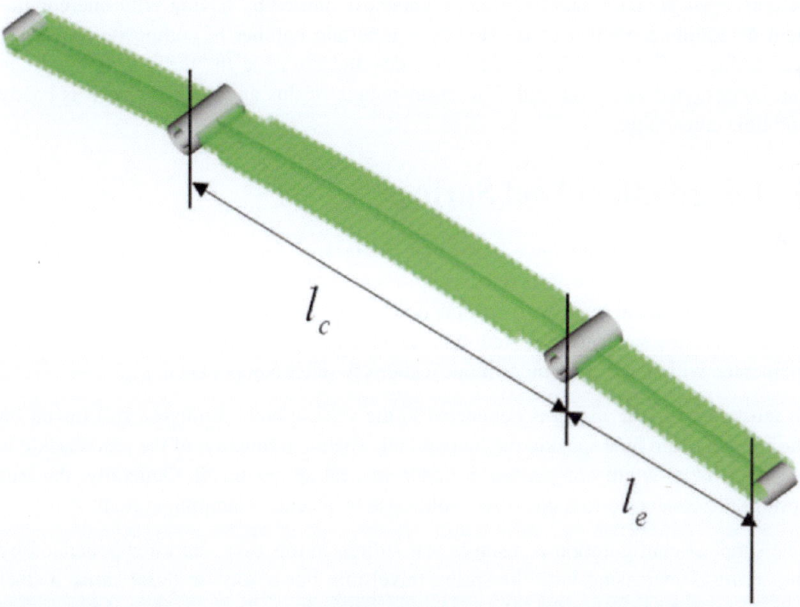

Figure 4: Spread between the inner connectors (l_c) and length of the end sections (l_e)

The design of the inner connectors is complex as it is subjected to multiple sometimes conflicting requirements. In vertical direction the inner connectors should have a high stiffness and be capable of taking a high preload. In lateral direction as well as in rotation around the global vehicle x-axis the inner connectors should have a very low stiffness to avoid undesired internal forces as well as to minimize parasitic spring rates.

There are few challenges in laying out a leaf spring suspension. One of them is package. When standard coil springs and a stabilizer bar system are replaced by a lateral composite leaf spring, package can become challenging. Driven by the longitudinal position of the interfaces between the ends of the lateral composite leaf spring and the wheel carriers or suspension control arms, the longitudinal position of the spring is often constrained. In addition, the center part of the leaf spring is exposed to a pronounced vertical motion, which needs to be package protected. Adjacent systems and components are body as well as frame structures and panels, driveline parts and exhaust pipes.

The tuning work on a transverse composite leaf spring suspension might also represent a challenge. The freedom of independently tune parallel and roll motion by mean of coils springs and a stabilizer bar is somehow limited by having only one component determining both of them. However, a certain number of composite transverse leaf springs, offering different characteristics and covering the desired tuning range can be designed and produced. The disadvantage of this approach is that it is costly and time consuming.

3 Longitudinal Leaf Springs

Longitudinal leaf springs have been used extensively for a long time, especially on commercial vehicles and light and heavy trucks. For these applications, the substitution of conventional steel with composite material can lead to very big weight savings. In case the suspension system will not change, the composite springs can be implemented with no or only minor modifications to other components.

In most cases a leaf spring is connected to the vehicle body through a soft mount on the front end and by a shackle on the rear end. The main function of the rear shackle is to allow for a length compensation under vertical spring travel. Generally, the leaf springs are connected to a transverse solid axle by a center clamping system.

The simplest configuration is a single leaf spring. In this case, the force generated by the spring is increasing with the spring travel in a linear way. In these cases, a steel leaf spring can be substituted by a composite leaf spring.

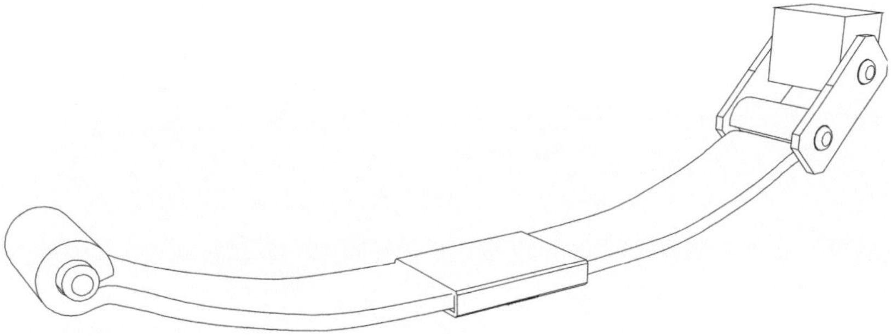

Figure 5: Single leaf spring with linear vertical force versus travel characteristic

In many cases a non-linear force versus wheel travel characteristic is desired in vertical direction in order to improve the vehicle dynamics performance and/or the ride comfort of a particular vehicle. Typically, the spring system is designed to deliver low spring rates for light loading conditions from rebound towards curb and higher spring rates for heavy loading conditions towards full jounce.

Longitudinal leaf springs essentially generate a linear force versus travel characteristic over the wheel travel in case the longitudinal length variation of the leaf spring under bending motion is unconstrained. In turn, a non-linear force versus travel characteristic can be achieved by constraining the length variation of the composite leaf spring under bending.

A second alternative to achieve a non-linear force versus travel characteristic for a suspension system with longitudinal leaf springs is the combination of a main leaf spring with an additional helper leaf spring. In this case, the main leaf spring accounts for the primary rate and the main leaf spring plus the additional helper leaf spring accounts for the secondary rate. Typically, such two-stage leaf spring systems are laid out in a way that for low vertical wheel forces the additional helper leaf spring is not engaged. This means that only the main leaf spring determines the total rate of the leaf spring system. For high vertical wheel forces the additional helper leaf spring engages at a specific ride height respectively wheel travel. In this condition, both the main leaf spring and the additional helper leaf spring determine the total rate of the leaf spring system.

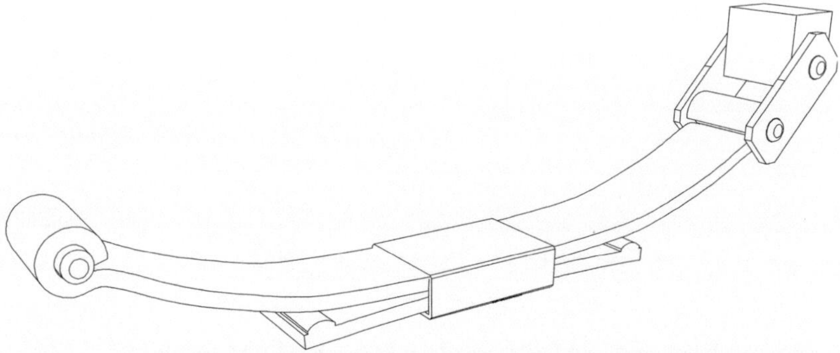

Figure 6: Longitudinal leaf spring with additional helper leaf spring and progressive force versus travel characteristic

A third variant to realize a longitudinal leaf spring suspension system with a non-linear progressive force versus travel characteristic is a combination of longitudinal leaf springs and additional bumper elements. Like for the system explained above with main leaf springs and additional helper leaf springs, the leaf spring suspension system with additional bumper elements generates a first rate at low vertical wheel forces with the main leaf springs only. For high vertical wheel forces, the additional bumper elements engage and the second stage rate is generated by the leaf springs and the bumper elements.

Next to deliver vertical stiffness as one main function of a longitudinal leaf spring, there are many other requirements to be fulfilled like for example durability, strength, lateral stiffness and wind up stiffness. In addition, the leaf spring needs to be packaged in a specific vehicle environment, considering all clearance requirements.

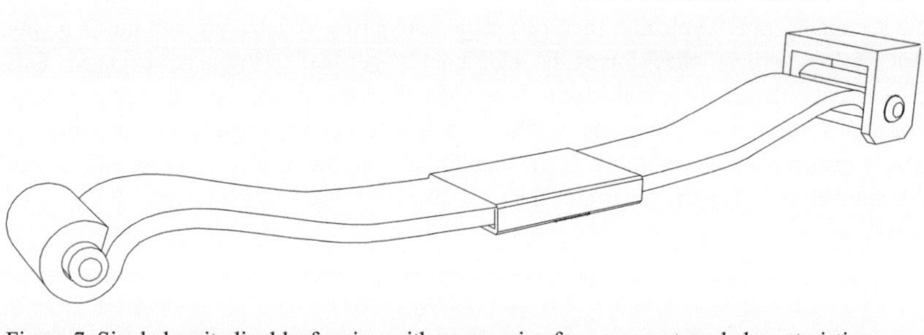

Figure 7: Single longitudinal leaf spring with progressive force versus travel characteristic

The package can become soon a challenge, especially when the composite leaf spring needs to replace a steel leaf spring, keeping all the other parts and the system unchanged. In particular as the composite material stiffness is much lower than the steel material stiffness, the spring dimensions need to be bigger. In vertical direction, this is usually not a big problem, but in lateral direction a wider spring may require a new clamp group and may generate other package issues with adjacent components.

The extensive use of CAE tools especially in early development phases is very important to meet several attribute requirements and guarantee an efficient development process of composite components.

Simple analytical tools allow engineers to come up with basic design solutions upfront like for example the shape of a generic longitudinal leaf spring to achieve the required stiffness. A quick assessment of maximal stresses reached on the component can be determined at the same time.

The higher complexity of composite materials in comparison to steel materials and the stringent demand of strong and durable components challenge engineers to achieve best-performing solutions combining lightweight and robustness. The advantages of an extensive implementation of CAE analyses, thanks to their continuously increasing reliability, are a reduction of expensive physical prototypes and the related efforts in testing. Hardware is still required but more and more to confirm manufacturing process robustness and to finally evaluate the fulfilment of the designed characteristics.

As the main function of a spring is to supply the suspension with a certain stiffness value, numerical evaluations to assess spring kinematic and compliance within the given design space are needed, but differently than in strength or fatigue analyses, the modelling effort can be limited. Since the spring working range requests for high deformation capacity as high strength properties to withstand multiple loads acting in different directions, extensive strength and fatigue analyses are needed.

The nonlinearity of the problem, the multitude of possible failure mechanisms and the sensitivity of strength and toughness to impurities and defects from the manufacturing processes makes accurate modelling very challenging.

Aside from compression forces acting on the leaf spring in the center clamping area, the main load experienced by leaf springs is a bending deflection due to the vertical travel. One side of the spring deflecting over the travel is loaded under tension and the other one under compression.

Following the beam theory and assuming that the material is homogenous without irregularities or defects, the maximum stresses or strains are reached on the outer surfaces of the component. Including these aspects into a CAE analysis can play a key role for the design of such components in order to ensure a durable part. This approach requires high time-consuming computations.

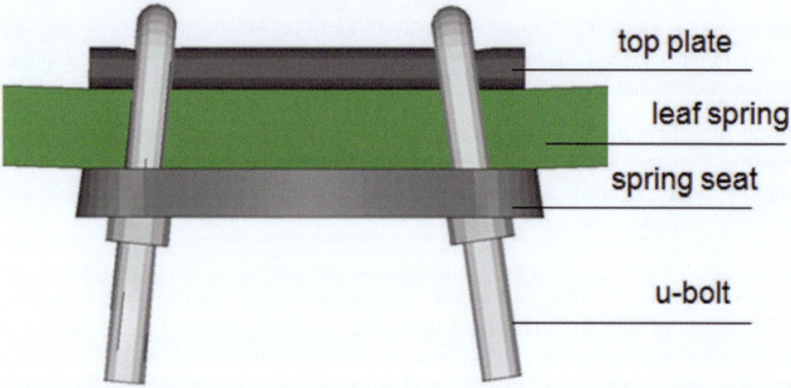

Figure 8: Clamp group

The connection between axle and leaf spring is typically realized by a clamp group which consists of a stamped or forged spring seat and a top plate which is fastened on the spring by U-bolts. The torque applied to the U-bolts is translated into a certain pressure distribution on the upper and lower surface of the spring to realize a strong and robust joint between the longitudinal leaf spring and the transverse solid axle.

An accurate and detailed analysis of such a joint is mainly required because of two main aspects: fatigue and creep of the composite material. The vertical compression acting on the outer surfaces of the spring by the upper and lower plates combined with the cyclic bending load result in a bi-axial loading condition which is fatigue relevant. On the upper side of the leaf spring a combination of cyclic tensile stresses in fiber direction in combination with compressive stresses acting primarily on the matrix can be observed. On the lower side of the leaf spring compressive stresses acting both in fiber and in matrix directions occur.

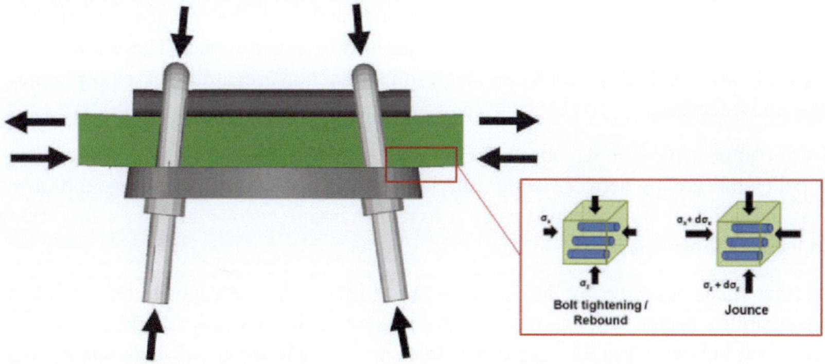

Figure 9: Loads within clamp group

This particular combination of compressive stresses in combination with possible weakening effects of the matrix due to material temperature sensitivity can lead to observable failure modes.

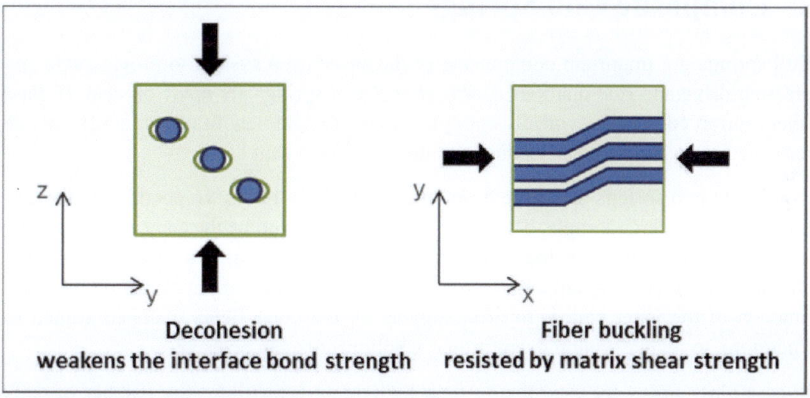

Figure 10: Compressive stresses

Compression in fiber direction can lead to fiber buckling issues in fiber reinforced plastic materials. The fiber buckling is resisted by the shear strength of the matrix. Introducing a normal compression force, a detaching effect between fibers and matrix can be induced and this can negatively influence the shear strength of the matrix. This determines a lower limit for the admissible axial compressive stress. Additionally, fiber misalignments can cause a reduction of the maximum shear strength.

Besides fatigue issues, high compressive stresses acting on the composite material in directions normal to the fiber orientation can lead to creep effects. These effects can cause clamp load losses which can result in malfunctions regarding assembly robustness and integrity.

Since compressive stresses normal to the fiber orientation are unavoidable in conventional clamp groups between composite leaf springs and solid axles, the goal of a robust and efficient clamp group design is to realize homogeneous pressure distributions on the component surfaces.

Another focus point during the development of longitudinal composite leaf springs is the durability performance of the leaf spring components. Various manufacturing processes are known to produce composite leaf springs for automotive applications. Next to filament winding processes, prepreg processes as well as resin transfer molding (RTM) processes are available on the market. Various failure modes like for example micro cracks in the matrix as well as delamination effects between individual layers within a component must be avoided for automotive suspension leaf springs as safety critical chassis components.

4 Composite Coil Springs

Coil springs are important components of the suspension system to provide safe and pleasant drive and ride qualities. Conventional coil springs are made of steel. If glass fiber reinforced plastic (typically epoxy) is used a weight benefit of about 50% can be reached. One example of this solution is the Audi A6 Avant Ultra.

Composite coil springs are made from layers of endless fibers wrapped around a center core with a certain angle. The number of layers as well as the fiber angles are determined in order to meet the spring requirements (stiffness, load, travel, side load, etc.) and the stress or strain targets. As a coil spring mainly works in torsion the main function of the inner core is to offer support for the outer fibers wrapped around it. Therefore, to reduce weight even further, a hollow core can be used.

In case glass fibers are used due to their high strain capabilities the number of coils can generally be reduced compared to an equivalent steel coil spring developed based on the same set of requirements.

Different manufacturing processes are possible to produce a composite coil spring. Even though manufacturing processes are not the primary focus of this paper, some general principles and differences between them will be presented.

Figure 11: Steel coil spring (left) versus composite coil spring (right)

The first step is the production of the composite rope. The different layers are placed around a central core with different angles relative to the rope main direction. The core can be made of different materials and can be either permanent or temporary, e.g. it can stay in the final product or it can be melted before the coil spring is finalized. The impregnation of the fibers to form the composite material can be done before (pre-impregnated fibers), during (in-line impregnation) or after (resin injection) this step. Depending on impregnation strategy, the rope manufacturing step will slightly vary, especially in terms of the supporting equipment. The rope length and thickness depend, as mentioned above, on the desired spring characteristics.

A following step consists in coiling the rope around a mandrel to generate the final spring shape. This step must be done until the resin is not yet cured, to allow the possibility of bending the rope. To guarantee flexibility in the spring shape (e.g. pigtail design) the mandrel can be melted at the end of the process to allow the spring to be removed.

The next step is the curing of the material. This is done in an oven under a certain temperature for a certain period of time.

The final step, after the matrix is completely cured and the mandrel is removed, is the machining of the coil spring like cutting the spring to the right length, eliminating imperfections and excessive resin or reworking the spring ends.

Some geometric and functional checks will be performed at the end of the process on each spring to check the quality of the products.

In terms of benefits of the composite coil springs weight reduction is for sure the most important. Nevertheless, other advantages make these products very interesting. An example is the lack of corrosion and in general a safer failure mode compared to steel springs. A composite coil spring will not break suddenly in two pieces as a steel coil spring might do under certain conditions but will have a progressive failure, leaving the driver the time to recognize that something is wrong long before a catastrophic failure.

Another important benefit is the road noise performance. Compared to an equivalent steel coil spring, a composite coil spring generally shows a reduced level of noise. This has three main root causes. The composite coil spring provides approximately twice as much damping than the steel spring. Although the level of damping is rather low in both cases, there is a broadband improvement in the interior noise as well as in the acceleration responses. A second reason is the modal response of the composite springs. Some fundamental modes are shifted up by 15-20 Hz when compared to the modes of a steel spring. This alone is unlikely to be the reason for an improved performance, because the coil spring modes were not adapted to the modes of the vehicle at the strut, suspension and body. Finally, the moving masses are very much reduced, so that similar vibration levels would contain much less vibrational energy in the spring.

The modal analysis has been conducted in different conditions, e.g. 'free-free', where the springs are suspended in free length with bungee cords.

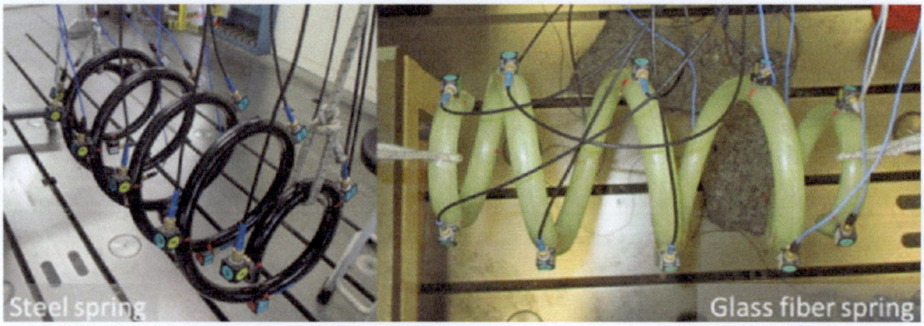

Figure 12: Springs in 'free-free' condition

In all tested conditions each spring is instrumented with approximately 3 tri-axial accelerometers per coil for a total of 17 accelerometers for the steel spring and 14 accelerometers for the glass fiber spring. This layout is sufficient to measure with the desired accuracy the spring modes up to a frequency of $f \approx 130$ Hz. The springs have been excited with impulsive signals using a common instrumented hammer.

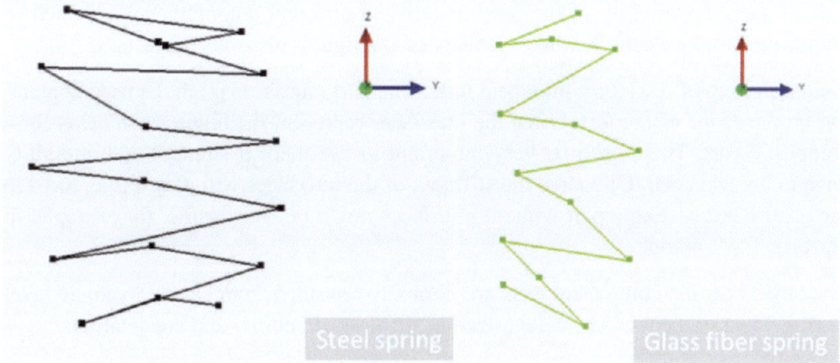

Figure 13: 1st bending mode, 'free-free' condition – comparison between steel and composite coil spring

5 Composite spring tests

One of the most important parts of the development process to implement a composite spring into a vehicle is testing. This is the time when designs, assumptions as well as material and manufacturing process are verified.

The test program is of course specific to the technology (transverse leaf spring, longitudinal leaf spring or coil spring) and to a certain extent to the application (e.g. race car, passenger car or commercial vehicle). However, many tests are common or at least very similar.

The material is normally tested first. The results give a good reference for comparison to existing databases and can be used as benchmark later in the development process. If a material change is desired or needed most of the material tests are performed on coupons specifically prepared for the test to execute. The advantage of this approach is that the specimen geometry can follow existing standards making some comparisons more meaningful. An example of tests done with specifically prepared coupons are mechanical tests like for example tension, compression or bending tests to check the strength and stiffness of a specific material.

Other tests are done with material coming from a real part. The advantage of doing this is that the influence of the manufacturing process on the material quality is considered. Some examples are the fiber volume content, void content or glass transition temperature analyses.

Testing the compatibility of the used material to different automotive fluids such as engine oils, brake cleaners, wheel cleaners or UV light is also very important.

Another group of tests are component tests. The part interfaces might be used depending on the scope of the test. Often the interfaces represent the highest risk areas for a composite part. Therefore it is very important to use them if strength and durability have to be evaluated. Of course the stiffness of the interfaces will play a role, too. On the other hand, testing a part without interfaces might be worthwhile, for example, to correlate CAE models.

The targets for the component tests are normally cascaded from existing vehicle level targets. Most of them can be categorized in stiffness, durability and creep targets.

To test the durability a cycling input is given to the spring. Typically the full range of suspension motion is supposed to withstand for a lifetime representative amount while rate and load are checked. These values have to stay within a certain tolerance to pass the test. Visual appearance is also evaluated during and after the test. Even if the function of the part is maintained it is done to avoid customer dissatisfaction. The durability tests can be done in different conditions to represent the worst case scenario in terms of customer usage. To reproduce very high or low temperatures and/or high humidity levels a climate chamber is often used.

Creep tests are static tests where a constant load or displacement is applied for a certain period of time. Similar to the durability tests rate and load are checked. For composite springs these tests are extremely important to check any possible sag which would cause the vehicle to lose ride height. Similar to durability tests creep tests are done at different environmental conditions.

When enough confidence is reached on component level vehicle tests can start. One of the first tests normally executed is the so-called ‘K&C test’ where the kinematics and the compliance of the suspension are analyzed. This test offers a good opportunity to compare and correlate suspension and vehicle models.

Regarding attribute performance road noise and vehicle dynamics tests are typically performed if possible as A to B comparisons to a standard steel spring version of the composite spring application.

Finally, full vehicle durability testing is extremely important to verify if the lifetime targets have been properly cascaded during the design phase and if the solution developed is robust to the different noise factors that are present in a proving ground test. For chassis components this test is a repetition of a series of tests, each of them provoking high stresses.

If the material and the parts survive the full test program the confidence level that the customer expectations will be met is very high.

6 Summary / Conclusions

Three types of composite spring designs have been presented in this paper, namely transverse leaf springs, longitudinal leaf springs and coil springs. Each chapter gives a brief insight to the most critical items for each component.

Ford Motor Company and especially the Research and Innovation Center in Aachen, Germany have been studying and developing these concepts for several years, acquiring a lot of experience for what makes a good design, what works and what not.

It is believed that the applications of composite springs, which have already grown in the last few years, will continue doing so at even higher speed.

This is justified by the increased pressure to reduce vehicle weight, the downward trend of material costs, the competition to improve manufacturing processes and the amount of research that industries, research institutes and universities are dedicating to this subject.

Mubea chassis springs – development of GFRP leaf springs for best-in-class lightweight design and functional performance

Authors: Dr. Jochen Asbeck, Dr. Joffrey Stimpfl

P.E. Pfeffer (Ed.), *8th International Munich Chassis Symposium 2017*,
Proceedings, DOI 10.1007/978-3-658-18459-9_13

1 Introduction of Mubea

Mubea is the market leader in many segments of the spring industry and achieved a remarkable global presence in the past years. Mubea's growth is based on global presence and technological innovations. Over the past few years we have established a worldwide production network which we are gradually expanding. We make sure that our company-specific know-how paired with our knowledge of local culture are available at all these locations. Whether in Europe, America or Asia, Mubea supplies products to all of the automobile manufacturers and leading Tier 1 system suppliers worldwide. Our primary goal is to develop close ties with our worldwide customers in a sustainable way. On the one hand we are a global player and on the other hand we will continue to be a family-owned business with its own identity, high flexibility and long-term orientation.

Over the past 40 years, the company has focused on heavy-duty-impact springs for passenger cars. For about ten years now, Muhr and Bender have expanded their product portfolio to include innovative products in the body frame sector. The first successful step has been made with the development of the so-called "flexible rolling" (also known as Tailor Rolled Blanks©). This lightweight construction technology, developed by Muhr and Bender for industrial production, has established itself over the past few years as a standard solution for weight reduction of body frame components worldwide.

The acquisition of Mubea Carbo Tech in Salzburg in 2011 was a strategically sensible addition to Mubea's expertise in lightweight construction components. As a pioneer in the carbon industry, Mubea Carbo Tech has made a name for itself as the world's first serial manufacturer of monocoques and hybrid wheels. Mubea Carbo Tech has more than doubled its sales of Mubea Carbo Tech since the acquisition. From now on, we are concentrating on intelligent applications where carbon is used "in the right place" and taking full advantage of this unique material. We are convinced that in the coming years carbon will be successfully used in the body and chassis field. We have prepared a cost-effective production of these scopes by setting up a production site in the Czech Republic.

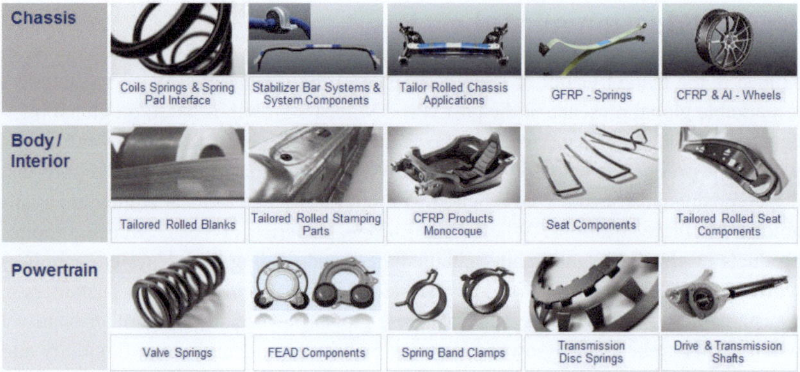

Fig 1: Mubea product portfolio

A major reason for the growth of the company over the past years is a multitude of innovations in the chassis. For axle springs, the use of the new generation technologies and SLM technologies has allowed approx. 15% weight reduction. For stabilisers, the worldwide triumph of the tubular stabiliser continues. With the recent build-up of our own tube plants, we remain faithful to our strategy to extend the value chain. At the same time, the global footprint of the chassis products was expanded by Turkey, Mexico and two locations in China. The company in Prostejov/Czech Republic has been expanded step by step as the world's largest factory for the production of axle springs and stabilisers with having today more than 1,100 employees. The location in Weißensee/Thüringen was strengthened by concentrating all chassis development activities in a new, generous development centre.

We believe it is possible that our innovations can reduce the weight of the axle springs products and stabilisers by a further approx. 15% in the coming years. For a medium-class vehicle, weight reductions in the range of 2 to 3 kg are hereby connected.

On the occasion of the IAA of 2011, we were able to present our customers for the first time with a leaf spring made of fibreglass – since then we have led this development to the standard production stage and are aiming to expand this enormous weight potential of up to 75% corresponding to approx. 50 kg per axle. Thus, we offer our customers a unique know-how combination with regard to springs <u>and</u> fibre composite materials. The first two series projects with SOPs in 2016 and 2018 will be industrialised at the Attendorn site, while further series projects will use the synergies with other fibre composite activities in Eastern Europe.

2 GRP leaf springs from Mubea

2.1 Product portfolio

GRP leaf springs are available in automotive vehicles as a version in the form of a transverse leaf spring or longitudinal leaf spring. The transverse leaf spring does not assume any wheel guiding tasks, but replaces the coil springs and, in the best case, the stabiliser as well. Fig. 2 shows a transverse leaf spring application in a Volvo platform in which the vertical dynamics are performed by a GRP transverse leaf spring. However, a stabiliser for providing the roll steering rate is still required in this example.

In addition, longitudinal leaf springs also take over the wheel guide. Especially when used on a driven rear axle, the requirements placed on the GRP material, the design and the technology are therefore higher. Figure 2 shows that since 2016 the Mercedes Sprinter has been fitted with a GRP longitudinal leaf spring for the driven rear axle for the first time. The manufacturer and supplier of the GRP leaf spring is Mubea.

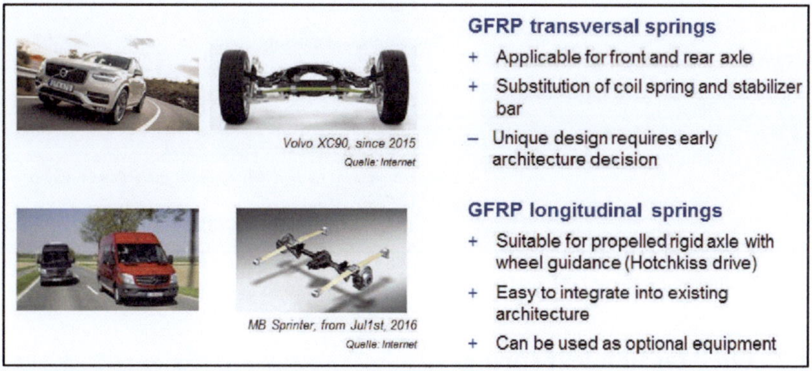

Fig 2: Successful market launch of GRP leaf springs

In principle, Mubea offers two types of longitudinal leaf springs for the rear axle of light commercial vehicles and pick-ups. Longitudinal leaf springs with a linear spring characteristic curve account for 90% of the applications in the EU and are designed in most cases as 1-layer steel leaf springs. The GRP longitudinal leaf spring is functionally identical but has a significantly lower weight with good damping behaviour. Furthermore, GRP springs never suddenly fail. This means that the axle remains guided even in the event of a failure.

On vehicles with higher payloads and with all pick-ups, longitudinal leaf springs with a progressive, multi-stage spring characteristic are used, which in the case of steel leaf springs consist of multi-layer steel blade packs. For this application, the GRP longitudinal leaf spring is designed as a tension leaf spring, which is further explained in chapter 2.2. Figure 3 shows an overview of the products and characteristics of the GRP leaf springs from Mubea.

The spring is connected to the chassis by using aluminium spring eyes with rubber bearings at the front and rear ends of the spring. When the shackle of the steel spring to be replaced is positioned under the steel spring, the rear as well as the front chassis connection point can also be maintained as a rule.

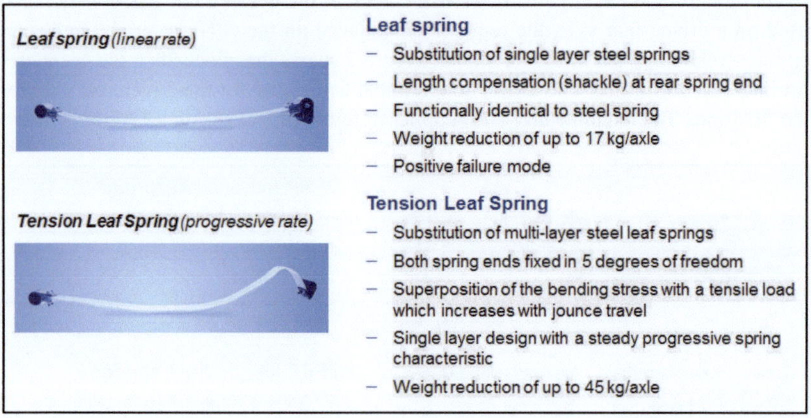

Fig 3: Product portfolio of GRP leaf springs

2.2 GRP tension leaf springs

A special innovation is the Mubea GRP tension leaf spring, in which not only steel is replaced by the lightweight GRP material, but a new Mubea-patented fibre-reinforced design is used. With the help of the tension leaf spring design it is possible, for the first time, to replace multi-layer steel springs for light commercial vehicles or pick-ups with single-layer GRP leaf springs. The novel design and the elimination of the length compensation by using the shackle lead to an increasing overlap of the bending portion with a pulling portion during compression. Thus, for the first time, a continuously progressive spring characteristic curve is possible despite a single-layer spring design.

The biggest customer benefit is the weight savings, which can be up to 45 kg at the rear axle when using a GRP tension leaf spring. There is hardly any other component in the chassis of an LCV or pick-up, which offers as much savings potential as the leaf spring. The continuous tension leaf spring design furthermore makes it possible for the first time to achieve a continuously progressive spring characteristic, despite a single-layer spring design, see Fig. 4.

This leads to a significant increase in driving comfort compared to a sudden change in the spring characteristic curve for multi-layer steel springs. The contact impact during the engagement of the second steel blade is then also no longer applicable. Furthermore, the contact impacts can lead to a noise generation. On the other hand, a change in the masses in contact leads to a suddenly changing natural frequency in the acoustic path.

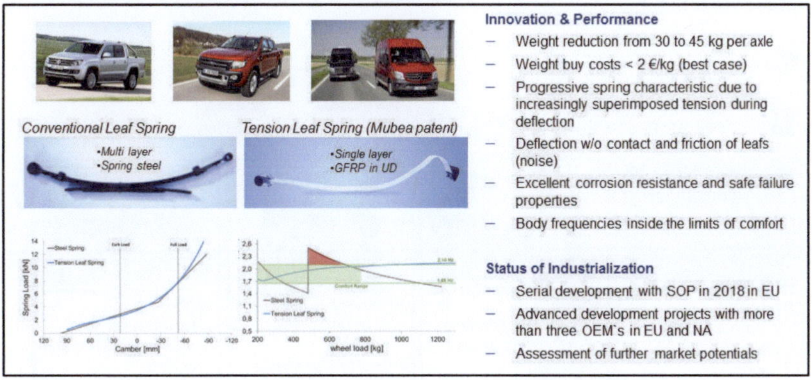

Fig 4: Comparison of spring characteristic and natural frequency of a BZF with a 2-ply steel spring

Overall the Mubea GRP tension leaf springs do not only offer a very high weight saving in light weight construction costs, but also a significant increase in driving comfort.

2.3 Production

The Mubea GRP leaf springs and tension leaf springs are produced in series by using the production technology of prepreg presses. The special feature here is the use of a strip-shaped semi-finished product, the so-called prepreg (pre-impregnated fibres). This results in some advantages with regard to the manufacturing process as well as the GRP spring product. On one hand, with continuous prepreg production, all the fibres in the spring are soaked and unidirectionally aligned (in one direction). On the other hand, the stretched and laterally untrimmed fibres (during pressing) fit optimally to the contour, despite the thickness and width variation of the spring geometry. Furthermore, the chemical meshing of the plastic takes place completely under high pressure. As a result, pore formation, microcracks or unevenness on the surface caused by material shrinkage or expansion differentials, are avoided during curing.

Already in 2010, with the construction of the GRP pilot plant, the prepreg pressing process was already defined as a manufacturing process for GRP springs at Mubea. The serial production of GRP leaf springs was then started at Mubea in 2016 at the Attendorn site. By 2018 the capacity will be expanded to a 6-digit annual volume.

The required process steps for the production of GRP springs using prepreg presses are shown below, see figure 5. This process has been successfully industrialised with the current serial production of the conventional GRP leaf spring for the Daimler Sprinter, see Fig. 6.

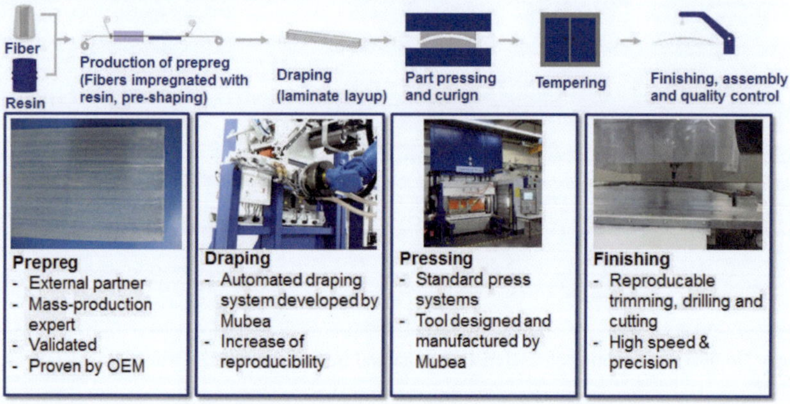

Fig 5: Process steps of the production of GRP springs using prepreg presses

Fig 6: Serial production of GRP leaf springs

The main advantages of the prepreg technology for GRP leaf springs compared to other manufacturing methods arise in the individual process steps as follows:

Production of semi-finished prepreg (**pre**im**preg**nated fibers)

- The optimised impregnation of the, during the process, individually spliced fibres allow for a very stable process management, by pressing into a resin applied to film with low viscosity and by temperature input. The process parameters are very controllable and enable fully automatic, continuous and economical production. The quality of the semi-finished product compared to other processes is very high.
- No stitching up or suturing is required to ensure the stability of the semi-finished product. Therefore there is no risk with regard to undulations or defects due to e.g. cross points of sewing threads.

Automatic robotic draping of prepregs using tape-laying head to a blank

- The process is a fully automated, reproducible, monitored length cut with exact positioning of the layers.
- A 100% inspection of the processed prepreg by an on-line surface inspection using a camera as well as monitoring the positional length and position.

Pressing and curing the prepreg blank into a spring

- The draped blank is narrower than the spring width. A wide width variation over the structural length is possible since no sewing threads hinder lateral movement of the fibres during pressing. This ensures a maximum transverse stiffness of the leaf spring through locally increased width without encroaching the installation space (packaging).
- The prepreg does not have to be trimmed laterally for a wide variation in the spring, there are no severed fibres on the highly loaded spring edges.
- The laminate quality is high, since pressing is already in the first phase high pressure. Cavities between the prepreg layers are securely closed and the remaining air is flushed out by squeezing excess resin.
- The complete curing also takes place under high pressure. Microcracks, cavities and surface irregularities on the surface due to material shrinkage are avoided by the sustained high pressure during curing

The described prepreg manufacturing process with the robot-guided draping and pressing in a multiple cavity with force-controlled position control is a unique feature of Mubea.

3 Summary

Over decades Mubea has developed into a world leader for chassis springs. Our product strategy, consisting of a combination of innovative lightweight construction combined with excellent robustness and function, the greatest possible value creation depth and global availability has made us the innovation driver for axle springs and stabilisers.

We will continue to see this path as a key success factor and will continue to focus on the continuous optimisation and further development of our product portfolios, such as the GRP leaf springs and related products presented in this article, which are shown in figure 7.

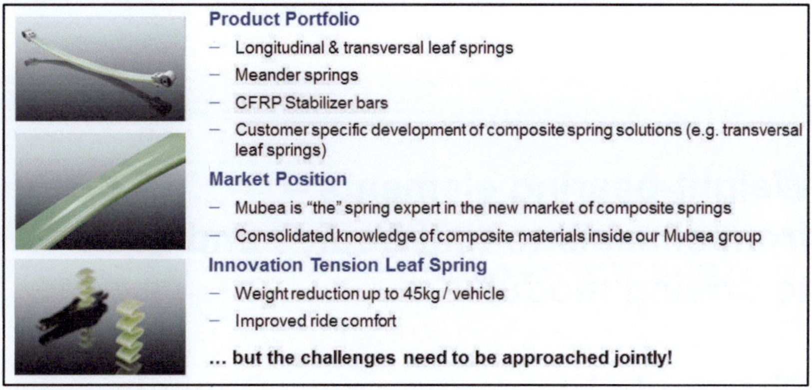

Fig 7: Summary

As a result, we are the best and most reliable partner to our world-wide chassis customers, because only together with our customers can we take advantage of the opportunities offered by the GRP technology for suspension springs and successfully apply them to series production. This technology defines new, significant light-weight built potentials that further improve the efficiency of our vehicles.

Weight-bearing elements – from chassis to unit bodies and back to driving modules

(Tragende Elemente – vom Chassis zur selbsttragenden Karosserie und zurück zum Fahrmodul)

AutoHistorica
Fahrzeugausstellungen und Automobilgeschichte

Erik Eckermann

P.E. Pfeffer (Ed.), *8th International Munich Chassis Symposium 2017*,
Proceedings, DOI 10.1007/978-3-658-18459-9_14

Weight-bearing Elements

From Chassis to Unit Bodies and back to Driving Modules

Creating a lighter weighing vehicle has been a challenge since horse drawn carriages and bicycles have been built. A lighter vehicle was the goal in order to keep costs down and to achieve better roadhandling and driving comfort. While the bicycle still has to rely on a frame the elimination of the coache's perch and the automobile's chassis was the start of the transition to self-supporting coach- and bodywork.

For centuries Gala- and coaches for other purposes were equipped with two *brancards* (bearers) or one perch connecting front and rear axles (Fig. 1). The invention of leaf or elliptical springs in 1805 and their installation between the axles and the box (Fig. 2) eliminated the need for bearers or perch. This reduced weight and step-up height, lowered the center of gravity and thus improved springing comfort, road handling, and safety.

Coming back to the bicycle, light construction was even more desirable because of restricted human power and endurance. Inspite of this nearly no improvements have been made since 1885 when Starley's safety bicycle appeared (Fig. 3). It weighed about 16 kg, just a few notches more than today's touring bikes. In the 19th century all bicycles used a frame consisting of wood, wrought iron, or steel tube. Today, frames are made of steel tubes, or, in special cases, alloy or „monocoque design" which usually is a curvaceous one-piece diamond frame.

So it seemed rather logical for the pioneers of the automobile to base their construction on bicycle or coach techniques. In 1886 Carl Benz choose a front-steered three-wheeled cycle (Fig. 4) popular in Britain and France, using a stationary engine common with many workshops of the time. The chain drive, tubular steel frame, wire spoke wheels, fork steering, and lack of weather protection all originated from the bicycle.

One year later Gottlieb Daimler and his companion Wilhelm Maybach installed an engine in an open carriage imported most likely from the US (Fig. 5). Though the engine towered above the floorboards between the bench seats in a rather ungainly and inorganic fashion, this nevertheless represented the world's first practical four-wheeled motorcar with the characteristics of a horse drawn vehicle: wooden body with underlock, fifth-wheel steering (kingbolt or turntabel steering), fullelliptics, wood spoke wheels, and blockbrakes.

There seemed to be a great uncertainty about a definite design pattern for the motorcar for Benz and Daimler and the following manufacturers in Europe and USA meandered between bicycle and coach techniques for the years to come. This found the attention of the *Scientific American* (January 21, 1899, p.39) wishing „…all hail to the automobile, and may some gifted genius soon arrive who will whip it into shape and make it presentable! All things are possible, even a good-looking horseless carriage“.

2

America's longing for a good looking automobile was fulfilled sooner than expected. After less than 10 months of construction work Wilhelm Maybach presented such a car in December 1900 (Fig. 6). For marketing reasons ist was not called ‚Daimler'. At this time Mercedes featured a lowered chassis, extended wheelbase, rear wheel drive and a high performance 4 cylinder-engine of 35 HP hidden under a bonnet and behind a honeycomb radiator, whose increased surface area was coupled with a reduction in water volume, radiator dimensions, and weight. This vehicle marked the beginning of the modern automobile.

Firms throughout the world copied the Mercedes form and function – and often learned the hard way that chassis, clutches, springs, and many other components could not cope with the side effects of more powerful engines. One example is the chassis; tubular frame, adopted from the bicycle technology, proved too weak and had to be replaced by U-channel frames following Mercedes practice (Fig. 7). Firms now faced the realization that they would have to invest in expensive stamping and shearing dies, buy their frames from suppliers, or give up auto making entirely. Mercedes not only unleashed the first great wave of capital investment in the then young auto industry; it also sealed the fate of manufacturers that had inadequate capital.

The ladder frame à la Mercedes dominated chassis construction for passenger cars up to the 1930s, for buses up to the 1950s, and for trucks still today. However, passenger car manufacturers experimented with other types of frames rather early striving for advantages in production engineering, economics, and handling characteristics.

Edmund Rumpler, for instance, was an aircraft engineer that was forbidden to continue aircraft design according to the Treaty of Versailles after the First World War, applied his aerodynamic knowledges to his Teardrop Car of 1921 (Fig. 8). With its chassis layout dictated by the teardrop bodywork it carried, the Rumpler Teardrop Car displayed a break with traditional ladder frame practice. Instead, it had a deep frame, the double floor holding two spare tires, springs, linkages, power unit, and silencer.

In the Lambda of 1923 (Fig. 9) Vincenzo Lancia replaced the customary frame longitudinals with a pair of deep-section sheet steel members, joined to each other by the dashboard, seat consoles, and rear transom. The front and rear suspension members are carried by fabricated steel structures, and as there is also a wooden floor the Lambda may be regarded as a semi-unit body.

Another turning away from the ladder type frame is the backbone chassis appearing in greater numbers with the Tatra 11 of 1923 (Fig. 10). Designed by Hans Ledwinka the central tube rigidly connected power unit and rear axle housing incorporating swinging halfshafts. Ledwinka later constructed trucks (Fig. 11) with central tube chassis and swing axles arguing that the reduced unsprung masses of independently suspended wheels in connection with a twist-resisting backbone would save the road surface.

3

Ledwinka however was by no means the inventor of the central tube chassis. The English Rover 8 HP dated 1904 as well as the 1909 Simplicia (Fig. 12) built by the construction firm and occasional auto-maker Lacoste & Battmann in Paris featured such a frame. The tubular backbone measured about 140 mm in diameter and took the drive shaft which needed no cardan joint in this configuration. Cardan joints were subject to wear and tear up to the 1930s and thus not liked too much by the engineers.

In 1934 Ledwinka managed to launch the Tatra 77 (Fig. 13), an aerodynamic limousine of everyday utility seating up to 6 persons. It was based on a backbone chassis and box design with welded-on sheet panels and a rear fork to hold the aircooled V8 engine. Because of its futuristic shape and unconventional engineering lay out experts and the public believed in a new epoch of automobile design, not only for the 77-followers cultivated backbone chassis, swing axles, aircooled, rear mounted engines and streamline but also Ferdinand Porsche's KdF-Wagen which was the immediate forerunner of the Volkswagen Beetle.

Citing the ‚Handbook of the KdF-Wagen' (Fig. 14) which was published during the war, the chassis is described as ‚torsion stiffened centre frame' with welded-on sheet metals. One has to add that the transition between backbone frame with sheet panels on the one hand and a platform frame with backbone on the other hand is flowing thus making a clear classification more difficult. As for the rear engined models, Volkswagen was talking of a central tube chassis, while Porsche described the chassis of its 356-model (Fig. 15) as a platform frame with box bearers or just as a box frame.

Another version of the backbbone chassis could be seen in the X-frame introduced by Mercedes in the 1930s (Fig. 16). Two oval-shaped tubes are nearly touching each other in the middle of the car making separate body supports obligatory. However, in the case of a side impact there was nearly no passenger protection.

Swing axles and independent wheel suspensions instead of live axles plus hydraulic brake systems instead of mechanical ones paved the way for lowered ladder-type frames (Fig. 17) which brought down the body and centre of gravity.

Though aluminum was applied as early as 1899 by body makers, followed by engine and gearbox designers around 1910, aluminum chassis seldom reached production status. In the United States it remained restricted to 16 so-called Pomeroy-cars built by Alcoa and Pierce Arrow in 1925/26, and a V16-Peerless prototype of 1931 which fell victim to the world economic crisis.

In France the multi-talented engineer Dimitri Sensaud de Lavaud built two or three passenger cars (Fig. 18) with aluminum platforms and other sophisticated technical solutions in 1926. His ‚Nouvelle Conception de la Voiture Automobile‘ caused a lot of attention but never came into production, quite in contrast to the front driven Amilcar Compound (Fig. 19), of which nearly 700 units could be sold between 1938 and May/June 1940 when German troops occupied France. The Amilcar‘s alloy structure consisted of side members incorporating sills, firewall, A-pillars, and the windshield frame. After the war, Jean-Albert Grégoire‘s semi self-supporting construction emerged with Panhard’s Dyna-models and with the Hotchkiss Grégoire built from 1950 to 1957.

4

Meanwhile, the body design has changed from composite construction which is sheet metal panels over wooden frame to all-steel bodies which were available in the United States as early as 1912. Initially introduced to cut costs the all-steel body offered an advantage nobody thought of at first: combining the floor sheet metal with the frame would create a twist resisting floor panel eliminating a separate chassis and, welded together with the body, would end up in a unitary construction – with additional adantages: low unit cost per copy on long runs, less weight, and greater structural strength.

After having applied for patents in 1928 Budd Manufacturing Co in Philadelphia launched such a prototype in 1931 (Abb. 20) which eventually became the Citroen 7 CV Traction Avant, though slightly altered. The Citroën went into production in 1934, its semi-unit body consisting of a basic frame of stamped steel members, encompassing rocker panels and firewall, with pickup points for the rear axle and front powertrain assembly.

The first volume-production automobile with a pure-blooded unit body was the GM-based Opel Olympia of 1935 (Fig. 21). The body consisted of sheet steel structural members and stampings only, with the roof and roof pillars also sharing structural loads. In comparison to its forerunner in chassis/body-configuration the Olympia saved about 11% or 110 kg on weight.

Unit body construction is not restricted to a certain material. The long-forgotten German firms Slaby-Beringer 1919 and DKW 1932 used chassis-less bodies made of plywood, as well as the English Marcos 1960 (Fig. 22). However, wood and its derivatives is not able to cope with the requirements of modern cars. For a long time steel sheet members and stampings dominated body construction although today steel is often supplemented or replaced by other materials.

For instance steel was replaced by aluminum in the SHW-car of 1924 (Fig. 23), designed by Wunibald Kamm. This car had a pure aluminum monocoque, but only three cars have been built. Some 106 experimental cars have been produced along Alcan-principles by European and American companies between 1983 and 1995, and only Honda, Jaguar, and Lamborghini were ready for series production. Honda produced more than 10.000 units each of NSX and Insight, Jag and Lambo stayed below 500 each, always in aluminum sheet construction.

To cut costs Alcoa and Audi worked on aluminum space frames (ASF) covered with alloy-panels. First such a full-aluminum car was the premium-segmented Audi A8 of 1994 (Fig. 24), followed by the compact-class A2 six years later. Todays ASF-cars show different materials on spots better suited for steel, stainess steel or RPM/carbon.

5

In the 1950s some sports and racing cars were equipped with space frames (Fig. 25) made of steel tubes welded together in such a way that the tubes were mainly used on pull and pressure, less on bending stress. These ‚birdcage‘ frames evolved into monocoques, first made of alloy (Fig. 26), later of carbon. There were also some attempts in self supporting RPM-bodies (Fig. 27), but as production costs ran too high they remained prototypes. In fact, Lotus could sell more than 1000 units from its Elite model (Fig. 28) from 1957 to 1963, but its fiberglass body was reinforced by metal in the door post/windscreen area and for mounting front suspension, engine, and radiator.

Though the unitary construction has been employed for decades and by all car companies it is inevitably not the ideal solution for the cars to come. As the motors for battery-driven cars must not be installed in the front (or rear) but preferable near the wheels and as the batteries are to be located as low as possible between the front and rear wheels, some sort of tub with higher sides may be more useful, embracing all electrical and mechanical parts and clusters. Picking up this theme Rumpler‘s Teardrop Car of 1921 should be mentioned, see Fig. 8, and in more modern times, General Motors‘ concept car HyWire of 2002. In the HyWire, a space frame 30 cm deep includes fuel cell, hydrogen-tanks, and E-motor. Instead of steering linkages, hand brake lever, pedals, or other mechanical connections by-wire technology for steering, braking, and driving is employed. The body also offers weather protection. BMW (Fig. 29) followed with the i3 electric model in 2013, featuring an aluminum platform (Drive Module) with a carbon body (Life module) on top.

Back to the roots? Even when coachbuilder Kellner's 1925 advertising-virgin (Fig. 30) whispered in her lover's ear „…the chassis as per your choice, but the body only from Kellner“ it remains to say that an attractive chassis could fill one with enthusiasm as well.

Erik Eckermann

Tragende Elemente

Vom Chassis zur selbsttragenden Karosserie und zurück zum Fahrmodul

Schon bei Kutschwagen und Fahrrad, den beiden Vorläufern des Automobils, strebte man neben Leichtbau Kosteneinsparungen, rationelle Fertigung, sichere Straßenlage und erhöhten Fahrkomfort an. Während das Fahrrad auch heute noch auf einen Rahmen angewiesen ist, ersetzte man diesen schon beim Kutschwagen, später beim Auto durch einen selbsttragenden Aufbau, womit die Kriterien zumindest zum Teil eingelöst werden konnten.

Jahrhundertelang waren Gala- und andere Kutschen entweder mit zwei Schwellern oder mit einem Langbaum gesegnet, die Vorder- und Hinterachse miteinander verbanden (Abb. 1). Nach der Erfindung der Elliptikfedern 1805 (Abb. 2), die zwischen den Achsen und dem nunmehr selbsttragenden Wagenkasten angeordnet waren, konnten Schweller und Langbaum entfallen, womit sich neben reduziertem Gewicht auch noch andere Vorteile ergaben: Preisgünstigere Herstellung durch weniger Materialeinsatz, erhöhter Federungskomfort und tiefer gelegter Schwerpunkt, dadurch sicherere Straßenlage, geringere Kippgefahr und verringerte Einstiegshöhe. Das sind Kriterien, die auch heute noch gelten.

Auch im Fahrradbau war und ist das Gewicht Kriterium, denn hier muss der Mensch selbst die Kräfte aufbringen, um sich voranzubringen. Erstaunlicherweise sind hier kaum Verbesserungen festzustellen: Das Gewicht des Nieder- oder Sicherheitsrads von Starley 1885 (Abb. 3), das etwa 16 kg wog, wird von modernen Tourenrädern nur wenig unterboten. Allen Fahrädern ist ein Rahmen gemeinsam, der im 19. Jahrhundert aus Holz, Schmiedeeisen oder Stahlrohr bestand, heute aus Stahlrohr oder Aluprofil, in Sonderfällen aus Leichtmetall- oder Kunststoff-Schalen oder Carbon.

So erschien es den Autopionieren nur logisch, sich beim Bau ihrer Motorwagen am Fahrrad oder am Kutschwagen zu orientieren. Carl Benz wählte die Fahrradbauweise: Er hängte 1886 einen Stationärmotor und andere Organe in ein Vordersteuer-Dreirad Abb. 4), einem in England und Frankreich weit verbreiteten Dreispur-Fahrrad mit zwei Sitzen. Es wies mit Stahlrohrrahmen, Gabellenkung, Stahlspeichenrädern, Kettenantrieb und fehlendem Wetterschutz die typischen Merkmale eines Fahrrads auf.

Gottlieb Daimler und Wilhelm Maybach hingegen versahen 1887 einen vermutlich aus den USA importierten vierrädrigen Kutschwagen mit Motor und Antriebselementen (Abb. 5) und übernahmen dessen typische Kennzeichen wie hölzerner Aufbau mit Durchlauf und zwei Sitzbänken, Drehgestelllenkung, Vollelliptikfedern, Holzspeichenrädern, Klotzbremsen und hohe Bauweise.

In der Folgezeit wechselten Benz, Daimler und andere Motorwagenhersteller mehrmals von der Fahrrad- auf die Kutschwagenbauweise und umgekehrt. Die Gründe für die mäandernde Entwicklung mögen in der Unsicherheit über die zukünftige Auslegung eines neuen technischen Gegenstands gelegen haben, möglicherweise auch in der Hoffnung, an die jeweilige Klientel – Radfahrer, Kutschwagenbesitzer – verkaufen zu können.

2

Bis zur Jahrhundertwende befand sich die Branche im Experimentierstadium, das Wilhelm Maybach mit dem im Dezember 1900 herausgebrachten Mercedes (Abb. 6) beendete. Mit dem vor dem Motor angeordneten Bienenwabenkühler und der anschließenden Motorhaube, mit Hinterradantrieb, verlängertem Radstand, niedrigem Fahrgestell und (später) gleichgroßen Rädern löste sich das Automobil von Technik und Formvorstellungen des Kutschwagen- und Fahrradbaus. Die Maybach'sche Fahrzeugauslegung wurde zur Standardbauweise

Voraussetzung dafür war auch die Auslegung des Chassis, das bisher entweder aus Holzrahmen mit Eisenverstärkungen oder aus Stahlrohren bestand. Die aus dem Fahrradbau stammenden nahtlosen Stahlrohre waren durch Hartlötung miteinander verbunden, wogen zwar wenig und ließen sich preiswert herstellen, doch hielten sie höheren Belastungen nicht Stand. Sie konnten auch nicht ohne weiteres verstärkt werden, weil man an vorhandene Rohrquerschnitte gebunden war. Rahmenträger mit offenen Querschnitten dagegen, z.B. U-Profile, eignen sich wegen höherer Trägheits- und Widerstandsmomente besser als Träger mit geschlossenem Profil, wie es der Rohrrahmen darstellt, doch waren die vom Hoch- und Tiefbau her bekannten Eisenträger für Automobile viel zu schwer. Die Lösung bestand aus einem U-Profil-Rahmen, hergestellt aus gepresstem Stahlblech, bestehend aus zwei Längsträgern und einigen Querverbindungen (Abb. 7). Der leiterförmige Rahmen verengte sich wegen des Lenkeinschlags und zur Aufnahme der Motorpratzen im vorderen Bereich. Damit konnte auf den bisher erforderlichen Hilfsrahmen im Fahrgestell verzichtet werden, was Material, Kosten und Gewicht einsparte.

Der Leiterrahmen à la Mercedes beherrschte die Fahrgestellkonstruktionen für Personenwagen bis in die 1930er Jahre, im Omnibusbau bis in die 1950er Jahre und im Lastwagenbau bis heute. Doch tauchten gerade im PKW-Bau schon früh andere Lösungen auf, von denen sich ihre Anwender wirtschaftliche, fertigungstechnische und fahrdynamische Vorteile erhofften.

So verließ Edmund Rumpler (Abb. 8) den Grundriss eines angenäherten Rechtecks des üblichen Leiterrahmens und passte die Längsträger der spindelförmigen Karosserie seines Tropfenwagens von 1921 an. Zudem bildete er die U-Profil-Träger hochbordig aus, so dass eine Art Kellergeschoss entstand. Es war gegen den Staub der Straße mit einem Schutzblech nach unten abgekapselt und nahm neben zwei liegenden Ersatzrädern diverse Gestänge, Längsblattfedern, Antriebseinheit und Auspufftopf auf.

Auch Vincenzo Lancia (Abb. 9) trennte sich von der herkömmlichen Fahrgestellbauweise und ersetzte beim Lambda von 1923 die U-Profil-Träger durch hochbordige Stahlbleche, von Armaturenbrett, Heckwand und Sitzkästen querversteift. Die Stahlbleche reichten bis zur Gürtellinie und waren mit Aussparungen für Türen und zur Gewichtserleichterung versehen. Lancia erreichte dadurch das für ein Zweiliter-Auto erstaunlich geringe Gewicht von 780 kg. Karosserieboden aus Holzbrettern und Stahlrohr-Hilfsrahmen für Vorderradaufhängungen und Hinterachse sprechen gegen eine selbsttragende Karosserie, die dem Lambda zugesprochen wird.

3

Beherrschte während der 1920er Jahre die von Wilhelm Maybach mit dem Mercedes 1900 vorgegebene Standard-Bauweise mit Leiterrahmen den Automobilbau, verblüfften europäische Firmen immer wieder mit unkonventionellen technischen Lösungen. So stellt der Mittelträgerrahmen eine radikale Abkehr vom Leiterrahmen dar. Er besteht aus einem zentral angeordneten Längsträger mit meist rohr- oder kastenförmigem Querschnitt und benötigt angeschweißte Querträger zur Aufnahme des Aufbaus.

Protagonist des Mittelträgerrahmens war Hans Ledwinka, der als Chefkonstrukteur der Tatra-Werke 1923 den Typ 11 herausbrachte (Abb. 10). Merkmale dieses leichten, anspruchlosen Kleinwagens waren neben dem Zentralrohrrahmen, der Antriebsblock und Hinterachse starr verband, luftgekühlter Boxermotor und Schwingachse hinten. Die Zentralrohrbauweise in Kombination mit Schwingachsen wendete Ledwinka später auch bei Liefer- und Lastwagen an (Abb. 11) mit dem Argument, die geringeren ungefederten Massen einzeln aufgehängter Räder an einem verwindungssteifen Mittelträger schonten den Oberbau der Straßen.

Ledwinka war keineswegs der Erfinder des Zentralrohrrahmens. Dieser tauchte schon 1904 beim englischen Rover 8 HP auf, gefolgt 1909 vom Simplicia der Pariser Konstruktions- und gelegentlichen Autofirma Lacoste & Battmann (Abb. 12). Das etwa 140 mm starke Zentralrohr des optimistischerweise auch ‚Sans Chassis' genannten Wagens nahm, wie später Ledwinkas Tatra 11, die Antriebswelle auf, die bei dieser Auslegung kein Kardangelenk benötigt. Das Kardangelenk galt bis in die 1930er Jahre als verschleißfreudiges Bauteil, von den Konstrukteuren gern vermieden.

Mit dem Typ 77 gelang es Ledwinka 1934 (Abb. 13), eine alltagstaugliche und formal befriedigende Aerodynamik-Limousine für vier bis sechs Personen herauszubringen. Sie baute auf einem nun kastenförmigen Mittelträger mit angeschweißten Bodenblechen auf, in dessen hinterer Gabel ein luftgekühlter V8-Motor montiert war. Wegen seiner futuristischen Form und der unkonventionellen Technik erregte der 77er ungeheures Aufsehen, Fachleute und Publikum wähnten sich am Beginn einer neuen Automobilepoche. Denn nicht nur die Tatra-Folgemodelle kultivierten Zentralträgerrahmen, Schwingachsen, luftgekühlten Heckmotor und Stromlinie, sondern auch Porsches KdF-Wagen. Der KdF-Wagen ist der unmittelbare Vorläufer des Volkswagens Typ 1 (Käfer).

In der Kriegsausgabe des ‚Handbuchs zum KdF-Wagen' (Abb. 14) wird dessen Unterbau als ‚Drehungssteifer Mittelrahmen' mit am Mittelrohr angeschweißten Bodenblechen beschrieben. Dazu ist zu bemerken, dass die Übergänge zwischen Zentralträgerrahmen mit Bodenblechen einerseits und Plattformrahmen mit Mittelträger andererseits fließend sind und eine eindeutige Zuordnung erschweren. So sprach das Volkswagenwerk für seine Heckmotor-Modelle von Zentralrohrrahmen, Porsche beim Typ 356 (Abb. 15) von Plattformrahmen mit Kastenträgern oder einfach vom Kastenrahmen.

4

Eine Sonderform stellt der von Mercedes in den 1930er Jahren entwickelte X-Rahmen dar (Abb. 16). Er wird definiert als zwei sich etwa in Fahrzeugmitte fast berührende (Ovalrohr-)Träger in X-Form. Auch hier tragen angeschweißte Querträger den Aufbau, der bei einem Seitenaufprall in Richtung Fahrzeugmittellinie zurückweicht und somit so gut wie keinen Insassenschutz bieten kann.

Schwingachsen und Einzelradaufhängungen statt Starrachsen und hydraulische Bremskraftübertragung statt Gestänge oder Seilzug erlaubten den Übergang vom üblichen Hoch- auf den Niederrahmen (Abb. 17). Dank des tiefliegenden mittleren Teils des Leiterrahmens konnten Karosserie und damit Schwerpunkt abgesenkt werden.

Während Aluminium schon seit 1899 im Karosseriebau und seit etwa 1910 im Motoren- und Getriebebau verwendet wurde, blieben Fahrgestelle aus Aluminium die ganz große Ausnahme. Bekannt geworden sind 16 sogenannte Pomeroy-Wagen, gebaut 1925/26 von Alcoa und Pierce Arrow in den USA, sowie ein Prototyp von Peerless 1931 mit V16-Motor, der wegen Weltwirtschaftskrise und Produktionsaufgabe nicht mehr in Serie gehen konnte.

In Frankreich baute der vielseitige Ingenieur Dimitri Sensaud de Lavaud zwei oder drei Fahrzeuge (Abb. 18) mit Alu-Plattformrahmen und anderen außergewöhnlichen technischen Lösungen. Seine ‚Nouvelle Conception de la Voiture Automobile' erregte zwar erhebliches Aufsehen, kam aber nicht aus dem Prototypen-Stadium hinaus. Dagegen schaffte der kleine Amilcar Compound (Abb. 19) den Sprung in die Serie: Zwischen 1938 und dem Einmarsch der deutschen Truppen in Frankreich im Mai/Juni

1940 konnten immerhin knapp 700 Exemplare abgesetzt werden. Beim Amilcar bestand die Struktur des Vorderwagens aus einem Alu-Gussstück, das Spritzwand, Instrumententafel, Windlauf und die bis ins Dach reichende A-Säulen mit Frontscheibenrahmen einschloss. Die halb-selbsttragende Bauweise nach Jean-Albert Grégoire tauchte nach dem Zweiten Weltkrieg bei den Panhard-Dyna-Modellen und beim Hotchkiss Grégoire wieder auf, gebaut von 1950 bis 1957.

Inzwischen hatte sich auch auf dem Karosseriesektor ein Wandel vollzogen insofern, als die seit 1912 in den USA ausgelieferten Ganzstahl-Karosserien die herkömmlichen, mit Stahlblechtafeln überzogenen Holzgerippe-Konstruktionen mehr und mehr verdrängten. War die Ganzstahl-Karosserie einst eingeführt worden, um kostengünstig größere Stückzahlen am Band produzieren zu können, ergab sich mit der Zeit eine weitere konstruktive Möglichkeit, an die man anfangs noch gar nicht gedacht hatte: Die Kombination von Karosserieboden und Fahrgestell zu einem verwindungssteifen Bodenblech, das einen separaten Fahrzeugrahmen überflüssig machte und, verschweißt mit dem Karosseriekörper, eine selbsttragende Karosserie ergab.

5

Einer ersten, 1928 von der Budd Manufacturing Co in Philadelphia patentierten Versuchskarosserie in selbsttragender Bauart folgte 1931 ein Prototyp (Abb. 20), der mit Änderungen 1934 als Citroën 7 CV Traction Avant in Serie ging. Er besaß noch einen Hilfsrahmen für den Antriebsblock. Eine reinrassige selbsttragende Stahlkarosserie in Schalenbauweise und damit Verzicht auf einen eigentlichen Rahmen brachte Opel 1935 mit dem Olympia heraus (Abb. 21), ebenfalls basierend auf Vorarbeiten von Budd und General Motors. Im Fall Opel konnten etwa 11% oder 110 kg Masse gegenüber dem Vorgänger in Rahmenbauweise eingespart werden.

Die selbsttragende Karosserie ist nicht an einen bestimmten Werkstoff gebunden. So

sind rahmenlose Holzkarosserien auch im Autobau versucht worden, z.B. Slaby-Beringer 1919, DKW Sonderklasse 1932 oder der Sonderfall Marcos 1960 (Abb. 22), dessen Rahmen und Karosseriestruktur aus Sperrholz bestand. Doch Holz stellte sich als überforderter Werkstoff heraus. Jahrzehntelang, noch heute, herrscht die selbsttragende Stahlblechkarosserie in Schalenbauweise vor, auch wenn heute Stahl stellenweise durch andere Werkstoffe ergänzt oder ganz ersetzt wird.

Zum Beispiel durch Aluminium. Der von Wunibald Kamm konstruierte und von den Schwäbischen Hüttenwerken 1924 herausgebrachte SHW 4/20 PS (Abb. 23) mit selbsttragender Alu-Karosserie blieb mit drei gebauten Exemplaren ebenso im Prototypenstadium stecken wie die Versuchsausführungen von einigen europäischen und amerikanischen Autoherstellern, die in den Jahren 1983 bis 1995 mit einem von Alcan neu entwickelten Tiefziehverfahren experimentierten. In die Serie dagegen schafften es selbsttragende Alu-Karosserien bei Honda, Jaguar und Lamborghini, jeweils in Blechschalenbauweise.

Als kostengünstiger erwies sich die Space Frame-Technik, unter der man ein raumumschließendes Fachwerk oder Rahmentragwerk versteht, von Audi 1994 mit dem A8 (Abb. 24) in der Oberklasse und mit dem A2 sechs Jahre später in der Kompaktklasse eingeführt. Heutige Alu-Tragwerke sind, wie auch die Stahlblechkonstruktionen, an bestimmten Stellen durch dafür geeignetere Werkstoffe ergänzt.

Im Sport- und Rennwagenbau sind ab 1952 Gitterrohrrahmen (Abb. 25) eingesetzt worden, bei denen zu Dreiecken zusammengeschweißte Rohre meist nur auf Druck oder Zug, nicht jedoch auf Biegung beansprucht werden. Aus den Gitterrohrrahmen ging die Monocoque-Bauweise hervor, zunächst aus Leichtmetall (Abb. 26), später aus CFK. Selbsttragende Karosserien aus GFK sind versucht worden (Abb. 27), konnten sich jedoch wegen zu hoher Herstellungskosten nicht durchsetzen. Das in mehr als 1000 Exemplaren gebaute Lotus Elite-Coupé (Abb. 28) bestand zwar aus miteinander verklebten GFK-Hohlprofilen, einer abschließenden Bodenwanne und einem Dach in Sandwich-Bauweise, besaß zur Versteifung jedoch zwei eingeklebte Stahlprofile im Motor- und A-Säulen-Bereich.

6

Eine Abkehr von der selbsttragenden Karosserie, die heutzutage auf veränderbaren Plattformen aufgebaut ist, wird möglicherweise der Elektro-Antrieb bringen. Beim 2002 vorgestellten Konzept-Auto Hy-Wire von General Motors nimmt ein 30 cm hoher Fachwerkrahmen den Antrieb, bestehend aus Brennstoffzelle, Wasserstofftanks und E-Motor, auf, die Karosserie bietet lediglich Wetterschutz. Lenkgestänge, Handbremshebel, Pedalerie oder sonstige mechanische Verbindungen bestehen nicht, Antrieb, Lenkung und Bremsen funktionieren elektrisch über eine Nabelschnur (by wire-Technik). Ähnlich der in Serie gegangene BMW i3 von 2013 (Abb. 29): Hier besteht das Fahrmodul aus einer Aluminium-Plattform und das aufgesetzte Karosserie-Modul aus CFK.

Rückkehr also zur Fahrgestell-Bauweise? Auch wenn die Werbe-Jungfrau der Berliner Karosseriefirma Alexis Kellner (Abb. 30) Mitte der 1920er Jahre ins Telefon hauchte: „…das Chassis von wem du willst: die Karosserie aber nur von Kellner“, bleibt festzuhalten, dass auch ein attraktives Chassis begeistern kann.

Erik Eckermann

Weight-Bearing Elements

From Chassis to Unit Bodies and back to Driving Modules

Tragende Elemente

Vom Chassis zur selbsttragenden Karosserie und zurück zum Fahrmodul

AutoHistorica

Fahrzeugausstellungen und Automobilgeschichte

Erik Eckermann

Ca. 1810 Dick & Kirschten

Coupé mit Langbaum und C-Federn

Coupé with pearch and C-springs

Abb. 1

AutoHistorica

Fahrzeugausstellungen und Automobilgeschichte

Abb. 2

Ca. 1890 Mayer
Selbsttragendes Coupé mit Vollelliptikfedern
Selfsupporting Coupé with fullelliptics

Abb. 3

1885 Starley
Sicherheitsfahrrad mit Stahlrohrrahmen
Safety Bicycle with tubular steel frame

1886 Benz-Motorwagen

Vordersteuer-Dreirad mit Rohrrahmen und Motor

Front-steering tricycle sociable with tubular steel frame and engine

Abb. 4

1887 Daimler/Maybach-Motorkutsche

Selbsttragende Américaine mit Motor

Selfsupporting Américaine with engine

Abb. 5

Abb. 6

1901 Mercedes 35 PS
Viersitziges Phaeton auf U-Profil-Pressstahlrahmen
Four seater Phaeton on U-section pressed steel frame

Abb. 7

Ca. 1909 Adler
Vorn eingezogener, ungekröpfter Leiterrahmen mit Traversen
Unraised ladder-type chassis with inswept front and cross members

Abb. 8

1921 Rumpler Tropfenauto/Treardrop Car
Spindelförmiger, hochbordiger Stahlblechrahmen
Spindle-shaped pressed steel frame with high webs

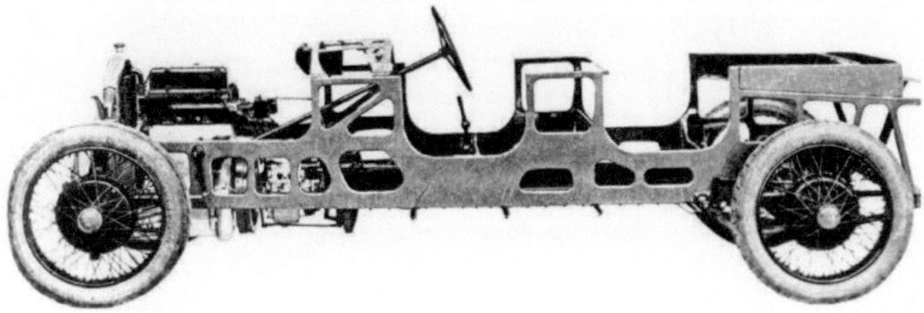

Abb. 9

1923 Lancia Lambda
Bis Gürtellinie hochgezogene Stahlstruktur mit Querversteifungen
Deep stamped steel structure with transverse members

1923 Tatra 11
Zentralrohrrahmen mit angeflanschtem Antriebsblock und Hinterachse
Central tube chassis with bolted-on drive unit and rear axle

Abb. 10

AutoHistorica
Fahrzeugausstellungen und Automobilgeschichte

1928 Tatra 24 Truck-Chassis
Zentralrohrrahmen mit Schwingachsen
Central tube chassis with swing axles

Abb. 11

AutoHistorica
Fahrzeugausstellungen und Automobilgeschichte

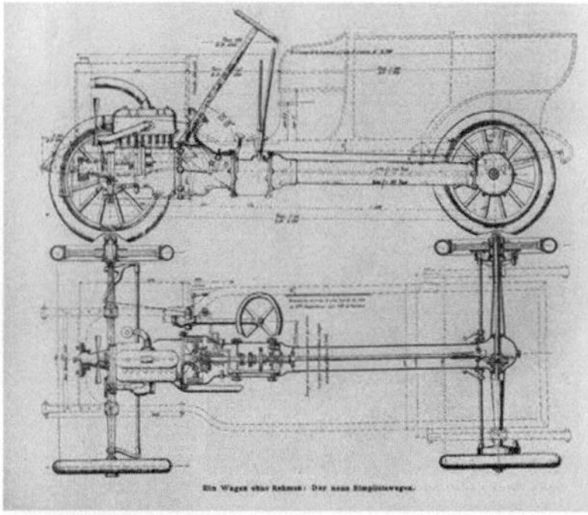

1909 Lacoste & Battmann Simplicia

Abb. 12

Zentralrohrrahmen mit Starrachsen

Central tube chassis with rigid axles

1937 Tatra 87

Abb. 13

Zentralkastenrahmen mit angeschweißten Bodenblechen

Central box frame with welded-on sheet metal

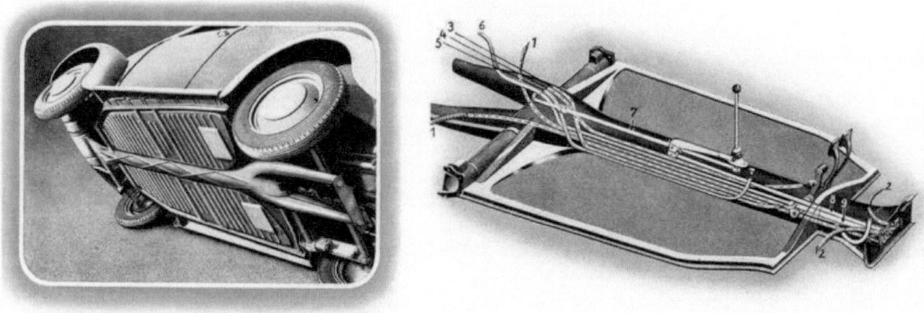

Abb. 14

Ca. 1940 KdF-Wagen
Zentralrohrrahmen mit angeschweißten Bodenblechen
Central tube chassis with welded-on sheet metal

Abb. 15

1955 Porsche 356 A
Plattformrahmen mit Kastenträgern
Platform frame with longitudinal box-shaped bearers

1949 Mercedes-Benz 170 S

Abb. 16

X-förmiger Ovalrohrrahmen mit Auslegern

X-shaped oval tube frame with body supports

AutoHistorica

Fahrzeugausstellungen und Automobilgeschichte

1933 Mercedes-Benz 380

Abb. 17

Niederrahmen mit Einzelradaufhängung

Low-slung frame with independent wheel suspension

AutoHistorica

Fahrzeugausstellungen und Automobilgeschichte

Abb. 18

1926 Sensaud de Lavaud

Aluminium-Plattformrahmen mit integrierten Tunnel und Schwellern

One-piece alloy casting with integrated tunnel and floor wells

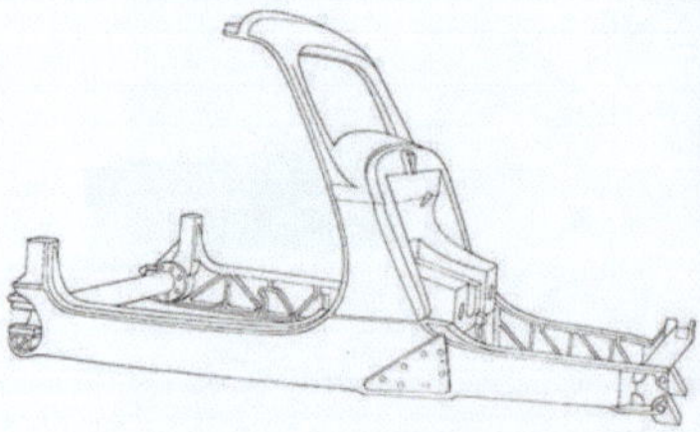

Abb. 19

1938 Amilcar Compound

Alu-Guss-Vorderwagenstruktur inklusive Schweller und A-Säulen

Frontend alloy structure incorporating sills, firewall and pillars

Abb. 20

1931 Budd Prototyp
Niedrige Silhouette dank selbsttragender Karosserie und Frontantrieb
Low silhouette thanks unit body and front drive

Abb. 21

1935 Opel Olympia
Selbsttragende Karosserie in Schalenbauweise
Unit body consisting of sheet steel structural members

Abb. 22

1960 Marcos GT

Rahmen-Karosserie-Verbund aus Sperrholz mit B-Säule ins Dach

Plywood frame/body-unit with B-pillar extension into roof

AutoHistorica
Fahrzeugausstellungen und Automobilgeschichte

Abb. 23

1924 SHW 4/20 PS

Selbsttragende Karosserie aus vernieteten Alu-Tafeln

Unit body construction consisting of riveted alloy-sheets

AutoHistorica
Fahrzeugausstellungen und Automobilgeschichte

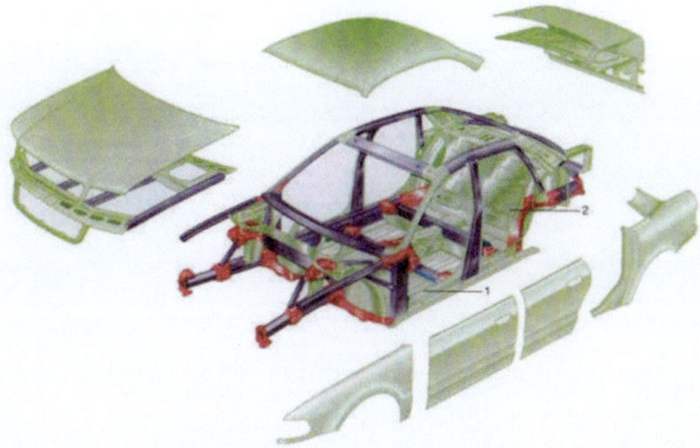

1994 Audi A8

Selbsttragende Voll-Aluminium-Karosserie in Fachwerk-Bauweise

Alu Space Frame unit body construction

Abb. 24

AutoHistorica

Fahrzeugausstellungen und Automobilgeschichte

1954 Mercedes-Benz 300 SL

Gitterrohrrahmen aus verschweißten Stahlrohren

Space frame consisting of welded steel tubes

Abb. 25

AutoHistorica

Fahrzeugausstellungen und Automobilgeschichte

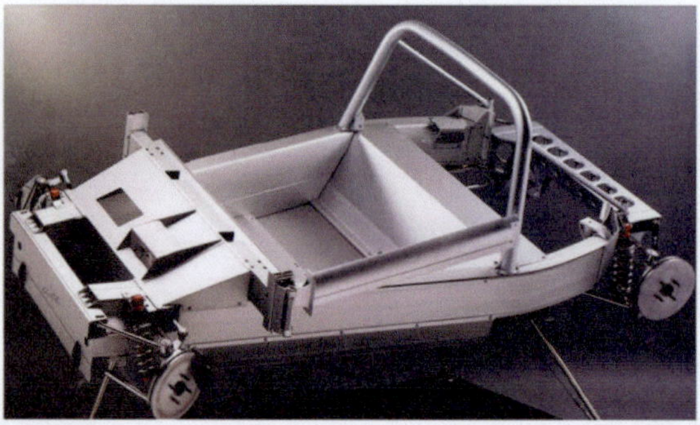

Abb. 26

1995 Lotus Elise
Aluminium-Wanne mit Stahl-Hilfsrahmen
Alloy honeycomb monocoque with steel auxiliary frame

Abb. 27

1967 Bayer K 67
Selbsttragende Bodengruppe aus Polyurethanschaum mit GFK-Aufbau
Driving module consisting of polyurethan foam, GRP-body

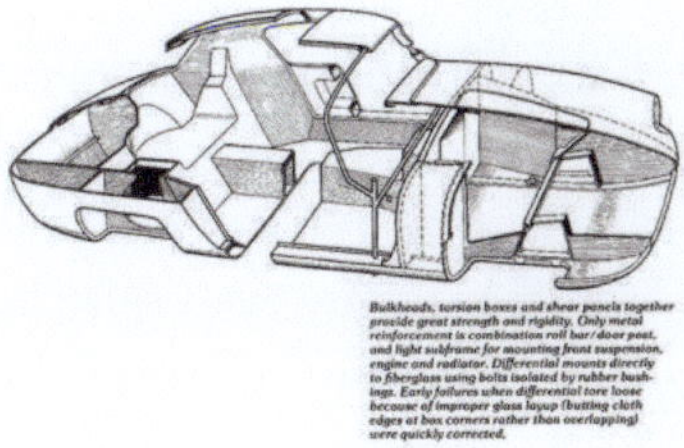

Abb. 28

1957 Lotus Elite

Selbsttragende GFK-Karosserie mit Stahlverstärkungen

Selfsupporting GRP-body with auxiliary steel sections

AutoHistorica
Fahrzeugausstellungen und Automobilgeschichte

Abb. 29

2013 BMW i3

Alu-Fahrmodul, CFK-Karosserie-Modul

Alloy driving module, carbon life module

AutoHistorica
Fahrzeugausstellungen und Automobilgeschichte

1925 Kellner (Coachbuilder) Advertisement

Abb. 30

2015
Auto und Karosserie
Springer Vieweg Verlag
ISBN 978-3-658-07425-8
€ 129,99

Abb. 31

PARALLEL STRAND II

VIRTUAL DEVELOPMENT METHODS

Explicit model predictive control of semi-active suspension systems using Artificial Neural Networks (ANN)

Dipl.-Ing. Ronnie Dessort
Dr.-Ing. Cornelius Chucholowski

TESIS DYNAware GmbH

P.E. Pfeffer (Ed.), *8th International Munich Chassis Symposium 2017*,
Proceedings, DOI 10.1007/978-3-658-18459-9_15

Abstract

An optimal nonlinear control method for a semi-active suspension system subject to state and input constraints is presented and compared with a passive system and a combined Skyhook/Groundhook controller. The model-based controller synthesis considers nonlinearities in the suspension system in terms of the variable kinematic transmission factor, frequency dependent vertical tire behavior as well as asymmetric continuously variable damper characteristics. A tool-based parameter identification process using virtual suspension and tire test rig is demonstrated. By applying the method of nonlinear programming, this approach incorporates inequality constraints regarding the restrictable maximum spring deflection as well as the actuating variable due to the operating range of the damper. The solutions of individual optimization problems are used to create a state-dependent control law, which is integrated in the feedback loop. Initial conditions are therefor uniformly distributed in a pre-defined state space region. Finally the controller table is determined by artificial neural networks (ANN) processing the optimal control of each initial value problem. The definition of the objective function enables focusing on either ride comfort, ride safety or a trade-off of these criteria. The performance of the proposed approach is tested and analyzed under various road conditions in both quarter car and full vehicle simulations using a physically based tire model.

1 Introduction

The main goal of a suspension system is to isolate passengers from any road induced chassis vibrations by simultaneously ensuring a safe road handling characteristic and taking constructional as well as system input limitations into consideration. Due to their energy efficiency, semi-active suspension systems consisting of continuously variable dampers are primarily integrated in modern vehicles. These electrohydraulic devices achieve performance benefits in both ride comfort and ride safety compared to passive elements by deploying specifically tailored control laws, [18]. A challenging task is to consider the dissipativity condtion and the working range of the damper directly in the optimal control problem in contrast to linear control approaches producing a clipped system input.

In general, the presented control approach was initially published in [29]. This contribution is focused on a further development towards increased computational efficiency during controller synthesis and online execution as well as demonstrating a virtual development process.

2 System Modeling

This section shows the quarter car modeling approach used by the proposed controller and its parametrization process utilizing virtual component test rigs. Furthermore, the goals of vertical dynamics control along with the corresponding performance metrics are explained. For controller evaluation purposes the selected benchmark data regarding stochastic road profile and comparative state-of-the-art controller is presented.

2.1 Quarter Car Model

For a model-based controller design, a sufficient modeling depth should be deployed. Although it is a very abstract form, while considering only vertical dynamics a quarter car model as shown in Figure 1 can be utilized, [1]. Insights based on this model also hold true for the complex full vehicle system. Basically the model generation procedure in this domain comprises of two tasks: physical or (semi-) empiric depiction of real system behavior by time-based ordinary differential equations as well as quantification of properties describing particular system components.

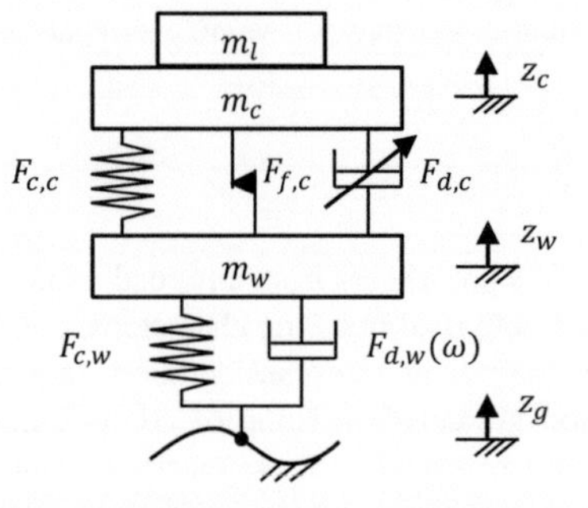

Figure 1: Nonlinear quarter car model

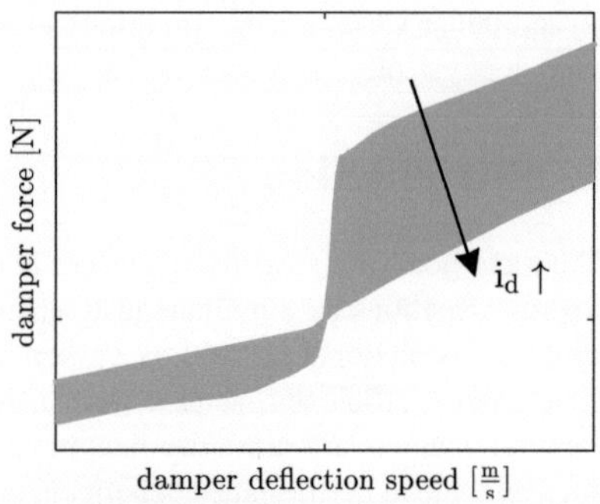

Figure 2: Characteristic damper force map

The implemented quarter car model consists of the chassis mass (sprung mass) m_c, representing a quarter of the unloaded vehicle body, the wheel mass m_w as well as an additional mass m_l considering passenger weights and arbitrary loading. Generally, independent spring and damper elements connect chassis and wheel by its individual kinematic relationship. The latter transforms forces from the respective plain (sp=spring plain, dp=damper plain) to the inertial wheel plain. The spring force $F_{c,c}$ comprises of the impact of the main spring as well as end stops for compression and rebound. Figure 2 depicts the characteristic force map of the continuously variable damper, which is controlled by the damper current i_d as system input. A first-order time delay $T_{d,c}$ can be applied on the static damper force $F_{d,st,c}$ to take hysteresis effects into account, [28]. Furthermore friction in the suspension is also considered by means of the force $F_{f,c}$. The wheel is attached to the ground by using a single contact point model, which consists of a parallel spring force $F_{c,w}$ and damper force $F_{d,w}$. Basically the tire damping effect is small compared to the hydraulic damper, but the tire has a complex dynamic behavior, [20]. This influence is also considered by a first-order lag element comprising of inflation pressure (p_t) dependent time constant $T_{d,w}$ and viscous damping coefficient d_w. Finally, a state space model of the quarter car is defined as follows:

$$\begin{aligned}
\dot{x}_1 &= x_2 \\
\dot{x}_2 &= -\frac{1}{m_c + m_l} \cdot [F_{cw}(\boldsymbol{x}) + m_l g] - \frac{1}{m_w} \cdot [-F_{wg}(\boldsymbol{x}) + F_{cw}(\boldsymbol{x})] \\
\dot{x}_3 &= x_4 - z \\
\dot{x}_4 &= -\frac{1}{m_w} \cdot [F_{wg}(\boldsymbol{x}) - F_{cw}(\boldsymbol{x})] \\
\dot{x}_5 &= \frac{1}{T_{d,c}} \cdot [F_{d,st,c}^{dp}(\boldsymbol{x}, u) - x_5] \\
\dot{x}_6 &= \frac{1}{T_{d,w}(p_t)} \cdot [d_w(p_t) \cdot (x_4 - z) - x_6]
\end{aligned} \tag{1}$$

where the first four states denote $x_{1,.,4} = [z_c - z_w, \dot{z}_c - \dot{z}_w, z_w - z_g, \dot{z}_w]$ and $x_{5/6}$ defines the dynamic damping forces between chassis and wheel or wheel and ground, respectively. The system output is given by

$$\begin{aligned}
y_1 &= x_1 \\
y_2 &= -\frac{1}{m_c + m_l} \cdot [F_{cw}(\boldsymbol{x}) + m_l g] \\
y_3 &= -F_{wg}(\boldsymbol{x})
\end{aligned} \tag{2}$$

representing the suspension deflection, vertical chassis acceleration and dynamic wheel load. In total the resulting force between chassis and wheel is computed according to

$$F_{cw}(\boldsymbol{x}) = \left[F_{c,c}^{sp}\left(z_{cw}^{sp}(x_1)\right) + x_5\right] \cdot i_{dyn}(x_1) + F_f \cdot \tanh(k_c x_2), \tag{3}$$

whereas the dynamic force between wheel (w) and ground (g) is derived from

$$F_{wg}(\boldsymbol{x}) = F_{c,w}(x_3) + x_6. \tag{4}$$

Wheel lift-off is considered by limiting F_{wg} to the negative static wheel load.

Especially in the field of real-time vehicle dynamics simulation pursuing the target of control units validation, most OEM's apply an already established model parametrization process. The main focus in this respect is usually on the chassis and tire component. If no K&C measurements are available, a valid approach is to use complex multibody systems as a master to derive a parameter set for real-time capable full vehicle simulation, [6]. In this place the parameter identification process under these conditions is based on performing various static and dynamic test cases on virtual component test rigs in order to generate the required quarter car data set. The utilized simulation framework DYNA4 supports this workflow with required functionality.

Firstly, supposing an already existing arbitrary full vehicle parametrization typical for the SUV class, the goal of using a suspension test rig is mainly to determine the kinematic axle behavior. While parameters like nonlinear spring stiffness and characteristic damper force map can be adopted from the full vehicle setup, the kinematics analysis concerns the static transformation of spring/damper deflection in inertial z-direction (wheel plain) to its actual deflection in the spring/damper plain. The corresponding local derivatives yield a variable transmission factor applied on velocities and forces. Therefor quasi-static parallel wheel lift tests on front and rear axle are conducted as depicted in Figure 3.

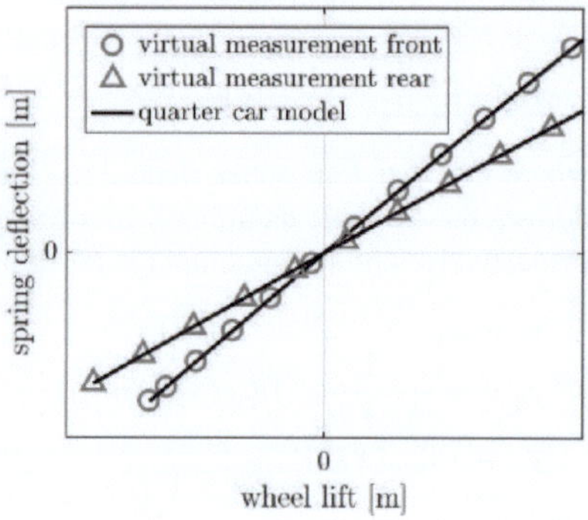

Figure 3: DYNA4 suspension test rig (left) and quarter car kinematics fitting result (right)

Furthermore it is necessary to identify possible load distributions for various passenger configurations, since the proposed control method is able to adapt its strategy regarding this feature. Figure 4 shows the derived working area for left/right symmetrical loading caused by different configurations of passenger weights and trunk load:

- A: front passengers only
- B: configuration A plus rear passengers
- C: configuration B plus second rear passengers
- D: configuration C plus trunk load

The front axle is therefor loaded in a range of [0; 60] kg, whereas the rear axle load varies approximately between [0; 180] kg.

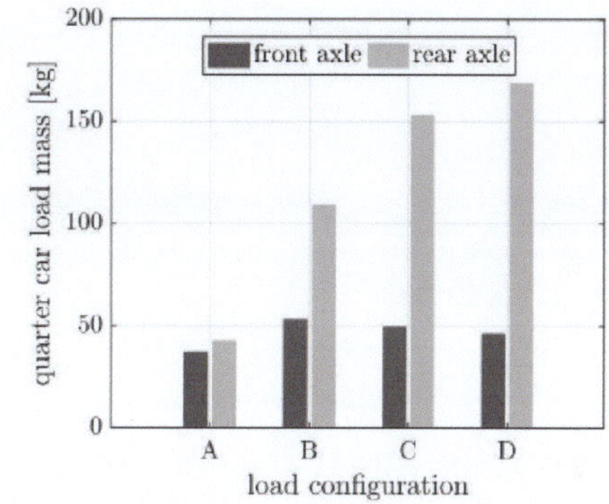

Figure 4: Full vehicle load (passengers and trunk) with corresponding load distribution (right)

Secondly, a further important model part is the tire and its vertical characteristics. With respect to handling dynamics semi-empiric tire models like TMeasy5 fulfill requirements with a proper grade of complexity, [7]. Usually the vertical tire force is depicted by means of a parallel spring and damper element. This is a fairly simple abstraction compared to a physically based tire model like FTire [27], which is used in analogue as a more complex master model. Along with an exemplary data set related to the utilized vehicle, FTire is simulated on a virtual tire test rig in static and dynamic test cases. As it can be seen in Figure 5, a quadratic function depicts the static spring stiffness very well (top left). After conducting a sine sweep with constant excitation speed in the given frequency range, the stated first-order lag element of the dynamic tire damping force is fitted to the resulting frequency response (right). The gain also shows a good regression with a slight deviation in the phase. The lower left diagram of Figure 5 depicts a linear influence of the inflation pressure on the damping coefficient in the range of [25;125]% of the LI specification. Here the tire behavior is linearized around the operating point given by 50% LI load.

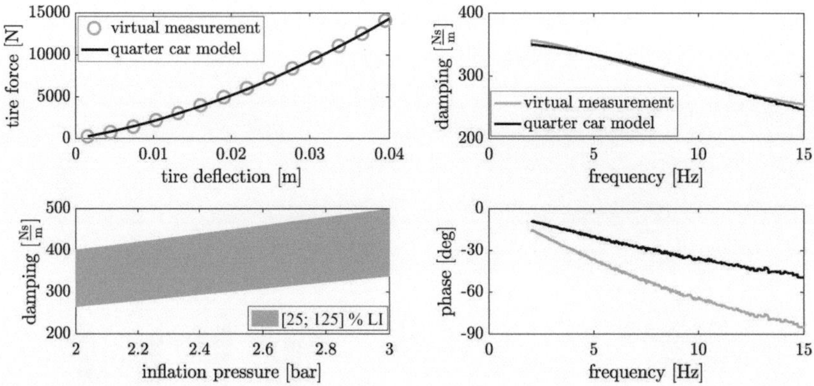

Figure 5: Static (top left) and dynamic (right) tire fitting (at 2.5 bar, 50% LI load)

2.2 Performance Measurement

The goals of vertical dynamic suspension design are, in general, defined by optimizing both ride comfort and ride safety under restriction of the available suspension deflection. A widely used quantity to describe these objectives is defined by the root mean square (RMS) value

$$\|\chi\|_{rms} = \sqrt{\frac{1}{T_1 - T_0}\int_{T_0}^{T_1} [\chi(t)]^2 dt}, \quad (5)$$

where χ denotes either the vertical chassis acceleration $\ddot{z}_c$ as a measure for ride comfort or the dynamic wheel load F_{dyn} to indicate the ride safety performance index.

Studies regarding the impact of mechanical excitation on the body of sitting humans described in VDI 2057 [2] and ISO 2631-1 [3] have shown, that their sensitivity for vibrations is frequency dependent and in a range of 4 to 8 Hz, is particularly distinct, [4]. Taking this behavior into account, a shaping filter stated in [2] is adopted and can be incorporated into the system by modeling it as a fifth-order linear transfer function, [5]. Furthermore, the maximum of the absolute value of the vertical chassis acceleration $\max(|\ddot{z}_c|)$ has to be considered since it mainly represents the influence of singular events like bumps or potholes.

In terms of ride safety, the RMS value of the dynamic wheel load should not exceed the limit defined by

$$\left\|F_{dyn}\right\|_{rms} \leq \frac{F_{stat}}{3} = \frac{(m_c + m_l + m_w) \cdot g}{3}, \tag{6}$$

[4]. In addition the minimum dynamic wheel load $\min(F_{dyn})$ is an important quantity as it indicates a possible loss of contact between tire and road.

Finally, to evaluate the effectiveness of different controller setups a performance gain is used and defined by

$$P = \left(\frac{\alpha_{controller}}{\alpha_{passive}} - 1\right) \cdot 100\ [\%], \tag{7}$$

where α denotes any of the above mentionend quantities used as objective system characterization. As it can be seen in eq. (7), every utilized controller is compared to a passive reference system, where a negative percent value indicates an improvement.

2.3 Benchmark Data

Real roads exhibit stochastic unevenness rather than harmonic sinusoidal wave forms causing the vehicle system to roll, pitch and heave oscillations. Empirical studies have shown that the power spectral density (PSD) decreases linearly regarding the spatial frequency in a double logarithmic diagram, [19].

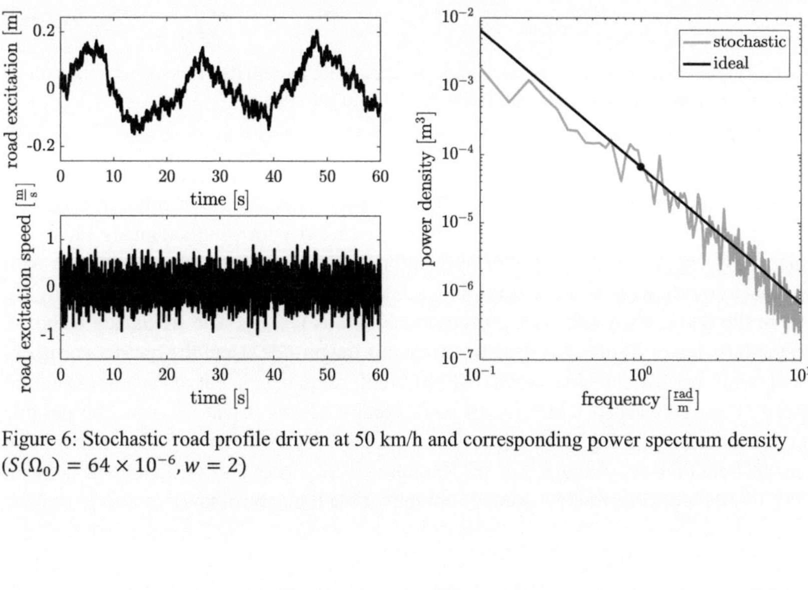

Figure 6: Stochastic road profile driven at 50 km/h and corresponding power spectrum density ($S(\Omega_0) = 64 \times 10^{-6}, w = 2$)

This statistical relationship can be described by means of two parameters, namely roughness measure and waviness. The latter assumes the value of two on an average, which is also taken for a norm road profile. Figure 6 shows on the left side the further on utilized road excitation or its speed used as model disturbance input, respectively. On the right side the mentioned linear dependency of a synthetic road is verified by using the standard waviness and a power spectral density of 64 x 10^{-6} m^3 at 10^0 rad/m. The road profile depicted in Figure 6 is used as benchmark excitation to evaluate the subsequently discussed damper systems.

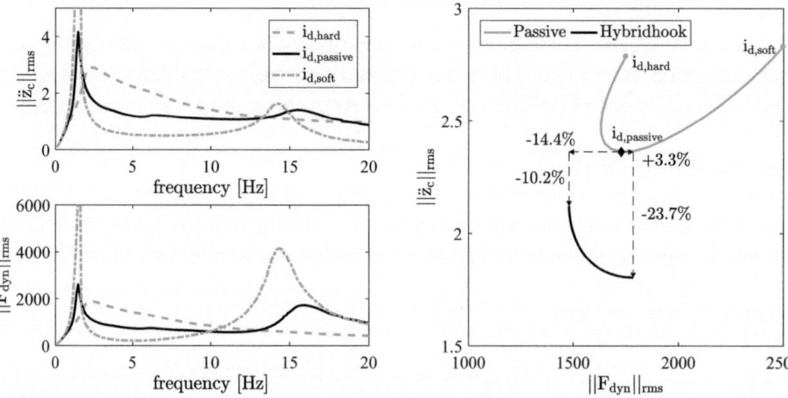

Figure 7: Front axle frequency response of passive reference system (left) and carpet plot with combined Skyhook and Groundhook (Hybridhook) control (right)

Two different damper controller setups are considered for benchmark purposes. Firstly, a constant damper current is determined to serve as a passive reference system. Here the focus is on providing maximum ride comfort with simultaneously satisfying ride safety characteristic. Therefor simulations based on the stated road profile with various currents were evaluated to obtain the carpet plot depicted in Figure 7 (right), where the determined reference current is marked as $i_{d,passive}$. The frequency response on the left side of Figure 7 is used to verify this magnitude. Here the quarter car is excited with permanent sine waves of different frequencies but constant excitation speed. Then converged RMS values were computed. As it can be seen, the passive reference system is a good compromise between ride comfort and ride safety. The second benchmark system is set to a combined Skyhook and Groundhook control, [18]. Firstly a desired damper force is computed via the control law

$$F_{d,desired} = d_{sky}(q) \cdot \dot{z}_c - d_{gnd}(q) \cdot \dot{z}_w, \tag{8}$$

where $d_{sky}>0$ and $d_{gnd}>0$ indicates the skyhook and groundhook damping coefficient and q serves as a fading factor between ride comfort and ride safety. This force is then transformed to the applicable damper current by means of an inverse characteristic damper force map.

3 Controller Synthesis

In Model Predictive Control (MPC) the control action is obtained by solving a finite horizon open-loop optimal control problem at each sampling step. Each optimization yields a sequence of optimal system input, but only the first action is applied to the process, [17]. According to system complexity this online optimization might only be applicable to rather slow processes. Explicit model predictive control (eMPC) tries to overcome this limitation by solving optimization problems offline for states within a given region. Usually this results in constant feedback gains valid for particular polyhedral regions in the state space hypercube, [16]. The main drawback is the exponentially increasing number of regions while enlarging the prediction horizon, which causes more time to perform an online region searching heuristic. More detailed explanations can be found in [17].

The proposed optimal control approach can be categorized as eMPC, since it aims on solving optimization problems offline. But instead to only consider linear models or compute different linear feedback gains, this method enables the utilization of MPC in its general idea even for high dynamic systems with both low online computational effort and low memory usage.

For the nonlinear dynamic semi-active quarter car model defined by eq. (1) and (2) the optimal control problem can be formulated as

$$\begin{aligned} \min_{u(t)\in[0;t_e]} \; & J = q \cdot \| \ddot{\bar{z}}_C \|_{rms} + (1-q) \cdot \left\| \bar{F}_{dyn} \right\|_{rms} \\ \text{s.t.} \quad & \dot{\boldsymbol{x}} = f(\boldsymbol{x}, u) \\ & \text{with } \boldsymbol{x}_0 = \boldsymbol{x}(0),\ \mathbf{x}(t_e) \text{ free} \\ & x_1(t) \leq -x_{1,min} \\ & u_{min} \leq u(t) \leq u_{max}, \end{aligned} \tag{9}$$

where u_{min} and u_{max} represent the lower and upper input constraints on the damper current and $x_{1,min}$ ensures a maximum spring compression due to constructional limitations. The cost function consists of the sum of normalized RMS values for vertical chassis acceleration and dynamic wheel load, whereby the factor q enables a fading from ride safety to ride comfort and vice versa. For arbitrary initial state conditions $\mathbf{x}_0$ and prediction horizon t_e the optimal control problem (9) is solved numerically. Then the first control input is stored along with each initial state and is used to train a metamodel afterwards, which is finally integrated in the feedback control loop. Here the latter is chosen to be an artificial neural network. Figure 8 gives a short overview of the workflow described in the following subsections.

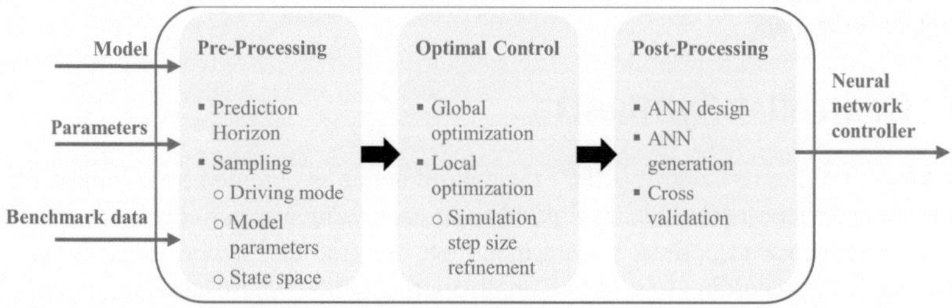

Figure 8: Workflow for offline model predictive control of semi-active suspension systems

3.1 Pre-Processing: Design Variables

In order to enable a fast computation of the overall optimization loop, special attention was paid to the sampling procedure. A three-staged approach is based on an arbitrary sampling of driving mode (fading from pure ride safety to ride comfort), where each factor again consists of a closed sampling region for the two considered model parameters load mass and inflation pressure. The last stage covers the system states. The main goal is considering only those optimization problems relevant because of their actual occurrences. Instead of distributing initial state conditions in the entire state space, a more proper subspace is determined. This approach assumes, that with respect to the given system dynamics, a wide region of the state space might never be reached or only with a low probability, respectively. As shown in Figure 9, this subspace is approximated by a hyperelliptic hull of states resulting from a simulation of the passive quarter car model excited by the benchmark road profile stated in section 2.3. In general, the surface of an arbitrary oriented hyperellipsoid centered at **c** is defined by the solutions **x** to the equation

$$(\boldsymbol{x}-\boldsymbol{c})^T \cdot \boldsymbol{A} \cdot (\boldsymbol{x}-\boldsymbol{c}) = 1, \tag{10}$$

where **A** is a positive definite matrix, [30]. Thus one has to find such a matrix minimizing the four-dimensional hyperellipsoid volume

$$\begin{aligned} &\min_{\{\boldsymbol{A};\boldsymbol{c}\}} J = \frac{\pi^2}{2} \cdot \prod_{i=1}^{4} r_i \\ &\text{s.t.} \quad \lambda_i > 0, \quad i = 1, \dots, 4 \\ &\qquad (\boldsymbol{x}_k - \boldsymbol{c})^T \cdot \boldsymbol{A} \cdot (\boldsymbol{x}_k - \boldsymbol{c}) \le 1, \quad k = 1, \dots, N \end{aligned} \tag{11}$$

where r_i denotes the specific radius of an elliptical semi-axis defined by the squared reciprocal eigenvalue λ_i of matrix **A**, [30]. Therefor all eigenvalues have to be real and greater than zero. Furthermore it has to be ensured that all N reference samples are located within this hyperellipsoid. This optimization problem can be successfully solved using the NOMAD solver [8] incorporated in the OPTI Toolbox [9]. Setting **A** to be symmetric yields the property of orthogonal diagonalization, i.e. all eigenvectors $\mathbf{v}_i, i \in [1;4]$ are mutually perpendicular and therefore an orthogonal basis. Finally, initial states are uniformly distributed within [10] and on the surface [11] of a unit hypersphere and afterwards transformed to the hyperelliptical region as depicted in Figure 9 via

$$\boldsymbol{x}_{e,i} = \boldsymbol{V} \cdot \left[\mathrm{diag}(\boldsymbol{r}) \cdot \boldsymbol{x}_{s,i}\right] + \boldsymbol{c}, \tag{12}$$

where $\mathbf{x}_{s,i}$ defines the i-th sample in the unit hypersphere and $\mathbf{x}_{e,i}$ the related sample in the hyperellipsoid.

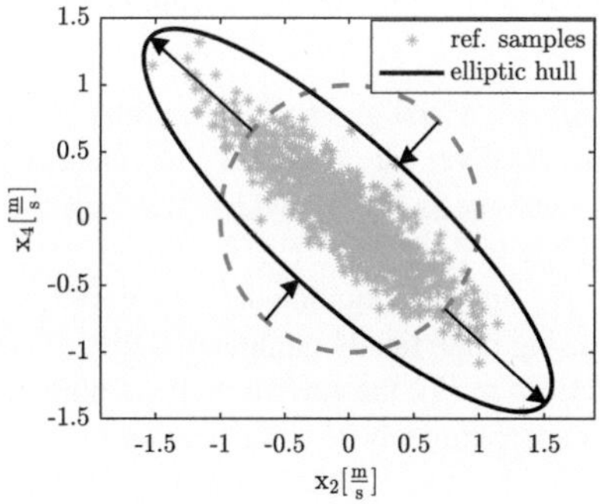

Figure 9: Elliptical state space sampling (projection of reference samples into x_2-x_4-plain)

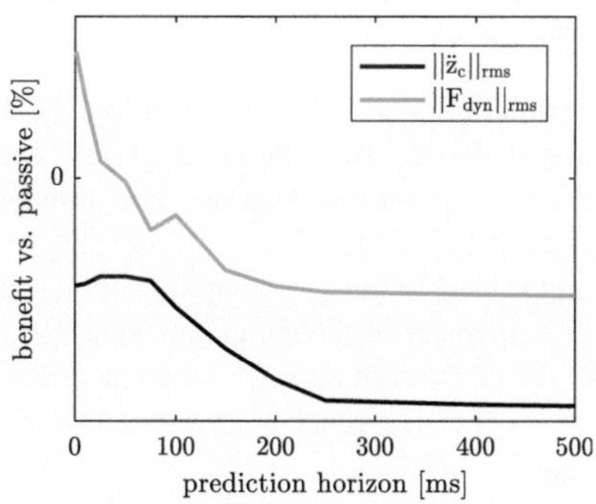

Figure 10: Estimated impact of prediction horizon on controller performance

Before starting to solve optimal control problems numerically, a fixed prediction horizon has to be defined. This quantity also scales the computational effort and simultaneously influences the controller performance in the end. Figure 10 depicts a rough estimation, where the benefit towards the passive damper system dependent on the control time span shows an almost saturated performance gain beginning at approximately 250ms. This is derived by computing controllers with this proposed method as look-up tables based on k-nearest-neighbor interpolation [12] for one particular physical setup (curb load, medium inflation pressure) and focusing on either ride comfort or ride safety.

3.2 Two-level Optimization

Nonlinear optimal control problems usually cannot be solved analytically. Using indirect methods the solution can be found solving a boundary-value problem, [13]. In contrast, with partially discretization of the system dynamics, the direct method of Nonlinear Programming [13] is used to find the optimal control solution minimizing the measure J while keeping the state and input constraints. The state at an arbitrary time is only dependent from its initial condition and the applied control sequence. Thus, by discretizing the system input and numerically solving the ordinary differential equations (1) via

$$\boldsymbol{x}_{k+1} = \boldsymbol{x}_k + \Delta t \cdot \boldsymbol{\phi}_k(\boldsymbol{x}_k, u_k), \qquad k \in \left[0; N = \frac{t_e}{\Delta t}\right], \tag{13}$$

where $\boldsymbol{\phi}_k$ is defined by any explicit Runge-Kutta method [14], the time-continuous optimal control problem (9) can be transformed to a static nonconvex optimization problem.

In order to prevent being tacked to a local minimum, a first guess of the optimal solution is derived by applying a global optimization solver like DIRECT [15] incorporated in [9] on the problem. The simulation step size is therfor wider lowering the computational effort, but still focusing on a stable integration. Next, this rough solution is used to run a local optimization routine, e.g. interior-point method, by iteratively refinement of the simulation step size. By using analytically computed derivatives of the discretized objective and constraint functions in (9), the current implementation achieves convergence to an optimal solution in approximately 3s on an average[1].

3.3 Post-Processing: Metamodel

In the next step one has to define a proper modeling technique to depict the gathered optimal control solutions with adequate matching. Although the concept of artificial neural networks is already well-known, the availability of sufficient computational power and user-friendly software libraries has contributed to common dissemination. The following metamodel is derived by utilizing the Neural Network toolbox from The Mathworks, Inc.

Once neural networks were chosen as the method to solve this function approximation or scattered data interpolation problem, respectively, a set of metamodel hyperparameters have to be defined. These hyperparameters comprise of different types of network settings. Aiming at good generalization properties and prevention of data overfitting, the training method is particularly set to Bayesian regularization.

1 Intel Core i7-7700K 4.2 GHz, 16 GB RAM

In Figure 11 the analysis results of different network architectures with respect to model performance and computational effort is depicted. Here the former is measured by the mean squared error (mse) between model output and target data, whereas the latter is characterized by the time needed to finish one iteration of the underlying optimization process (time per epoch). For the sake of simplicity, the variations consist of different numbers of hidden layers, where each hidden layer again always contains the same number of neurons. In general, the computation time per epoch increases exponentially with raising model complexity (either due to more hidden layers or more neurons per layer, notice logarithmic abscissa scale on the right side). On the left side of Figure 11 it can be seen that adding more neurons to a small hidden layer amount causes an improvement in the fitting result. With increasing number of layers however, this effect is reversed. The carpet plot on the right side shows the evaluation data of each configuration. Thereby efficient or dominant combinations, i.e. there is no point in the lower left region related to a specific sample, are marked with a diamond symbol, respectively. According to the analyzed space, a network consisting of three hidden layers and 20 neurons per layer (marked as black triangle) achieves the lowest mse value, while still requiring an acceptable amount of time per epoch.

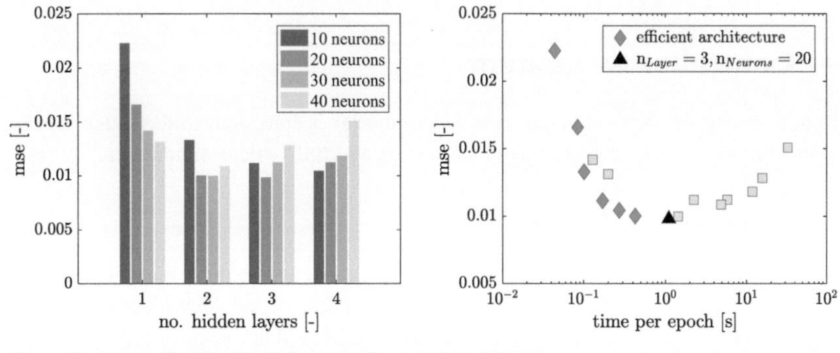

Figure 11: Pareto-front to determine neural network architecture

While using this optimal neural network architecture, an analysis of different training data set sizes has shown that a further performance gain can hardly be achieved by increasing the amount of training samples. Due to moderate computational effort along with satisfying fitting performance, a state space sampling subset of only 1000 samples seems to be sufficient for this kind of regression problem.

Figure 12 shows the final neural network architecture obtained, where each neuron consists of a sigmoid activation function. This is also assigned to the output layer, which implicitly ensures output values in the valid range of possible damper currents.

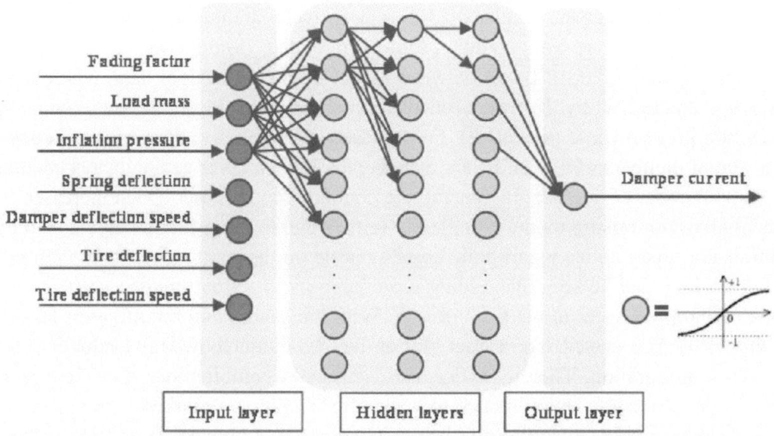

Figure 12: Utilized neural network architecture (three hidden layers, 20 neurons per layer, fully connected) with seven inputs (classified according to driving mode, system parameters, states) and one output (damper current)

4 Controller Evaluation

In this section the derived neural network controller is compared to the stated benchmark damper systems in both quarter car test rig and full vehicle simulations.

4.1 Quarter Car Test Rig

In order to evaluate the proposed optimal nonlinear control method, both this strategy (NLP) and the combined Skyhook/Groundhook (here called Hybridhook HH) are compared to the performance of the passive reference system. Each of the control approaches are utilized in a safety (sf) and comfort (cf) oriented configuration. The stated benchmark road profile serves in each test case as quarter car system excitation.

Table 1 gives an overview of the obtained simulation results and the achieved performance regarding objective measures. The configuration NLP_{sf} decreases the RMS value of dynamic wheel load by 20.2 % compared to only 14.4 % achieved by HH_{sf}. Moreover NLP_{sf} reduces the minimum wheel load and thus the possibility of wheel lift-off by 26.2 %. Both these controllers also improve even ride comfort almost equally, but NLP_{sf} ensures a further reduction of the maximum absolute chassis acceleration by 28.9 %. Focusing on ride comfort, the new control approach NLP_{kf} significantly reduces both comfort weighted and absolute chassis acceleration RMS value by approximately 34 % compared to 23.7 % resulted from HH_{kf}. This is accompanied by only a slight deterioration regarding ride safety measures and equally sized for both control concepts. Overall, all controllers enhance the induced wheel travel with a slight benefit towards the HH approach.

Table 1: Simulation results of front axle quarter car model on benchmark road profile

Quantity	Passive	HH_{sf}	HH_{kf}	NLP_{sf}	NLP_{kf}
$\|\ddot{z}_c\|_{rms}$ in $\frac{m}{s^2}$	2.36	2.12	1.80	2.10	1.56
Benefit vs. passive		-10.2 %	-23.7 %	-11.0 %	-33.9 %
$\max(\|\ddot{z}_c\|)$ in $\frac{m}{s^2}$	10.59	8.68	7.42	7.53	7.27
Benefit vs. passive		-18.0 %	-29.9 %	-28.9 %	-31.4 %
$\|\ddot{z}_{c,comf}\|_{rms}$ in $\frac{m}{s^2}$	2.12	1.96	1.62	1.91	1.40
Benefit vs. passive		-7.5 %	-23.6 %	-9.9 %	-34.0 %
$\|F_{dyn}\|_{rms}$ in N	1729	1480	1786	1380	1769
Benefit vs. passive		-14.4 %	+3.3 %	-20.2 %	+2.3 %
$\min(F_{dyn})$ in N	-5661	-4460	-5619	-4172	-5838
Benefit vs. passive		-21.1 %	-0.6 %	-26.2 %	+3.3 %
$\|z_c - z_w\|_{rms}$ in cm	2.1	1.6	1.8	1.7	1.8
Benefit vs. passive		-23.8 %	-14.3 %	-19.0 %	-14.3 %
$\min(z_c - z_w)$ in cm	-6.5	-5.6	-5.2	-5.5	-5.6
Benefit vs. passive		-13.8 %	-20.0 %	-15.4 %	-13.8 %

Further on the importance of model parameter based controller adaptivity is shown in Figure 13. In this contribution the controller synthesis considers the impact of various load mass and inflation pressures, see Figure 12. Especially on the rear axle, a huge spread exists regarding the static wheel load, whereas the inflation pressure is specified to be in a range between [2;3] bar. On the left side of Figure 13 the influence on ride safety of NLP_{sf} used as fixed (curb load, medium inflation pressure) or adaptive controller is depicted. In this place even an adaptation on load mass cannot ensure a relatively constant performance with respect to passive damping. This is completely contrary while focusing on ride comfort verified with NLP_{kf}. On the right side a significant necessity is shown for taking the load mass during controller development process into account. Hence it is possible to achieve an almost constant performance gain in ride comfort over a wide loading range. Overall it can be seen that a varying load mass influences the controller performance much more than a change in the tire stiffness and damping behavior induced by a higher or lower inflation pressure, even though the latter shows a slightly higher sensitivity in the context of ride safety.

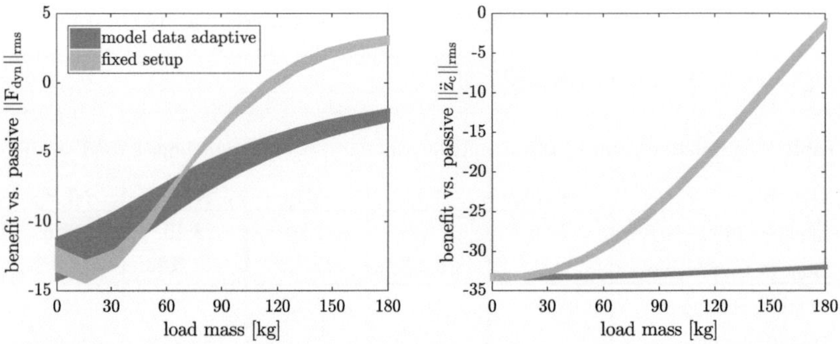

Figure 13: Influence of model parameter variation on rear axle controller performance regarding ride safety (left) and ride comfort (right)

4.2 Full Vehicle Simulation

Finally all damper systems are directly and unchanged incorporated into the full vehicle simulation, which is carried out by using the physically based tire model FTire available in the simulation framework DYNA4. The road is changed to a realistic Belgian block (curved regular grid (crg) data, [21]) type rather than the so far used synthetic profile, see Figure 14.

Table 2: Simulation results of full vehicle model driving on Belgian block

Quantity	Passive	HH_{sf}	HH_{kf}	NLP_{sf}	NLP_{kf}
$\left\Vert\ddot{z}_{c,driver}\right\Vert_{rms}$ in $\frac{m}{s^2}$	2.40	2.21	2.08	2.22	1.88
Benefit vs. passive		-8.0 %	-13.4 %	-7.6 %	-21.7 %
$\left\Vert\ddot{z}_{c,driver,comf}\right\Vert_{rms}$ in $\frac{m}{s^2}$	2.00	1.82	1.72	1.82	1.57
Benefit vs. passive		-9.2 %	-14.0 %	-9.3 %	-21.7 %
$\left\Vert F_{dyn}\right\Vert_{rms}$ in N	3601	3111	3726	3000	3781
Benefit vs. passive		-13.6 %	+3.5 %	-16.7 %	+5.0 %

Analogue to the quarter car test rig simulations, the same controller configurations are used here as well. As shown in Table 2 the performance evaluation is compressed to ride comfort measured at the driver seat position and ride safety as a result of averaging the RMS values of all four wheels. Basically the insights derived from the quarter car tests can be adopted. The neural network controller considerably outperforms the benchmark strategy in both ride comfort (-21.7 % compared to approximately -14 %) and ride safety (-16.7 % compared to -14.0 %). Both control concepts were however not able to transfer their test rig performance gain onto the full vehicle setup. As the proposed method is a model-based approach, a different or neglected system behavior might yield such deviations. These can be caused by a different kinematic behavior or an additional spring force coming from the anti-roll bar due to opposite wheel lift. Furthermore, different movement patterns occur, i.e. simultaneous heave, roll and pitch. In the synthesis process, the tire rotation is neglected and only vertical excitation during standstill is assumed.

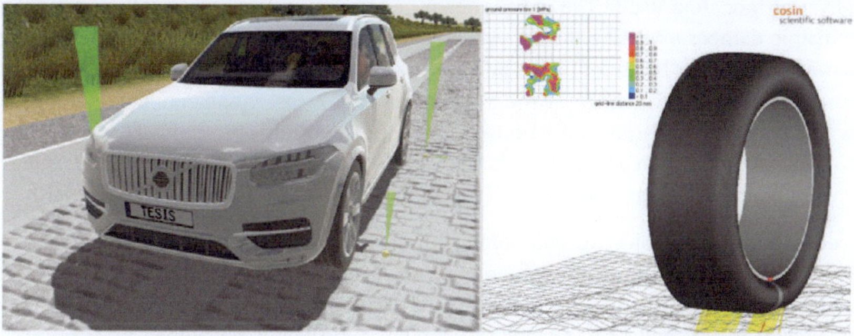

Figure 14: DYNA4 full vehicle simulation (left) on Belgian block using FTire (right)

5 Summary and Outlook

An optimal nonlinear control approach based on explicit model predictive control is shown, applied to semi-active suspension systems and evaluated in full vehicle tests using the simulation framework DYNA4. A quarter car model is implemented considering nonlinearities resulting from axle kinematics, overall spring stiffness and continuously variable damper force. The complex tire characteristic is depicted by a nonlinear spring and first-order delayed damper element. All parameters are derived by performing various static and dynamic test cases on virtual suspension and tire test rig. Furthermore the controller synthesis process is described in detail. Basically, optimal control problems have to be solved for multiple initial value problems, which are uniformly distributed in a pre-defined closed state space region. Thereby only the first control move in the damper current is stored along with the initial state condition as well as the current vehicle properties (i.e. loading and inflation pressure) and the intended driving mode (fading between ride safety and ride comfort). Then artificial neural networks are chosen as modeling technique to obtain a function approximation of this scattered result data set. After defining some general settings to avoid overfitted data during the training phase, a proper network architecture is manually derived by systematically varying the number of hidden layers and its contained neurons. For evaluation purposes an adaptive combined Skyhook and Groundhook controller is adjusted. The proposed method is tested in both quarter car test rig and full vehicle simulations. For both environments the new control concept significantly outperforms the benchmark in terms of ride safety (reduced RMS value of dynamic wheel load) and ride comfort (reduced RMS value of absolute and filtered chassis acceleration) improvement compared to a passive reference system.

Future work might comprise of the integration of dynamic neural networks in the physical modeling itself. The work already published in [22] can be adopted for the dynamic damper modeling. Moreover, a deployment of this concept to depict the dynamic vertical tire behavior has to be examined. It should always be possible however to obtain partial derivatives of the system enabling a fast optimization. Thinking of wider system boundaries, the proposed approach might also be applicable to an antiroll bar control or an integrated chassis control concept, respectively. As all states cannot be measured, an observer [23] has to be deployed and tested.

Looking into the subject of autonomous driving, the amount of data exchanged between vehicles and their environment will be constantly increasing. Methods such as reinforcement learning provide techniques for online based system optimization. As it is shown in [25] and [26], an online learning controller is developed by a reward-inaction based algorithm. Considering cloud-based computation units [24], this can be used for a fine-tuning process of application parameters during customer's real driving. An identification of optimal controller settings for particular road segments are therfor possible.

References

[1] J.Y. Wong. Theory of ground vehicles, John Wiley & Sons, 2008

[2] VDI 2057 Blatt 1. Einwirkung mechanischer Schwingungen auf den Menschen – Ganzkörper-Schwingungen, 2002

[3] ISO 2631-1. Mechanical vibration and shock Evaluation of human exposure to wholebody vibration, 1997

[4] M. Mitschke and H. Wallentowitz. Dynamik der Kraftfahrzeuge. Springer Verlag, Berlin, 4. edition, 2004

[5] G. Koch, K. J. Diepold, and B. Lohmann. Multi-objective road adaptive control of an active suspension system. In H. Ulbrich and L. Ginzinger, editors, Motion and Vibration Control, pages 189–200. Springer, 2009

[6] S. Schmeiler, K. Köstlin. Using Adams as master model for ECU system simulation. Global Vehicle Dynamics Conference 2016, MSC Software, 2016

[7] R. Dessort, C. Chucholowski, G. Rill. Parametrical Approach for Modeling of Tire Forces and Torques in TMeasy 5. Proceedings of the 16th Stuttgart International Symposium, 2016

[8] S. Le Digabel. Algorithm 909: NOMAD: Nonlinear Optimization with the {MADS} algorithm. ACM Transactions on Mathematical Software, 37(4):44:1-44:15, 2011

[9] J. Currie and D. I. Wilson. OPTI: Lowering the Barrier Between Open Source Optimizers and the Industrial MATLAB User. Foundations of Computer-Aided Process Operations, Georgia, USA, 2012

[10] https://de.mathworks.com/matlabcentral/fileexchange/9443-random-points-in-an-n-dimensional-hypersphere, accessed on April 16, 2017

[11] G. Marsaglia. Choosing a Point from the Surface of a Sphere. Ann. Math. Stat. 43, 645-646, 1972

[12] D. T. Larose. Discovering Knowledge in Data: An Introduction to Data Mining. John Wiley & Sons, Inc., Hoboken, NJ, USA, 2004

[13] D. Kirk. Optimal control theory. Dover Publications, 1998

[14] J.C. Butcher. Numerical Methods for Ordinary Differential Equations. John Wiley & Sons, 2008

[15] M. Björkman, K. Holmström. Global Optimization Using the DIRECT Algorithm in Matlab. Advanced Modeling and Optimization Volume 1, Number 2, 1999

[16] L. Csekő, M. Kvasnica, B. Lantos. Analysis of the explicit model predictive control for semi-active suspension. Period. Polytech. Elec. Eng., Vol. 54, No. 1-2 (2010), pp. 41-58

[17] A. Alessio, A. Bemporad. A Survey on Explicit Model Predictive Control. Nonlinear Model Predictive Control, LNCIS 384, pp. 345–369

[18] S. Savaresi, C. Poussat-Vassal, C. Spelta, O. Sename, L. Dugard. Semi-Active Suspension Control Design for Vehicles. Elsevier, 2010

[19] B. Heißing, M. Ersoy, S. Gies. Fahrwerkhandbuch. Springer, Berlin, 2013

[20] M. Mitschke, H. Wallentowitz. Dynamik der Kraftfahrzeuge. Springer, Berlin, 2004.

[21] http://www.opencrg.org/

[22] J.C. Tudón-Martínez, R. Morales-Menendez. Nonparametric Modeling of an Automotive Damper Based on ANN: Effect in the Control of a Semi-active Suspension. Volume 577 of the series Studies in Computational Intelligence, pp. 295-309

[23] N. Pletschen; K.J. Diepold. Nonlinear State Estimation for Suspension Control Applications: A Takagi-Sugeno Kalman Filtering Approach. Control Engineering Practice 61, 2017

[24] Z. Li, T. Chu, I. Kolmanovsky, X. Yin, X, Yin. Cloud Resource Allocation for Cloud-Based Automotive Applications. Cornell University Library, https://arxiv.org/abs/1701.04537, 2017

[25] M.N. Howell, G.P. Frost, T.J. Gordon, Q.H. Wu. Continuous action reinforcement learning applied to vehicle suspension control. Mechatronics, 7 (3). pp. 263-276. ISSN 0957-4158

[26] M.J.L. Boada, B.L. Boada, B. Munoz, V. Diaz. Neural control for a semi-active suspension of a half-vehicle model. International Journal of Vehicle Autonomous Systems 3(2), 2005

[27] M. Gipser. FTire and Puzzling Tire Physics: Teacher, not Student. 4th Tyre Colloquium, University of Surrey, 2015

[28] G. Koch, S. Spirk, B. Lohmann. Reference model based adaptive control of a hybrid suspension system. Proceedings. of the IFAC Advances in Automotive Control, 2010

[29] E. Pellegrini, S. Spirk, K.J. Diepold, R. Dessort, B. Lohmann. Nonlinear control of a semi-active suspension system considering actuator and state constraints using methods of optimal control. Proceedings of the American Control Conference, 2013

[30] K. Shah. EE263: Introduction to Linear Dynamical Systems. Lecture Slides 20, http://ee263.stanford.edu/lectures/ellipsoids.pdf, accessed on April 24, 2017

Using virtual prototypes for a cross-domain increase in efficiency in the development process

Martin Elbs, General Manager Sales & Business Development

Alexander Frings, Product Manager Engineering Services

IPG Automotive GmbH

P.E. Pfeffer (Ed.), *8th International Munich Chassis Symposium 2017*, Proceedings, DOI 10.1007/978-3-658-18459-9_16

1 Introduction

One of the primary goals of the traditional vehicle development process is the development of components which fit the respective vehicle optimally but which are also reusable in as many other models of the OEM as possible in order to allow for savings in terms of further development costs.

Furthermore, until now, each department usually comprised experts who were primarily specialized in their specific fields but hardly carried out any cross-domain work.

As part of the increasing networking of today's vehicle systems, the development departments of OEMs and suppliers are now facing new challenges. It is becoming increasingly difficult to adopt the traditional approach to development which will become impossible to realize in the future. Here, the so-called virtual prototype offers a solution approach which takes the difficulties of today's vehicle development into account and which redefines the development process.

This paper provides a detailed definition of a virtual prototype and an illustration of the essential advantages it offers in today's and future development processes. An exposition is then provided of the process that is necessary for building virtual prototypes and their cross-domain use throughout the entire development process. Furthermore, an overview of best practice examples of renowned OEMs and suppliers is given as well as a detailed description of quality enhancement and numerous potentials for savings in the development process.

Fig. 1 The virtual prototype

2 Complexity Drivers

The considerable increase of complexity in vehicle development over the past years can be traced back to a variety of reasons. As part of the electrification and hybridization of powertrains, the systems are tailored to the specific vehicle with significantly more precision than it was previously the case with traditional powertrains. The requirements need to be defined exactly. On the one hand, this requires the consideration of the overall system, on the other, it severely limits the reusability in other vehicles of the OEM.

Automated driving contributes another significant part to the increased development effort. The technical overlap of the subject areas of vehicle dynamics, advanced driver assistance systems and powertrain causes major difficulties for developers since none of these areas can be considered in isolation – for automated driving, for instance, both lateral and longitudinal dynamics are essential.

The testing of individual functions is only possible in the context of the whole vehicle since the functioning of the respective control unit depends on the signals provided by other components as well as the behavior of the vehicle within its environment. Only in the context of the whole vehicle can complex systems and their interplay be recognized, analyzed and optimized. Moreover, the mode of operation of advanced driver assistance systems, such as an emergency brake or lane keeping assist system, acts on vehicle dynamics as well as the powertrain and therefore cannot be tested independently either.

Nonetheless, the networked systems must be tested as early as possible – under realistic conditions and in the context of the whole vehicle. This is further complicated by the fact that a variety of development departments and partners need to be included in the process in time.

The lacking availability of real prototypes is another impeding aspect. This problem applies both to later stages within the development process in which prototypes are usually on hand as well as the number of available real prototypes that is simply too limited to ensure access for all development engineers at the required time. Despite the lack of real prototypes, there is a mounting pressure to shorten development cycles and avoid errors in the constantly growing diversity of variants.

3 Virtual Prototype – Definition

When discussing virtual prototypes, this paper focuses predominantly on prototypes for the development of electronic and mechatronic functionalities of automotive electronics. Nearly all automotive software development and test activities can be supported by such prototypes. Component simulation models (including Simulink-based models) have been in use for these tasks for more than 25 years. Development of whole-vehicle virtual prototypes originated first in vehicle dynamics development.

In light of the high number of vehicle configuration variants and the interaction with active safety control functions, test cases for vehicle dynamics control have long reached a level of complexity that has resulted in an increasingly expensive testing of these functions in real-world test driving alone. The higher workload related to critical tests on real prototypes additionally tends to increase costs, while not providing sufficient reproducibility of the test conditions to determine and quantify functional improvements.

Virtual test driving allows for repeatable testing in combination with a considerable cost advantage and reduces the risks involved dramatically. With virtual test driving, millions of kilometers can be driven overnight in automated test scenarios. Negative effects of changes to the software or parameters can be discovered in parallel with software development. Therefore, this approach first and foremost allows for appropriate quality control measures during systems development. It also enables a reduction of the number of real prototypes required and the corresponding time-to-market of a vehicle platform.

The virtual prototype is a fully parameterized full-vehicle model consisting of all components and systems relevant for automotive systems development. The model is optimized to reflect correct vehicle dynamics behavior. All relevant interdependencies and influences of the various vehicle systems can be simulated to varying degrees of fidelity, depending on the application context.

Appropriately detailed and parameterized virtual prototypes allow for a cross-domain use in diverse areas of application and significantly add to an increase in the efficiency of development processes. The application of virtual prototypes in the vehicle development process of an OEM is therefore best supported by a dedicated and centralized model development team that caters to a wide range of development departments as shown in Figure 2. Jaguar Land Rover has incorporated a Virtual Engineering division that produces virtual prototypes for the entire business with active models from internal and external suppliers integrated within the same. This enables any engineer within the business to set requirements, build and test their systems on a pre-validated virtual vehicle within a test driving environment.

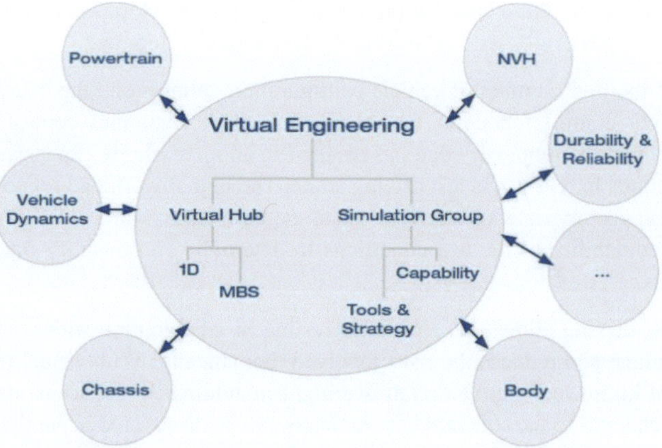

Fig. 2 Virtual prototypes at Jaguar Land Rover

Automotive systems development involves numerous stakeholders that include not only the OEMs but also various suppliers. Development of virtual prototypes accordingly combines systems of many departments involved both at OEMs and suppliers to provide simulation environments in an early stage of the development process.

This is the basis of scenario-based testing. In this approach, the virtual vehicle, driven by a virtual driver, is confronted with complex traffic situations in realistically modeled virtual worlds. Scenario-based testing describes the simulation of the complete system (within the vehicle as a full system) under realistic conditions (in freely definable scenarios).

The virtual prototype is available in early phases of development projects and can therefore be used throughout all stages of the development process – from the early testing of the control algorithm (proof of principle) with the model-in-the-loop method up to function and release tests of the system on a hardware-in-the-loop test bench (functional prototype) or in the real vehicle (vehicle-in-the-loop) itself.

In addition, access to the prototype is ensured for every engineer due to the ability to use a validated virtual prototype. These advantages form the foundation of a highly efficient development model for components and accommodates their interactions as well as interdependencies with each other.

ESC homologation is an area in which the use of virtual prototypes extends even to formal certification legislations of systems through simulation. The use of simulation leads to almost 80 % of the homologation tests being conducted using simulation, after 20 % of the tests are conducted and validated with the real vehicle. This leads to a large savings in effort, infrastructure, costs as well as a reduced number of prototype variants for a platform.

By using virtual prototypes, the PSA Group was able to reduce the number of real prototypes by approximately 30 percent. Stringent application of the process results in a significant reduction of time-to-market.

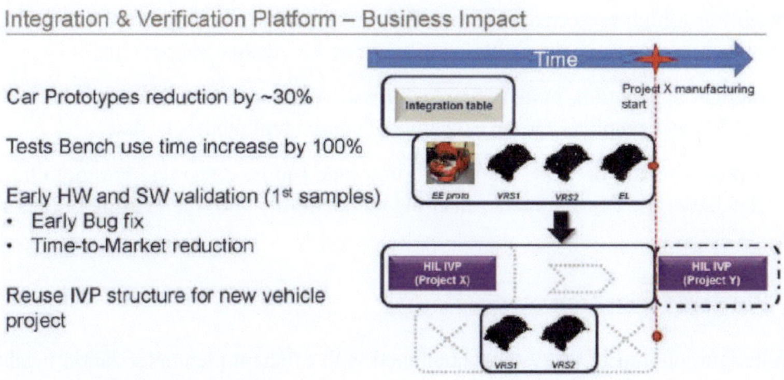

Fig. 3 Virtual prototypes at PSA

4 Building Virtual Prototypes

The following section illustrates the method of building a virtual prototype.

4.1. Definition of required model complexity

The foundation of successful application of virtual prototypes is a high-quality vehicle dynamics representation of the target vehicle. The following paragraphs will describe the required steps to efficiently develop high quality models.

Development processes built around the use of virtual prototypes require competence in three areas:

1. Early availability of representative models
2. Efficient parameterization methods using easily accessible data
3. High level of refinement in both modeling and parameterization

Applications of the virtual prototype determine the refinement required in both modeling and parameterization. Powertrain-related development activities such as the improvement of real driving emissions (RDE) require a specific focus on powertrain component models and integration interfaces to engine dynamometers. The development of functions for advanced driver assistance or automated driving functions requires models of a multitude of different sensors in application-specific model complexity apart from representative fidelity of the active systems involved.

Additionally, the magnitude of simulations to be performed will affect the required model resolution: Highly parallelized simulations of millions of kilometers of driving overnight on a high performance computer cluster requires a different modeling approach when compared to durability investigations for chassis components.

In a practical application, model complexity will also depend on the cooperation between OEMs and suppliers for the exchange of component models or data.

From a process point of view, the virtual prototype can be integrated into both traditional databases and PLM systems, ensuring compatibility with the existing infrastructure.

4.2. Parameterization

Such integration into PLM systems combined with efficient parameterization mechanisms for the virtual prototypes facilitates the automatic generation of vehicle variants: Various vehicle components from different sources parameterized in dedicated data files can be copied and subsequently combined to form a complete vehicle code.

As the vehicle development process progresses, measurements of real components and systems can validate and refine the initial data-based modeling of the virtual prototype. Such validation loops between (estimated) simulation and real-world measurement will also help improve the model development process in the long run.

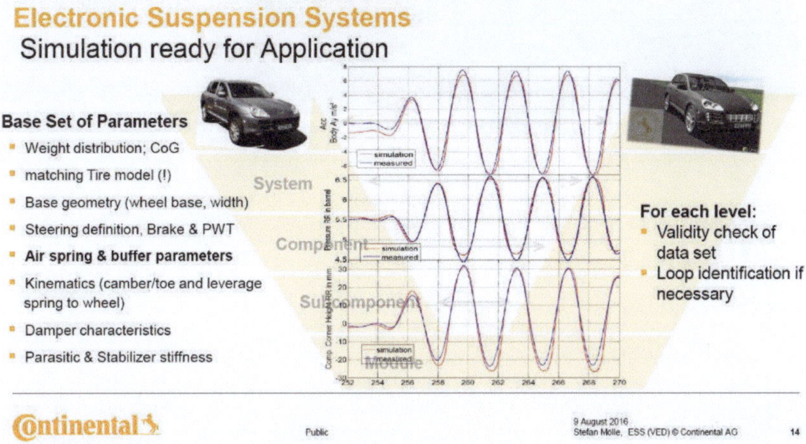

Fig. 4 Model parameterization for the development of electronic suspension systems at Continental

Continental serves as an example here. In the application of electronic suspension systems, e.g. air spring systems in customer projects, virtual prototypes play an important role. Parameterization is performed initially with vehicle parameters provided by the OEM customer and, over the course of the project, validated using data acquired in road test as well as on various test rigs (K&C, component test rigs).

Virtual test driving enables Continental to test and parameterize functions reliably and reproducibly at a very early stage of development.

4.3. Vehicle handling validation

The virtual vehicle model is validated by means of comparative measurements obtained in real-world test driving. Numerous test maneuvers are performed in simulation as well as in reality. A valid parameterization is achieved when the measurement data of the real vehicle is consistent with the measurement data of the virtual vehicle.

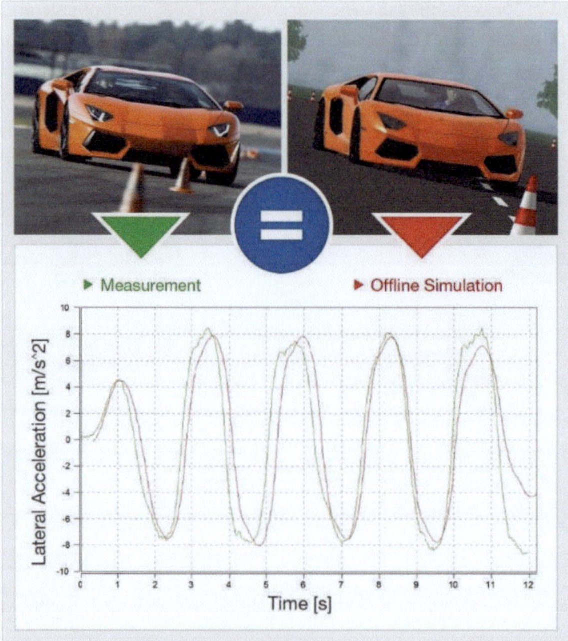

Fig. 5 Vehicle handling validation

4.4. Model integration

Early applications of the virtual prototype in the development process focus on development and testing of new functions and systems in the virtual world (model-in-the loop). Control functions and subsystems frequently come from various sources (suppliers and OEM). A well-structured integration environment is a prerequisite to manage the complexity of vehicle and component variants.

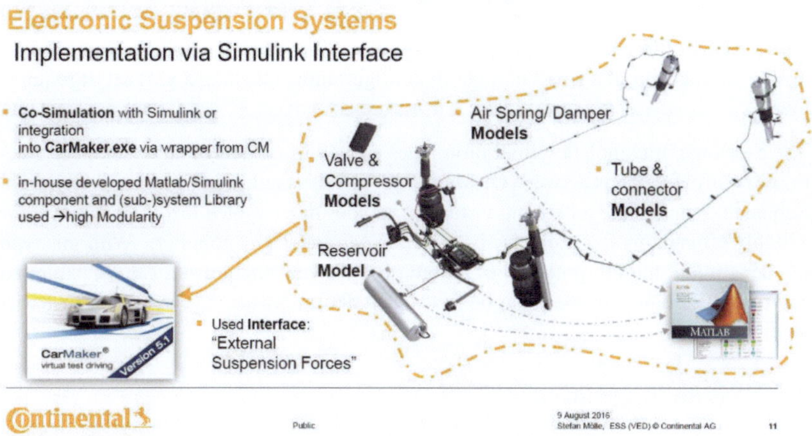

Fig. 6 Model integration of electronic suspensions systems at Continental

Typical paths for the integration of component models are C code, Simulink models or FMUs. Each development department needs to have access to the same validated models in order to synchronize simulation results at all times.

Another reference application is the development of a concept for the conversion of a vehicle equipped with an internal combustion engine to a hybrid system – without a real-world prototype as demonstrated by Bosch Engineering.

Fig. 7 Hybrid powertrain concept study by Bosch Engineering

With virtual prototypes, required design decisions can be made during the concept phase. Many different hybrid powertrain configuration variants can be evaluated to allow a pre-selection of a small number of configurations fulfilling a vast set of requirements.

The exchange of complete virtual prototypes as well as models of subsystems or electronic control functions between OEMs and suppliers is facilitated by model encryption. Submodels can be integrated into virtual prototypes in encrypted form. This allows for a highly efficient model exchange between OEMs and Tier 1 suppliers. With this type of cooperation, the full potential of automotive systems engineering can be exploited and the development process can be fundamentally optimized, e.g. by means of frontloading.

4.5. System integration

The integration of real vehicle components into the development process using virtual prototypes provides additional opportunities for optimization of the development process, thereby combining the benefits of testing real components with the flexibility of simulation. With the hardware-in-the-loop method, any real vehicle component can be tested while the vehicle itself as well as the entire environment are simulated.

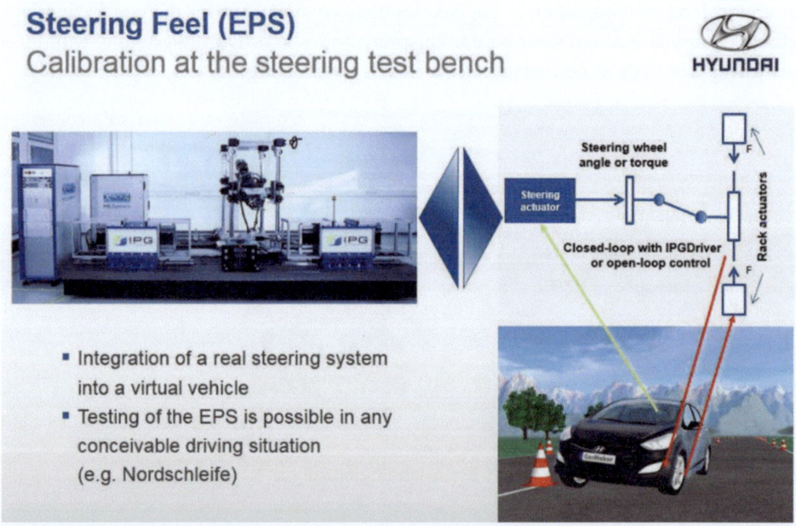

Fig. 8 Steering-in-the-loop for optimization of steering feel

As an impressive example of the far-reaching use cases of virtual prototypes, Hyundai in cooperation with IPG Automotive has proved the feasibility of virtually tuning the parameters of an electrically assisted power steering (EPS) system to improve steering feel in a production vehicle. Steering feel plays a major part in the perception of handling characteristics and is increasingly gaining importance as a customer-relevant criterion and as an element of the car manufacturer's brand DNA. The integration of virtual prototype models into a steering system test bench enables a reproducible and efficient tuning process in laboratory conditions.

In this case, the hardware-in-the-loop method offers distinct advantages in comparison to traditional test drives – full reproducibility as well as significant savings in costs.

While this example highlights very advanced applications of virtual prototypes in cutting-edge development processes, many examples demonstrate significant efficiency gains from using virtual prototypes. A relevant application case here is the testing of trailer stability control systems.

Fig. 9 Trailer stability control

Adaptation of trailer stability functions of an ESC system requires significant parameterization effort, e.g. depending on the position of the load in the trailer and the resulting change in the center of gravity. Simulation allows the elimination of time-consuming preparations in real-world testing. Different application and load conditions can be tested with very low effort. Adding to the savings in time, test scenarios that are difficult to represent and/or dangerous (e.g. swaying trailers) can thoroughly and repeatably be simulated.

5 Conclusion

The automotive systems engineering approach is required to manage increasingly networked development processes. Virtual prototypes are a key component to take the development process to the next level.

This work has provided a detailed demonstration of the process that is necessary for building virtual prototypes and their cross-domain use throughout the entire development process. Furthermore, an overview of best practice examples as used by renowned OEMs and suppliers was briefly described.

The use of virtual prototypes offers several essential advantages that may contribute fundamentally to the optimization of vehicle development. Particularly, the early availability of virtual prototypes to all development engineers results in significant savings in time and costs. Development efficiency can be increased considerably.

The vehicle of the future in which electronic controls autonomously manage the complete transportation from point A to point B will not be developed and tested without the application of virtual prototypes throughout the whole development process.

REFERENCES

(1) Steffen Schmidt, The future is connected – which possibilities are available already today?, Baden-Baden Spezial, 2016

(2) Building Virtual Prototypes – with Our Engineering Services, IPG Automotive GmbH, Karlsruhe, 2016

(3) IPG CarMaker at Jaguar Land Rover, Robert Chase, apply & innovate, Karlsruhe, 2016

(4), (5) CarMaker @ Electronic Suspension Systems Dipl.-Ing. Stefan Mölle, apply & innovate, Karlsruhe, 2016

(7) Durchgängige Simulation zur Elektrifizierung des Antriebs, Christian Appel, Dr. Stefanie Freudenstein, Christian Temmen, MTZ 02/2014

(8) Improvement of Steering Feel Virtual Approach with HiL, Dipl.-Ing. Axel Honisch, Dipl.-Ing. Michael Lugert, Dipl.-Ing (FH) Timo Schöning, Dipl.-Ing. Stephan Hakuli, ATZ 05/2015

Optimizing passive vehicle dynamics for active safety and autonomous driving

Marco Fainello, Chief Technical Officer
Danisi Engineering S.r.l., Italy

P.E. Pfeffer (Ed.), *8th International Munich Chassis Symposium 2017*,
Proceedings, DOI 10.1007/978-3-658-18459-9_17

Abstract

This paper explores the possibility of increasing global vehicle performance by reducing its passive stability margins and exploiting the presence of autonomous drive.

The design of vehicle handling has historically considered a human driver as its target controller. This in turn mandates generous stability margins are built into the vehicle's passive dynamics for safety reasons. This approach represents a trade-off between open-loop performance and closed-loop stability.

This investigation, made at a simulation level using a Driver-in-Motion simulator, consists of a closed-loop evasive manoeuvre. The vehicle's passive dynamics are modified at both architectural and tuning levels, and the resulting performance of an autonomous and human driver are compared.

Grip vs Stability

The passive grip and stability limits of a vehicle depend primarily on the cornering stiffness' of the front and rear axles. These are designed into the vehicle using tools such as tyre sizing, suspension kinematic and compliance, elastic and geometric load transfer. The cornering stiffness of a vehicle is not a constant parameter but rather something that varies in function of the vehicle's dynamic state.

This paper represents the stability of a vehicle setup using the static margin [1] as calculated from a steady state lateral ramp steer manoeuvre. The static margin is observed to vary as the vehicle's state transitions from its linear range to its saturation range as the front and rear cornering stiffness' change as shown below.

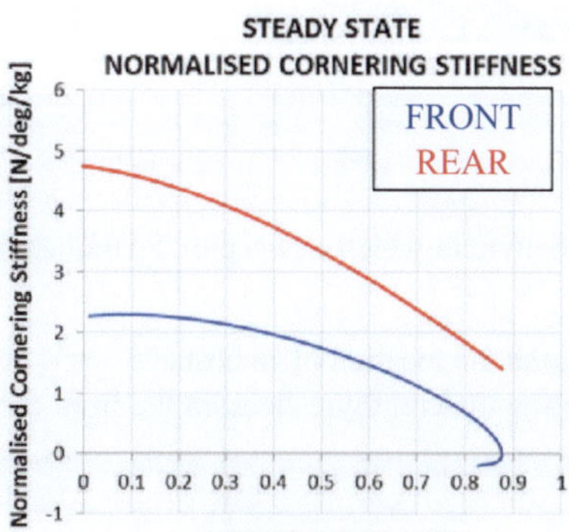

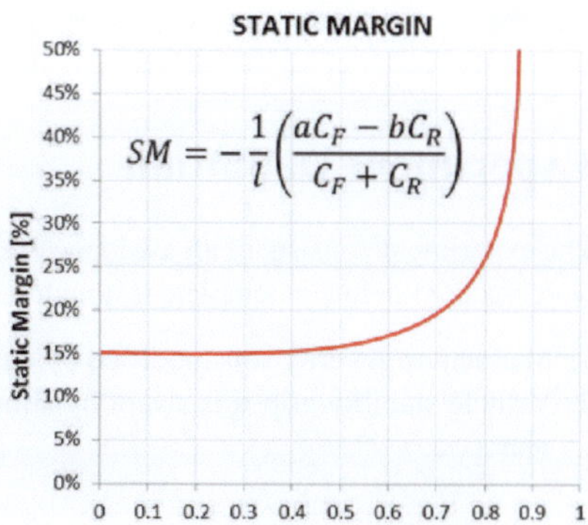

The trade-off between absolute grip and stability is apparent above in that the rear axle doesn't ever reach its saturation state (cornering stiffness = 0) , thereby constraining the static margin to be always positive (i.e. stable).

One can increase the global grip limit by rebalancing the vehicle to use more rear axle grip but this comes at the expense of rear cornering stiffness and hence stability as expressed using the static margin metric.

Below, an example is shown which demonstrates how a higher peak grip can be of assistance in maintaining a closed loop trajectory on a dual lane road. The blue vehicle was simulated with a static margin of 15% and the red vehicle had a reduced static margin of -2%. Both vehicles saturated the front axle during the manoeuvre but the red vehicle was able minimise its path deviation and remain in its assigned traffic lane due to its higher peak grip capacity.

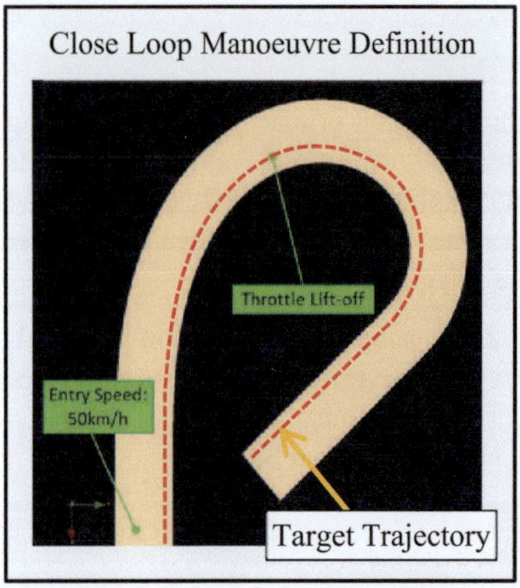

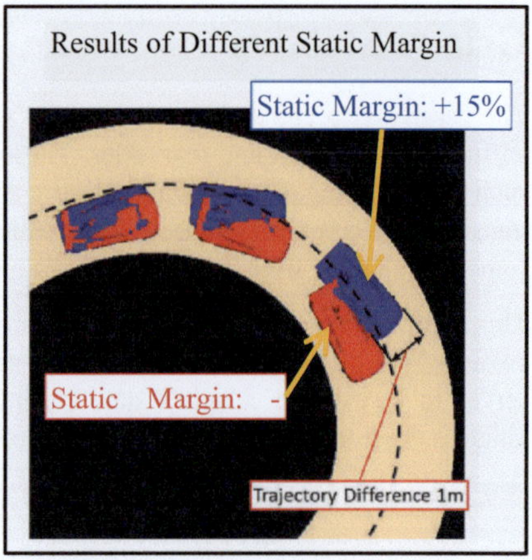

Experiment Definition

The experiment consists of an avoidance manoeuvre in which a sweep of 2 parameters were made to define a complete test matrix.

1. Manoeuvre severity was modified by changing the location of an obstacle
2. Vehicle stability/grip was swept by simulating different tyre sizes on the front and rear axle.

Manoeuvre

To evaluate the human-vehicle limit for controllability and stability, an avoidance manoeuvre was built using a dynamic driving simulator. The manoeuvre is based on a traditional lane change scheme but with the a few important modifications.

1. For better repeatability a graphical blocker (basically a 2m high wall) was added after the first gate to inhibit the driver from anticipating the steering or throttle release actions.
2. The manoeuvre in the simulator was randomly set to be to the left or the right – again to avoid driver anticipation.
3. The lane widths are not ISO standards but are sized to recreate an avoidance manoeuvre on a standard 3.75m wide highway lane
4. The "avoidance length" was lengthened to incorporate the delay of the driver's reaction time due to the graphical blocker
5. The "recovery length" was swept from a high value to a low value in order to find the point at which the driver is unable to control the vehicle.

The manoeuvre layout is shown below:

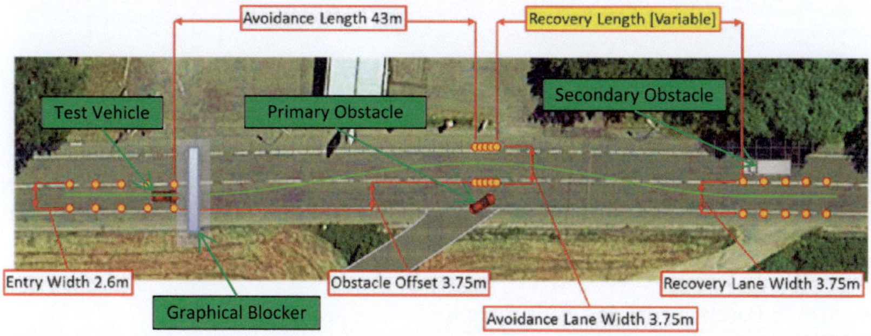

The execution of the manoeuvre was done such that for each nominal recovery length – the driver has 10 attempts to perform the manoeuvre. The driver is then scored based on the number of manoeuvres completed successfully (i.e. no cones hit).

Vehicle Setup

The baseline vehicle is an E segment series production sedan that is validated for transient and combined use on the driving simulator. Its stability and grip was swept from understeering (static margin +15%) to neutral (static margin -2%) by using a tyre size modification tool which is based on the Pacejka formulation (MF-Tyre 6.1). The vehicle setup sweep progressively reduced the rear axle grip and cornering stiffness and simultaneously increased those at the front.

Below one can see the relationship between the peak attainable lateral acceleration and the static margin. As the static margin is reduced – the peak lateral acceleration increases. It was found that about 10% extra lateral grip performance could be obtained by reducing the static margin from 15% to -2%.

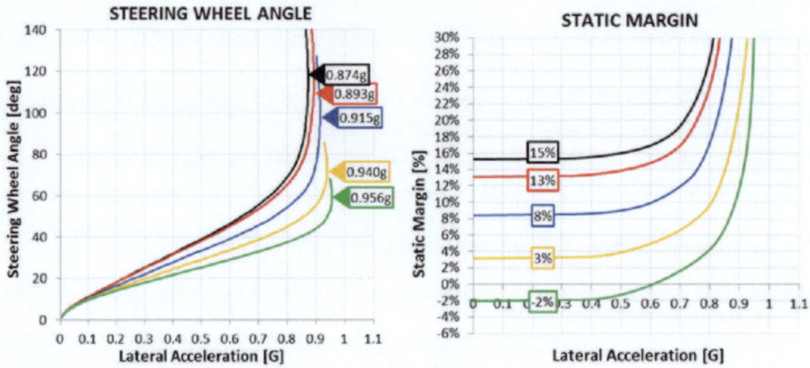

Complete Test Matrix

The complete test matrix then consisted of a sweep of the manoeuvre severity (recovery length) and the vehicle stability (static margin) for a total of 60 manoeuvres for each driver:

5 vehicle setups were selected:

- Step 0: Baseline vehicle / static margin of 15% / peak Ay of 0.874g
- Step 2: Setup variation / static margin of 13% / peak Ay of 0.893g
- Step 4: Setup variation / static margin of 8% / peak Ay of 0.915g
- Step 6: Setup variation / static margin of 3% / peak Ay of 0.940g
- Step 8: Setup variation / static margin of -2% / peak Ay of 0.956g

A total of 6 different geometries of the lane change manoeuvre were used:

- Recovery length: 35m
- Recovery length: 32m
- Recovery length: 29m
- Recovery length: 26m
- Recovery length: 23m
- Recovery length: 20m

3 drivers were used in the study – one professional, one 'enthusiast' and one 'everyday' driver. All 3 have significant simulator experience which was necessary to exclude the effect of the simulator learning curve on the results.

Simulator Implementation

In the driving simulator the graphical environment was implemented as shown below whereby cones were placed at the desired locations depending on the required manoeuvre severity. The graphical blocker was added as a signboard which covered the full width of the highway and is around 2m high. The obstacle consisted of a barrier object placed adjacent to the avoidance lane marked by the cones.

The driver was manually required to maintain 100km/h and release the throttle as soon as he passed through the graphical blocker.

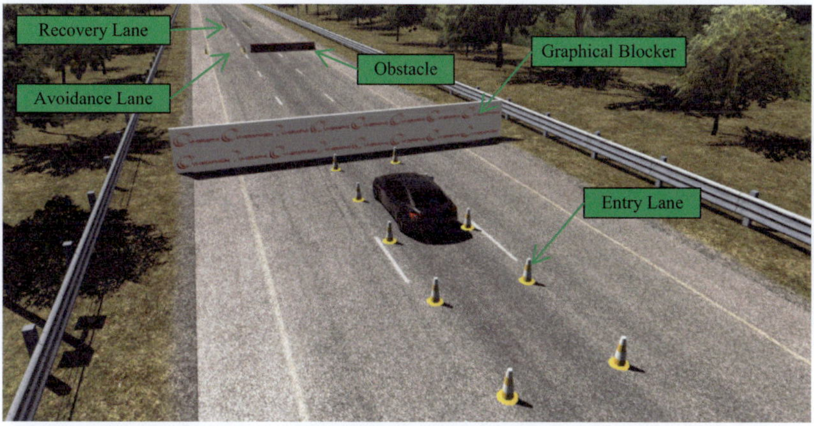

Robotic Controller Implementation

The manoeuvre was implemented as a closed-loop path following task with a robotic steering controller. The success of the manoeuvre was assessed automatically using a trajectory analysis tool that takes the vehicle trajectory and the cone locations.

The robotic controller was implemented in co-simulation with the real-time vehicle model and consists of a model-based predictive type controller (MPC). The MPC controller uses a simplified vehicle model that reflects the chosen vehicle setup and it simulates the vehicle's response for multiple possible input conditions along a predictive 'event-horizon' before choosing the one which best follows the target trajectory.

Results

Human Driver

The human driver was scored based on the percentage of manoeuvres that were passed for each combination of static margin and manoeuvre severity. The pass-rate ranged from an average of 80% (combined from the 3 drivers) for low manoeuvre severity and high static margin – down to 7% for the case of high manoeuvre severity and low static margin.

A correlation coefficient of around -0.7 was found between the recovery length and static margin (expressed as percentage points). In other words – every 1% decrease of static margin increases the required recovery distance by 0.7m in the lane change manoeuvre at 100km/h.

Subjectively, the drivers described the performance of the neutral steer vehicle (static margin = -2%) as extremely unstable in a transient condition (i.e. the second part of the manoeuvre) and very slow to respond to steering input. The professional driver quickly learnt to anticipate the steering input before the change of trajectory was actually required – however once the vehicle started to respond it was almost impossible to make any changes to its trajectory. The non-professional drivers found it a lot more difficult to anticipate the steering by the right amount such that the vehicle arrives at a safe location for both the primary and secondary obstacles and as a result they demonstrated a zero success rate with the severe manoeuvre and neutral steer vehicle.

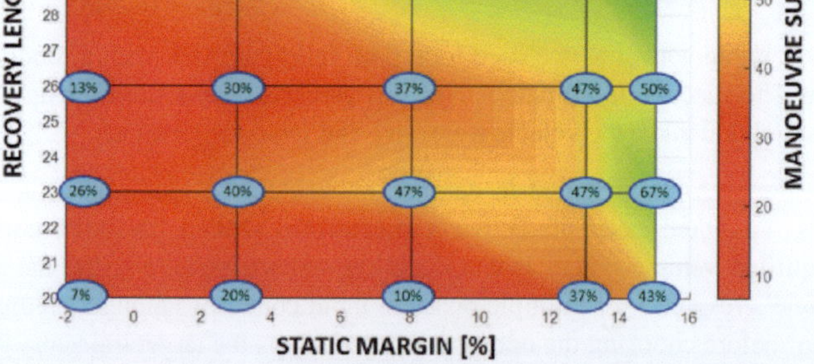

Robot Controller

The robot controller was able to successfully complete all combinations of the manoeuvre severity and vehicle static margin however it was noted that the control task became increasingly difficult for the vehicle setups closer to neutral steer due to the inherent response delays that accompany a vehicle with a low static margin. This delay was found to be much larger than the event horizon of the MPC controller and as such it was not able to follow the trajectory in a robust way. To successfully pass the most severe manoeuvre with the neutral steer vehicle it was necessary to modify the target trajectory with a variable offset to induce steering commands much earlier.

The final configuration of the controller + trajectory environment resulted in the following behaviour for the most severe manoeuvre with the neutral steer vehicle.

Robot Controller Commands

Recovery Length 20m // Static Margin -2%

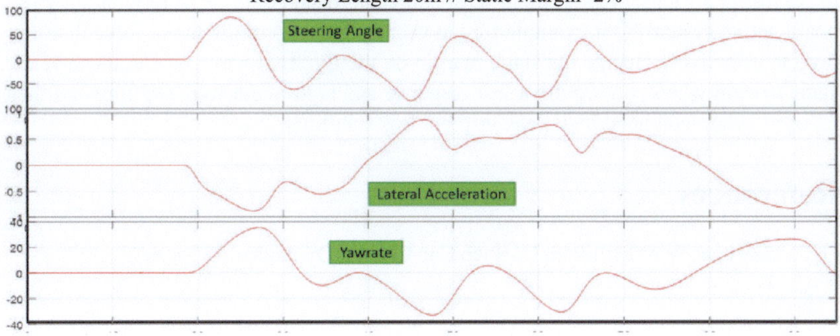

It can be seen that there are effectively 8 distinct steering inputs (peaks in the steering angle graph) which correspond to the 4 changes of direction required in the lane change manoeuvre. On further analysis it was understood that every change of direction was executed in an action-reaction manner whereby the first input created the change in direction and the second input damped the otherwise unstable response.

Conclusions

This investigation explored the theme of exploiting extra grip by tuning a vehicle's passive vehicle dynamics with less stability in order to assess if such a concept could be feasible to use in an autonomous drive application.

It was shown that a neutral steer vehicle with less stability and more grip is able to avoid a potentially dangerous loss of trajectory in a low frequency / low dynamic cornering manoeuvre. This behaviour was also able to be replicated by a human driver.

However in a dynamic condition, such as a lane change manoeuvre, the human driver was not able to control the neutral steer vehicle due to its lack of manoeuvre stability.

One could conclude from this analysis that to exploit the maximum performance envelope of an autonomous vehicle it is not enough to simply reduce the static margin because it was seen that the delays introduced into the vehicle response are too large to control in a robust way. A more reasonable approach could be to rely on other onboard active systems (for example torque vectoring or electronic brake force distribution) to effectively change the static margin on the fly. That way you can increase the static margin when you require a fast response and high manoeuvre stability and then reduce it when you need high peak grip but less stability.

References

[1] Milliken, W & Milliken, D, Race Car Vehicle Dynamics, 1995, SAE

OBJECTIVE ASSESSMENTS

Objective assessment of the dynamic system behavior of multi-chamber air springs

M. Eng. Emre Boyraz, Dr.-Ing. Christoph Kandler,
Dipl.-Ing. Markus Gantikow, Dr. Ing. h.c. F. Porsche AG

Prof. Dr.-Ing. Dr. h.c. Dieter Schramm, University Duisburg-Essen

P.E. Pfeffer (Ed.), *8th International Munich Chassis Symposium 2017*,
Proceedings, DOI 10.1007/978-3-658-18459-9_18

Abstract

Porsche aims to offer its customers sports cars which are well-suited for everyday driving. Therefore, special emphasis is placed on the realization of functions, which enable best in class driving behaviour with regard to performance and ride comfort. This can be achieved by the consistent use of highly developed mechatronic subsystems to achieve a wide spread between performance and ride comfort. A major proportion of this spread can be accomplished by introducing multi-chamber air springs. The synergistic interplay, consisting of consequent combination of mechanical, electrical and information technology, enables to fully exploit the high potential of these systems. However, along with these benefits, multi-chamber air springs add system complexity, posing new challenges for chassis development.

This paper presents a new method to efficiently describe the dynamic system behaviour of multi-chamber air springs by using chassis test rigs. Based on objective vehicle measurements, the proposed method offers considerable potential to strengthen the early development phase and to reduce development time and costs.

1 Introduction

Porsche aims to add even more vehicle performance with each new model line, but at the same time maintaining their typical character. There are various requirements to be met to offer Porsche customers the driving experience they expect from their vehicle. To achieve the desired driving behaviour, the characteristics of many subsystems need to be fine-tuned to optimize the overall function of the system. This demands an iterative development process and a high degree of testing requirements. Therefore, a reliable assessment of distinct vehicle properties is necessary for a target oriented development process. Objective methods for evaluating the dynamic system behaviour on chassis test rigs are proposed here to strengthen subjective tuning during the development process as depicted in FIG. 1. This provides the potential to control the increased degree of complexity and to reduce development costs.

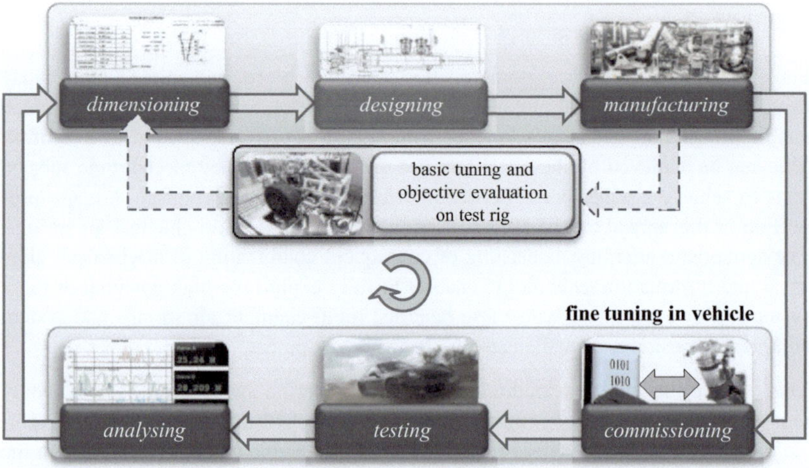

Figure 1: Iterative development process [1]

Porsche aims at developing best in class performance with a wide spread of ride comfort. Especially multi-chamber air springs allow the chassis developers to optimize the driving behaviour, as depicted in FIG.2. The spring rate of an air spring can adapt to the actual driving situation by using dynamically switchable chambers. The multi-chamber air spring, being analysed in this paper, has three chambers. The lowest spring rate is provided by actuating all three chambers. By deactivating one or two of these chambers and thus by increasing the total volume, the spring rate increases. For a more detailed description of these design features we refer to reference [2].

To further maximise the spread between performance and ride comfort, a precise understanding of the dynamic system behaviour of multi-chamber air springs is beneficial. Based on an objective assessment of a full vehicle, relevant parameters of the air spring can be identified and optimized by using chassis test rigs.

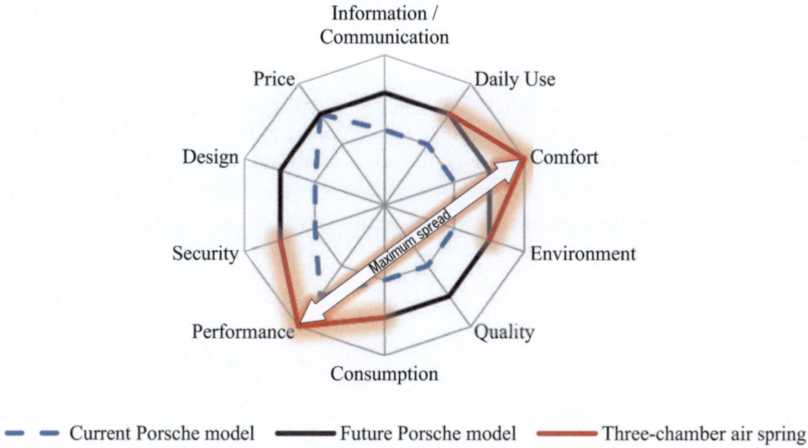

Figure 2: Full vehicle targets [2]

2 Objective assessment of full vehicle

The objective assessment of vehicles on roads regarding ride comfort represents the key factor as part of an efficient vehicle development process. Thus, the requirement for an obvious and reproducible description of the driving behaviour of vehicles is becoming more and more pronounced. The ride comfort of a vehicle can basically be divided into the following categories [3].

- "Body control": periodic vibrations in the range of the vehicle body's natural frequencies.
- "Solidity": periodic vibrations in the range of the subsystem's natural frequencies.
- "Stochastic vibrations": irregular or constrained vibrations and absorption.
- "Single obstacle": impulse and damping behaviour.

Further evaluation methods for the axle and individual chassis components can be derived based on a driver-independent description of the driving behaviour of vehicles. Thereby, each chassis system can be designed optimally on individual component test rigs, which are available in the early development phase and can be adapted to one another on composite test rigs. This chapter describes a method according to [3], which allows an objective assessment of the ride comfort of a full vehicle. This method is used to derive an evaluation method for an air spring test rig.

2.1 Methodology for evaluation

To quantify the subjective comfort criteria, objective methods are used. Based on road measurements taken with a full vehicle, objective parameters can be derived from measured data and an effect on the driver's perception can be established [4]. For this purpose, acceleration sensors are used to measure the mechanical vibrations during driving. The vehicle, which is exemplary analysed in this paper, was equipped with four acceleration sensors on its chassis domes, one on the driver's seat rail and four on wheel carriers. FIG. 3 shows the described measurement setup.

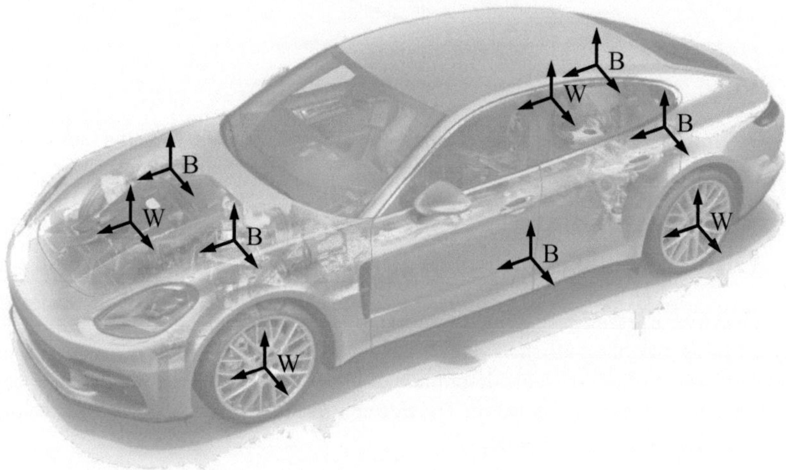

Figure 3: Measurement setup for ride comfort (B: body, W: wheel)

For the assessment of the chassis dependent ride comfort, it is reasonable to evaluate the vehicle body accelerations [5]. The body control describes the vibration in the range of the vehicle body's natural frequencies. To apply ride comfort measurement, it is useful to excite the vehicle in the primary movement direction of the vehicle body. The modal movement can be simplified with the input vector $\vec{u}$ and the road height velocities v_{Road} at each wheel as follows.

$$\begin{array}{llllll} & FL & FR & RL & RR & \\ u_{Heave} = \{ & 1 & 1 & 1 & 1\} & v_{Road}(t) \quad (1) \\ u_{Pitch} = \{ & -1 & -1 & 1 & 1\} & v_{Road}(t) \quad (2) \\ u_{Roll} = \{ & 1 & -1 & 1 & -1\} & v_{Road}(t) \quad (3) \end{array}$$

The measurements for this paper were carried out on a comfort test road on the Porsche proving ground in Weissach. To evaluate the ride comfort, the spectra of the vertical body accelerations were calculated by first transferring the five body acceleration signals into the frequency range and then averaging them. To make sure that all relevant body modes are excited in the whole frequency range, a comfort measurement consists of several individual measurements at different speeds. The speed is kept constant during one measurement but is increased step by step for the following measurements. Afterwards, the synthesis from the individual measurements is calculated by weighting with the corresponding reciprocal value of the velocity and subsequently averaging them. This makes it possible to obtain a most evenly distributed modal excitation spectrum and to compare different vehicle variants with one another [6]. FIG. 4 shows an example of a ride comfort measurement.

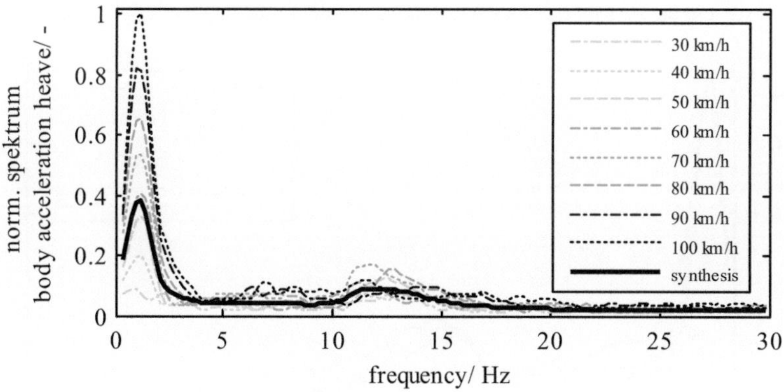

Figure 4: Normalised spectra of measurements on the road with different speeds

Two characteristic maxima of the body acceleration spectrum of the result synthesis are displayed in FIG. 4 in the ride comfort measurement. The first peak at approx. 1 Hz is caused by the natural frequency of the body. The natural frequency of the wheel is responsible for the maximum at approx. 12 Hz. The range between body and wheel natural frequencies is known as isolation range. This range has a great influence on the ride comfort of the vehicle occupants according to [7] and [8]. In these publications, a frequency factor is used for reflecting the frequency-dependent sensitivity of human vehicle occupants to vertical mechanical vibrations. This factor is largest in the isolation range because some natural frequencies of human organs are within this region [4]. Consequently, the influence of the multi-chamber air spring on the isolation range is being analysed in this paper. The mean value is taken as the characteristic value and used for the comparison of the variants in the frequency range between 4 Hz and 10 Hz.

2.2 Influence of different chassis modes

Porsches mission is to offer its customers the sportiest vehicle combined with a high level of comfort in each segment. There are three chassis modes for the driver to choose from. In *Normal Mode*, the mechatronic chassis systems are set in such way, that the mechanical vibrations acting on the driver are kept as low as possible in the area of most sensitive human perception. This improves the well-being of the occupants. In this case, all three air chambers of the air spring are activated and the force request in the damper is set low by a small damping constant in the controller. At the other extreme, in *Sport Plus Mode*, the driver experiences the best performance at an adequate of comfort. In this case, two chambers of the air spring are deactivated and the force request of the damper is maximising by a large damping constant. Measurements were carried out to analyse the different chassis modes. FIG. 5 shows the normalised body acceleration on comfort road in the vertical direction as a function of the frequency.

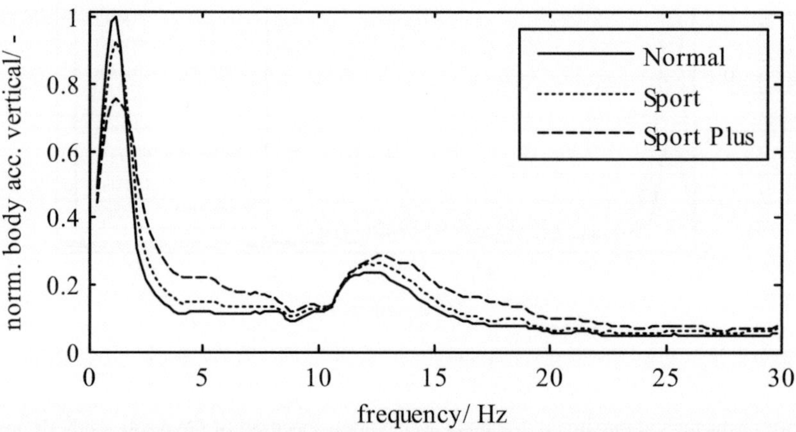

Figure 5: Normalised vertical body acceleration of three chassis modes

The comparison of the measurements shows, that the accelerations in the natural frequency of the body in *Normal Mode* are greatest. In contrast, it is obvious, that the accelerations in the Sport Plus mode are greatest around the natural frequency of the wheel and in the isolation range. With the frequency weighting curve, according to [8], a weighting of the measured accelerations can subsequently be made, in order to make a subjective assessment of the ride comfort of the vehicle. FIG. 6 shows the frequency weighting curve for the seated posture with the measuring points on the seat in the vertical direction.

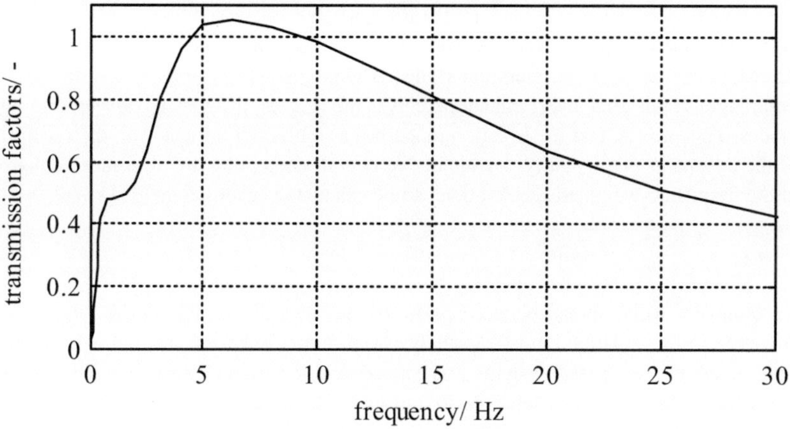

Figure 6: Frequency weighting curve for the seated posture [8]

The frequency-weighted acceleration $a_w(t)$ at the point of entry can be used to make a statement about the subjective comfort perception of the human being. For this purpose, the accelerations occurring at the seat rail during the comfort measurement were recorded to analyse the three chassis modes and weighted with the frequency weighting curve. FIG. 7 shows the normalised acceleration of the seat rail in the vertical direction on the left and the described weighted acceleration is plotted on the right.

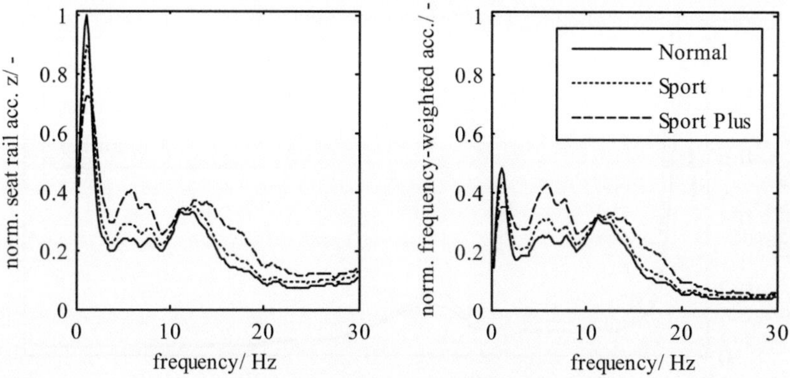

Figure 7: Normalised seat rail and frequency-weighted acceleration

Based on the frequency-weighted acceleration, it can be observed that comfort perception is the best in *Normal Mode*. In *Sport* or *Sport Plus Mode*, the comfort perception decreases due to higher acceleration values. The largest deviation occurs in the isolation range. In this area, the occupant perceives the greatest difference between driving modes. This proves, that the isolation range has a significant influence on the comfort of the occupants. To quantify the subjective overall perception, the root-mean-square of the frequency-weighted acceleration can be calculated according to [7] [8], by

$$a_{wT} = \sqrt{\frac{1}{T}\int_0^T a_w{}^2(t)\mathrm{d}t} \ . \tag{4}$$

To achieve a wider spread between performance and ride comfort in the future, the deviation in the isolation range was analysed more precisely. For this purpose, an influence analysis of chassis systems was carried out by parameter variation. The aim is to reduce the occurring accelerations to improve ride comfort.

2.2.1 Influence of air spring stiffness

To determine the influence of the stiffness of a multi-chamber air spring on the isolation range, further ride comfort measurements were carried out. For this purpose, the number of switched chambers was varied. The damping constant of the damper was set to the minimum value and kept constant to have the smallest possible influence of the damper. FIG. 8 shows the results of the investigation. It can be seen, that the stiffness of an air spring has no significant influence on the isolation range. Therefore, the air spring stiffness is not the cause for the occurring deviation from FIG. 7.

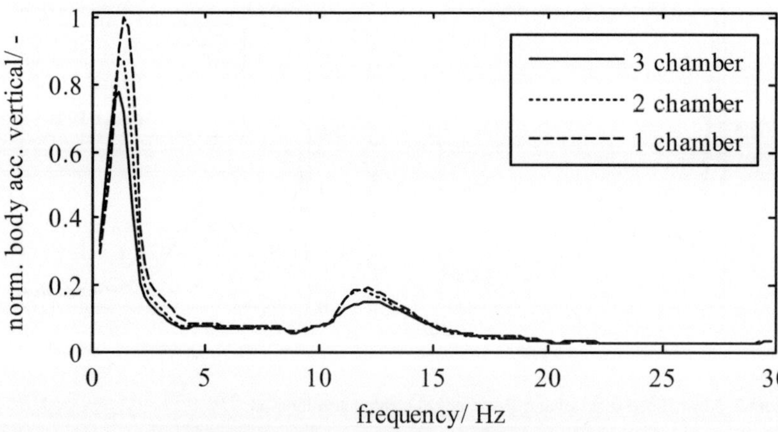

Figure 8: Normalised body acceleration vertical of three air spring modes

2.2.2 Influence of damping constant

The influence of the damping constant of the damper on the isolation range was investigated afterwards. This factor was varied by the damping controller. In this investigation, only one chamber of the air spring was activated. With reference to FIG. 9, where the modal acceleration of the body and the frequency-weighted acceleration on the seat rail is shown, it is recognizable, that the isolation range is deteriorated by a hard damper. Due to the high damping constant of the hard damper it can be stated that the damping characteristics of the damper have a large influence on the isolation range and is significant for the deviation from FIG. 7. In addition, it should be noticed, that a hard damper causes smaller accelerations in the range of natural frequency of body control. This favours the body control and thus the performance of the vehicle as well.

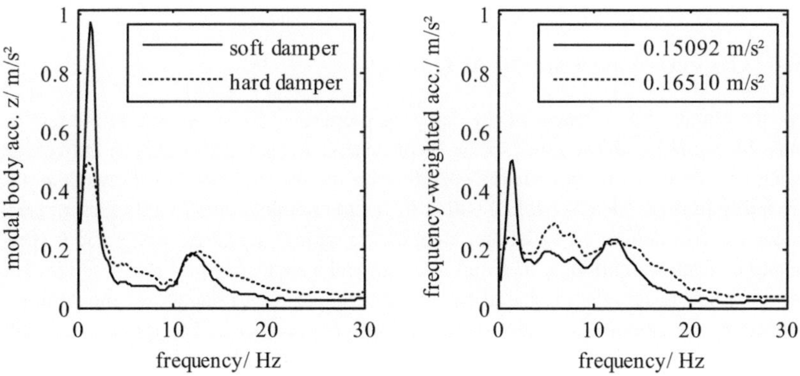

Figure 9: Modal body acceleration and frequency-weighted acceleration of seat rail

3 Damping constant of air spring as a comfort criteria

The influence analysis shows, that the damping characteristic of the damper has a significant influence on ride behaviour. A large damping constant improves on the one hand the performance by an improved body control. On the other hand, the ride comfort is deteriorated by higher accelerations in the isolation range. This interaction is described in the literature as "invariant points" and it is the reason that the damping constant of the damper cannot be designed as small as desired [9] [10].

The optimisation of the dynamic characterisation of an air spring is another possibility to improve ride comfort. In [1], it has already been explained, that an air spring has in addition to the springy also frictional and damping properties. The damping characteristic of the air spring can be reduced by a targeted change of its geometric parameters. A distinct description of the dampening characteristic of air springs, by using chassis test rigs, can support the development process by a corresponding target area definition.

This chapter presents the influence of damping characteristics of air springs for the ride comfort of a full vehicle. For this purpose, three different air springs were measured on the test rig according to method [1] and in the full vehicle afterwards. The investigated air springs have the following geometric parameters:

	Rolling piston diameter	Number of threads in bellows
• design A:	111 mm	650
• design B:	113 mm	750
• design C:	111 mm	750

3.1 Objective assessment at test rig

For the objective assessment of the damping properties, the air spring on the test rig must be excited with the same spring displacement, which also occurs in the vehicle when driving on a comfort road. Depending on the driving speed of the vehicle, the air spring in the vehicle is excited with a different combination of amplitudes and frequencies. To examine the isolation range, the air springs are tested with various combinations from an amplitude of 1 mm to 7 mm and a frequency of 4 Hz to 10 Hz. The resulting spring force is measured by a load-cell. For the evaluation of the damping properties the damping constant of the air spring is required. For this purpose, the damping work is calculated by numerical integration of the force-displacement hysteresis curve:

$$W_D := \int_0^T F\,ds \tag{5}$$

The damping work describes the dissipated energy of a system. Considering the quotient of damping work to the excitation amplitude and the averaged speed, a damping constant of an air spring can be calculated:

$$d_F = \frac{W_D}{2 \cdot s \cdot \bar{v}} \tag{6}$$

The calculated damping constant is suitable to describe the damping characteristics of a nonlinear system. For the comparison of the three air springs, the spring damping constant can be illustrated as a function of the excitation amplitude and frequency, as depicted in FIG. 10.

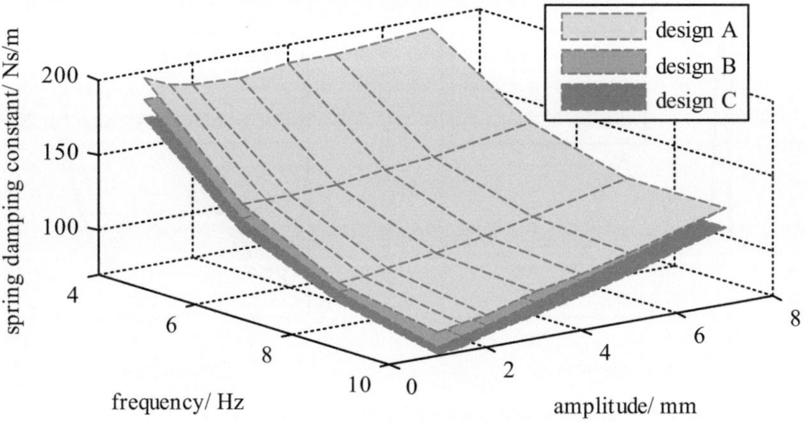

Figure 10: Spring damping constant depends on frequency and amplitude

It is recognizable that the three air springs have different damping characteristics due to their design features. The air spring "design A" has the largest damping characteristics over the entire characteristic map. For this reason, it can be assumed that this air spring causes the greatest accelerations at the body in the isolation range on the vehicle body. In contrast to this the air spring "design C" due to the lowest values would have to cause minor accelerations and thus promote ride comfort.

3.2 Influence on ride comfort

To investigate the influence of the damping behaviour of an air spring on the driving comfort, the three air springs were also measured in the full vehicle. The analysed air springs are developed for the rear axle of the vehicle. For this reason, only the air springs of the rear axle were varied during the measurement and of the front axle was not changed. The frequency-weighted acceleration at the seat rail can be used to assess subjective perception. The result is shown in FIG. 11.

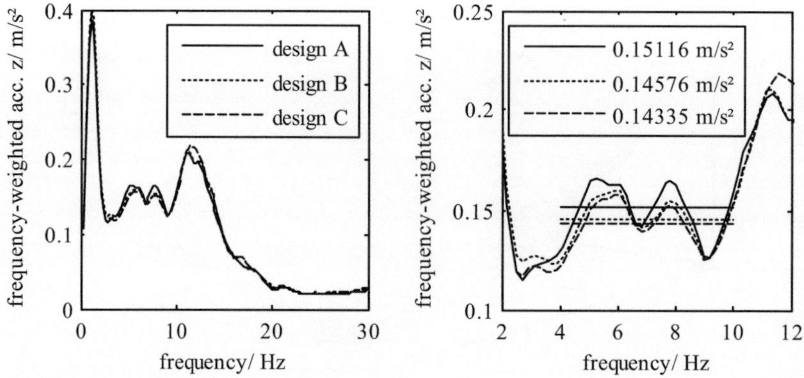

Figure 11: Seat rail frequency-weighted acceleration

It can be seen, that the analysed air springs of the rear axle cause different accelerations on the seat rail in the isolation range area. As already indicated by the test rig measurements and by the calculation of the spring damping constants, the air spring "design A" causes the largest accelerations due to its great damping characteristic. A root-mean-square of 0.15116 m/s² in the range of 4 Hz until 10 Hz can be determined on the seat rail. In comparison, the air spring "design C" has the smallest spring damping constant by its geometrical feature and causes the smallest effective value of 0.14335 m/s². This corresponds to an improvement of 5.45% compared to the air spring "design A".

Thus, it becomes clear, that only a small change in a geometrical parameter of the air spring affects the full vehicle behaviour and thus the subjective perception of the occupants. The ride comfort can be improved by reducing the damping behaviour of the air spring. Furthermore, it was shown that the determined spring damping constant on the test rig correlates with the root-mean-square value of the full vehicle measurement.

4 Summary

Due to the increasing demands on driving behaviour, Porsche is increasingly using mechatronic systems. To achieve the desired driving behaviour during the development of the vehicle, the characteristics of many mechatronic systems need to be fine-tuned to optimize the overall function of the system. This demands an iterative development process and large amount of testing activities is required. Objective Methods provide the potential to straight support the subjective tuning process by corresponding target area definition.

In this paper it has been shown, that the damping property of an air spring has a relevant influence on the ride comfort of a vehicle and the subjective perception of the passengers. For this purpose, the damping property has been quantified by the calculation of the spring damping constant at the test rig, using three different air spring designs. Afterwards, the air springs have been measured in the full vehicle. It has been found that a smaller spring damping constant of an air spring favours the isolation range and thus improves the overall riding comfort.

With this knowledge, a preliminary examination can be carried out in the early development phase in the future. The number of tests in the vehicle can be reduced by means of a pre-selection of air spring variants. This provides a potential to strengthen the development process by improving system understanding and reducing development time and costs. In addition, the presented method can contribute to achieve a maximum spread between performance and ride comfort.

Bibliography

[1] E. Boyraz, C. Kandler, M. Gantikow and D. Schramm, "Investigation of methods for the objective assessment of the dynamic system behaviour of air springs," in *17. Internationales Stuttgarter Symposium*, Stuttgart, Springer Fachmedien Wiesbaden, 2017, pp. 593-609.

[2] M. Gantikow , E. Boyraz, N. Kallert and R. Legierski, "The new multi-chamber air spring by Porsche – future innovation in chassis mechatronics and integration," in *7th International Munich Chassis Symposium 2016*, Munich, Springer Fachmedien Wiesbaden, 2016, pp. 437-456.

[3] M. Eisenbarth, G. Boisdequin and F. Gauterin, "Consistent assessment of chassis relevant ride comfort on roads and test rigs," in *14. Internationales Stuttgarter Symposium*, Stuttgart, Springer Fachmedien Wiesbaden, 2014, pp. 41-53.

[4] B. Heißing, M. Ersoy and S. Gies, Fahrwerkhandbuch, München, Lemförde, Wolfsburg: Vieweg + Teubner Verlag, 2011.

[5] M. Kosfelder, "Die Objektivierung des Schwingungskomforts von Kraftfahrzeugen auf der Grundlage globaler Bewegungsformen," in *14. Aachener Kolloquium Fahrzeug- und Motorentechnik*, Aachen, fka – Forschungsgesellschaft Kraftfahrtwesen mbH, 2005, pp. 1631-1644.

[6] T. Henn, M. Eisenbarth and D. Schramm, "Model-Based Calibration of an Adaptive Shock Absorber by Objective Comfort Criterions," in *24th Aachen Colloquium Automobile and Engine Technology*, Aachen, fka – Forschungsgesellschaft Kraftfahrtwesen mbH, 2015, pp. 807-818.

[7] ISO 2631-1, *Mechanical vibration and shock – Evaluation of human exposure to whole-body vibration – Part 1: General requirements,* Geneva, Switzerland, 1997.

[8] VDI 2057-1, *Einwirkung mechanischer Schwingungen auf den Menschen Ganzkörper-Schwingungen,* Düsseldorf, 2002.

[9] J. Hedrick and T. Butsuen, "Invariant properties of automotive suspensions," in *Proceedings of the Institution of Mechanical Engineers, Part D: Journal of Automobile Engineering*, California, Massachusetts, Sage Publications, 1990, pp. 21-27.

[10] D. Karnopp, "How significant are transfer function relations and invariant points for a quarter car suspension model?," in *Vehicle System Dynamics*, California, Taylor & Francis Group, 2009, pp. 457-464.

Subjective testing of a torque vectoring approach based on driving characteristics in the driving simulator

Fridrich, Alexander (IVK)

Nguyen, Minh-Tri (IVK)

Janeba, Anton (IVK)

Dr. Krantz, Werner (IVK)

Neubeck, Jens (FKFS)

Prof. Dr. Wiedemann, Jochen (IVK/FKFS)

P.E. Pfeffer (Ed.), *8th International Munich Chassis Symposium 2017*,
Proceedings, DOI 10.1007/978-3-658-18459-9_19

Abstract

The increasing electrification of the vehicle's drive train leads to the development of new active systems to provide specific driving characteristics. In this context torque vectoring constitutes an influential system. Within this study a control approach based on a specification of driving characteristics is presented. This approach does not influence the vehicle behavior by changing single control parameters, but uses specific vehicle characteristics as the target behavior. The target behavior, for example, is specified by a reference model which shows the desired characteristics.

This work shows the implementation of such a torque vectoring controller. The structure comprises a summary of several approaches from literature which promises to generate different driving characteristics in terms of safety and agility. A control allocation is integrated to handle the desired target behaviors and ensures an optimal distribution of the drive torques. The calculated distribution of the torques results in a necessary yaw moment to meet the desired vehicle characteristics. Additional weighting factors enable an emphasis between different vehicle characteristics.

This torque vectoring control approach is integrated within an overall vehicle simulation environment. In order to evaluate the subjective impression of the vehicle's target behavior, the whole simulation environment is implemented in the Stuttgart Driving Simulator. This implementation makes it possible to test the controller functions with regard to the driver's impression. Since the driver is in the loop, changes to the controller are immediately perceptible and the development engineer has the ability to tune the model parameters while driving. Moreover, the subjective evaluation data generated in the driving simulator can be used to establish a correlation to objective data from the simulation of the vehicle with torque vectoring control.

Introduction

In times of the vehicle's electrification the development of the driving dynamics becomes more and more sophisticated. The integration of electric engines allows a wide range of new possibilities to optimize the vehicle dynamics. As a consequence, new active systems are realized to generate desired driving characteristics in any driving situation. Among other active systems torque vectoring is a powerful system to realize this purpose [17]. This work shows the development and structure of such a torque vectoring controller. The implementation of the torque vectoring control approach in this paper uses tire longitudinal forces induced by the vehicle's wheel individual motors which is referred to as *Direct Yaw Moment Control (DYC)* [13]. The aim of the torque vectoring system is to generate a desired target vehicle behavior which is specified by the driving characteristics of a reference model. The reference vehicle is a single-track model which represents vehicle lateral dynamics in the horizontal plane, expressed by the states yaw rate $\dot{\psi}$ and vehicle side slip angle β [12]. The modification of the axle cornering stiffnesses leads to the desired lateral dynamics behavior and allows the specific parameterization which generates the desired agility or safety.

A distinguished feature of Direct Yaw Moment Control is its ability to generate a desired yaw moment even if the lateral forces are nearly saturated [17]. This is an advantage in contrast to other active systems to influence vehicle lateral dynamics like active steering [10]. From Shibahata et al. [13] it is known that vehicle stability in a certain driving state is not only determined by the yaw rate but also by the vehicle side slip angle and its time derivative. Since the states $\dot{\psi}$ and β are always coupled within the vehicles differential equations in the horizontal plane, one cannot be controlled independently [10]. Therefore in the control approach described in this paper both a feedforward and feedback strategy is proposed: The feedforward controller aims to tend the vehicle side slip angle towards a small magnitude whereas the feedback controller computes a yaw moment to track a desired yaw rate. The aggregation of the two control outputs results in a total yaw moment and guarantees a stable tracking of the driving characteristics-based reference model.

As a tool to handle the overactuated Direct Yaw Moment Control system discussed in this paper, control allocation strategies are applied. Advantages are to integrate not only primary control objectives like the generation of a certain vehicle yaw moment or longitudinal force but also to consider further secondary objectives. These secondary objectives comprise, e.g., actuator limits or the tire workload usage which describes the ratio between the effective and the maximum horizontal tire force. The implemented allocation strategy in this paper is optimization-based and refers to existing allocation schemes like in [10, 15]. The torque vectoring control system is implemented in the Stuttgart Driving Simulator (Fig. 1). This allows evaluating the control functions in virtual test drives. Compared to real track testing quick changes between controller configurations are possible which also supports the process of subjective evaluation. This advantage makes the Stuttgart Driving Simulator an effective development tool for analyzing active system like the torque vectoring approach mentioned above.

Fig. 1: Stuttgart Driving Simulator

For a real world implementation of the control approach presented here, many control strategies are suitable. Examples are adaptive controllers, model predictive controllers and sliding mode controllers [10]. They are not chosen here because the intention in this work is to point out the advantages of using virtual road testing for a driving characteristics-based controller development. For this purpose robustness is less important and an easy to use control concept with respect to implementation and parameterization is best suited. De Novellis et al. show in [1] that PID controllers are proper solutions for the formulated request.

Target Vehicle Model

The implemented vehicle model calculates the vehicle movements which are rendered by the motion system of the simulator. The rendering of motion has a significant influence on the driver's impression in the simulator. To achieve an optimal impression, the model requires a suitable complexity to describe all necessary vehicle movements [11]. For this investigation an IPG CarMaker vehicle model is used. It consists of a rigid vehicle body which is linked to four unsprung masses. Kinematics, elastokinematics and suspension force elements base on characteristic diagrams. The wheel forces are calculated by Magic

Formula 6.1. Additionally, the model uses the mechanical module of the Pfeffer steering model to describe spring and damping of the non-linear steering system.

The calculated vehicle movements are realized without scaling factors by the simulator's motion system. The motion cueing algorithm only uses tilt coordination technique at quasi stationary cornering. This motion cueing setup guarantees a nearly identical realization of the computed vehicle motion and the simulator motion and produces the most realistic impression of the vehicle behavior.

To provide the individual wheel torque distribution, the torque vectoring controller based on Matlab/SIMULINK is implemented by the CarMaker Plug-in model interface and replaces the standard drive train. The four calculated wheel torques are directly applied on the drive shafts. Using this interface developed controller functions can be implemented and verified in the virtual test drive. The effects of several control algorithms are immediately perceptible and assessable by the driver.

Torque Vectoring Controller

In the following the holistic control structure of the implemented torque vectoring system is shown which is explained in detail in the subsequent description.

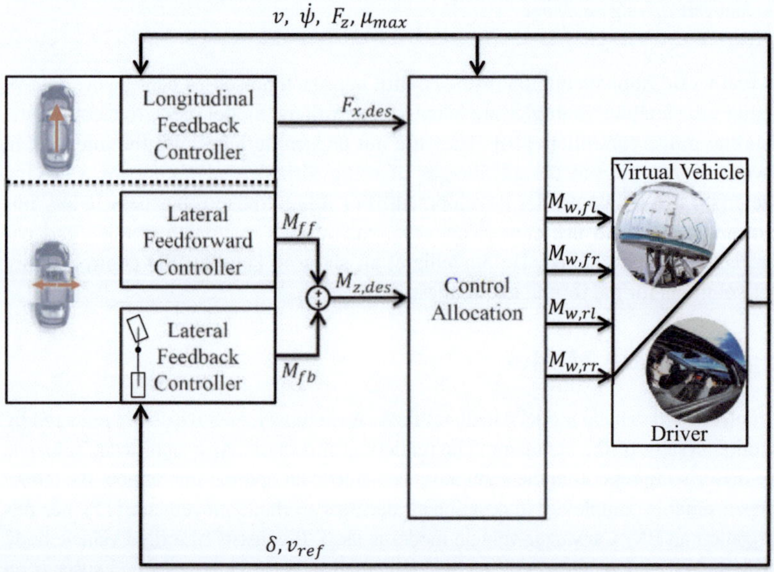

Fig. 2: Overall control structure of the implemented torque vectoring system

Design of feedforward controller

The design of the feedforward controller is based on a linear single-track model. The single-track model's transient behavior has to match the target vehicle's transfer functions with regard to the yaw rate and the side slip angle. To achieve these requirements the parameters of the single-track model are identified by an evolution based optimization algorithm technique. According to [6] and [14] the Laplace transformation of the differential equations for the vehicle side slip angle in the center of gravity for a linear single-track model can be concluded as

$$\beta(s) = \frac{(e_1 s - e_1 a_{22} + e_2 a_{12})\delta(s) + b_2 a_{12} M_{ff}(s)}{(s - a_{11})(s - a_{22}) - a_{21} a_{12}}. \tag{1}$$

In equation (1) the Laplace variable is denoted s, the front steering angle δ and the feedforward yaw moment M_{ff}. The left over factors are mathematical combinations of parameters of the single-track model, i.e.:

$$\begin{aligned} a_{11} &= -\frac{c_{sf}+c_{sr}}{mv},\ a_{12} = -1 - \frac{c_{sf}l_f - c_{sr}l_r}{mv^2}, \\ a_{21} &= -\frac{c_{sf}l_f - c_{sr}l_r}{\theta_z},\ a_{22} = -\frac{c_{sf}{l_f}^2 + c_{sr}{l_r}^2}{v\theta_z}, \\ e_1 &= \frac{c_{sf}}{mv},\ e_2 = \frac{c_{sf}l_f}{\theta_z},\ b_2 = \frac{1}{\theta_z}. \end{aligned} \tag{2}$$

The equations (2) contain the vehicle mass m and the moment of inertia about the inertial vertical axis θ_z, the vehicle's absolute velocity v, the distance from the center of gravity to the front and rear axle l_f and l_r as well as the cornering stiffness of the front and rear axle c_{sf} and c_{sr} of the single-track model.

Now, the feedforward yaw moment gain G_{ff} is introduced, according to

$$M_{ff}(s) = G_{ff}\delta(s). \tag{3}$$

Inserting equation (3) into equation (1) and taking the condition of a vanishing vehicle side slip angle in steady state as a basis, the yaw moment gain can be expressed as

$$G_{ff} = c_{sf}\frac{c_{sr}l_r l_f - l_f mv^2}{mv^2 + c_{sf}l_f - c_{sr}l_r}. \tag{4}$$

Regarding the last two equations, the feedforward yaw moment demand gets:

$$M_{ff} = c_{sf}\frac{c_{sr}l_r l_f - l_f mv^2}{mv^2 + c_{sf}l_f - c_{sr}l_r}\delta \tag{5}$$

Design of driving characteristic-based feedback controller

A PI-feedback controller is used to control the yaw rate of the target vehicle. The following equation represents the calculation of the virtual control output M_{fb} which is a demand for the later described control allocation system to realize the desired yaw movement $\dot{\psi}_{des}$ [2]

$$M_{fb} = K_P(\dot{\psi}_{des} - \dot{\psi}) + K_I \int_0^t (\dot{\psi}_{des} - \dot{\psi})\, d\tau. \tag{6}$$

With the design of the PI-controller in equation (6) an easy to implement and parameterize controller is used to track a desired yaw rate with a zero steady-state error. The desired yaw rate is a result of the subsequently explained reference yaw rate generator. Since the two states of the vehicle's horizontal movement – the yaw rate and the vehicle side slip angle – are not independently controllable with torque vectoring actuators, in this paper the feedback loop only controls $\dot{\psi}$. A small β is ought to be achieved via the feedforward control output M_{ff} which is a method among further others, e.g. depicted in [1]. Both, the feedback control output M_{fb} and the feedforward output M_{ff} are added to get the desired yaw moment which is transferred to the control allocation system, i.e.

$$M_{z,des} = M_{fb} + M_{ff}. \tag{7}$$

The generation of the desired yaw rate is based on the work of Graf [4]. This method uses differently parameterized axle cornering stiffnesses of the linear single-track model. The variation of axle cornering stiffnesses results in different self-steering behavior. Graf [4] divides the self-steering behavior in three sections depending on the lateral acceleration. The first section ($a_y \leq 4$ m/s²) represents the agility of the vehicle. In this section the axle cornering stiffnesses are determined in a way to obtain a ration between rear and front axle cornering stiffness of 1.5. The third section (6 m/s² < a_y < 7 m/s²) is responsible for the stability of the vehicle. Here the ratio of axle cornering stiffness is increased to 2.0 for stability reserve aspects. For lateral acceleration states in between (4 m/s² < $a_y \leq 6$ m/s²) a weighting factor scheme defines a convex interpolation between the cornering stiffnesses of the linear and nonlinear driving states [4]. It has to be taken into account that this process is primarily defined by the steady-state self-steering behavior. Nevertheless, the changing of the model's axle cornering stiffness also affects the dynamic behavior of the reference model [4,6].The real-time integration of the differential equations of the parametrized and state-dependent reference single-track model results in the desired yaw rate $\dot{\psi}_{des}$.

The control structure and the driven maneuvers (see section "Virtual Test Drive and Results") require a constant vehicle velocity. Therefore, keeping the focus on the lateral dynamics of the controlled closed-loop system, in the conducted simulator experiments a simple PI-controller is used to keep the vehicle velocity constant.

Control Allocation

A useful tool to combine different demands of overactuated systems is control allocation. With control allocation techniques it is possible not only to instantiate primary but also secondary objectives and to meet actuator restrictions. Control allocation strategies separate the actuator choice problem from the controller design problem [5]. A way to formulate both the primary and secondary objectives in one mathematical description is a scalarized, weighted sum optimization problem [5], like in equation (8). The primary objectives describe the concordance between the virtual control output $v_c(t)$ and the true control output $u(t)$. The secondary objectives can be characterized by p-normed differences of the true control output $u(t)$ and a desired control output $u_d(t)$, i.e.,

$$u_w = arg\ \min_{\underline{u} \leq u \leq \bar{u}} \left(\gamma \left\| W_v \left(B_{eff} u - v_c \right) \right\|_p^p + \| W_u (u - u_d) \|_p^p \right). \tag{8}$$

Equation (8) supposes a linear relation between the true and the virtual control output via the control effectiveness matrix B_{eff} [5]. By choosing the constant γ large, the focus on the primary control objectives can be emphasized. In the same way the weighting matrices W_v and W_u provide the opportunity to balance the optimization problem.

To meet real-time demands, the control allocation problem is formulated as a convex optimization problem and further a convex quadratic one, i.e.

$$\min_{-\mu_{max} F_z \leq F_{x,opt} \leq \mu_{max} F_z} \left(\left\| W_{xz} \left(B_{xz} F_{x,opt} - \begin{pmatrix} F_{x,des} \\ M_{z,des} \end{pmatrix} \right) \right\|_2^2 + F_{x,opt}{}^T\, Q\, F_{x,opt} \right), \tag{9}$$

$$B_{xz} = \begin{bmatrix} \cos(\delta_{fl}) & \cos(\delta_{fr}) & 1 & 1 \\ -\cos(\delta_{fl}) \frac{d_f}{2} & \cos(\delta_{fr}) \frac{d_f}{2} & -\frac{d_r}{2} & \frac{d_r}{2} \end{bmatrix}. \tag{10}$$

Where $W_{xz} \in \mathbb{R}^{2x2}$ is the positive semidefinite diagonal weighting matrix of the vehicle longitudinal force and the yaw moment, $B_{xz} \in \mathbb{R}^{2x4}$ is the effectiveness matrix with steering angles at front left an front right wheels δ_{fl}, δ_{fr} as well as the vehicle's front and rear track width d_f, d_r. The term $F_{x,des} \in \mathbb{R}^{1x1}$ comprises steering angle caused contributions of lateral tire forces to the vehicle's longitudinal motion. The secondary objective of the control problem is depicted through the second quadratic term in equation (9) which contains the elements of the positive semidefinite matrix $Q \in \mathbb{R}^{4x4}$ as

$$q_{ii} = \frac{w_{\mu,i}{}^2}{\left(\mu_{max,i} F_{z,i}\right)^2} \text{ and } q_{ij} = 0 \text{ for } i = [fl, fr, rl, rr],$$

to denote the corresponding wheel. The denominators of q_{ii} are composed of the maximum friction coefficient $\mu_{max,i}$ and the vertical tire load $F_{z,i}$.

The secondary objective describes the tire workload usage on every wheel of the vehicle. The priority for minimizing the wheel individual tire workload usage can be manipulated by the corresponding weighting factors $w_{\mu,i}$. The set of feasible control outputs is chosen as the highest transmittable horizontal friction force between the road and the tire.

The outstanding characteristic of convex functions like equation (9) is that any locally optimal solution is also globally optimal, with an amount of zero, a unique or an infinite numbers of solutions [5]. To solve the custom problem stated in equation (9) in real-time, the software tool CVXGEN is used. It is based on a standard primal-dual interior-point method with Mehrotra correction to solve the optimization problem efficiently, robustly and reliably, guaranteeing accurate results within 5-25 iterations [8].

Equation (9) concerns multi-objective optimization and therefore results depend on how the weighting factors in W_{xz} and Q are chosen. Since the objective function comprises elements of different magnitudes an efficient normalization method according to Mausser et al. [9] is implemented. This method depicts approximations of Pareto Minima and Maxima for each element of the cost function by finding the extrema of a random sample within the problem's constraints and uses the absolute value of the difference between the individual Pareto extrema to normalize. Since studies show that the best results in weighted sum optimization problems are achieved by the basis of a convex combination of cost functions [7], the weighting factors in equation (9) sum up to 1.

The resulting longitudinal forces of the optimization process $F_{x,opt,i}$ define the moments acting on the every i-th wheel of the driving simulator vehicle, i.e.

$$M_{w,i} = F_{x,opt,i} r_{dyn}, \tag{11}$$

where r_{dyn} describes the dynamic radius of the wheel.

Virtual Test Drive and Results

The test drive takes place on a virtual Autobahn track with constant velocities of 80 km/h and 180 km/h. Within two lanes maneuvers like weave tests, lane changes and ramp steers are performed to evaluate the agility and the steering behavior of the vehicle. To analyse the influences of different control configurations in the overall vehicle model the following variants are defined:

Tab. 1: Vehicle control configurations

Variant	Long. Control	Lat. Feedforward Control	Lat. Feedback Control
A	ON	OFF	OFF
B	ON	ON	OFF
C	ON	OFF	ON
D	ON	ON	ON

Vehicle Configuration A induces wheel individual different torques only for keeping the vehicle velocity constant which does not result in a yaw moment around the inertial vertical axis. Therefore a transfer behavior almost identical to the base vehicle with standard drive train becomes apparent (Fig. 3). Like in every vehicle configuration the criteria of minimal tire workload usage are aspired. Fig. 7 shows the steering wheel angle versus the lateral acceleration which results from a quasi-stationary ramp steer maneuver with constant velocity. The thin line of variant A reveals a non-linear steering behavior at lower lateral accelerations. This is caused by the elasticity of the steering system.

The driving behavior of vehicle configuration B is subjectively more predictable and thus more controllable than variant A. The amplitude response of the vehicle side slip angle decreases, whereas the phase response increases in absolute value (Fig. 3). A less agile driving behavior arises since the yaw rate amplitude response is decreased and its phase response is increased, too. In Fig. 7 the dashed line of variant B shows an increasing effect of the non-linear steering behavior at lower lateral acceleration compared to variant A. This is due to the additional yaw moment of the feedforward control. In this exceptional example, the effect occurs in such a way that the vehicle's reaction leads to negative lateral acceleration which is contrary to the steering direction. Furthermore, the gradient of the steering wheel angle due to lateral acceleration increases in comparison with variant A.

Based on the feedback controller a variant C with higher agility compared to variant A is generated. For this variant the axle cornering stiffnesses of the reference single-track model are defined according to Graf [4] that the peak response time is reduced about 20 % compared to the base vehicle. As a consequence the controller forces the vehicle to track the reference model's yaw rate which is shown in Fig. 5. As expected, the vehicle C has a high magnitude of yaw rate while phase shift between steering wheel angle and yaw rate decreases (Fig. 5). The natural coupling between the side slip angle and the yaw rate also leads to a higher magnitude gain of the side slip angel's frequency response and a decreasing phase shift between the side slip angle and the steering wheel input compared to the base vehicle (Fig. 6). The subjective evaluation of variant C shows better agility than variant A. Especially at on-center handling the lateral vehicle reaction due to steering is too sensitive and devaluates the straight running behavior. This result is explained in Fig. 7. The non-linear behavior of the steering wheel angle at small lateral accelerations is eliminated by the feedback controller. The controlled vehicle hereby shows a linear self-steering behavior like the reference single-track model. This means that the driver needs less steering wheel input to achieve a certain lateral acceleration compared to variant A.

Vehicle configuration D is effectively a combination of the variants B and C which represents the characteristics of both configurations. The transfer behavior of the variants C and D in Fig. 5 and Fig. 6 shows a similar steady state behavior of the yaw rate and the side slip angle. The same assertion can be stated with regard to the self-steering gradient (Fig. 5). In comparison to vehicle configuration C, the amplitude amplification of D decreases at higher steering frequencies. This refers to the influence of the feedforward yaw moment. In the subjective evaluation the driving behavior of D shows a good controllability with a still high agility.

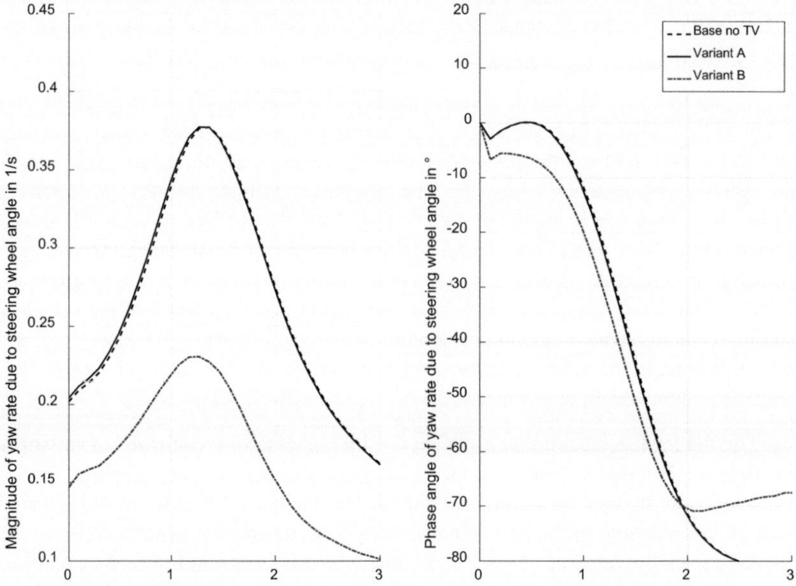

Fig. 3: Frequency response of the yaw rate of the base vehicle and the variants A and B generated by sine sweep maneuver with 180 km/h and maximum steering wheel angle at steady state lateral acceleration of 2 m/s².

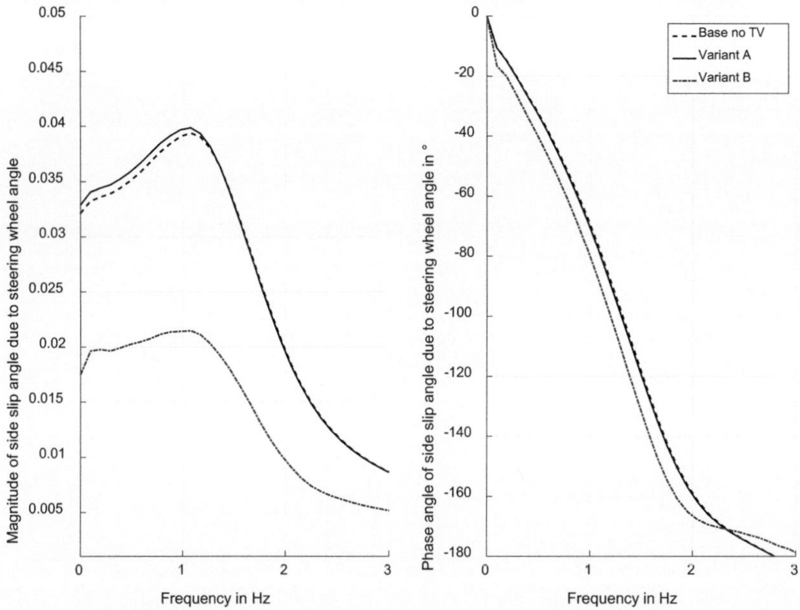

Fig. 4: Frequency response of the side slip angle of the base vehicle and the variants A and B at vehicle CG generated by sine sweep maneuver with 180 km/h and maximum steering wheel angle at steady state lateral acceleration of 2 m/s².

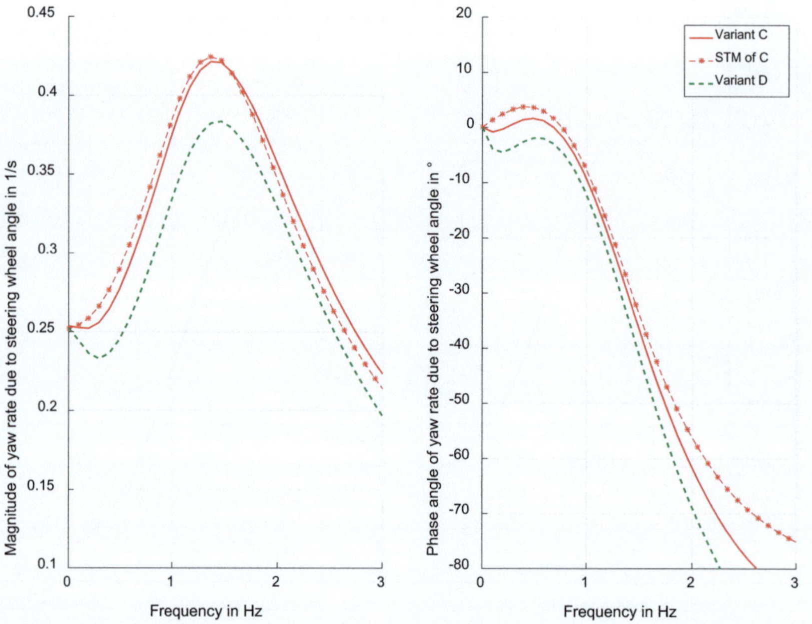

Fig. 5: Frequency response of the yaw rate of the variants C and D generated by sine sweep maneuver with 180 km/h and maximum steering wheel angle at steady state lateral acceleration of 2 m/s².

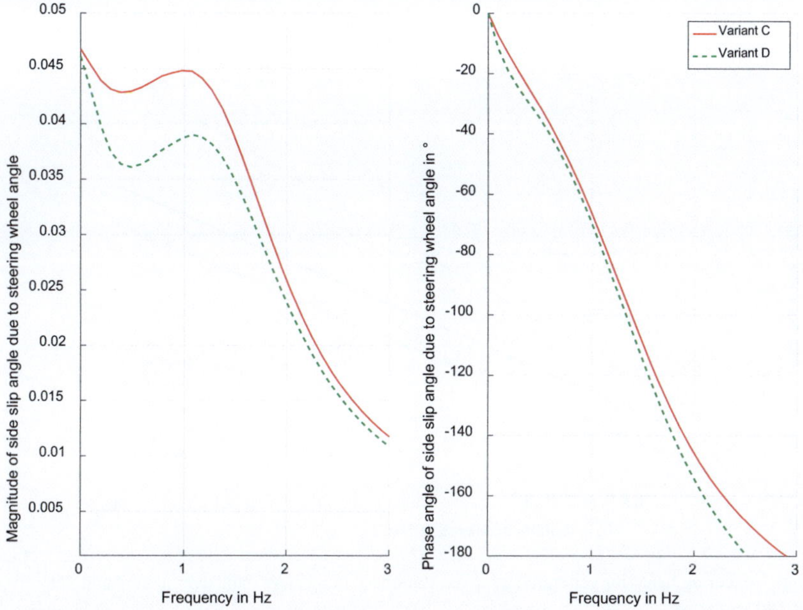

Fig. 6: Frequency response of the side slip angle of the variants C and D at vehicle CG generated by sine sweep maneuver with 180 km/h and maximum steering wheel angle at steady state lateral acceleration of 2 m/s².

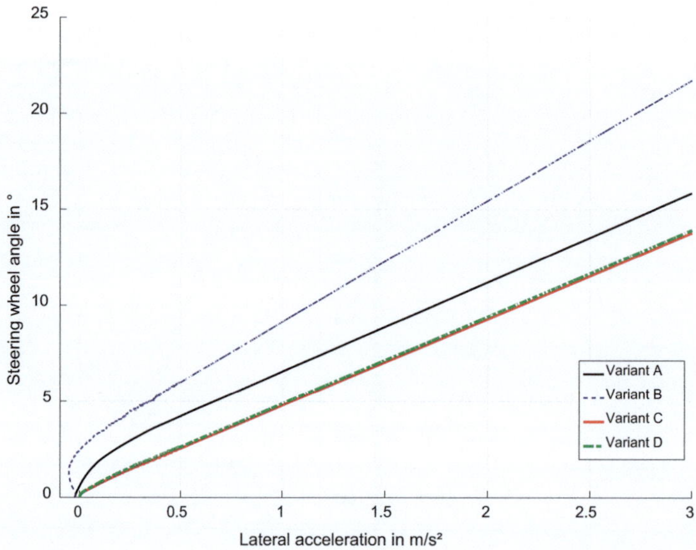

Fig. 7: Steering behavior generated by quasi stationary ramp steer maneuver at constant velocity of 180 km/h

Conclusion

This works illustrates the implementation of a driving characteristics-based torque vectoring system in the Stuttgart Driving Simulator. Test runs in the driving simulator generate subjective driving impressions of the torque vectoring controlled overall vehicle. The effect of the feedforward control is validated in the virtual test run. According to the preliminary results, the feedforward controller induces a reduction of the absolute side slip angle. Smaller angles cause a better controllability [6]. The PI feedback yaw rate controller realizes the tracking of reference model's driving characteristics. In this example the reference model is specified with a higher agility. This more agile behavior is subjectively confirmed in the simulator. The control allocation with the input of the combined yaw moment shows that the desired characteristics of both feedforward and feedback can be considered. With the help of subjective evaluation more detailed knowledge about pre-described control system is acquirable. This knowledge will have direct influence on the further development of the torque vectoring controller.

Literature

[1] De Novellis, Leonardo; Sorniotti, Aldo; Gruber, Patrick; Pennycott, Andrew: Comparison of Feedback Control Techniques for Torque-Vectoring Control of Fully Electric Vehicles. IEEE Transactions on Vehicular Technology, Volume 63, Number 8 (2014), pp. 3612-3623

[2] Dodds, Stephen J.: Feedback Control: Linear, Nonlinear and Robust Techniques and Design with Industrial Applications. Springer, London, 2015

[3] Esmailzadeh, Ebrahim; Goodarzi, Avesta; Vossoughi, Gholamreza: Optimal yaw moment control law for improved vehicle handling. Mechatronics, Volume 13, Issue 7 (2003), pp. 659-675

[4] Graf, Michael: Methode zur Erstellung und Absicherung einer modellbasierten Sollvorgabe für Fahrdynamikregelsysteme. Dissertation, Technische Universität München, 2014

[5] Härkegard, Ola: Backstepping and Control Allocation with Applications to Flight Control. Doctoral Thesis, Linköping University, 2003

[6] Heißing, Bernd; Ersoy, Metin; Gies, Stefan: Fahrwerkhandbuch: Grundlagen, Fahrdynamik, Komponenten, Systeme, Mechatronik, Perspektiven. 4. Auflage, Springer Vieweg, 2013

[7] Marler, R. Timothy; Arora, Jasbir S.: Function-transformation methods for multi-objective optimization. Engineering Optimization, Volume 37, Number 6 (2005), pp. 551-570

[8] Mattingley, Jacob; Boyd, Stephen: CVXGEN: A Code generator for Embedded Convex Optimization. Optimization Engineering, Volume 13 (2012), pp. 1-27

[9] Mausser, Helmut; Grodzevich, Oleg; Romanko, Oleksandr: Normalization and Other Topics in Multi-Objective Optimization. Proceedings of the Fields-MITAACS Industrial Problems Workshop, Toronto, 2006

[10] Mousavinejad, Eman; Zhu, Yong; Vlacic, Ljubo: Control Strategies for Improving Ground Vehicle Stability. 10th Asian Control Conference (ASCC), pp. 1-8, Kota Kinabalu, 2015

[11] Nguyen, M.-T.: Subjective Perception and Evaluation in the Virtual Test Drive, 17th Stuttgart International Symposium, 2017

[12] Riekert, Paul; Schunck, Theo: Zur Fahrmechanik des gummibereiften Kraftfahrzeugs. Ingenieur Archiv, Volume 11, Number 3 (1940), S. 210-224

[13] Shibahata, Yasuji; Shimada, Kazuhiko; Tomari, Tatsuhiro: Improvement of Vehicle Maneuverability by Direct Yaw Moment Control: Vehicle System Dynamics, Volume 22 (1993), pp. 465-481

[14] Shino, Motoki; Nagai, Masao: Yaw-moment control of electric vehicle for improving handling and stability. In: Society of Automotive Engineers of Japan JSAE Review, Volume 22 (2001), pp. 473-480

[15] Sun, Jinlin; Ding, Shihong; Zhang, Shengdao; Zheng, Wie Xing: Nonsmooth stabilization for distributed electric vehicle based on direct yaw-moment control. Proceedings of the 35th Chinese Control Conference, Chengdu, 2016

[16] Zanten, Anton T. van; Erhardt, Rainer; Pfaff, Georg: VDC, The Vehicle Dynamics Control System of Bosch. SAE Technical Paper 950759, 1995

[17] Zheng, Shuibo; Tang, Houjun; Han, Zhengzhi; Zhang, Yong: Controller design for vehicle stability enhancement. Control Engineering Practice, Volume 14, Issue 12 (2006), pp. 1413-1421

Objectification of steering feel and application in the context of virtual steering feel tuning

Stefan Grüner, Thomas Werner, Bartosz Käpernick

P.E. Pfeffer (Ed.), *8th International Munich Chassis Symposium 2017*, Proceedings, DOI 10.1007/978-3-658-18459-9_20

1 Introduction

Today steering feel tuning is typically a purely subjective process. In practice, this is an iterative procedure involving subjective tuning by tuning experts from the steering manufacturer, and subjective assessment by experts from the vehicle manufacturer. Due to the nature of subjective processes, in general, this often becomes a time consuming undertaking with moving targets.

The scenario above is in clear contrast to the efforts and advancements that are made in many other engineering disciplines. They move from real world testing to simulation-based development methods. The simulation-based methods render development more independent of sample availability and increase time efficiency. The key to these developments is twofold. First, it is the ability to simulate the relevant scenarios in sufficient detail. Second, it is the quantitative description of the desired state in terms of quantities that can be deduced from simulation results.

Looking at the two aspects with respect to steering feel tuning, this comes down to vehicle dynamics simulation with detailed steering gear models including the ECU (Electronic Control Unit) software and an objective description of steering feel. As for the first, .i.e. the simulation, it is clear that much progress has been made over the last years, and the needed simulation environments are becoming state of the art. Major simulation environment suppliers have established themselves in the industry. With respect to objective description of steering feel, the situation is considerably different. There are many metrics proposed in the literature, e.g. (1), and supported by in-vehicle testing and assessments. However, the subject is far from being standardized.

This paper aims at presenting our findings from steering feel objectification based on different OEM points of views and the deduced results that eventually point to the potentials for full Steer-by-Wire steering systems.

The paper is organized as follows. Section 2 gives a self-contained introduction to human perception of steering feel and the psychological background with respect to subjective steering feel assessment. On this basis, section 3 gives the results of a purpose designed subjective steering feel assessment campaign, performed recently. Section 4 then gives details about the performed objectification of steering feel and relates it to the results given in section 3. With this background, section 5 discusses the implications for stationary vehicle simulators in the context of steering feel assessment. This discussion eventually moves the interest to a process for steering feel tuning which is fully simulation based, as sketched in section 6. In this context a more detailed inspection of the results from section 4 points to an increased steering feel tuning potential for Steer-by-Wire (SbW) steering systems. Section 7 summarizes the main results and gives a perspective on future work in the context of the present paper.

2 Steering Feel and Human Perception

Before going into the details of the subjective assessment campaign and the objectification, this section begins by giving a common understanding of steering feel perception as well as steering feel assessment.

2.1 Definition of Steering Feel

As a starting point, a reasonable definition of the notion "steering feel" needs to be given. In (1) such a definition of steering feel is presented. It is given in Fig 1 and shall be explaind here in some detail to keep the paper self-contained.

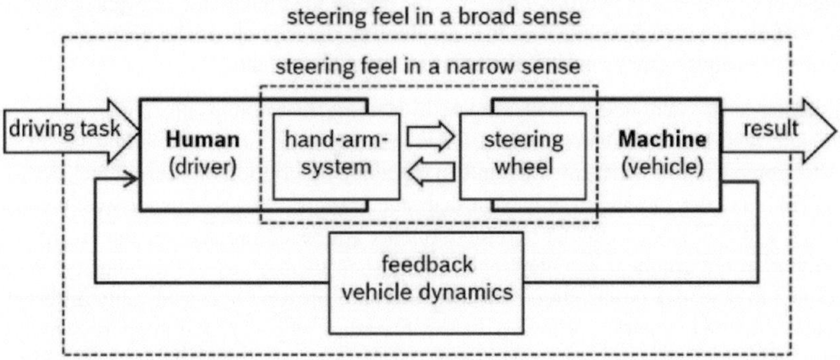

Fig. 1 Definition of steering feel: steering feel in a narrow sense and steering feel in a broad sense and their interrelation.

In (4) the authors propose a two layered definition of steering feel: steering feel in a narrow sense and steering feel in a broad sense.

2.1.1 Steering Feel in a narrow sense

The inner layer is the steering feel in a narrow sense and focuses on the interaction of the hand-arm system on the human side and the steering wheel on the machine side. In this setting steering feel is completely captured by steering wheel angle and steering wheel torque. Hence any steering feel characteristics related to the steering feel in a narrow sense may be deduced from analysing the relation between angle and torque. This means that steering feel in a narrow sense is completely unrelated to the vehicle, i.e. vehicle reactions to the driver inputs are irrelevant for the steering feel in a narrow sense.

2.1.2 Steering Feel in a broad sense

The steering feel in a broad sense encapsulates the steering feel in a narrow sense and moves the matter to a wider context. It does not only account for the hand-arm system and the steering wheel, but brings in the vehicle and its reaction to the driver input. In order to understand the meaning of the diagramm given in Fig. 1, it is convinient to take the external input driving task as a starting point. Based on the driving task compared to the present vehicle state, the human driver generates a steering input. This in turn results in a vehicle reaction which is perceived by the driver, which is then again put in relation to the driving task and so an.

Therefore, in summary, steering feel is the holistic perception of vehicle lateral dynamics, i.e. the sum of the visual, vestibular, haptic and acoustic sensing of a human driver while steering a vehicle.

2.2 Physiological aspects: human perception paths

From the preceeding paragraphs it should be clear that steering feel is about human perception. In Fig. 2 a human driver in a driver's seat holding a steering wheel is sketeched together with the human perception paths.

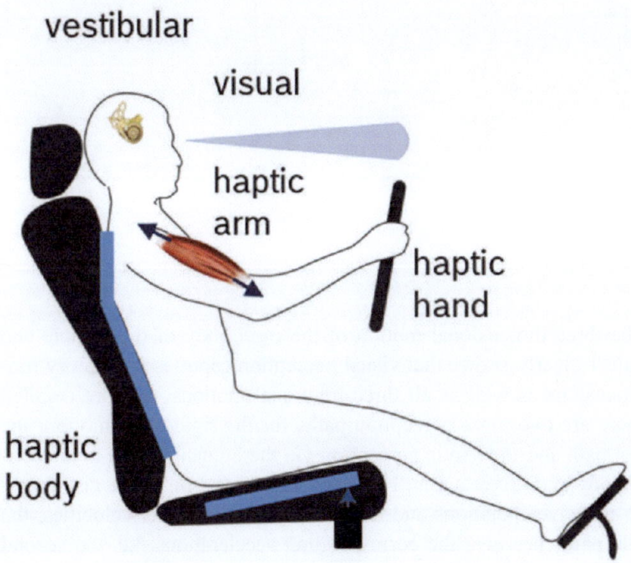

Fig. 2 Human perception path with relevance to vehicle guidance.

Several works exist in the literature that may facilitate a better understanding of human perception while performing lateral control of a vehicle (e.g. (1), (2), (3) and the references therein). A review of these helps in answering the question as to by what perception paths human perceives translatory and rotatory motion in all three space dimensions as well as torque and angle while operating a steering wheel. The results of the literature review are compiled in Table 1.

Table 1 Human ability to perceive 3d motion in general and steering wheel torque and angle while operating a steering wheel.

Motion		visual	vestibular	haptic			perception threshold
				hand	arm	body	
longitudinal	position/velocity	++					90 mm / 0.06 m/s
	acceleration		+		o	++	0.02-0.8 m/s² 1)
lateral	position/velocity	++					75 mm / 0.054 m/s
	accelleration		++		o	+	0.05-0.1 m/s²
vertical	position/velocity	++					215 mm / 0.145 m/s
	acceleration		++		o	+	0.02-0.05 m/s² 1)
yaw	angle/velocity	++					4 ° / 0.6-0.9 °/s
	acceleration	+	++				0.05-5 °/s²
roll	angle	++	+				-
	velocity/acceleration	+	++				0.2 °/s / 0.1-0.2 °/s²
pitch	angle	++	+				-
	velocity/acceleration	+	++				0.4 °/s / 0.1-0.2 °/s²
steering wheel	angle	+		o	++		0.2-2 °
	torque level/vibration			+	++		0.5-1.5 Nm / 0.05 Nm

++ … primary perception path,
+ … secondary perception path,
o … subordinate perception path.

Focusing first on the three dimensional motion of the rigid body of the vehicle and thus the driver, Table 1 clearly shows that visual perception captures translatory motion in all three dimensions as well as all three rotational motions. A more detailed look reveals that there are two more perception paths for the rigid body motion: the vestibular perception path and the haptic perception via the human body, i.e. the contact points between body and driver's seat. However, in comparison to the visual perception path, which perceives positions and their time derivatives, i.e. velocities, the vestibular and haptic paths perceive the corresponding accelerations, i.e. the second order time derivatives. From this fact, one may infer that the vestibular and haptic perception paths are faster in comparison to the visual path.

It is remarkable to note that for the translatory motion this is a strict separation in the sense that the visual path only perceives positions and velocities while the vestibular and haptic path only perceives accelerations. For the rotational motion, this is no longer true. While it is again the visual path acting as perception path for absolute values and first order time derivatives, the vestibular and haptic paths now take the role of a secondary perception path and vice versa for accelerations.

The dependencies sketched in the two preceding paragraphs will become relevant in the upcoming discussion of stationary vehicle simulators for steering feel tuning and assessment.

2.3 Psychological aspects: steering feel assessment

Before starting with the discussion of the psychological aspects of steering feel assessment, it is worthwhile to point out that despite the fact that the notion “steering feel” contains the word “feel” it is not a feeling from a psychological perspective. In contrast to well-known feelings such as e.g. fear, anger, joy, happiness and the like, steering feel is a perception, specifically, the holistic perception, of vehicle lateral dynamics characteristics.

On this basis, it is straightforward to realize that steering feel assessment is the quantitative rating of the perception of vehicle lateral dynamics characteristics. This is illustrated in Fig. 3 given below.

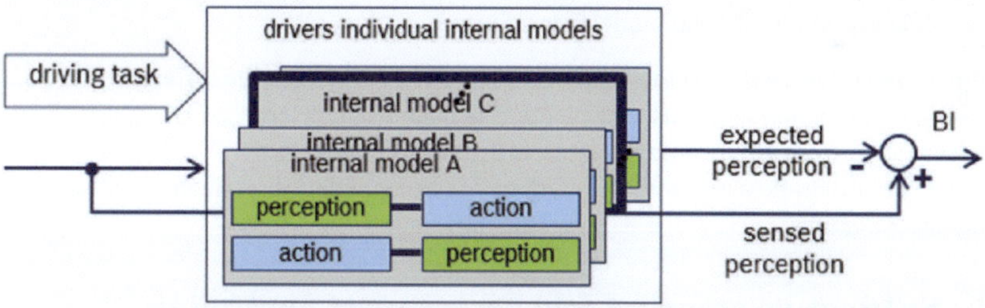

Fig. 3 Subjective steering feel assessment: Rating the difference between expected perception as expected on basis of some internal model and sensed perception.

The baseline for this quantitative assessment is the expected perception. It is produced by some internal model of the driver. The internal model can be thought of as being composed of two major parts. Based on the individual perception of the vehicle lateral dynamics and the driving task the driver performs some action, i.e. the model has mappings from perceptions to actions. In addition to that, the driver expects some reaction of the vehicle lateral dynamics to his actions; hence, the model also includes mappings from actions to expected perceptions. These two classes of mappings open the possibility that a perception is mapped directly to an expected perception. The act of steering feel assessment is hence the rating of the difference between this expected perception and the sensed perception.

In the Fig. 3 some important aspects of the internal models are given. The first important one is the fact that the internal models are individual models. Every driver has his very own internal models. This in turn means that it is unlikely that any two drivers will give the very same assessment of steering feel for some vehicle. Almost always, will there be differences, in the details at least. The second important aspect is the existence of a whole set of internal models. When assessing steering feel, the driver automatically chooses one model out of his set of models. Consider as an example situation booking a rental car. While booking the rental car, the process of selecting the internal model gets started. It is influenced by objects on the way to the rental station, by reception of the key, the visual impression while approaching the car, the weather, having plenty of time or being late and the like. All that builds up expectations, i.e. selects the internal model. Upon starting the trip, the way the car is rated, the underlying baseline, is given by the selected model.

The process of reselecting the internal model is initiated by large differences between expected perception and sensed perception, e.g. if some changes to a new vehicle it feels at first somehow different while after some time this perception has vanished. This can be interpreted as an outcome of the fact that some other internal model has been selected or the internal model was adjusted.

In the subjective assessment campaign presented in the sequel, the difference between expected perception and sensed perception is quantified in terms of BI values. These values range from 10, meaning perfect, to 1, meaning intolerable. A rating of 5 corresponds to borderline cases while a rating of 7 is acceptable for production without further investigations.

3 Subjective steering feel assessment campaign

The aim of the performed subjective steering feel assessment campaign is twofold. On the one hand, it serves to investigate in what aspects the steering feel of a range of production vehicles from different brands differ. On the other hand, it is to be shown if these aspects are the same no matter from what perspective the assessments were performed.

3.1 Structure of the assessment campaign

The vehicles to be investigated are made up by six vehicles that are currently supplied with steering gears produced by the company the authors are employed by. In the company's costumer oriented vehicle testing department, vehicle tuning and assessment experts are available that have many years of tuning and assessment experience with the relevant OEMs. Hence, they are considered valid proxies for real OEM tuning and assessment experts. In the sequel, therefore, OEM is written as a short form for these tuning and assessment experts. No assessments were done with assessment experts from the OEMs. However, for some of the OEMs, crosschecks with OEM tuning and assessment experts were performed. These crosschecks confirmed them as valid proxies, i.e. their internal models for steering feel may be assumed to be sufficiently close to those of the real OEMs.

To get a detailed understanding of the steering feel, an assessment sheet with more than 100 assessment criteria has been developed together with steering feel assessment experts. These criteria can be grouped to form steering feel attributes, e.g. precision. To facilitate orientation, the attributes can be clustered with respect to the driving situation, e.g. handling.

3.2 Overall Results and First Insights

In the matrix given in Fig. 4 the overall average of the assessments is given. Each entry in this matrix corresponds to the average assessment result given by a specific OEM for a specific vehicle. The entries on the main diagonal are the self-assessments, i.e. the assessments the OEMs gave of their own vehicles.

Comparing the self assessment for each vehicle, i.e. the result shown on the main diagonal, to the other element in that column, it can be seen that these self assessments are always at least as good as the best off-diagonal elements. This shows that OEMs like their own steering feel more than the other OEMs like the steering feel of the competitor's vehicle. On the other hand, comparing the diagonal elements to the off-diagonal elements in a row, the results are somewhat mixed. For most OEMs, any other steering feel is inferior to their own steering feel. A very clear example for this is given in the first row. Just in the row below, however, the situation is different. Here the steering feel of the vehicle represented in the first column is rated better than the own vehicle, and the vehicles given in column five and six are only 0.1 marks behind the self assessment.

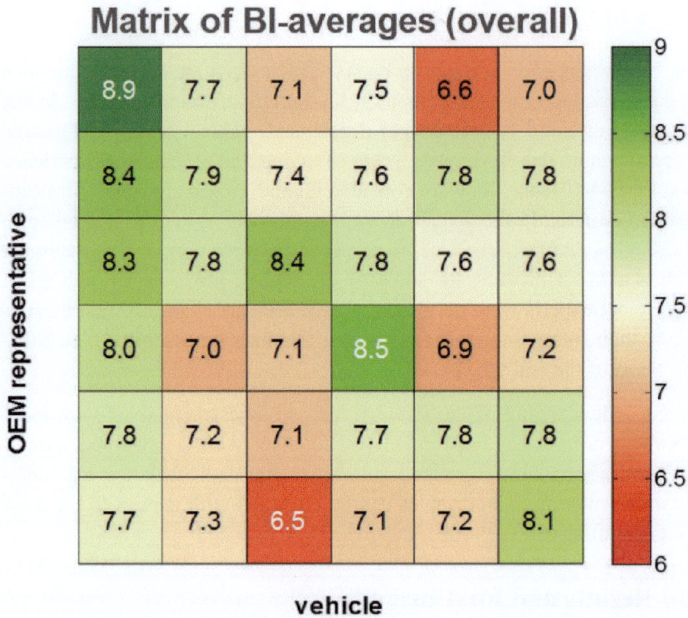

Fig. 4 Overall average result of subjective assessments for all vehicles as given by all OEMs.

3.3 Detailed Results

The foregoing results of the overall average analysis raises the question: what is the origin of these differences? In order to get some insight, a systematic analysis on the level of the individual criteria was performed automatically. This analysis was focused on two specific situations. The first on, called OEM-high, is defined as follows: The self assessment is significantly higher than the average of all third party vehicle assessments. This is shown in Fig. 5 as an example.

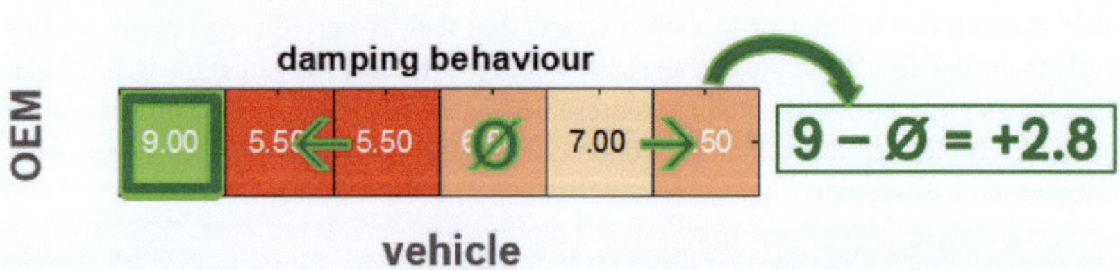

Fig. 5 OEM-high: Self assessment is significantly higher than the average of all third party vehicle assessments.

The interpretation is straightforward since this indicates situations where the OEM is pleased with his own implementation of the attribute and considers his implementation superior to all other implementations of this attribute. Here and in the remainder, "significantly higher" means that the difference is considerably higher than the overall variance in the assessments given.

The second, called OEM-low, is defined as follows: The self assessment is significantly lower than at least one third party vehicle assessment. This is shown in Fig. 6 as an example.

Again, the interpretation is straightforward. The OEM is sensitive to this attribute, but in contrast to OEM-high, now a third party vehicle showed a superior implementation of this attribute.

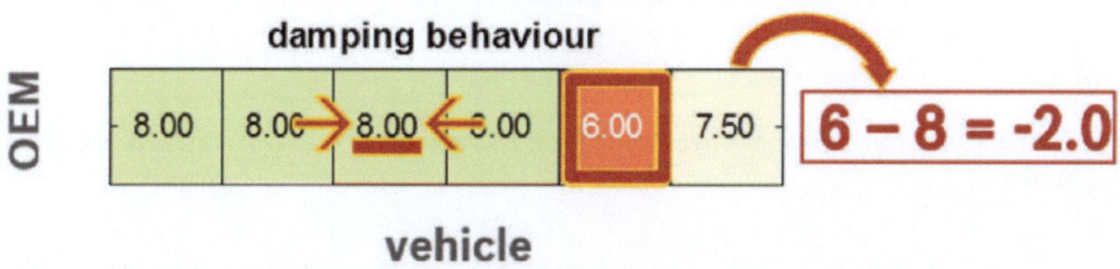

Fig. 6 OEM-low: Self assessment is significantly lower than at least one third party vehicle assessments.

Note that both OEM-high and OEM-low only take into account attributes for which the OEM assessments show a variance that is well above the overall variance of the assessments. Hence, these are steering feel attributes with a distinctive perception by the OEM.

3.4 OEM Specific Steering Feel DNA

Collecting all of these characteristic assessments identifies the attributes that are relevant to the OEM#s specific steering feel. This collection may thus be interpreted as the OEM#s steering feel DNA. These results are given in the matrix shown in Fig. 7. Positive numbers correspond to OEM-high, while negative numbers correspond to OEM-low assessments. Empty elements represent attributes without any characteristic assessments.

The a priori expectation off this matrix was to find some attributes that are relevant to all OEMs and just a few OEM specific attributes. However, looking at Fig. 7 the first clear observation is that there are no such common attributes. At most 4 OEMs share a characteristic attribute while some are shared by only two OEMs or are even specific for a single OEM.

This goes in line with the observation that some OEMs have only a small number of characteristics that make up their steering feel DNA, while others have a quite detailed steering feel DNA with many characteristics.

A further interesting observation is that there are some OEMs that only have OEM-high characteristics. This can be interpreted as the result of having a distinctive target steering feel while at the same time putting the required effort into the steering feel tuning to achieve the target steering feel.

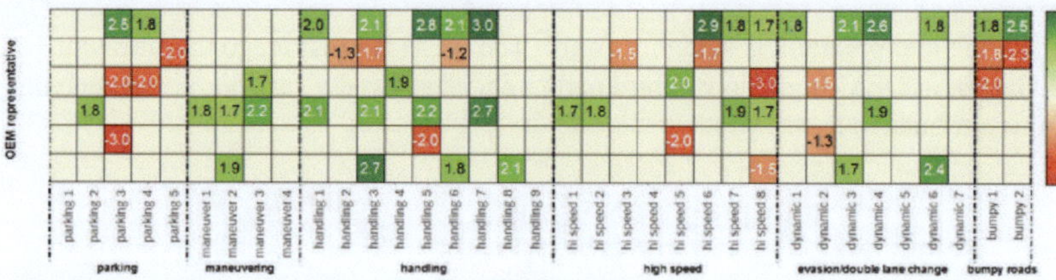

Fig. 7 Characteristics in the subjective steering feel assessments as seen from different OEM perspectives.

In turn, many OEM-low assessments do not indicate that the steering feel of the vehicle is bad. It is rather a manifestation of the fact that the OEM has a detailed target for that steering feel attribute, and some other vehicle showed that steering feel attribute much closer to the target than the OEM's own vehicle. It is well possible that this is at the price of being worse with respect to some other steering feel attributes than the OEM's own vehicle.

In summary, this analysis clearly shows that there is no common set of steering feel attributes that may be used to characterize the steering feel DNA of all OEMs. Quite to the contrary, the analysis emphasizes that the steering feel DNA is a very OEM specific topic. Before discussing the objectification of steering feel in the next section, it shall be noted that a similar analysis of the subjective assessment results from a vehicle point of view was performed. These results shall be only summarized for brevity here. The main insight is that while an OEM is happy with his implementation of steering feel and considers it superior to the steering feel of other vehicles, this opinion may not be shared by other OEMs. At the same time, it reveals that some OEMs are supercritical as they assess their own vehicle worse than all the other OEMs. Again, this emphasizes the key finding of the analysis: the steering feel DNA is highly OEM individual.

4 Objectification of Steering Feel

The short analysis of subjective steering feel for different OEMs and the analysis of the human perception paths in the preceding sections set the stage for the attempt of describing the steering feel by objective metrics with focus on capturing the OEM-specific steering feel DNAs. To be fully utilizable in virtual engineering applications, an objectification of steering feel must be able to quantify the differences between the OEMs with respect to their individual subjective steering feel DNAs.

4.1 Acquisition of Measurement Data by Open-Loop Driving Maneuvers

Based on the human perception paths and the perception limits as given in Table 1, suitable measurement devices were chosen to capture the rigid body motion of the vehicle as well as the steering wheel angle and torque.

One way of defining open-loop maneuvers could be to derive them from the maneuvers that are performed during the subjective assessments. However, these maneuvers differ from OEM to OEM at least in the details. Hence there is no advantage in using this approach to be expected compared to the use of standard open loop maneuvers as given in Fig. 8.

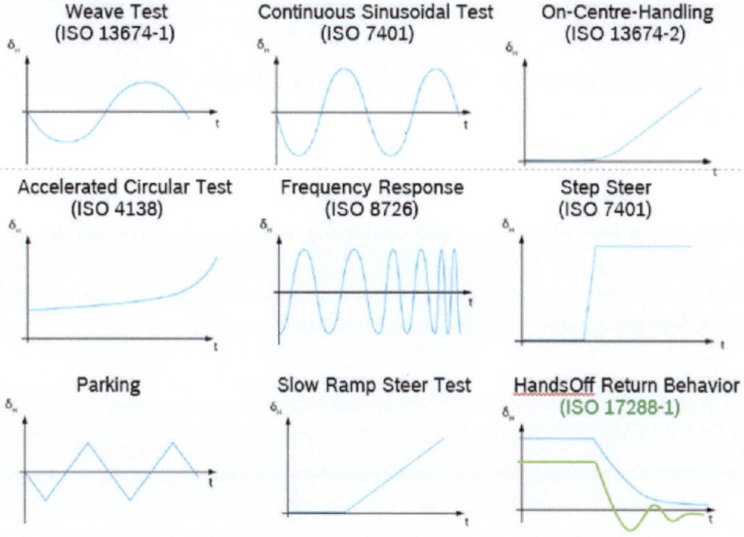

Fig. 8 Open-loop driving maneuvers used to capture objectively the steering feel.

Most of the maneuvers are defined in the ISO norms as noted in Fig. 8. However, some additional maneuvers were included, as the ISO norms do not consider these aspects. The three maneuvers are given in the lower column of Fig. 8. The first one addresses parking by steering from lock to lock at walking speeds down to standstill. The other one is the hands-off return behavior at low speeds. The ISO maneuver with the same name is defined at much higher vehicle speeds and starts at much lower steering angles. The third one is the slow-ramp steer which was performed over a wide range of vehicle speeds and is, in contrast to on center handling, continued up to the vehicles individual stability limit at each individual speed.

4.2 Metrics to Objectively Capture Steering Feel

From section 2 on the definition of steering feel and human perception it becomes evident that it is the relation of two quantities to one other that make up steering feel. There are in general three possible combinations: an action with respect to a perception, an action with respect to some other action and a perception with respect to some other perception.

Hence, for the following analysis, all combinations of measured signals were plotted against each other and the resulting curves were quantitatively described by metrics. The metrics were chosen in such a way that from their numeric values a detailed reconstruction of the curves is possible. Hence, the metrics capture all the information that can be deduced from the curves themselves.

4.3 Correlation Analysis: Mapping the Objective Metrics to Subjective Assessments

With this massive number of metrics, which is in the range of tens of thousands values, a correlation analysis was performed. The correlation analysis was set up to identify metrics that show a reasonably good correlation with the subjective assessments given by an individual OEM for the range of vehicles. The fitting function is chosen to be quadratic in nature. The idea behind this choice is the fact that all expected relations can be captured by this function, as shown in Fig. 9.

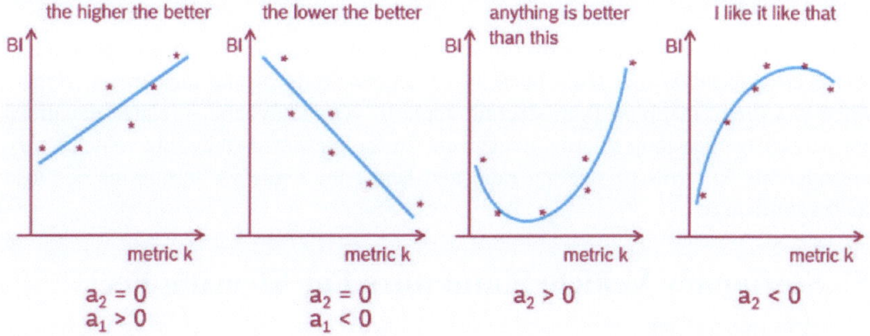

Fig. 9 Potential shapes of the chosen mapping function *BI = a2*metric^2 + a1*metric + a0*. The points correspond to the pair (measured metric for a vehicle | BI given by OEM for that vehicle).

Further to this, a correlation analysis of the BIs with respect to all pairwise linear combinations of metrics was performed. However, it was only in very rare cases that this linear function of two metrics produced slightly better results than the quadratic function of one metric, as given above.

4.4 Interpretation of the Results of the Correlation Analysis

Based on a preliminary analysis of the found correlations some interesting observations can be made. The first one, and maybe the most encouraging one, is the observation that most of the subjective steering feel DNA can be quantified by metrics. Many subjective attributes even can be quantified by more than one metric, and within such a group of metrics typically not all metrics are correlated with one another. On this basis, however, it is surprising to see that even for OEMs sharing a subjective attribute in their DNA according to Fig. 7 it is seldom the case that they also share the relevant metrics. This points to the need for an OEM specific set of metrics to quantify each OEMs steering feel.

The next observation is based on the nature of some of the metrics that have strong correlations with subjective attributes: they are metrics deduced from plots of two vehicle lateral dynamics related quantities, e.g. lateral acceleration vs. roll angle. Hence, these metrics cannot be influenced by tuning the steering feel via the steering; they have to be addressed by the chassis design.

The last observation to be mentioned here is the fact that metrics from the combination of steering wheel torque vs. steering wheel angle do not contribute to any significant correlations. At first, this may seem odd as many textbooks mention this relation.

However, taking into account that the vehicles investigated differ considerably in terms of steering gear ratio makes the situation more transparent. For a given vehicle, the steering gear ratio is fixed quantity or depends only on the rack travel. Hence, there is a direct mapping from steering angle to vehicle dynamics related quantities for particular open-loop steering maneuvers. In the presented study, the vehicles vary considerably in terms of steering ratio and hence the angle vs. torque metrics lose their significance.

5 Stationary Vehicle Simulators For Steering Feel Assessment

The results of the correlation analysis mentioned before point towards the possibility to end up with a fully virtual steering feel tuning process in the future. However, until then quite a bit of research is needed and hence it is worthwhile to look for a short-term alternative. This alternative should not need the real vehicle and at the same time not rely on objectifying the steering feel. A fair solution for this task is a stationary driving simulator.

5.1 Stationary Driving Simulators

The notion of a "stationary driving simulator" refers to a driving simulator where the driver is not moving in space but is seated on a stationary seat. Typically, a steering wheel together with a set of pedals to accelerate and decelerate a simulated vehicle is provided. The perception of the moving vehicle is produced by means of high quality computer animated graphics, ideally in combination with artificial sounds simulating the engine, wind and tire noises. Hence, the key components in this setup are the vehicle model and the computer graphics. In both fields, impressive progress has been made over the last few years and both sets of components have now reached a level of high maturity. Especially for the graphics, the recent development of virtual reality glasses gives a visual impression that is close to being perfect.

To give even more detail to such simulators they may be built up as a steering in the loop test bench with an actual steering gear that is exposed to real rack forces. In this setup, the loads on the rack are produced by some load simulating test bench equipment such as e.g. linear motors.

5.2 Perception of the Steering Feel in Stationary Driving Simulators

The simulator setup as described in the previous section shall now be discussed in some detail in the light of the discussion on steering feel, its perception via the human perception paths and the psychology of steering feel assessment as given in section 2.

First of all the steering feel in a narrow sense can be perfectly simulated in such a stationary vehicle simulator. Based on the finding of section 4 especially vehicle dynamics related perceptions, e.g. lateral acceleration, have to be considered, i.e. the perceptions related to the steering feel in a broad sense. Looking at the human perception paths as given in Table 1, the visual perception is the main perception path for the rigid body motion and thus the lateral vehicle dynamics. Again, this perception path can be perfectly simulated by highly quality computer graphics.

However, we believe that it is necessary to take a closer look at the situation with focus on what is going on during steering feel assessment. As pointed out in section 2 this is about expected perception due to some internal model and sensed perception of the driver. Within these models are, among others, perceptions related to other perceptions, and this is the point where things get critical for stationary driving simulators. While the visual path is relevant for the perception of rigid body dynamics, it is also the vestibular perception, which plays a role here. However, the vestibular perception acts on the level of accelerations while the visual one is slower as it is focused on changes in the position and the positions as such. Hence, it is reasonable that a sensed vestibular perception of an acceleration creates the expected visual perception of a change in position. In a stationary driving simulator, the driver senses no acceleration but only changes in position and hence there is a systematic mismatch. This mismatch becomes most relevant as motion sickness or the so-called driving simulator sickness, i.e. people start feeling sick after driving for a short time in such a driving simulator. Over time, most people accommodate to this situation, but this is nothing else as a modification of their internal models. This raises the general question to what extend stationary driving simulator results on steering feel are valid. To overcome this systematic problem non-stationary driving simulators may be considered, but in comparison to stationary driving simulators, the effort is immense.

6 Sketch of a Fully Simulation Based Steering Feel Tuning Process

In order to overcome both, the potential loss of validity of stationary simulator experiments due to driver internal model adaptation and the difficulties of non-stationary driving simulators it is worthwhile to think about a steering feel tuning process that is fully simulation based.

In Fig. 10 the main functional blocks of such a simulation-based steering feel tuning process is sketched.

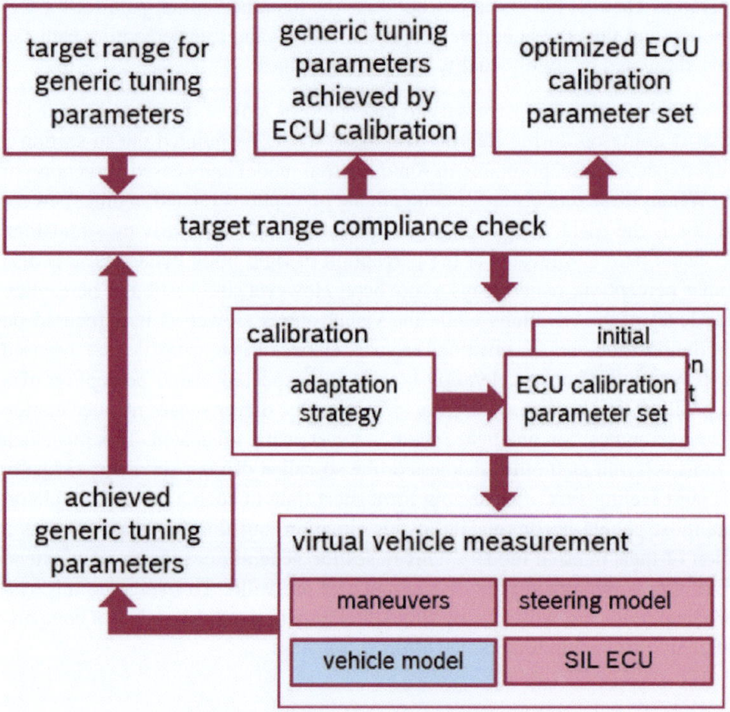

Fig. 10 Overview of a fully simulation based steering feel tuning process.

In Fig. 10 the metrics used to define the desired steering feel are called generic tuning parameters as they take on the role of traditional ECU tuning parameters. The iterative process starts with a virtual vehicle measurement, i.e. the open-loop maneuvers defined earlier are simulated with a detailed model comparison a vehicle model combined with a detailed mechanical model of the steering together with a Software-In-the-Loop (SIL) model of the real ECU code. From the simulated measurements the metrics are computed, i.e. the achieved generic tuning parameters. These achieved generic tuning parameters are now compared to the target range for the generic tuning parameters. Based on the observed differences, the actual ECU parameters are adapted and the virtual vehicle measurement is performed again. This process is repeated until the difference does not reduce any further.

Such a simulation based tuning process relies on the quality of the used models. Hence, care has to be taken with all models used in this process: vehicle model, model of the mechanical parts of the steering gear and the SIL model.

With such a simulation based tuning process it even seems feasible to systematically include the steering ratio in a rigorous way in the tuning process and thus to fully exploit the potentials of a Steer-by-Wire steering system. This is supported by the observation from the correlation analysis that with steering torque (but without steering angle) only approximately 50% of the DNA can be captured. Including both torque and steering angle either in combination with one another or in combination with some vehicle dynamics related quantities, again most of the OEM specific steering feel DNA can be captured by metrics. In addition to the general feasibility of such a process, the potentials for Steer-by-Wire steering feel design with such a process will be investigated in the future.

7 Conclusions and Outlook

Based on a detailed definition of steering feel, the necessary knowledge about human perception and the psychological process of steering feel assessment, the results of a subjective steering feel assessment campaign from different OEM perspectives was analyzed. It revealed that the steering feel DNA is strongly OEM-specific, not only in terms of target values but also in terms of the relevant attributes. Moreover, the results of a correlation analysis were presented. The analysis revealed that most of the DNA can be described quantitatively; however, the relevant metrics are again OEM specific. In this light, the use of stationary vehicle simulators was discussed. Their stationary nature leads to serious doubts with respect to the validity of the obtained results. Hence, in the last section, a tuning process was proposed that is fully simulation based.

Future work will focus on proving the feasibility of such a fully simulation based tuning process, and its application to Steer-by-Wire steering systems will be investigated.

References

(1) Harrer, Manfred; Pfeffer, Peter: Steering Handbook. 1st ed. 2017. Berlin, Heidelberg: Springer (2016).

(2) Decker, Medon: Zur Beurteilung der Querdynamik von Personenkraftwagen. PhD Thesis, University Munich (2009).

(3) Wolf, Hagen J.: Ergonomische Untersuchung des Lenkgefühls an Personenkraftwagen. PhD Thesis, University Munich (2009).

Model-based safety validation of the automated driving function highway pilot

Presenter:

Halil Beglerovic,
Research Engineer / Research and Technology

Co-Authors:

Abhishek Ravi,
Application Engineer / Calibration Applications

Niklas Wikström,
Simulation Engineer / Powertrain and Hybrid Systems

Hans-Michael Koegeler,
Principal Engineer / Calibration Applications

Andrea Leitner,
Project leader / Research and technology

Jürgen Holzinger,
Project manager ADAS

AVL List GmbH

P.E. Pfeffer (Ed.), *8th International Munich Chassis Symposium 2017*,
Proceedings, DOI 10.1007/978-3-658-18459-9_21

Abstract

In the development cycle of an Advanced Driver Assistance System (ADAS) emphasis is always placed on passenger safety, which directly correlates to adequate and extensive testing and validation procedures. However, testing and validation of ADAS systems is not a simple task and cannot be performed in a similar manner as conventional testing approaches. The main challenges is the vast amount of scenarios and environment parameter variations that might occur during operation. Current validation and testing procedures mostly rely on real world tests conducted on roads; however, because of the cost and complexity, these tests are not exhaustive and compromises on the scenario types and the number of considered parameters are made. In this paper we propose a model-based validation approach performed in a SiL (Soft-ware in the Loop) environment to validate an ADAS system under various conditions and the proposed methodology is presented on a Highway Pilot case study.

Preliminaries

In this document we are going to use the following terminology:

- Test scenario: *A generic traffic situation i.e. crossing, roundabout, highway driving etc. (including variation points for certain environment parameters).*

  For example: Driving on a high way within one lane

- KPI: *Key Performance Indicators.*

- Test case: *traffic scenario with defined discrete parameters – e.g. the number of other traffic participants or the number of lanes + defined KPIs in order to assess the function. Continues variables like starting velocities and positions of the traffic participants are still given as valid ranges only. For example: Driving on a 2 lane road behind a proceeding vehicle in a distance between 80 to 130 m with velocities of the proceeding vehicle between 40 and 150 km/h with defined KPIs like minimum clearance distance happening in the next 5 km of driving.*

- Test run: *Test case with defined values for all (also continuous variable) environment parameters. For example: Driving on a 2 lane road behind a proceeding vehicle in a distance of 80 m with a defined velocity profile of the proceeding vehicle stepping from 80 to 50 km/h, during light rain and slippery road with defined μ of 0.5. The KPIs remains like given in the test case.*

- ADAS: *Advanced Driver Assistance System. Typically below SAE-level 3*

- ADF: *Automated Driving Functions.*

- ENABLE S3: *European Initiative to Enable Validation for Highly Automated Safe and Secure Systems.*
- ACC: *Adaptive Cruise Control,* LKA: *Lane Keep Assist,* LCA: *Lane Change Assist.*
- ESP: *Electronic Stability Program,* EBA: *Emergency Brake Assistant.*
- Highway Pilot: *Combination of ADAS functions like ACC, EBA, and others for level 4 automation on highways.*
- MIL: *Model in the Loop,* HIL: *Hardware in the Loop,* VIL: *Vehicle in the Loop,* SIL: *Software in the Loop.*
- SUT: *System Under Test.*
- DOE: *Design of Experiments.*

1 Introduction

Technologically the development of automated driving functionalities is satisfactorily understood and witnessed by more than 1 million test kilometres already travelled by automated cars on public roads. The ever increasing demand and technological improvements in functionalities are leading to greater safety, lesser accidents as well as more efficient and environmentally friendly traffic. ADAS systems, and in the upcoming years ADF, have slowly become an irreplaceable part of the everyday driving experience. Nevertheless, Watzenig et.al. [1] state that new validation methodologies, procedures, and laws are needed in order to successfully incorporate emerging technologies into traffic and thus improve safety, reduce emissions, provide traffic flow optimization and enhanced mobility. Some steps towards this goal have already been taken. The EU made legal obligations on new passenger cars to include certain safety-related ADAS systems (EPS, EBA) and the level of automation will increase in the following years.

However, demonstrating the reliability, safety, and robustness of the technology in all conceivable situations, e.g. in all possible traffic situations under all potential road and weather conditions, has been identified as the main roadblock for product homologation, certification and thus commercialization. Winner et.al. [2] as well as Wachenfeld et.al [3] predict that more than 100 million km of road driving would be required to statistically prove that the automated vehicle is as safe as a manually driven car.

This is not feasible with current verification methods as it would require several years of testing. OEMs currently mainly rely on proving ground or public road testing in order to validate their systems because of the lack of alternatives. The test scenarios are usually taken from collections generated by engineers, which include the complete scenario description together with the expected response of the ADAS system. However, even when utilizing these collections of tests one cannot prove that the ADAS system will not fail in a test scenario that was not previously covered. In addition, proving ground and real world testing is associated with high costs, low reproducibility and long validation times. Especially reproducibility in a real world setup is challenging because of the difficulty to reach correct initialization, exact traffic behaviour, similar environmental influences, and so on. Furthermore, safety is a very important aspect and further limitations arise because some test cases could be dangerous or even impossible to be carried out by human drivers. All these limitations add up and influence the overall time needed to successfully validate an ADAS function.

Taking further into account the high number of vehicle variants and software versions, it becomes obvious that new approaches are required to validate automated vehicles within a reasonable time period at reasonable costs. New approaches are needed to reduce the effort required by today's state-of-the-art practices by orders of magnitude in order to become economically acceptable.

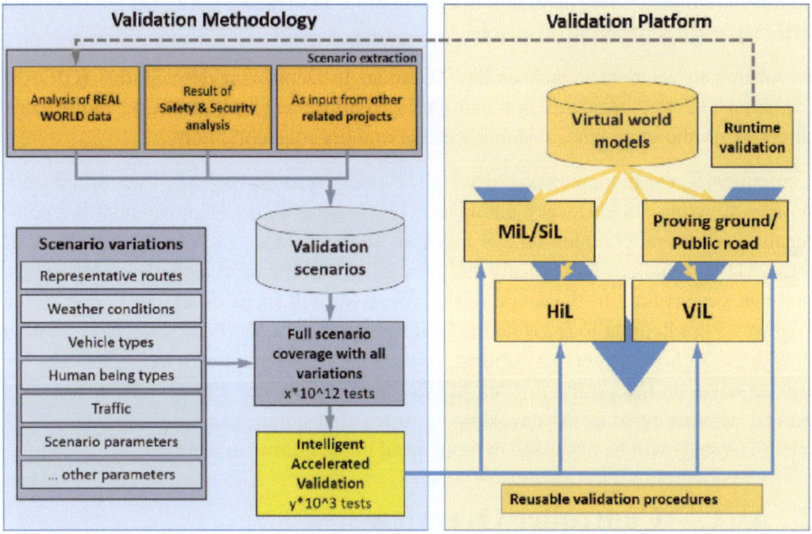

Figure 1: Modular Validation and Verification Framework envisaged in the ENABLE-S3 project [4]

Figure 1 shows the approach of a European research project that was started in May 2016 called ENABLE-S3 [4]. The main objective of the ENABLE-S3 project is to develop a modular framework for validation and verification of automated systems.

Because of the large scope and complexity the problem has been split into two parts. The validation methodologies on the one hand side describe the necessary steps and re-search on data acquisition and storage, scenario and metrics selection, as well as test optimization methods. The validation platform on the other side focuses on reusable validation technology bricks, which are able to seamlessly support various development stages (MIL, HIL, VIL, and Proving ground / Public road). By combining both parts and their respective technology bricks the project aims to achieve testing with at least 50% less effort compared to the actual established testing methods which are estimated with 10^{12} test runs.

This paper will show the first validation approach by selecting one of maybe 1000 relevant traffic scenarios for a Highway Pilot. By systematically varying scenario characteristics it is expected to achieve a better coverage compared to the classical approach. However, this approach is only feasible if adequate virtualization tools and MIL/SIL environments are available.

According to accident databases like "German In-Depth Accident Study" (GIDAS) [5], most highway accidents in recent years have occurred when driving in the same direction on the same lane, at sunny weather and dry road conditions.

Consequently, the first automated driving use cases focus on Highway Pilot functionality [6] – allowing a SAE level 4 automation [7]. In general, the Highway pilot is a combination of several complex ADAS functions such as: ACC, LKA, and LCA. Each of these ADAS functions individually and their integration presents a challenge for validation and verification. In the scope of this paper we will focus on the ACC part of the Highway Pilot in order to focus on the "following behind a car in the same lane" scenario with the highest accident probability. To keep the confidentiality of manufacturers, we put ourselves in the position of an ADAS-Function supplier and develop it from the ground up starting from the development to the first tuning and validation. The ACC control strategy will be discussed in more detail in the following section.

2 ACC – Controller Overview

The objective of the investigated control strategy is to allow fuel-efficient, comfortable and safe longitudinal speed profiles for the ego vehicle on a highway road, presenting the first and fundamental part of a Highway Pilot. To this end, a model predictive control (MPC) strategy is employed which adaptively controls the acceleration of the vehicle based on information from the prediction models. These models recurrently predict the motion of the ego vehicle as well as a preceding target vehicle over a prediction horizon of 20 seconds. This information is then used as inputs to a quadratic programming optimization problem, which aims to

$$\begin{array}{ll} \text{Minimize} & \boldsymbol{x}^T Q \boldsymbol{x} + \mathrm{q}^{\mathrm{T}}\mathbf{x} \\ \text{Subject to} & \text{Vehicle Constraints} \\ & \text{Traffic Speed Limits} \\ & \text{Headway Limits} \end{array} \tag{1}$$

The optimization problem is solved every 500ms to obtain desired vehicle accelerations. In equation (1), $\boldsymbol{x}$ denotes the state of the system, which includes the ego vehicle's longitudinal headway distance, speed, acceleration and mechanical jerk. The cost matrix $\boldsymbol{Q}$ and cost vector $\boldsymbol{q}$ describe the weightings between the different state variables. These parameters were tuned in an initial phase for a single route with fixed traffic and environment conditions and will be discussed in detail in the next section.

Figure 2: The minimum inter-vehicle distance is given by two headway time limits: the first soft limit may be violated at a cost and the second hard limit that must never be violated.

The vehicle is allowed to adapt its headway distance to the preceding vehicle within a predefined, velocity-dependent range. This is referred to as a flexible spacing policy. Figure 2 shows how the minimum distance is defined by two headway limits: a hard and a soft limit. The hard limit must never be violated, whereas the soft limit can be violated but at a cost. This cost grows quadratically approaching the hard limit.

The maximum distance, illustrated in Figure 3, is also composed of two components. The first component is a predefined limit, dominated either by a fixed distance or a velocity-dependent distance based on the preceding vehicle's velocity. The second component is proportional to the preceding vehicle's current speed, scaled by a 'catch-up' factor defined to be greater than one. The first component dominates when the ego vehicle is within the desired headway range, whereas the second component dominates when the vehicle is far behind. Neither of the constraints are allowed to conflict with the speed limit on the route, and lagging behind the limits is associated with a travel time cost.

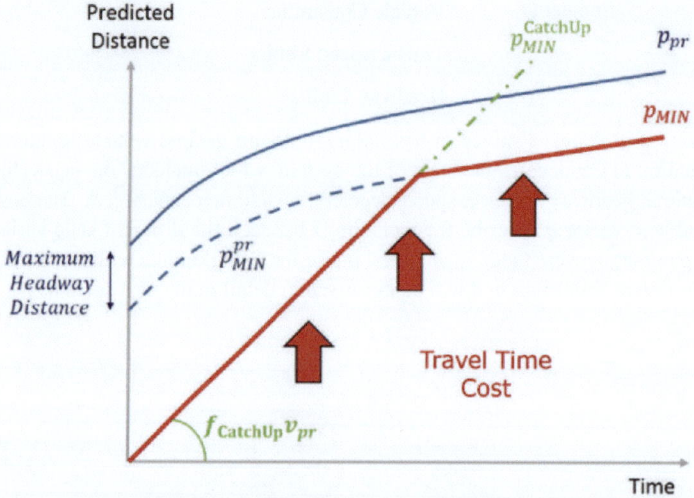

Figure 3: The maximum inter-vehicle distance $\boldsymbol{p_{MIN}}$ is defined by two limits: one limit dominates when the ego vehicle is located far behind the preceding vehicle ($\boldsymbol{p_{MIN}^{CatchUp}}$), while the other dominates after a predefined headway range has been reached ($\boldsymbol{p_{MIN}^{pr}}$).

3 Model-based Tuning

In order to reach a target behaviour of the ACC, or in more general terms the System under Test (SUT), the relevant influencing parameters have to be determined and varied systematically. Typically the first development environment in the development process is a Model in the Loop System (MiL), where the ACC function is simulated together with the environment.

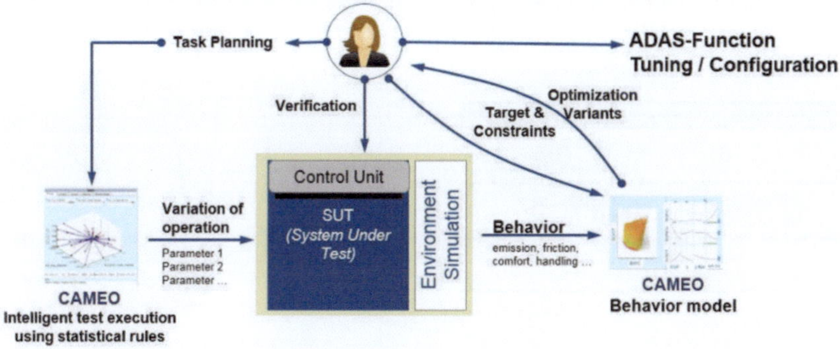

Figure 4: Model-based tuning of a System under Test (SUT)

In general, this tuning task is carried out by an expert operator, who based on his experience varies the most relevant parameters until the target behaviour is met. The results from such a subjective tuning process depends on: start values, step size, and number of considered parameters, and so on. Without a good understanding of the underlying correlations between the parameters, a perceived optimum behaviour could easily represent a local optimum. To improve that, a statistical Design of Experiment (DoE) based tuning methodology is presented in Figure 4, which tries to provide a correlation of the input parameters and the Key Performance Indicators (KPIs) using the least possible number of tests. Using an automated test procedure, all the relevant input parameters are varied simultaneously to identify the system behaviour. The observed behaviour of defined Key Performance Indicators (KPIs) is used to build behaviour models, which allows predictions of the system behaviour in the whole space defined by the variation parameters and some limited extrapolation depending on the model quality. The method assures that a global optimum is found, the parameter space is optimally covered, and the efficiency as well as traceability of the tuning-decisions is ensured. This statistical approach helps to understand the parameter influence and correlations, which is very beneficial in cases with a large number of input parameters. In such cases, the parameters with lower influence can be kept constant or excluded which reduces the tuning effort. Furthermore, the behavioural models built between the inputs and KPIs can be used for optimization tasks and trade-off analysis providing an additional tool for the adequate tuning of parameters.

The development environment for the SUT consisted of dynamic vehicle simulation program, Matlab-Simulink for the development of the ACC and AVL Cameo for the test automation, tuning and optimization.

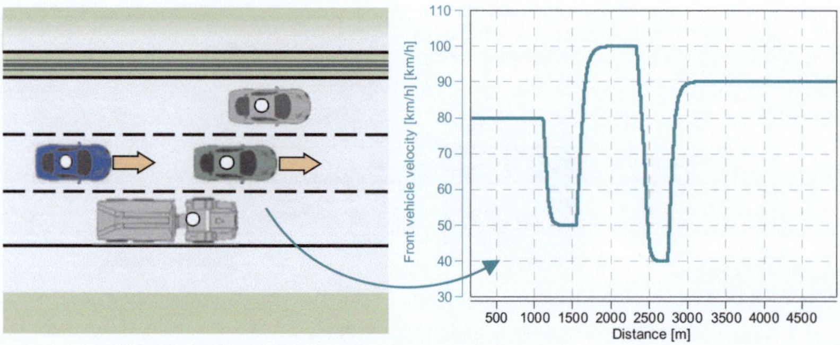

Figure 5: a) Scenario selection, b) Reference maneuver

3.1 Scenario Selection and Reference Maneuver

Figure 5 shows the scenario selected for the tuning of the ACC function. It consist of an ego vehicle (blue) following a preceding vehicle (green) using a model predictive ACC. The preceding vehicle has a predefined behaviour as shown in Figure 5 b) and the controller is tuned to satisfy several KPIs with respect to the preceding vehicle behaviour. The tuning task represents the first effort conducted for the new controller and it is placed early in the V development cycle.

3.2 Tuning Targets (KPIs)

For the tuning task of the ACC controller the following KPIs have been chosen:

- Safety – The controller should not violate the hard constraints regardless of the maneuver. As shown in Section 1 the controller has a hard constraint for the headway time. This constraint must not be validated in order to leave enough time for the vehicle to come to a complete stop in case of an accident.
- Fuel Performance – The controller should minimize the fuel consumption measured in liters per 100km and the influence of sudden changes in the preceding vehicle's velocity or maneuvers should be kept low. The fuel consumption when the ego vehicle perfectly follows the preceding vehicle with a fixed headway time is considered as the baseline.
- Driver Comfort – The controllers should maximize the driving comfort by reducing the integrated jerk. Similarly as above, the behavior of the preceding vehicle will be smoothed out and the controller will try to minimize braking while maintaining a safe and smooth ride.

3.3 Design Variables Sensitivity Analysis

As we are dealing with a new controller function without a priory knowledge of the behaviour, we have choosen 10 parameters and conducted a sensitivity analysis. Figure 6 shows the Relative Significance Indicator (RSI) for the 4 KPIs with respect to each of the 10 investigated tuning parameters. To calculate this the sensitivity algorithm analyzed the contribution of each individual variation channel to improve the statistical quality of the behavior model by reducing the root mean square error (RMSE). It lies between 0 and 1, denoting, how strong the remaining RMSE is reduced, when taking the corresponding variation parameter as additional input parameter to explain the observed behavior. A Robust Neural Network (RNN, [11]) is used as modeling base, which also considers nonlinear influences and interactions correspondingly. We conclude that only the first 5 parameters have a considerable influence on the overall behavior of the system. Therefore, for further tuning the rest of the parameters were kept constant.

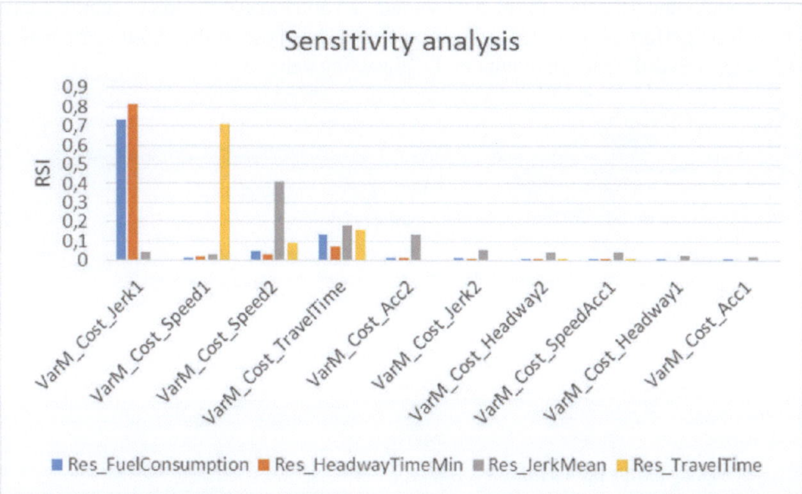

Figure 6: Parameter sensitivity analysis

3.4 Tuning Results

The tuning of the controller has been conducted with an interactive DoE procedure (COR DoE, [9]). The start design of 40 test runs for the 5 dimensions was automatically extended, in the area of interest, to 520 test runs in order to get reliable models in the area of interest, which is not known upfront. Without describing here the well-known “model based tuning process” in detail [8, 9, 10], Figure 7 shows the resulting tuned ACC-behaviour: The upper two signals represent the preceding and ego vehicles velocities respectively. The preceding vehicle’s velocity profile was fixed for the whole tuning task and the ego vehicle’s ACC controller was tuned in order to assure that the safety, fuel consumption and comfort KPIs are met. The next two signals represent the road elevation profile and the clearance between the ego and the preceding vehicle. We can see that the controller settings are chosen in such a way that robustness is assured and outside disturbances coming from the road elevation are successfully handled. Finally, the two remaining signals represent the accelerator and brake pedal position percentages, and we notice that the controller settings are minimize the braking energy and jerk leading to more efficient and comfortable ride.

For the evaluation of the tuning, we compared the behaviour and consumption of the ego vehicle with respect to the preceding vehicles’ velocity profile. The controller was able to achieve approximately 18% decrease in fuel consumption whilst still maintaining a smooth and safe ride. In the next section, a validation of the tuned control function will be performed in order to determine the robustness of the chosen parameters with respect to different behaviours of the preceding vehicle.

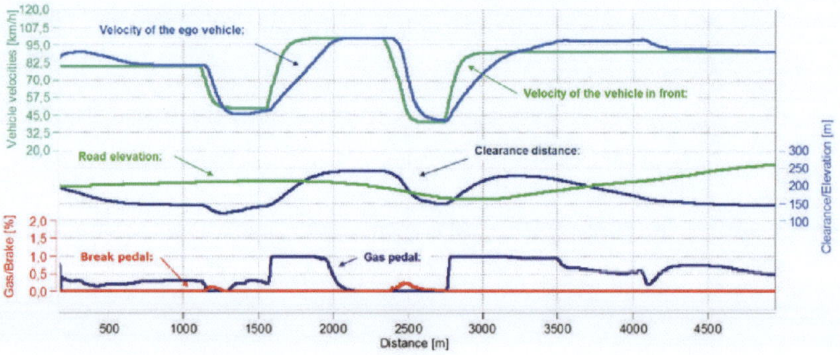

Figure 7: Tuning results for the reference scenario

4 Model-based Validation

After successfully tuning the SUT, a validation procedure is needed in order to prove that satisfactory behaviour and robustness is assured in a wider range of maneuvers. Figure 8 shows an overview of the validation task and we notice a considerable similarity to the tuning task discussed previously. In this paper, we aim to prove that the same principles of model-based tuning can be transferred to the model-based validation of Automated Driving Functions. The main difference between the tuning and validation is that now the external or environmental parameters are varied in order to validate the internal or controller parameters. Similar to the tuning task, a parameter space coverage is determined using DOE after which behaviour models are built for all KPIs. The most important benefits of this approach are [9]:

1) Compared to “Full factorial grid approach” the behaviour of safety-critical KPIs can be predicted in with an order of magnitude less simulation runs, giving complete results in the investigated ranges.
2) Interactions and significance of the parameter changes are fully considered when searching for safety-critical situations.
3) The highly automated process is much faster than a manual one and the human interaction happens on the level where decisions are made. In addition, a new interactive DOE procedure is available [11], which fills in points automatically after an initial sampling in order to improve the accuracy of the models in regions of high interest.

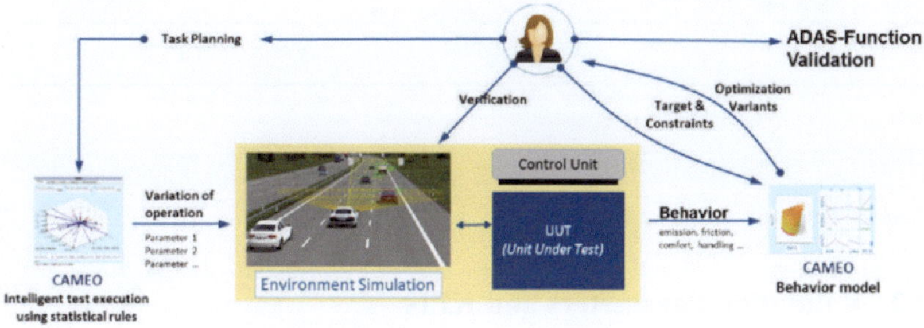

Figure 8: Model-based validation of a System under Test (SUT)

By selecting appropriate KPIs, it is possible to locate safety-critical behaviour regions in-side the parameter space by optimizing the models built in the previous step. Faulty behaviour of an ADF in a test run can be defined as the behaviour occurring when a critical criteria is not satisfied, e.g. the minimum clearance distance was breached. For a function developer it is of high interest to investigate the region in more detail and to get a better understanding of the faulty behaviour. This can be done automatically using the interactive DOE procedure.

4.1 Scenario Selection for Validation

The scenario selected for the validation of the Highway Pilot can be seen in Figure 9. In this scenario the ego vehicle (blue) is driving with the desired cruising velocity and a vehicle (green) is cutting in from the right side with a much lower speed. After cutting in, the preceding vehicle's velocity is changing following a sinusoidal trajectory where the mean velocity, amplitude, and period are varied as validation parameters. The length of the test run is fixed to 3km. In addition, this type of scenario represents the second most likely accident type defined by the "German In-Depth Accident Study" (GIDAS) [5].

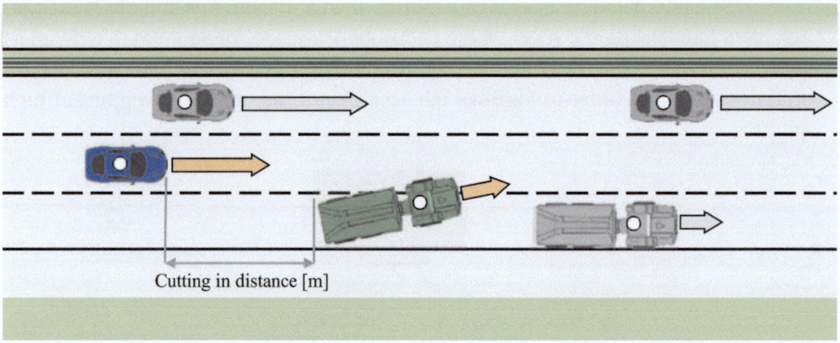

Figure 9: Selected scenario for validation

4.2 Validation Parameters and KPIs

The parameters of the scenario can be split in two parts. First, the parameter set by the driver:

- ACC target velocity (target cruising velocity) – If no preceding vehicle is detected, the ACC will accelerate to the target velocity.
- Maximum headway time to next vehicle – If a preceding vehicle is detected the ACC is going to keep the headway time below the maximum value, i.e. keeping a steady but safe distance from the vehicle in front.

Second, the environment parameters shaping the test case to a defined test run:

- Distance to the preceding vehicle – Distance between the ego and preceding vehicle at the cutting-in time instant.
- Road friction coefficient

- Amplitude, frequency, and mean velocity of preceding vehicle – In order to validate the ACC under a wide range of behaviors the preceding vehicles´ velocity was changed following a sinusoidal trajectory where the amplitude, frequency and mean are defined as input parameters for the validation.

Table 1 gives an overview of the validation parameters and their corresponding ranges for the selected example. The scenario was simplified for a better understanding in the following way: The cutting-in distance was fixed to a reasonable borderline value (for shorter distances the truck driver would be responsible for an accident). We did not consider variations in the road friction and the road gradients turned out to have no influence as long as the road friction is sufficient.

Table 1: Validation parameters

	Start velocity of the maneuver	fixed values		150 km/h
Selected by the Driver	ACC target velocity			150 km/h
	Maximum time gap			2 s
Maneuver - variation parameters		min	max	Start value
	Distance to the incoming vehicle	fixed values		80 m
	Road friction			1
	Road gradient			0%
	Average traffic velocity	40 km/h	140 km/h	80 km/h
	Sine velocity amplitude	- 20 km/h	20 km/h	0 km/h
	Sine velocity time period	10 km/h	40 km/h	10 km/h

Figure 10 shows one example taken from the test runs. The ACC set velocity was 150km/h and at time instance of 40s the ego vehicle's sensor detects a vehicle cutting in with a velocity of 65km/h. Finally, the velocity of the preceding vehicle is changing following a sinusoidal trajectory with an amplitude of 20km/h and a period of 25 seconds.

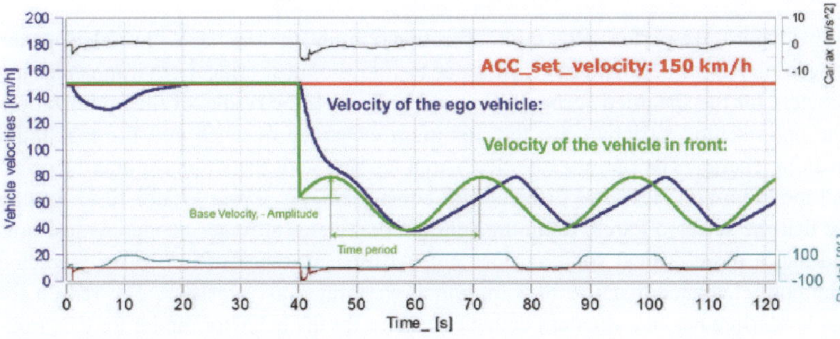

Figure 10: Example validation test run

For the validation, the following KPIs were considered:

- Minimum clearance distance between ego and preceding vehicle.
- Minimum headway time – representing the minimum time that the ego vehicle would need to reach the preceding vehicle's current position.

The KPI values for the example test run are shown in Figure 11.

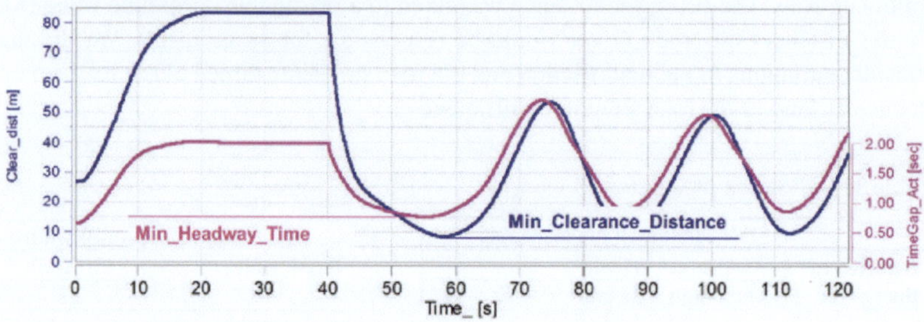

Figure 11: Clearance distance and time gap KPIs for the example test run

4.3 Validation Results

In order to validate the ACC for this specific scenario we first need to define a criterion that denotes a critical situation. The *"Safe distance between vehicles"* published by the *Conference of European Directors of Roads* [12] gives an overview of existing laws regarding the minimum clearance distance and headway time. In the publication we find that the minimum headway time defined by the Austrian law is 0.4 seconds regardless of the scenario. Following the regulations as guidelines we define that a dangerous test run ranges between 0.4s – 0.8s headway time and clearance distance ranging from 3 m to 15m, where scenarios with lower values represent crashes.

In this regard, we performed a multi objective optimization on the KPI models, in the dangerous ranges defined above, in order to find the most critical environmental parameters for the specified scenario. Figure 12 a) shows the results of the optimization. The optimization variables are the minimum clearance distance and the maximum breaking energy. The optimization leads to a test run with the strongest braking effort and the minimum clearance distance still considered safe. From Figure 12 a) we can see that the selected Pareto optimum solution dominates all other dangerous test runs and that a more critical solution does not exist except those with a lower predicted minimum clearance distance of 3 m, which were rated as "crashes". The reason for this borderline lies in a standard deviation of 1 m for the behavior model for the "minimum clearance distance". From Figure 12 b), where a verification simulation for this setting is shown, we determine that the minimum clearance distance and time for the worst test run are 2.2 m and 0.45s respectively.

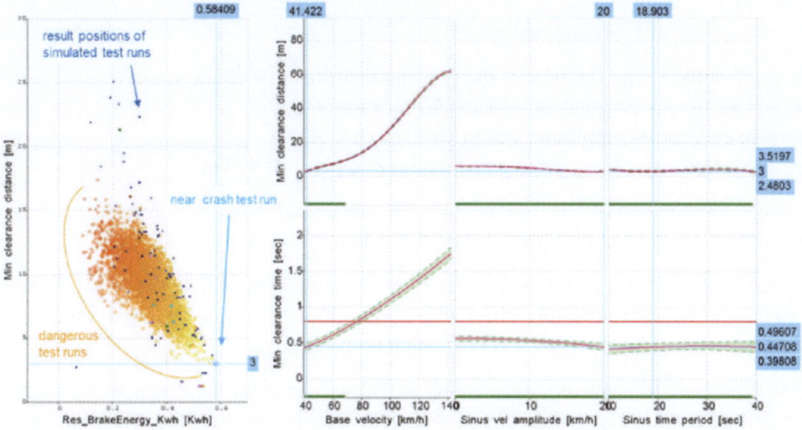

Figure 12: a) Pareto front for the critical scenario, b) Parameter influence on safety

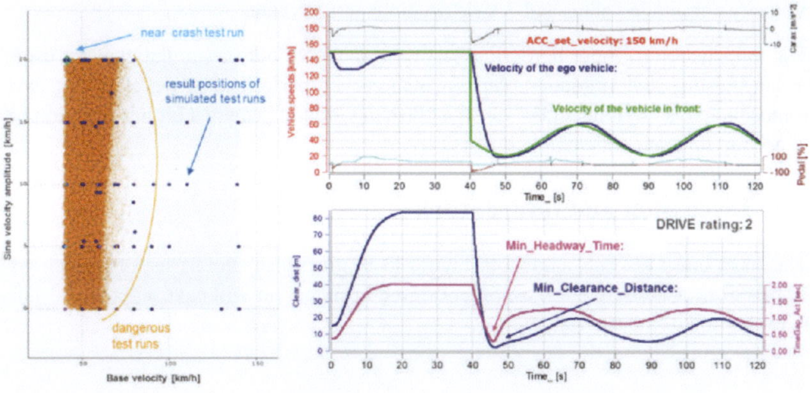

Figure 13: a) Dangerous test runs regarding the Sine parameters b) Most critical test run

In addition, Figure 13 a) shows all the dangerous test runs with respect to the sinusoidal velocity amplitude and the base velocity. We can see that these two parameters have the main influence on the minimum clearance distance especially when the amplitude of the sine is negative leading to lower velocities. This effect can be seen in Figure 13 b) where the most critical "near crash test run" is presented.

Thus we can state, that for the test run considering parameters in table 1 with an ACC target velocity of 150 km/h under

- dry conditions, on flat road, with a cutting in distance of 80 m,
- a sinus time period between 10 and 40 seconds
- with an velocity amplitude within +/-20 km/h
- a base velocity of the "cutting in vehicle" of more than 41 km/h,

for any combination of these parameters, the system will react safe in terms of "crash avoidance" (minimum clearance distance less than 3 +/-1 m is not reached).

Furthermore, as long as the base velocity of the cutting in vehicle is above 78 km/h (no dangerous test run results exists with higher velocity values) the legal limits of at least 0.8 seconds headway time is kept under all conditions.

One difference to a conventional testing is that we can draw these conclusions based on 210 simulation runs while considering 5 influencing parameters. Additionally, in this test case we did not take in consideration friction and road gradient; however, conventional testing methods would still needed 693 simulation runs, when changing each of the remaining 3 parameters in steps of 11 Average traffic velocity values, 9 Velocity amplitude values and 7 Velocity time period values.

The information of these borderline test runs, gathered out of simulations, are useful in later testing and validation stapes which can be performed in development environments like a "Driving cube" or on a "Proving ground" where a high integration between the function and vehicle is possible.

4.4 Discussion and future work

The "near crash test run" shown in Figure 13 b) gives also another aspect, which will be part of near future work: Passengers sitting in such a car will have a perceived safety evaluation of the test run.

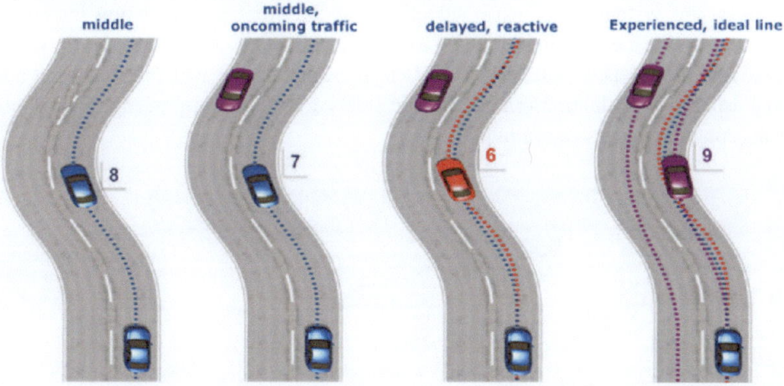

Figure 14: Rated passenger feeling, when driven in automated way, taking the other traffic participants into consideration [13]

In order to quantify the effect of the ADF to such subjective evaluation "perceived safety" and comfort metrics are introduced [13]. Figure 14 shows the perceived safety and comfort of passengers in a vehicle rated from 1 to 10 when driving on a rural road. This means that at any time in all possible situations the distances between the traffic partners must not be undershoot. But also for one and the same trace of an ego vehicle – keeping sufficient distances to other traffic participants from safety point of view – the perceived safety will depend strongly on the oncoming traffic.

Another important issue is that the most relevant scenarios and their respective test cases including KPI´s and realistically relevant parameter ranges for the function under consideration have to be selected.

The current tuning and validation example for the model predictive ACC was carried out on a single highway scenario and not all the critical parameters were considered. For a more extensive validation all possible scenarios and all relevant test cases need to be taken into account. In the end, all traffic situations have to be handled in proper way by the ADF. This immediately leads to the question of what it means "all" and what "to be handled in a proper way".

First, the most relevant scenarios and their respective parameters for this function have to be selected. In order to do so, probability distributions for the scenarios and their parameter value ranges have to be collected. Figure 15 shows the high level architectural framework elaborated within the ENABLE-S3 project. It provides a high level overview on the technology bricks required to validate an automated system and shows how the proposed approach of this publication (marked as "Test definition and control" + "Evaluation") is embedded in the overall big picture of automated system validation.

Together, the systematic scenario selection and the proposed validation method are going to provide a comprehensive and extensive validation. As a future work, we will investigate the scenario selection aspects in more detail and will focus also on additional aspects of evaluation (e.g. perceived safety for the driver/passenger to rate also the acceptance for these systems).

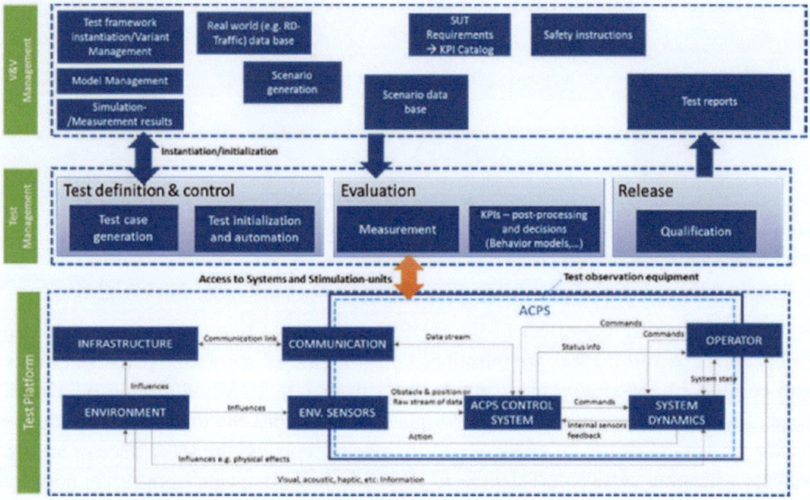

Figure 15: Validation and Verification framework of the ENABLE S3 Project

5 Conclusion

In this paper we presented a comprehensive introduction for the development, tuning and validation of a simple Automated Driving Function – ACC of a Highway pilot. The emphasis was put on the validation methodologies using as an example one test scenario. We have shown how principles of the well understood tuning task could be carried onto the validation. An ACC function was developed from the ground up presenting the first part of a Highway pilot. The tuning of the controller was done on a fixed scenario and the ACC was then validated on a test case, which was shaped in its appearance in 210 test runs. The resulting behaviour models allow us to specify areas for

- robustness and satisfactory response of the controller as well as
- areas, where legal limits are already violated for short time and further
- its borderline performance for the near crash situation.

We have shown that the ACC is able to stay above the absolute minimum safety margins for nearly all test runs in the specific scenario and in future work an additional perceived safety rating KPI will be taken into account.

It is planned to further develop this approach within the ENABLE S3 project. Thus a first methodical step towards a valid base for quantifying the remaining risk of a certain AD-driving function in its later lifetime is done. Such a quantification finally is needed in order to decide whether to accept the remaining risks or to request further testing to improving the functionality before allowing a usage in real traffic.

Literature

[1] Watzenig, Daniel, and Martin Horn, eds. *Automated Driving: Safer and More Efficient Future Driving*. Springer, 2016.

[2] Winner, Hermann, and Walther Wachenfeld. "Absicherung automatischen Fahrens, 6." *FAS-Tagung München, Munich* 2013.

[3] Wachenfeld, Walther, and Hermann Winner. "Die freigabe des autonomen fahrens." *Autonomes Fahren*. Springer Berlin Heidelberg, 2015.

[4] www.enable-s3.eu – European Initiative to Enable Validation for Highly Automated Safe and Secure Systems.

[5] www.bast.de – Bundesanstalt für Straßenwesen (BASt).

[6] Dr.-Ing. Armin Engstle, Dipl.-Ing. Helmut Assmayr, Dr. Georg Schwab, Softwareentwicklung für autonome Fahrfunktionen,

[7] www.sae.org – Society of Automotive Engineers

[8] Rainer, A., H. M. Koegeler, and D. Rogers. "Iterative DoE—Improved Emission Models and Better Optimisation Results Within a Shortened Measurement Time." *2nd Biennial International Conference on Powertrain Modelling and Control, Bradford, UK*. 2014.

[9] M.Sc. A.Ravi, Dr. H.M. Koegeler, M.Eng. Takayuki Miyata, M.Sc. A.Saroldi, *Tool Chain for development of ADAS systems*, CRF, Italy,

[10] Schick, Bernhard, et al. "Model-based steering ECU calibration on a steering-in-the-loop test bench." *6th International Munich Chassis Symposium 2015*. Springer Fachmedien Wiesbaden, 2015.

[11] Keuth N, Koegeler H, Fortuna T, Vitale G. *DoE and beyond : the evolution of the model based development approach how legal trends are changing methodology*. Conf. Des. Exp. Powertrain Dev., 2015.

[12] Conference of European Directors of Roads – CEDR, *"Safe distance between vehicles"*, 2010.

[13] Holzinger, Juergen; Schloemicher, Thomas; Bogner, Erik, *Objective assessment of comfort and safety of advanced driver assistance systems*, FISITA 2016 World Automotive Congress, Busan, Korea

CHASSIS.TECH SECTION

DEVELOPMENT METHODS

The basis for a process chain for functional component data

Michael Baumann, Dr. Christoph Böhm, Daimler AG

This manuscript is not available according to publishing restriction. Thank you for your understanding.

P.E. Pfeffer (Ed.), *8th International Munich Chassis Symposium 2017*, Proceedings, DOI 10.1007/978-3-658-18459-9_22

On the design of actuators and control systems in early development stages

M. Sc. Amir Zare, BMW Group

Prof. Dr.-Ing. Kai Michels, University of Bremen

Lucas Rath-Maia, BMW Group

Dr. Markus Zimmermann, BMW Group

P.E. Pfeffer (Ed.), *8th International Munich Chassis Symposium 2017*,
Proceedings, DOI 10.1007/978-3-658-18459-9_23

Abstract

Vehicle development faces new challenges due to increasing requirements on vehicle performance, e.g., on ride, comfort and safety. Adding mechatronic systems to the chassis helps to reach more ambitious design goals, however, it also adds to the complexity of the design process. The number of design variables increases significantly: in addition to the mechanical chassis, actuators of mechatronic systems and control logic parameters affect the overall vehicle performance. The complex interaction between many design variables poses a particular challenge. Engineers need to make important design decisions in early development stages, typically about the mechanical chassis, without knowing the final design, in particular the detail parameters of mechatronic systems.

The approach presented in this paper produces combined quantitative requirements on mechatronic systems and the mechanical chassis. As application example, a mechatronic rear steering system is considered. Requirements on the dynamics and overall performance of the actuator and its associated control logic parameters are derived from requirements on the overall vehicle performance. The derived requirements are ranges of permissible subsystem properties such as damping ratio, time constant or maximum rear angle. The permissible ranges are maximized in order to provide flexibility for actuator design and the application of control logic parameters in later development phases.

1 Introduction

In the last few decades, several kinds of methods and theories have been developed for modelling, design and control of large scale systems, composed of various subsystems [1, 2, 3].

Designing large scale systems can be accomplished by designing subsystems separately. In this case, each subsystem should be developed so that overall performances of the large scale system will be achieved. However, such an approach may lead to conflicts of goals regarding overall performances of a large scale system, since each subsystem attempts to accomplish overall performances and does not take into account contributions of other subsystems. In addition, dealing with various uncertainties of each subsystem adds to the complexity of the development process. In control engineering, this problem is addressed by different approaches, such as robust control design of large scale systems [4], centralized and decentralized control design of large scale systems [5]. However, there is a necessity in system engineering to propose a new method which considers not only control design but also the design of other subsystems with respect to uncertainties and interactions of subsystems on overall performances of the whole system. [6] present a new approach for designing interacting subsystems of a large scale system with respect to their uncertainties and overall performances. Overall performances are evaluated by defined quantitative requirements, typically objective requirements. Based on the V-model designing method, objective requirements are broken down into requirements adapted into subsystem performances.

Vehicles in general and their chassis in particular are examples of such large scale systems. Tires, suspension systems, steering systems and control systems with their associated actuators interact with each other and influence the vehicle dynamic behaviour significantly. Implementation of the simple but effective V-model approach with additional distinct steps, explained by [6] in the development of vehicle dynamics, gives engineers an opportunity to handle the complexity and uncertainties of the development procedure on the one side and designing subsystems simultaneously with respect to entire requirements of a chassis on the other side.

This paper is organized as follows. We introduce the idea behind the V-model and a new design approach based on three steps for the design of large scale systems in Section 2. In Section 3, we apply this methodology to design of vehicle lateral dynamic control systems and its associated actuators. Section 4 presents the extension of the methodology. Section 5 demonstrates the results of the application of the methodology to the rear steering system. Conclusions are drawn in Section 6. Future work is finally discussed in Section 7.

2 Methodology

The V-model is an effective way to decompose qualitative requirements of a large scale systems into subsystem requirements. However, it is not possible to evaluate quantitative subsystem requirements by this model. Therefore, additional steps have to be incorporated into this model to make it possible to pursue quantitative requirements of subsystems from overall requirements of a large scale system. In the first step, it should be clarified which design variables of each subsystem interact with objective quantities, possibly interact with each other and affect overall performances of a large scale system. This is done by creating a so-called dependency graph. In the second step, so-called bottom-up mappings, e.g., surrogate models have to be established, which evaluate the quantitative performances of a large scale systems in dependence on their design variables. Surrogate models must be as simple as possible and as complex as necessary. In third step, so-called top-down mappings are used: permissible ranges of design variables are derived from quantitative requirements on the large scale system. The preferred top-down mapping relies on the so-called solution space identification method [7]. Figure 1 shows the overview of all these three steps.

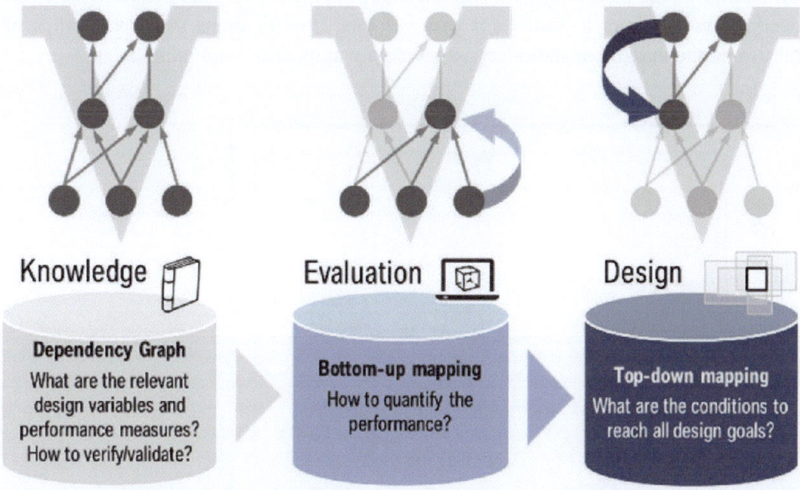

Figure 1 – Three steps for quantitative systems design

3 Application of the methodology to control and actuator design

The considered large scale system in this paper is a vehicle chassis. The subsystems to be designed are the lateral vehicle dynamics control system and its relevant actuator, a rear steering system.

3.1 Dependency graph: Design variables and objective requirements

To apply this methodology to the design of actuators and control systems requirements on the chassis with respect to ride, comfort and safety must be defined. Moreover, the requirements must be formulated objectively and associated to different driving manoeuvres [8].

For example, the weave manoeuvre is carried out with constant velocity and a sinusoidal excitation of the steering wheel with constant frequency. It is used for the determination of the stationary yaw velocity amplification [9]. Three requirements are specified with respect to stationary yaw velocity amplifications at the velocities v_1 and v_2 and the maximum stationary yaw velocity amplification, Figure 2.

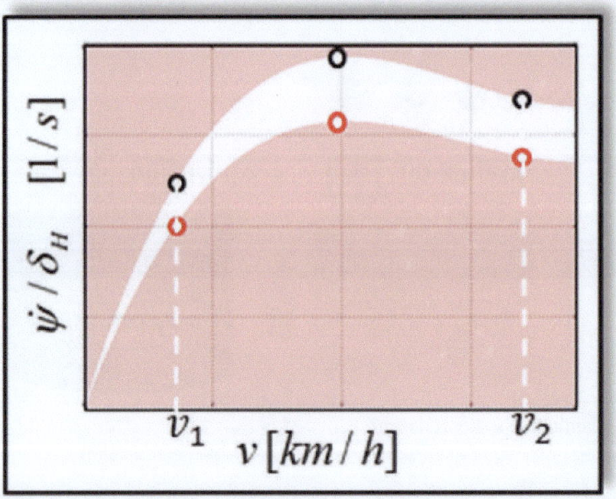

Figure 2 – Stationary yaw velocity amplification

Some of the objective requirements assessed in this paper are listed below.

Table 1 – Objective quantities

Objective Criterion	**Definition**
Maximum lateral acceleration	$a_{y,max} = \max\big(a_y(t)\big)$
Self-steering gradient at lateral acceleration at $\boldsymbol{a_{y,norm}}$	$\frac{d\,\delta_{norm}}{d\,a_y}$
Maximum stationary yaw velocity amplification	$\frac{\dot{\psi}}{\delta_{norm}}(v = v_{char})$
Stationary yaw velocity amplification at $\boldsymbol{v_1}$	$\frac{\dot{\psi}}{\delta_{norm}}(v = v_1)$
Stationary yaw velocity amplification at $\boldsymbol{v_2}$	$\frac{\dot{\psi}}{\delta_{norm}}(v = v_2)$

Furthermore, the most relevant design variables of the actuator and the associated control system are addressed in the following table.

Table 2 – Design variables

Design variables	**Definition**
Actuator damping ratio	ζ
Actuator time constant	T_w
Actuator time delay	T_d
Actuator no load speed	v_{noload}
Actuator maximum output force	F_{max}
Proportional term of the PID feedback controller	K_p
Frequency factor of the dynamic feedforward control	ω_F
Vehicle rear axle load ratio	HAL
Vehicle mass	m

An example of the dependency graph is shown in Figure. 3.

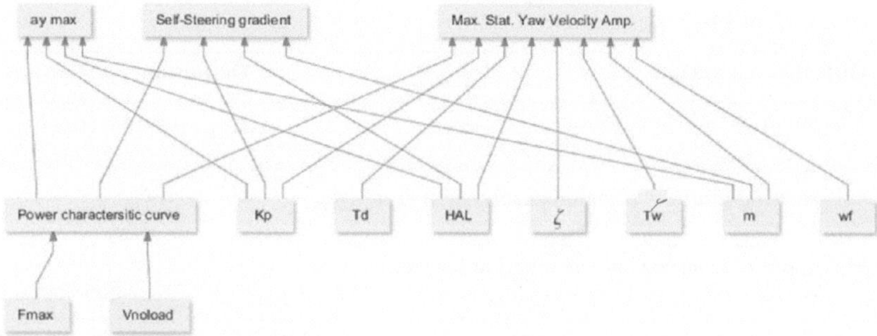

Figure 3 – Dependency graph

3.2 Bottom-up mapping: Model for Control System and Vehicle

For the bottom-up mapping, quantitative evaluation of inputs on outputs, a model is required which represents the physical relations between inputs and outputs, design variables and objective quantities respectively. Hence, we firstly pass on an overview about the vehicle model. It is based on a non-linear two-track model implanted at BMW in MATLAB®. This simulation tool also contains a driver, a traffic, a road model and enables the integration of any arbitrary control system, i.e. logic, actuator and sensor model [10]. Its non-linear two track model is composed of non-linear characteristics obtained from a multi-body simulation in ADAMS® and is able to simulate the vehicle dynamics at an adequate level. The magic formula is implemented in this model for the calculation of the tire dynamics.

Moreover, for the purpose of developing actuators or parametrizing logics of control systems, simplified models have to be developed. The simplified control system model considered in this paper has the structure shown in Fig. 4.

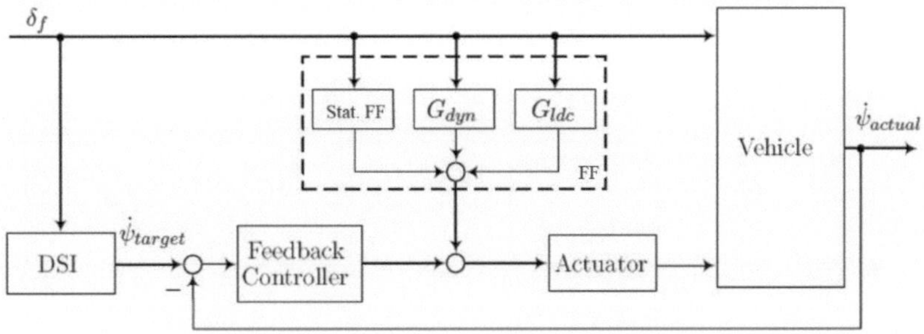

Figure 4 – Simplified control system structure

The control system includes feedforward control (FF) composed of stationary (Stat. FF), dynamic feedforward control (G_{dyn}) and disturbance correction feedforward control (G_{ldc}). The Actuator model is based on a combination of empirical and physical modelling (semi-physical), represented by a linear transfer function and the power characteristic cure respectively [11]. The order of the transfer function depends on the characteristics of the actuator, evaluated by the inspection of its component test rig tests. It can be assumed that the most active chassis actuators can be adequately modelled by a second order transfer function [12, 13]. This model is often used for the purpose of the interdependent development of logics and actuators in the early stages of the vehicle dynamics development. However, for the purpose of a simulative validation, once the logic and actuator are developed and delivered by suppliers, a more precise model is required, for example a second order LPV transfer function as follow

$$G_{Actuator(s,CF)} = \frac{K_p(CF)}{1 + 2\zeta\,(CF)T_w(CF)s + (T_w(CF)s)^2} e^{-T_d(CF)s}$$

where $K_p(CF)$ stationary gain, $\zeta(CF)$ Damping ratio, $T_w(CF)$ Time constant and $T_d(CF)$ time delay are dependent of counterinputs, in the case of this paper counterforces (CF), Fig. 5.

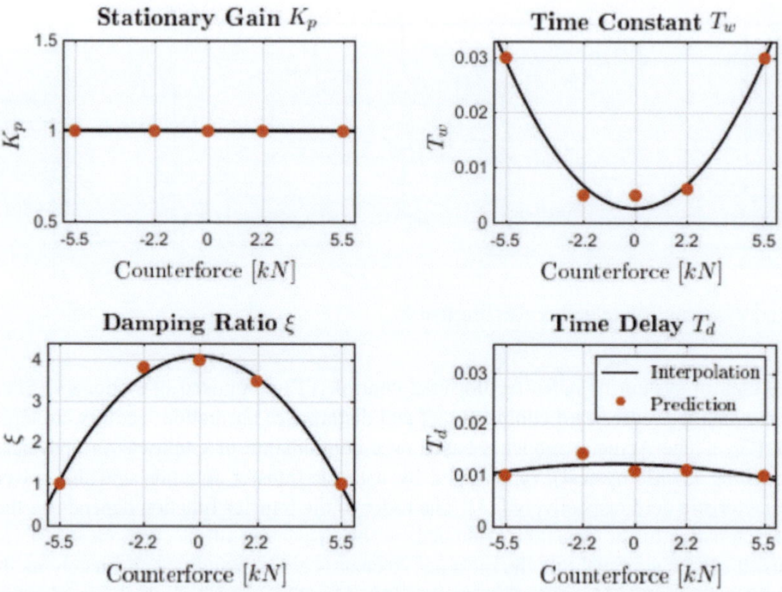

Figure 5 – Parameters of the actuator transfer function associated with different counterforces

Moreover, to extend the physical model to semi-physical, the power characteristic curve, force (torque)-speed curve, has to complement the physical model, Fig. 6

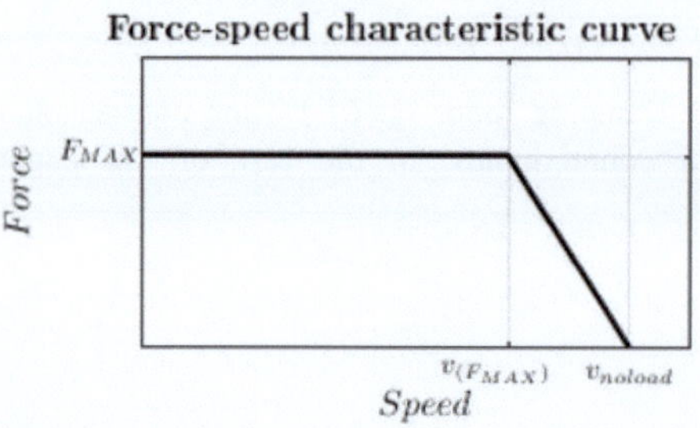

Figure 6 – Power characteristic curve

The authors will provide details of the modelling procedure of the logic and the actuator in a future article.

3.3 Top-Down Mapping: Solution Spaces

In the early development stages of vehicle dynamics, there exist different kinds of uncertainties. As a consequence, it is not desired to find one optimum solution which meets all the requirements sought by the designer in this stage. For example, figuring out just one virtual parameterization of the logic based on the simulation may not able to cope with the subjective driving experience. Furthermore, only one simulative actuator design may be not or highly expensive realized by a supplier. Hence, the approach to control uncertainties is to find a target region, a so called solution space, of all good designs for design variables, which fulfil all objective quantities. Besides, target intervals will be sought to realize independent target regions for design variable and the product of the intervals should be maximized to increase the design flexibility. As a result, the solution box calculated from the maximized intervals has the form

$$\Omega = I_{act,j} * \ I_{log,k} *$$

Where $I_{act,j} = [x_{l,act,j}, x_{u,act,j}]$ denotes an interval for the j-th variable of the actuator and $I_{log,i} = [x_{l,log,k}, x_{u,log,k}]$ for the k-th variable of the logic. It means,

$$\max_{\Omega \subseteq \Omega_{ds}} \mu(\Omega)$$

$$subject\ to \quad \underline{f_i}(X) \leq \underline{f_{c,i}}, \quad i = 1, \dots, n$$

Where $\underline{f}(X)$ and $\underline{f_c}$ represent vectors of vehicle performance measures and associated vehicle performance threshold values, respectively. n is the number of vehicle performance measures. The algorithm seeking the largest box in a solution space was presented in [7].

Fig. 7 represents a nominal solution space for actuator and logic design variables. The green dots express good designs and the bold box, so-called solution box, assigns each parameter a permissible interval. Coloured dots represent designs that violate one or several requirements. The solution box is allowed to have 5% violating designs.

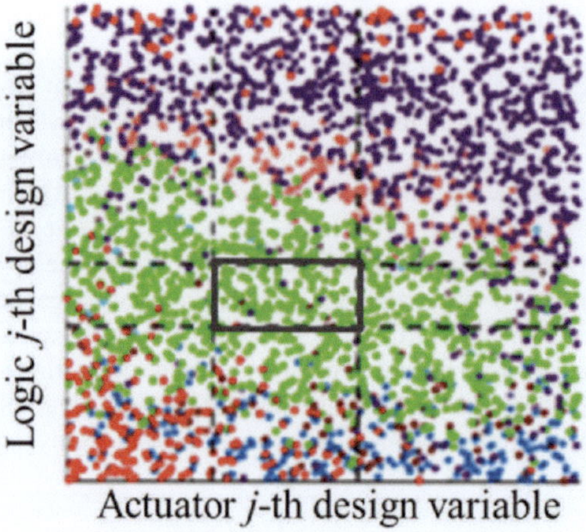

Figure 7 – A nominal solution space and a nominal solution box for design variables of an actuator and a logic of a control system

4 Extension of Methodology

4.1 Varying design variables

As mentioned in chapter 3.2, the actuator dynamic response depends on the counter-force (or countertorque) caused by forces acting on the tire. As shown in Fig. 5, the higher the counterforce, the worse is the dynamic response of the actuator. However, for developing the actuator dynamics and the virtual parameterisation of the control logic, the LTI model of the actuator is regarded.

It implies,

$$(\zeta, T_w, T_d, K_p) = \underline{X}$$

with $\underline{X}$ being constant. In this case, a solution space will be identified by using

$$\underline{Z} = \underline{f}(\underline{X})$$

where $\underline{Z}$ represents the vector of vehicle performances. Accordingly, the solution space will be defined as

$$\Omega_A = \left\{ \underline{X}_A \middle| \underline{f}(\underline{X}_A) \leq \underline{f_c} \right\}$$

It is assumed for the LPV model of the actuator in the design process that

$$\underline{\hat{X}}_A(F(t)) \in \Omega_A \quad \forall t \quad \square \quad \underline{Z} = \underline{f}\left(\underline{\hat{X}}_A(F)\right) \leq \underline{f_c}$$

where

$$\left(\zeta = \hat{\zeta}(\mathrm{F}),\ \mathrm{T_w} = \hat{\mathrm{T}}(\mathrm{F}),\ \mathrm{T_d} = \hat{\mathrm{T}}_\mathrm{d}(\mathrm{F})\right) = \underline{\hat{\mathrm{X}}}(\mathrm{F})\ ; \qquad \mathrm{F} = \mathrm{F(t)}$$

4.2 Additional Constraints

There is not only one particular solution box in the solution space. For instance, let us look at the solution space of two design variables, actuator transport delay T_d and natural frequency factor of the dynamic feedforward control ω_F Fig. 8.

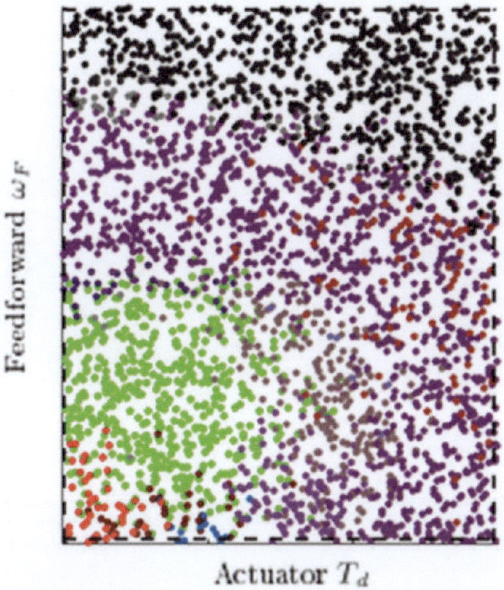

Figure 8 – Solution space of actuator transport delay and natural frequency factor of the dynamic feedforward control

As mentioned, the algorithm seeks a solution box with the maximum size and does not consider the meaning of design variables. For instance, the solution box illustrated in Fig. 9 restricts the lower bound of the transport delay interval. It means, the actuator transport delay must not begin from zero, which implies slow performance of the actuator to its input. But on the other hand, the flexibility for the parameterization of the natural frequency factor has been improved by extending its permissible interval with respect to this solution box. However, such a solution box is undesired, as we require a high-performance actuator and of course more flexibility for designing an actuator. Therefore, the algorithm has to be forced to avoid such effects. As a result, we fix the lower bound of the actuator transport delay in the design procedure in order to not allow the optimization algorithm to find a box which restricts the actuator performance. Based on the same reason, we also fix the lower bound of the time constant T_w of the actuator transfer function. Moreover, the lower bound of the damping ratio ζ has to be fixed to 0.707 in order to prevent a resonance magnification [14]. In order to prevent limiting the maximal actuator power, the upper bounds of parameters v_{noload} and F_{max} have also to be fixed.

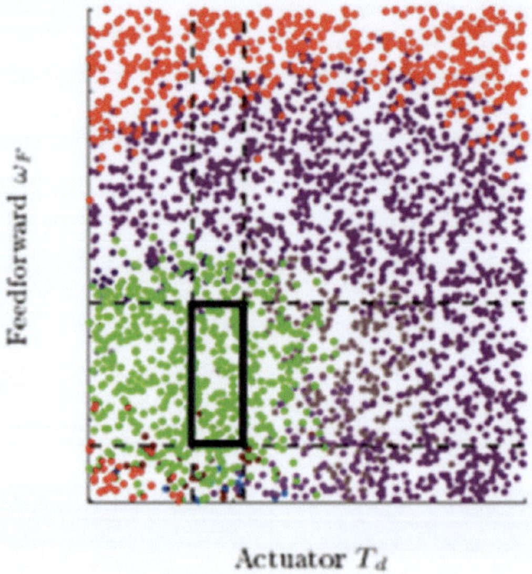

Figure 9 – Solution box with the largest interval for the natural frequency factor of the dynamic feedforward control

4.3 Balancing between actuator and logic design variables

As mentioned in the last section, the algorithm seeks a solution box with the maximum size. Yet, it does not consider the importance of design variables. It means, the algorithm can find a solution box which includes large intervals for logic parameters and small intervals for actuator parameters, which provides the manufacturer of the actuator with a restricted range to build an actuator. As a result, the actuator may be more expensive, heavier and consume more energy. Therefore, we should force the algorithm to find a solution box which fits larger intervals to the actuator design variables. The solution for finding such a solution box is to add weighting factors ω to the actuator and logic design variables as follow

$$\max_{\Omega \subseteq \Omega_{ds}} \mu(\Omega)$$

$$\mu(\Omega) = \prod_{i=1}^{n} (x_{u,act,i} - x_{l,act,i})^{\omega_{act,i}} * (x_{u,log,i} - x_{l,log,i})^{\omega_{log,i}}$$

Where Ω_{ds} denotes the design space, $\mu(\Omega)$ the volume of the box and n the number of design variables.

5 Results

In this chapter, we apply the abovementioned method on the design of variables of the logic and the actuator of a rear steering system. 20 design variables are considered. It means, the solution space has 20 dimensions and the solution box is visualized by its projection into two dimensional plots. The weighting factor is set to 5 for all actuator design variables and 1 for all logic design variables, $\omega_{act,i} = 5, \omega_{log,i} = 1$ respectively.

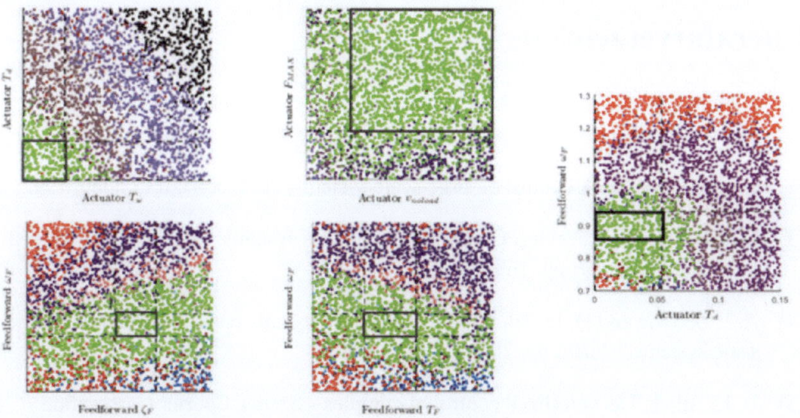

Figure 10 – Representation of the solution space into two dimensional plots

6 Conclusion

The paper shows how to formulate the quantitative requirements on actuator and vehicle lateral dynamic control systems. The simplified models of an actuator and logics have been demonstrated. Quantitative requirements have been formulated on a large scale system, i.e. a chassis. These requirements have been broken down into requirements which must be met by the actuator and the control system. Accordingly, solution intervals are derived by means of the numerical solution space algorithm. The intervals provide regions where all possible combinations of design variables fulfill the requirements on the large scale system.

7 Future work

In the product requirement document (PRD), the special requirements, which have to be met by an actuator delivered by a supplier, must be formulated with respect to different kinds of test rig tests. Accordingly, the evaluated permissible intervals for actuator design variables should be converted into requirements formulated on verification variables measured from the behaviour of an actuator with respect to test rig tests such as step response, phase response and actuator unit dynamic response tests. In this manner, the dependency graph will be extended. Moreover, the method can be applied on the parameterisation of more different control systems of lateral vehicle dynamics, if a chassis contains more than one control system and actuator.

Literaturverzeichnis

[1] J. Mohammadpour and K. M. Grigoriadis, Efficient modeling and control of large-scale systems, Springer Science & Business Media, 2010.

[2] J. Lunze, Feedback control of large-scale systems, Prentice-Hall London, 1992.

[3] D. Siljak and A. Zecevic, "Control of large-scale systems: Beyond decentralized feedback," *Elsevier,* pp. 169--179, 2005.

[4] A. I. Zecevic and D. D. Siljak, "Design of robust static output feedback for large-scale systems," *IEEE,* pp. 2040--2044, 2004.

[5] D. D. Siljak, Decentralized control of complex systems, Courier Corporation, 2011.

[6] M. Zimmermann, S. Koenigs, C. Niemeyer, J. Fender, C. Zeherbauer, R. Vitale and M. Wahle, "On the design of large systems subject to uncertainty," pp. 1--22, 2017.

[7] M. Zimmermann and J. E. Hoessle, "Computing solution spaces for robust design," *International Journal for Numerical Methods in Engineering,* pp. 290--307, 2013.

[8] D. Maeder, Simulationsbasierte Grundauslegung der Fahrzeug-Querdynamik unter Beruecksichtigung von Erfahrungswissen in der Fahrdynamikentwicklung, Verlag Dr. Hut, 2012.

[9] S. Botev, Digitale Gesamtfahrzeugabstimmung fuer Ride und Handling, VDI-Verlag, 2008.

[10] P. Kvasnicka, G. Prokop, M. Dorle, A. Rettinger and H. Stahl, "Durchgaengige Simulationsumgebung zur Entwicklung und Absicherung von Fahrdynamischen Regelsystemen," *VDI BERICHTE,* p. 387, 2006.

[11] K. Dettlaff, I. J. Wiedemann, U. Schaaf and I. Scharfenbaum, "The influence of the modeling depth of active chassis systems with respect to the development stage and their interaction with driving characteristics," in *6th International Munich Chassis Symposium 2015*, 2015.

[12] I. Scharfenbaum, A. Fratini and G. Prokop, "A Novel Method for the Development of an Idealised Active Roll Stabilisation System Model," in *2013 IEEE International Conference on Systems, Man, and Cybernetics (SMC)*, 2013.

[13] M. Greger, Auswirkungen einer variablen Momentenverteilung auf die Fahrdynamik, Technical University of Munich, 2006.

[14] K. Michels, Regelungstechnik – Vorlesungsmanuskript, Bremen, 2012.

Process design for a companywide geometrical integration of manufacturing issues in the early development phases based on the example of automotive suspension

Dipl.-Ing. (FH) Bastian Leistner

Univ.-Prof. Dr.-Ing. Ralph Mayer

Dipl.-Ing. (FH) Dirk Berkan

P.E. Pfeffer (Ed.), *8th International Munich Chassis Symposium 2017*, Proceedings, DOI 10.1007/978-3-658-18459-9_24

1 Introduction

The variety and complexity levels of chassis in one production plant increases with the rising number of different chassis concepts. Hence, the number of assembly requirements that are being addressed to the product and that should be considered during the development process of modern car chassis increases, as well. Furthermore, many functional requirements like driving dynamics and stability have to be applied. Most of them have plenty of geometrical impacts that have to be implemented while developing those chassis. On the other hand, more frequently used requirements of manufacturing and production issues with geometrical impact on the product should be considered.

Car inventory and population size [in Mio.] [a]

	2010		2030		2050	
	Car inventory	Population	Car inventory	Population	Car inventory	Population
World	842	6.909	1.382	8.309	1.810	9.150
Eu27	239	733	269	723	256	691
Latin Amerika	48	589	70	690	85	729
Africa	21	1.033	36	1.524	41	1.998
Germany	42	82	46	79	43	73
China	53	1.354	318	1.462	400	1.417
India	13	1.214	92	1.485	311	1.614
Japan	58	127	60	117	56	102
USA	210	318	227	370	202	404

Forecast worldwide Car-Inventory by type of drive [in Mio.] [b]

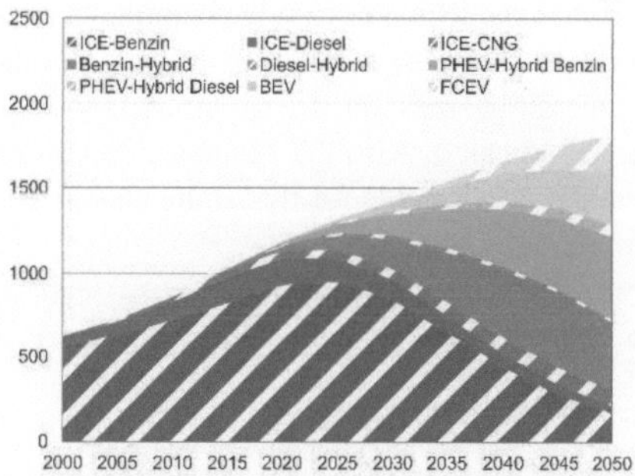

[a] Car inventory from IEA (2012), Basic data, population forecast from Eurostat (2010)
[b] IEA (2012), p. 443 Improved Case or 2 °C-Szenario: In year 2050 the world wide car inventory will be 1,8 Mrd. cars. In less optimistic 4 °C-Szenario are 2,3 Mrd. cars in 2050 predicted

Figure 1 – forecast of car inventory between 2010 and 2015[1]

Extra complexity is added due to the market shift in alternative drive train topology, as can be seen in figure 1. The number of battery electrical vehicles (BEVs) and plug-in hybrid electric vehicles (PHEVs) increases, while the quantity of sold cars with conventional internal combustion engines (ICEs) is nearly constant. In the past, the BMW Group produced all electrical vehicles such as BMW i3 or i8 (BEVs/PHEVs) in a separate assembly line in the plant in Leipzig. To address the rising demand on

1 Illustration based on: [11] p.24 – information based on [13]

BEV, the integration of these concepts on the same assembly line as well as the cars with internal combustion engine is necessary.

To achieve this, many assembly simulations need to be conducted whilst developing approaches for shared car architectures at the beginning of the product development process.

As most of the architecture relevant components are developed in the early phases of product development, many of these assembly simulations have to be completed nearly 70 to 40 months before start of production (SOP)[2]. Especially in these phases, many of the marginal conditions and development achievements are highly variable. Furthermore, many process-involved engineers have to supply these simulations with their expertise.

This paper serves an estimate to quantify, how probable and feasible the integration of a car concept in existing plants is, assuming as little changes as possible. Meanwhile, the changes which have to be completed at the chassis should be easily monitored and implemented. For example, the change of the geometry of a wishbone is supposed to create more space for the mounting tools. But not just the changes on the chassis have to be indicated. Also the necessary modifications in the production line should be revealed. An example for this is the change of the order of fitting or the usage of different mounting tools.

2 The challenge of assembly simulation in the early development phases

The first challenge, as mentioned in the introduction, is how to deal with the chassis technology mixture in the early development phases. While trying to give an answer to this question, there are a lot of other related questions raised. For instance, „How should the relevant data be stored and shared?" or "What kind of simulation has to be done at which time during the development process?". For the sake of delivering an answer to all these questions, it is necessary to understand what challenges development engineers and manufacturing planners are exposed to.

The main problem regarding the point of time in the product development process is when the first manufacturing simulations should be conducted. Since it is nearly 70 month before SOP, no hardware is available and all simulations need to be completed virtually. Therefore the usage of CAx-methodologies[3] is unavoidable.

2 Period of time refers to variable contents for developing car families

3 CAx – i.e. "Computer aided X"

One of the challenges is the allocation of the input data. Information outgoing from the chassis – for example the geometrical shape of individual chassis components – is as previously mentioned, highly variable. Meanwhile, information outgoing from the manufacturing process – for example the reference order of fitting – is not reliable because of the possibility of making an assessment based on current produced cars and existing plant layouts. Furthermore the order of fitting could be changed often.

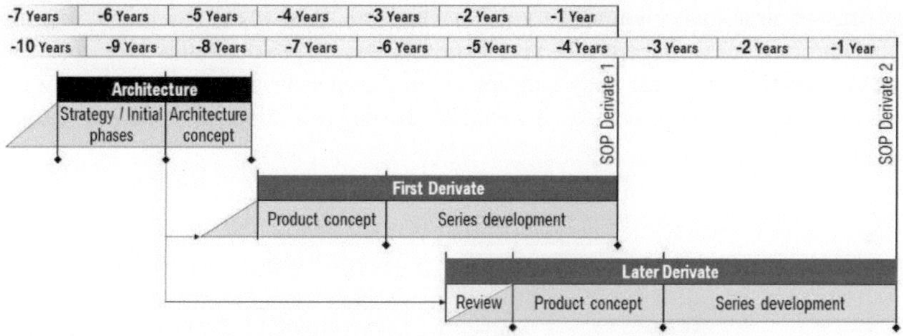

Figure 2 – Automotive product development process in terms of car families[4]

Another problem regarding to the point of time is when the manufacturing simulations get started in reality. There are two basic kinds of developing processes used by the BMW Group. The first one is called architecture development process and it is timely classified between 84 and 48 months before SOP. The second one is called derivative development process and is linked right after the architectural one (figure 2). In most cases, the developers and manufacturing planners start discussing the project when the project conditions are cleared because it would be inefficient to invest workforce of many hundred manufacturing planners in the early phases while their work would be required in the phase right before SOP for coordinating it. This results in the necessity of defining specifications of the car and the chassis. In this way, the guidelines based on the car family are defined as well and the capability of changing concept-relevant issues in a cost-efficient way is not possible anymore.

For example, the change of the concept of the marriage process in the series development phases of a derivate nearly 24 months before SOP would cause costs of several million euros. This concept is majorly defined by the "dresslevel"[5] of the pre-assembly peripheral of the chassis components – especially of the suspension strut, as can be seen in figure 3. If the suspension strut is mounted to the car body, it is much

---

4 Own illustration based on [12] p.9 and extended by expert knowledge

5 The "dresslevel" is a nomenclature to indicate different stages within the assembly

easier to transport the pre-assembled chassis to the marriage process station, because the position of the suspension strut is defined by the top mounting and the force of gravity. In addition, the parts of the trunk interior could be mounted in the steps before the marriage process but right after the fitting of the suspension strut because the accessibility for tools to mount the strut bearing is not necessary anymore. Besides, the suspension strut has to be guided by robot or by hand to the point of mounting to the chassis. If the suspension concept is implemented as a space functional axle with a suspension strut connection at the lower control arm, this way of guiding during the whole process of marriage can be very long. If a quality problem with the suspension strut connection at the car body gets noticed in the later development phases as a result of bad addition of tolerances, it would be possible that this concept will not work.

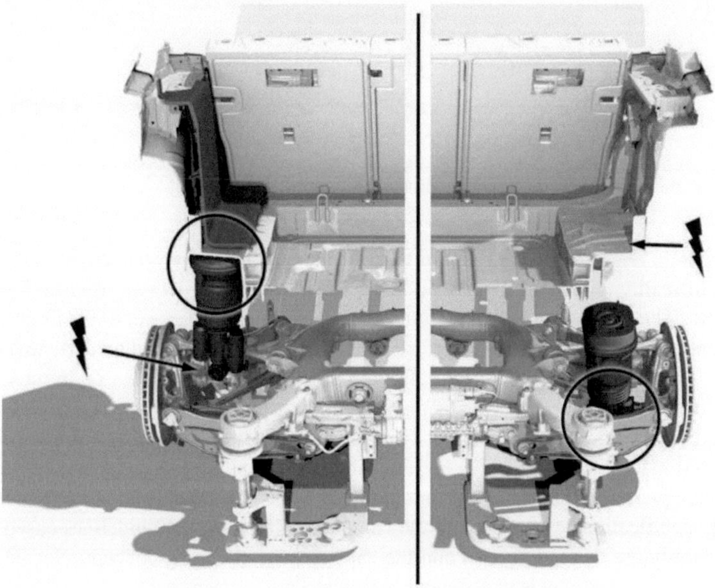

Figure 3 – Fitting process “Marriage” with two different orders of fitting

In this phase the manufacturing facility is already planned and appointed from the construction supplier. If the marriage concept should be changed now, the position of the whole marriage station in the production line has to be changed as well, because of the inaccessibility for the mounting tools for the trunk interior. This is just the result of changing the position of one single component in the order of fitting within the whole construction.

In order to avoid this adjustment, a possible solution would be to employ even more manufacturing planners in the architecture development and let them complete all the simulations they would normally do later in the derivate development. This approach is supposed to minimize the previously explained problem, but this process would be very inefficient. Most of the simulation results are not valid because of the variability of the input data. Hence, it is necessary to allocate as little manpower as possible in these simulations. For this purpose, the automatization of providing the input data and simulation results is necessary, too. This would enable the process engineer to rerun the simulations with minimal effort at any time there is new input data available.

In academic research and industry this approach of frontloading is not particularly new [1] [2] [3]. Also CAx-methodologies for simulating manufacturing and assembly processes got invented and executed [4].

For example MEIßNER and MÜLLER wrote in 2007 about challenges of integrating product development and production planning for products in series production [5]. Their focus was on actual conditions and used proceedings for an early phase's product validation. Later in 2008 MEIßNER made a proposal for a methodology which analyzes the accessibility of fitting tools [6]. With that methodology it is possible to estimate geometrical issues between existing production tools and products in an early development phases. The author defined the missing data structure the methodology requires as need for action, to get this approach prototypically implemented.

Simultaneously, RESCH and BAER described a methodology for a standardized representation of joints on the example of fixed connections [7]. Therefore, they analyzed the needed joint properties and elaborated a type of documentation, which allows an automatized validation. In 2010 WACK, BAER and STRAßBURGER indicated different challenges for the virtual validation of the production start-up [8]. As an example, they criticized the missing feasibility of simulating fitting sequences and therefore needed interfaces between simulation-involved tools. Later in 2011, WALLA, BAER and OVTCHAROVA presented a novel approach for developing a standardized and modularized production system [9]. There they used a kind of product structure that allows the developer to design a new product inside of the production environment. In this way, they specified a possible solution for the problem MEIßNER wrote in 2008. This connection between modularized production and the product structure has a problem described below.

For connecting a production module to a section of the product structure automatically it is necessary to identify the chosen section. Therefore, it is indispensable to standardize the product structure and the production modules. Another problem is that the production modules are regarded as invariable. So it is possible to value the product design concerning to the actual applied production environment, but it is not possible to value potential modifications inside the production modules. Interesting would be the optimum between changing the product and adapt the production, as well. Also the facility of simulating fitting sequences while developing new components would be attractive.

But the implementation of those operation models in an existing company combined with a data base for processes used in all development phases is not uncomplicated. In reality there are rather ridged corporate structures which is not necessarily bad. But over time every organization unit – for example product development or production – formed and optimized their own working processes usually isolated from each other. Now is the time to define a companywide process model for simulating and rating manufacturing processes for the early development phases – especially for the architecture development phases – which has an interface to manufacture planning tools and approval processes.

In the following paragraphs, the idea of a companywide work model and data exchange model necessary for this purpose will get explained.

3 Novel approach

In order to prove how probable and realistic the integration of a car concept in an existing plant is, it is necessary to combine the processes of organization unit product development and production much earlier in the development process in an efficient way. As the process is implemented in such early development phases all analyses, which have to be completed, have to be virtual. Therefore, the data structure for geometrical manufacturing simulations has to be designed in a way both departments can easily use. Because of the huge number of car projects and different plants, the structure and its data should be stored in an object-oriented manner. The objects which should be stored in this data structure are 3D-CAD-Models, review sheets, process information, and simulation results. Furthermore, the structure of this data structure has a logical connection to the manufacturing process. So it is possible to visualize the order of fitting especially for a defined car configuration on a manufacturing line directly in the CAD design tool.

While the development engineer is designing a chassis component, he is aware which other components are already mounted when this component will be added. So it is not directly essential to investigate which component gets mounted afterwards.

In the second step, the manufacturing planers can save their tools in the same data structure in the relevant fitting step. So the design space, which gets reduced by the mounting tools, is individual for each fitting step. In sum, the overall available design space increases – assuming the order of fitting does not change so often in this phase. This is a theoretical design space potential for the chassis development and needs to be further analyzed.

In the next stage of expansion, it is possible to execute fitting simulations directly and automatically with the data from the data base. So it is possible to simulate fitting processes – for example insert connection elements – and also fastening these components with the assigned tools (figure 4). The special feature is that this is not just possible in construction position. While simulating, it is possible to move all components of the chassis in a predetermined position, which is specific for this one fitting step. A good example for that is the process a production specialist calls "engaging". This will be further thoroughly explained in the following paragraph.

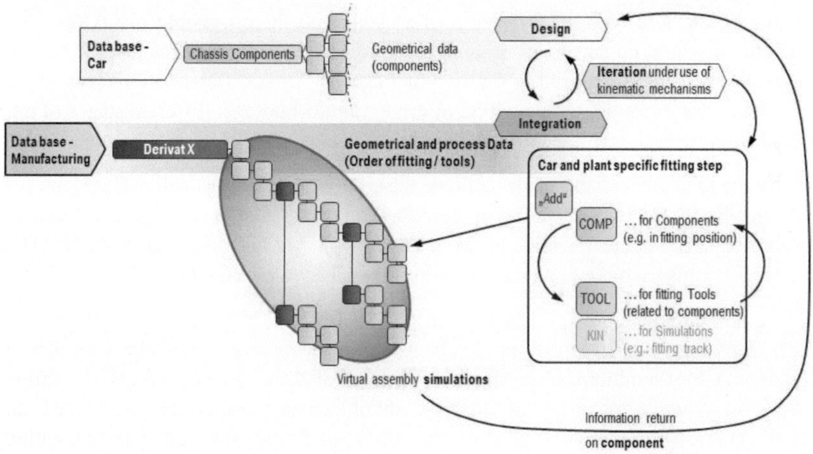

Figure 4 – Relation between different types of data structures for geometrical information

To give a future prospect – in the last step it is planned to combine this simulation structure with a special reporting tool. With this reporting tool it will be feasible to store all information abound what kind of input data from whom has ever been saved. This could be frozen every time a new phase in the development process gets started. So it would be possible to retrace all decisions have been done during the whole process of design matching.

4 The Details

The initial situation is that the task for developing a new architecture has been already finished. This is one of the earliest milestones in the architecture development process. The functional requirements each car has to achieve, have been approximately defined and a first indication of which derivate will be manufactured in which plant has been done.

A suitable type of chassis has to be chosen based on the functional requirements. As an example, the methodology for a double wishbone front axle with separate lower control arms will be presented. In the first step, the development engineer makes a design proposition with first geometrical forming. In the same step, the manufacturing engineer provides the information in which plant this kind of chassis with similarly characteristics is currently mounted. This information includes the order of fitting based on the parent model and is for this phase detailed enough. Because of the architecture development, it is not enough to do this for only one derivate. It is necessary for each car of this architecture and each possible manufacturing site. For example – for an architecture with 10 different car projects and three possible plants for each derivate, it would be necessary to generate 30 discrete data structures just for having an overview. This is exposed in figure 5 on page 10. Later differentiations have to be executed – for example, for each kind of drive train technology, different stages of optional equipment or left-hand drive/right-hand drive.

To handle this, it is indispensable to create these structures automatically. Therefore, the information for the manufacturing process should be available in an open and standardized format. This standardization is also needed for getting this information into a CAD-System – in this example implemented with Dassault Systèmes CATIA V5.

To explain the methodology it makes sense to look at the first step at just one car project. The whole data structure simply consists of basic structural elements – similar to the approach of modularized production [9]. Each of them represents a generic fitting step – for example mounting any lower wishbone to any axle carrier. Combined, all these generic fitting steps show a generic order of fitting. Recently after allocating concrete data out of a special car project, the order of fitting becomes car-specific. It means that the specific fitting step includes the car-specific wishbone. Important is that this specific wishbone has a logical connection to the CAD-Model of the car. This is explicitly needed to exclude redundancies in the data structure.

After specializing the order of fitting in the data structure, the opened CAD-Model has the appearance of a kind of stairs. If now the order of fitting should be adjusted as part of the development process, it would not be necessary to delete the information in the structure. It is easily feasible to change the content of two steps. The usability in the CAD-Model is also an important fact for doing a lot of iterations in this early phases of the development process.

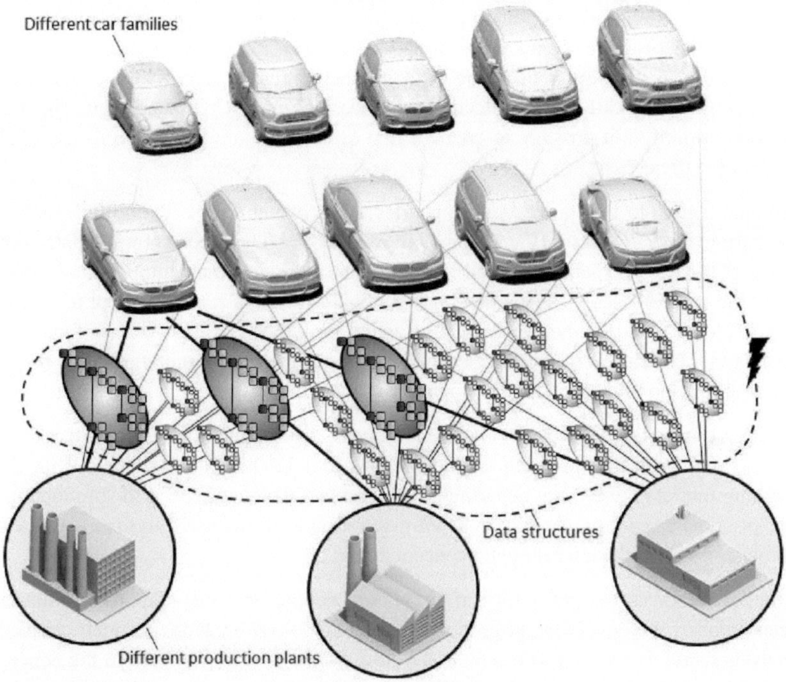

Figure 5 – Complexity of theoretical possible car-plan-configurations

So it is possible to create ready-made fitting sequences for components out of construction kits – for example wheel bearing, brake caliper and brake disc, which is also known as pre-assembling "cornermodule". So the developing engineer has the facility to use the so called "early-phases" fitting sequences for analyzing the wheel carrier concerning to manufacturing issues. In a later stage of expansion, these "early-phases" fitting sequences shall be available in a separate fitting order construction kit.

Looking at the whole architecture development, this kind of working model has the benefit that for each chassis type such a data structure could be easily created based on the concept-relevant connections – similar to the methodology of standardized connection documentation [7]. These concept-relevant connections are different for each type of chassis. A double wishbone front axle has different joints than an integral beam rear axle or a space beam axle. As an example – the concept-relevant joints of a double wishbone front axle with drained lower control arm level are illustrated in figure 6 on page 12. If each joint of a chassis type gets connected to a generic fitting step in a generic order of fitting, this order of fitting turns into a chassis-type-specific generic order of fitting. Only in combination with the car-specific geometrical data it becomes a car-specific order of fitting. This can be compared with a kind of manufactural construction kit for a selection of chassis types. Combined with the part construction kit, is it possible to create a first draft of a manufacturing process while developing the chassis.

Simultaneously to this process, the manufacturing planer inserts the correct tools to the correct step in the specific order of fitting. So the developer knows exactly how much of the design space is needed for fitting and screwing each component. Considering the early phases of development, it is detailed enough to use concept tools with less grade of detail or the tools from an actual produced car with similar functional characteristics. This is the connection between the organization unit "production" and the organization unit "product development".

Later on during the derivate development process, the manufacturing planer can insert the car and projected production line specific tool in the data structure. With this working model between the departments of "product development" and "production", it is possible to gain the level of concept maturation from the early architecture development phases to the later derivate concept phases.

Another reason is the possibility of simulating geometrical manufacturing issues in each point of the development process. The methodology includes geometrical clash analysis which is feasible to check all saved tools in the data structure in the concept design space. In this way, the time of each iteration step decreases. With the approval of a constant number of iteration steps while developing chassis components, the total time needed sinks. But this is not the desired result. A better effect of the methodology is the assumed increasing number of development iterations in the same development time and besides this, the increasing concept maturation.

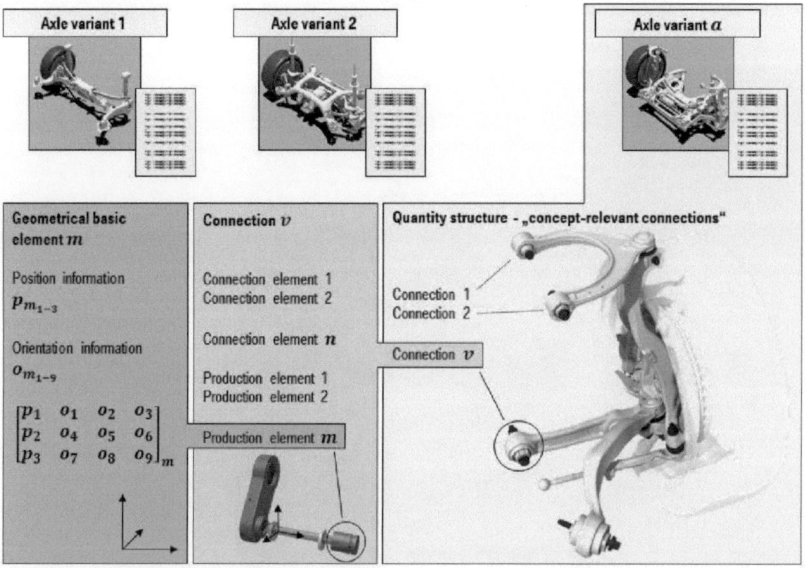

Figure 6 – concept-relevant connections on the example of a double wishbone front axle

A good example for this is the previously mentioned process of engaging. The peculiarity is that the so-called "dresslevel" of the axle – the level of actually fitted components – is high but not completely. When the totally mounted drivetrain is getting fitted to the preassembled chassis, the wheel carrier and the suspension strut have no geometrical defined position. For mounting the drive train, it is necessary to flap away the wheel carrier and suspension strut because the drive shaft needs a lot of space while fitting the drivetrain. While flapping out both components, it is possible to get the chassis in a position it would never get in in operating condition in the assembled car. So the angle in the ball joints between the control arms and the wheel carrier can be for a short period of time larger than the theoretical permissible maximum (figure 7 – page 12). The result is that the ball joint will be damaged before the chassis has ever been assembled to the car.

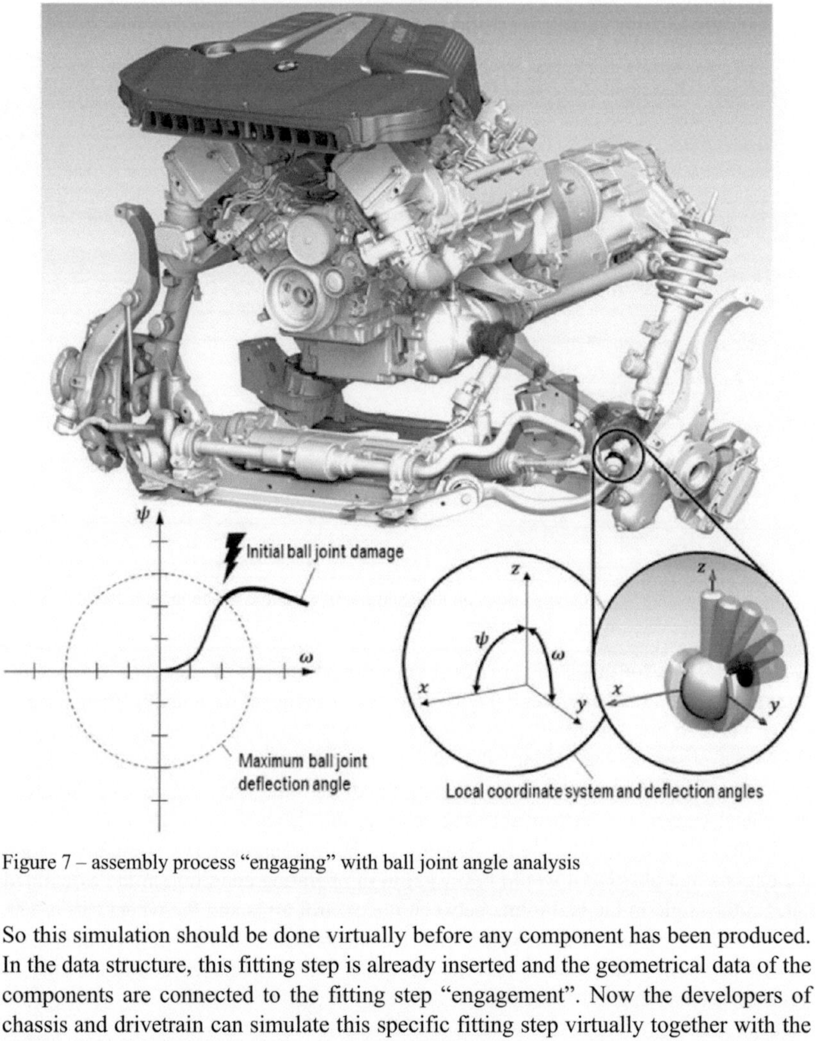

Figure 7 – assembly process "engaging" with ball joint angle analysis

So this simulation should be done virtually before any component has been produced. In the data structure, this fitting step is already inserted and the geometrical data of the components are connected to the fitting step "engagement". Now the developers of chassis and drivetrain can simulate this specific fitting step virtually together with the manufacturing planner this specific fitting step virtually. Therefore they do not have to create a new kinematic mechanism. This feature is directly implemented in the kinematic model of the chassis. So this functional model got enlarged by manufacturing functions with geometrical impact. The developer can simulate chassis with operating conditions and while manufacturing conditions in the same model with very fast iteration and changing steps.

In the next step, the manufacturing tools should be connected to the fitting step, as well. Now the order of fitting becomes car-specific and plant-specific – so the manufacturing process is explicitly described. All the geometrical simulations, which have been already conducted, can be saved in the data structure, too. The logical connection between fitting steps and the simulations, which have been done, is comprehensibly defined. With this approach it is possible to retrace sustainably while developing done decisions.

To depict this working model in a process diagrammatic it is necessary to choose a suitable type of process diagrammatic. In figure 8 is the introduced methodology in form of a cross functional process map visualized ([10] p.349 ff).

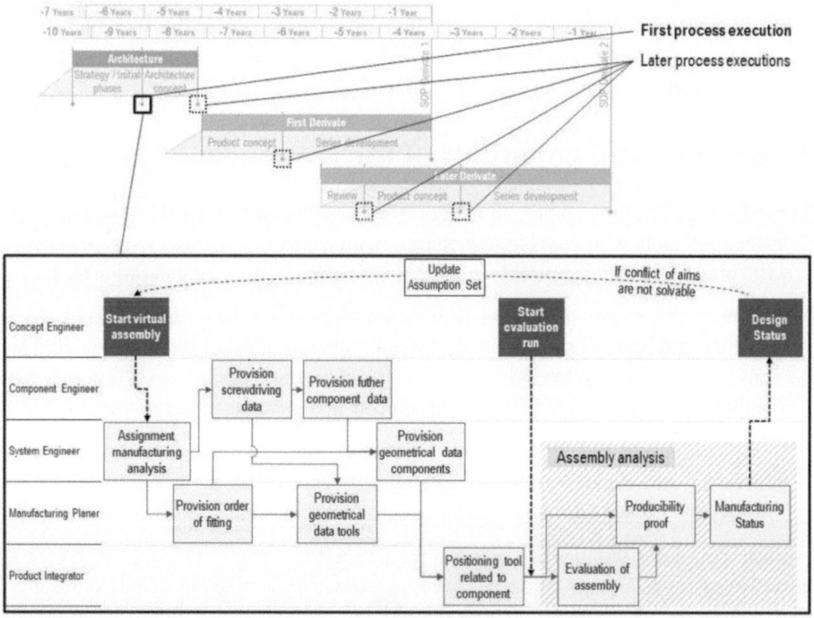

Figure 8 – process design for assembly analysis related to automotive product development

This is a first draft of a companywide working model for providing and using information for the simulation of manufacturing issues with prevalent geometrical impacts. As seen, the difficulty is to implement this in practical application. Therefore, the process has to be adapted so that the task of each process-involved engineer is included.

The process is companywide, because the process owner is located in the organization unit "Development" and some process-involved engineers are located in the organization unit "Production". Like in Figure 8 shown, the process owner is the so-called "Concept Engineer". He is responsible for the whole product concept, including the indemnification of producibility. The first part of the process is needed for the provision of many different kinds of information, for an example screwing data. All data should be saved in the previously explained data structure because the engineers and planers have the chance to adapt and correct their data due to providing them. This has to be an iterative process. These information has to be provided when the second part of the process starts. This process of assembly analysis accounts to persons out of the organization unit "Production". Nevertheless the result of this analysis is an important indicator for the concept maturity of the chassis concept and has to be committed to the concept engineer. By now it is to define this process accurate by introducing this method to all process-involved engineers to get a validation of the implementation in the early phases of the product development process.

5 Benefits and potentials of the approach

The main benefit of this approach is the clarity of collaboration between different organization units while developing chassis. Connected with a process-involved specific process chart it is possible to increase the efficiency of each iteration step.

Furthermore, the resulting increase in iteration steps in the early development phase improves the concept maturity by a constant level of effort. Therefore, the simulations have to be completed virtually and should be standardized. So it is possible to compare the results from different car projects in a meaningful way.

With standardized geometrical and functional simulations and identical post-processing and evaluation, the repeatability is feasible. So the transparency of assessment processes for manufacturing issues in early and later phases raises.

Further approaches will be researched in the future, for example a theoretical potential of design space as a result of sequential geometrical simulations of manufacturing issues with geometrical impact. If it is possible to fix the order of fitting in an earlier development phase, the overall available design space should theoretically increase.

The depicted working model has the following potentials and disadvantages which have to be further investigated.

To get the level of effort constant, it is necessary to automatize as many analysis steps as possible. Therefore, more standards and methods are needed. For example, a numerical representing of joints for an automatized clash analysis of mounting tools and joining elements should be developed.

Actually there is no alternative for a different order of fitting integrated. For Example, if a derivate gets produced in two plants, it is not possible to match the information between both production lines easily. It should be feasible to simulate the fitting sequences in both plants simultaneously in order to get a view on potential improvements in the production process and the product, as well. Therefore, an extension of this methodology is needed. Also the representation of market-specific extra equipment and their apportionment over the different plants is actually not implemented.

It is necessary to connect all data structures logically for developing whole car families. This could be implemented by adding a relation knowledge to each generic fitting step. So it is possible to get an overview about the complexity of the development system – especially in the period of time when new and old cars of one type are being produced at the same time.

6 Conclusion

The novel approach is a process design for a method of operation between different organization units with focus on virtual manufacturing simulations.

During the process of developing modern vehicles, many requirements regarding product functionality and process efficiency are to be implemented. Subsequently, this is due to constantly growing competitive pressure and evermore present regulatory framework.

A possible solution is the development of product families, vehicle architectures, which allow the acquisition of as many synergies as possible between single products for the sake of cost reduction in the production.

In order to establish a holistic evaluation of a product concept, it is necessary to analyze not only functional issues with geometrical impact, but also manufacturing engineering ones. While considering the range of requirements, it is noticeable that different requirements exclude each other depending on the process partner who imposed them.

Therefore, to support the process of finding an efficient and sustainable solution, it is imperative to connect the departments of product development and production planning. The following strategy, which outlines a modification of the product development, relates to the product development process and a consistent methodology.

This methodology enables the evaluation of product maturity in the early development phase, based on existing production and assembly structures which accelerates product maturity. The main aim of this new methodology is to allow the possibility of estimating product modification effort by changing the assumptions of manufacturing engineering.

Main constitutes of this methodology are:

- Structure of geometrical product data related to assembly processes and order of fitting
- Standardized definition of fitting elements for each chassis concept-relevant connection
- Working model and process design for a comprehensive provision of input data for virtual assembly simulations

Automotive chassis are convenient practical examples because the complexity of developing them is in geometrical and functional ways high. Also the specialty of manufacturing issues and the assembly of requirements in early development phases are extensive.

References

[1] K. Ehrenspiel, A. Kiewert, U. Lindemann und M. Mörtl, Kostengünstig Entwickeln und Konstruieren – Kostenmanagement bei der integrierten Produktentwicklung, Berlin, Heidelberg: Springer, 2013.

[2] M. Eigner, Frontloading – Ein Potential für die Optimierung des Produktentwicklungsprozesses, Universität Kaiserslautern, 2005.

[3] U. Seiffert und G. Rainer, Virtuelle Produktentstehung für Fahrzeug und Antrieb im KFZ, Wiesbaden: Vieweg + Teubner Verlag, 2008.

[4] J. Milberg, Moderne Methoden zur Montageplanung, München: Herbert Utz Verlag, 1998.

[5] R. Meißner und M. Müller, „Herausforderungen an die Integration von Produktentwicklungs- & Produktionsplanungsprozesse zur Entwicklung serienprozessfähiger Produkte," *18. Symposium "Design for X",* 11./12 Oktober 2007.

[6] R. Meißner, „Proaktive geometrieorientierte Montageabsicherung," *19. Symposium "Design for X",* 09./10 Oktober 2008.

[7] J. Resch und T. Bär, „Dokumentation und geometrische Absicherung von Verbindungen im Automobil-Produktentstehungsprozess," *19. Symposium "Design for X",* 09./10 Oktober 2008.

[8] K.-J. Wack, T. Bär und S. Straßburger, Grenzen einer digitalen Absicherung des Produktionsanlaufs, Karlsruhe: KIT Scientific Publishing 2010, 2010.

[9] W. Walla, T. Bär und J. Ovtcharova, „Impact of modularised production on product design in automotive industry," *International conference on engineering design, ICED11,* 15.-18 August 2011.

[10] S. Lunau und R. Meran, Six Sigma + Lean Toolset: Mindset zur erfolgreichen Umsetzung von Verbesserungsprojekten, Heidelberg: Springer, 2012.

[11] D. Kreyenberg, Fahrzeugantriebe für die Elektromobilität – Total Cast of Ownership, Energieeffizienz, CO2-Emissionen und Kundennutzen, Berlin: Springer Vieweg, 2016.

[12] J. Weber, Automotive development processes, Berlin, Heidelberg: Springer, 2009.

[13] DLR und Wuppertal-Institut, Begleitforschung zu Technologien, Perspektiven und Ökobilanzen der Elektromobilität – STROMbegleitung, Stuttgart, Wuppertal, Berlin, 2015.

DRIVING COMFORT

Experimental comparison of the analysis methods for vibration transmission in the chassis

Dipl.-Ing. Christian Schubert

Stephan Pries, M. Sc.

Dipl.-Ing. Martin Jaensch

Dr.-Ing. Kemal Çalışkan

Prof. Dr.-Ing. Ferit Küçükay

Institute of Automotive Engineering,
Technical University of Braunschweig

P.E. Pfeffer (Ed.), *8th International Munich Chassis Symposium 2017*,
Proceedings, DOI 10.1007/978-3-658-18459-9_25

Abstract

Vibration isolation is a crucial distinguishing feature in automotive development and corresponds directly with brand image and costumer complaints. Due to the continuous drop in drive noise by improved or even novel drive types the ride comfort and its enhancement is of growing importance. Regarding ride comfort, suspensions are often still tuned by expensive and time demanding expert evaluations with several road tests in late-stage development. To accelerate tuning processes and uncover further optimization potential of suspension components, it is necessary to detect the dominating vibration transfer paths between road and passenger compartment. In noise frequency range, extensive investigations of noise transmission in suspensions have been performed by means of different transfer path analysis (TPA) methods with differing quality and effort. However these methods have been sparsely investigated in the vibration (and harshness) frequency range.

This work focuses on detailed investigation of TPA methods applied to a suspension system in the low frequency range. A multi-link rear axle of a test vehicle was mounted in a quarter-car test-rig with fixed body frame and excited by the relative displacement between wheel-hub and body measured during a test on typical roads used for tuning processes. The test-rig is equipped with force sensors at several connection points between the suspension components and body frame. Measurement of the forces ensures a direct detection of the vibration transfer paths and thus serve as a reference for a comparative study of TPA methods in the low frequency range (up to 30 Hz).

1 Introduction

Optimizing the vibration comfort for passengers is an ongoing process and becomes more and more a crucial distinguishing feature in automotive development. Not least electric drives and, in particular, the increasingly automated driving result in the fact that the Noise Vibration Harshness (or short NVH) comfort is increasingly perceived by the vehicle occupants.

Regarding ride comfort, suspensions are often still optimized by expensive and time demanding expert evaluations with several road tests in late-stage development.

This is due to the high system complexity of the chassis components involved, since these hold, on the one hand, non-linear and time-varying transmission behavior (damper, rubber bearing, etc.) and, on the other hand, are often coupled in a complex geometry.

Although the characteristics of the individual components can be measured well on test-rigs and thereby modeled, the interaction of all components in the chassis is not possible to readily measure. Detecting (dominating) vibration transfer paths between road and passenger compartment allows to accelerate the tuning process and uncover further optimization potential of the components. In noise frequency range, many extensive investigations of noise transmission in chassis have been made in the last decades. Different transfer path analysis (TPA) methods with differing quality and effort were used. However these methods have been sparsely investigated in the vibration and harshness frequency range. This work focuses on detailed investigation of common TPA methods applied to a chassis system in low frequency range (up to 30 Hz).

Therefore this work is structured as follows:

1. Theory of transfer path analysis, especially in the context of an application to the vibration transmission within the chassis, and explanation of the methods used in this work.
2. Experimental investigation of the methods applied to a multi-link rear axle at a quarter-car test-rig.
3. Summary of findings in this work and outlook for further investigations.

2 Transfer path analysis methods

A transfer path analysis (TPA) is used to detect the vibro-acoustic energy transfer path of a vibrating system in operation and thus to identify the vibration transmission behavior of this system. The results of a TPA can be used to show the influences of system modifications and to obtain approaches for optimizing the transmission behavior. TPA is usually applied for noise treatments (structure and air-borne noise), but in same way they can be applied to vibration treatments [1,2].

In Figure 1 the transfer paths of a simple chassis for a comfort relevant vehicle vibration z_S is shown. For example, such a target vibration may be the vertical seat-rail acceleration at the driver's seat, which is known for a significant impact on the human vibrational receptance. The vibration system, which is exclusively excited by the road unevenness h, can be separated into an active system consisting of the chassis (indicated by *ch*), and a passive system consisting of the vehicles body (indicated by *bd*). In this, the vibrations on the passive side are exclusively caused by the vibrations of the active side. However, the vibrations on the active side are caused as well by the road excitation as by the coupled vibrations of the passive side. If relevant secondary excitations of the passive system are not taken into account, i.e. the active system is not sufficiently captured, an erroneous identification of the transfer paths could be the result. Therefore in this work the road unevenness is said to be the only source of excitation. The dividing line between the active and passive system can in principle be drawn as desired, but for the sake of clarity the chassis is to be regarded as an active system and the body as a passive system and thus as a receiver structure for the vibrations transmitted by the active system.

The contribution of the individual paths (in this case four paths) to the target response is then the result of a TPA:

$$z_S = z_{S,path\,I} + z_{S,path\,II} + z_{S,path\,III} + z_{S,path\,IV}. \quad (1)$$

A number of methods are presented in the literature to determine these path contributions, in which different weights between effort and accuracy are accounted for [e.g. 3 to 6].

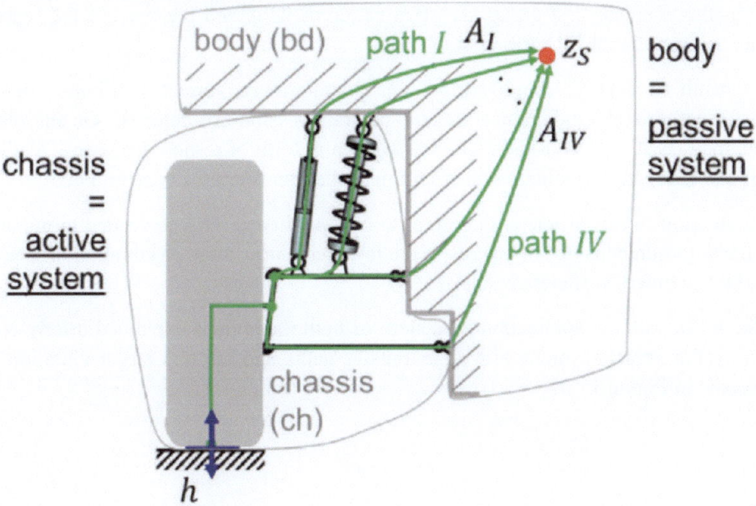

Fig 1: Basic concept of transfer paths applied to a chassis system only excited by road unevenness h with separation in active system (chassis) and passive system (body).

These methods for detecting the vibration transmission paths of a system in operation can basically be divided into two groups, as shown in Table 1:

- *Methods involving direct or indirect measurement of the forces and/or torques acting in operation and multiplication of these with the transmission behavior of the target structure*

  The forces are either measured directly through force sensors or calculated indirectly by using mount stiffness or matrix inversion of local inertances in cases where the application of force sensors is not possible due to weight or space restrictions. Depending on the number of interfaces in the system, this type of TPA can be very time-consuming and cost-intensive.

- *Methods involving measurement of the operational vibrations at so called indicator points and the target point to detect dominating transfer paths by using the transmissibility between indicators and target*

  In principle, different vibration variables (often accelerations, but also displacements and velocities) can be used as indicators. Since only operational measurements are required, this method holds significant time and cost advantages over the force-based methods.

The methods of the first group are often called as TPA and the ones of the second group as Operational TPA (or short OTPA).

As a result of a TPA, the physically, i.e. quantitatively, correct distribution of the transfer paths is obtained, since source and transmission are detected. On the other hand, the OTPA only considers the transmission in form of response-response evaluations and thus only a qualitative distribution of the transfer paths is determined.

The intention of this article is therefore, to assess whether TPA is needed to uncover the transmission behavior of chassis in the low frequency range of vibration or whether OTPA achieves sufficient results.

In the following, the application procedure of both methods is explained briefly considering the vehicle system with four transfer paths and the one target vibration z_S presented in Figure 1.

Table 1: Common methods for transfer path analysis applied to the vibrations transmission of a chassis.

	source	transmission	path contributions
TPA	operational in car 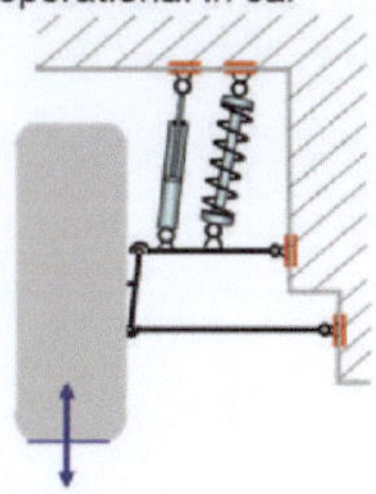	separate on test-rig 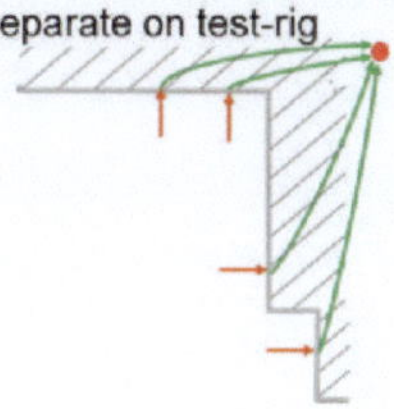	product of body-chassis interface loads and transfer functions of body
OTPA		operational in car 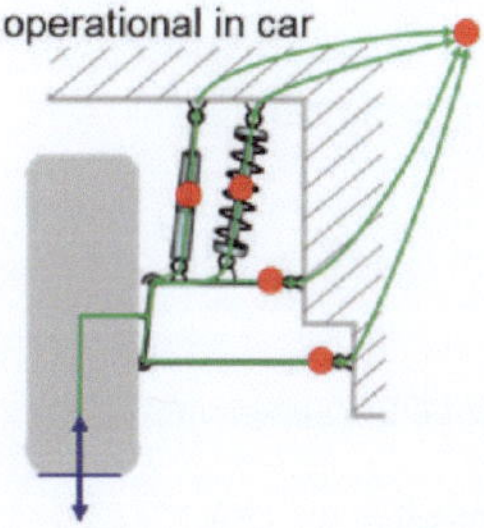	partial path dominations from transmissibility between target and indicator responses

TPA

Generally the TPA consists of three main steps:

1. Determine the interface loads F_i between chassis and body during operation along with the target response

$$\boldsymbol{F} = [F_I, F_{II}, F_{III}, F_{IV}]^T, \tag{2}$$

   a directly by force measurement, or

   b indirectly via mount stiffness properties and measurement of differential interface displacements, or

   c indirectly by measuring local inertances (i.e. Frequency Response Functions “FRF”) at the interfaces and multiplying the inverted inertance matrix with local measured indicator vibrations.

2. Determine inertances A^{bd} between interfaces and target using the isolated vehicle body, e.g. by impulse hammer, shaker or reciprocal measurements

$$\boldsymbol{A}^{bd}_{\boldsymbol{F}\to S} = [A^{bd}_{F_I\to S}, A^{bd}_{F_{II}\to S}, A^{bd}_{F_{III}\to S}, A^{bd}_{F_{IV}\to S}] \;. \tag{3}$$

3. Determine the path contributions from the interface loads and the inertances

$$z_S = \boldsymbol{A}^{bd}_{\boldsymbol{F}\to S}\, \boldsymbol{F} = A^{bd}_{F_I\to S} \cdot F_I + A^{bd}_{F_{II}\to S} \cdot F_{II} + A^{bd}_{F_{III}\to S} \cdot F_{III} + A^{bd}_{F_{IV}\to S} \cdot F_{IV} \;. \tag{4}$$

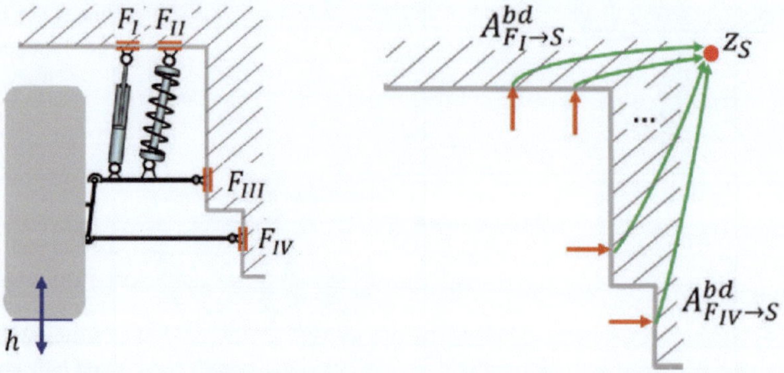

Force measurement during operation FRF measurement of the vehicle body

Fig 2: Measurements required for TPA.

OTPA

The OTPA method consists of three main steps as well, whereby only measurements in operation are required without separating the chassis from the vehicle body.

1. Measure vibrations w_i at path characterizing indicator points during operation along with target response.

2. Determine the transmissibility matrix from indicator and target responses:

$$\boldsymbol{H}^{ch\&bd}_{\boldsymbol{w}\to S} = z_S \cdot \boldsymbol{w}^{-1} \;. \tag{5}$$

Conditioning of the transmissibility matrix and thereby cross talk cancellation of the indicator responses is generally done by singular value decomposition or multiple coherence cancellation. Further information can be found in [7, 8].

In more detail, the transmissibility matrix is composed of the inertance matrix $\boldsymbol{A}_{h\to\boldsymbol{w}}^{ch\&bd} = \boldsymbol{w} \cdot h^{-1}$ between excitation h and indicators $\boldsymbol{w}$ and the inertance matrix $A_{h\to S}^{ch\&bd} = z_S \cdot h^{-1}$ between excitation h and target z_S:

$$\boldsymbol{H}_{\boldsymbol{w}\to S}^{ch\&bd} = A_{h\to S}^{ch\&bd}\left(\boldsymbol{A}_{h\to\boldsymbol{w}}^{ch\&bd}\right)^{+}, \qquad (6)$$

where the excitation is eliminated.

3. Determine the path contributions from the transmissibility matrix and the indicator vibrations

$$z_S = \boldsymbol{H}_{\boldsymbol{w}\to S}^{ch\&bd}\boldsymbol{w} \, . \qquad (7)$$

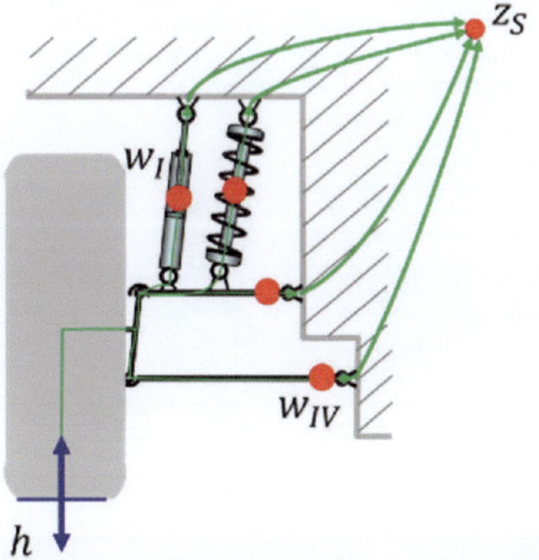

Fig 3: Measurements required for OTPA.

3 Experimental Comparison of TPA and OTPA

The multi-link rear axle shown in Figure 4 is used for investigation of the both analysis methods. An experimental vehicle with the specified rear axle is driven on typical roads used in suspension tuning process. In the next step, the rear axle of the experimental vehicle is built into a quarter car test-rig[1], which is shown in Figure 5 on the left side. One side of the rear axle together with the subframe is assembled to the test rig. The rear axle is equipped with the same sensor layout used in the road measurement.

The road unevenness profiles are then used to excite the left wheel of the rear axle via a single axis hydraulic cylinder. The vertical dynamic response of the axle at the test-rig is in good agreement with the axle response measured on the road. Based on the obtained results, it is concluded that the dynamic characteristics of the chassis system of the test vehicle are preserved at the quarter car test-rig. It should be noted that, in the quarter car test-rig the degree of freedom of the body frame is restricted to the vertical direction only (bounce). Thereby the pitch- and roll-dynamics are not present.

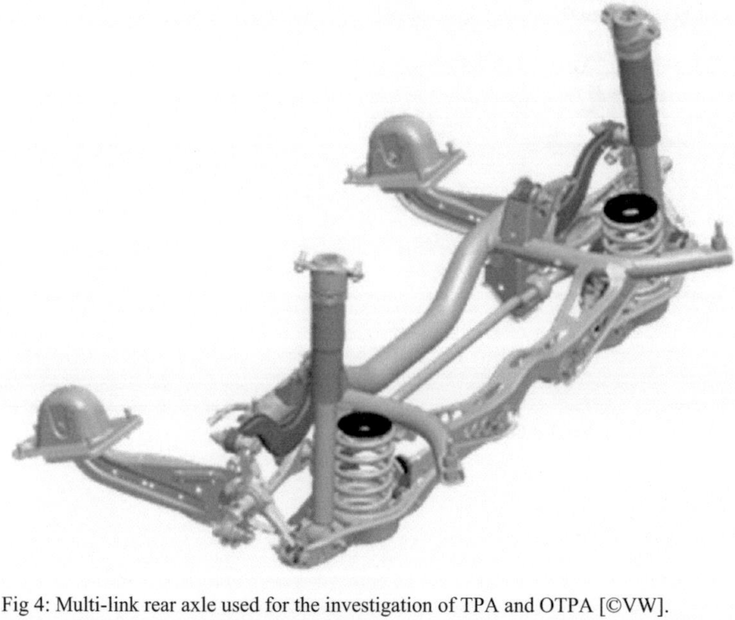

Fig 4: Multi-link rear axle used for the investigation of TPA and OTPA [©VW].

1 of the Institute of Automotive Engineering (IAE), Technical University of Braunschweig (TUBS)

In the following step, the body frame of the test-rig is fixed to the ground (i.e. bounce degree of freedom of the body is blocked) and all interfaces between rear axle and test-rig are equipped with force sensors, as shown in Figure 6. Measurement of all interface forces ensures a direct detection of the vibration transfer paths for TPA.

In addition to the force sensors, accelerometers are placed on all suspension components (longitudinal arm, upper transverse, lower transverse, tie rod, spring and damper) connecting the wheel carrier with the fixed test-rig frame and the axle subframe. Measuring accelerations of all components between active-system (=chassis) and passive-system (=body) provides sufficient indicators for the OTPA.

To eliminate the non-linear effects due to tire during the test-rig measurements and to allow an analysis of the vibration transfer characteristics of the axle only, the tire and wheel are replaced with a replacement adapter at the test-rig, as shown in Figure 5 on the right side. Consequently, to excite the axle similar to the road operation, the measured relative displacement h_{rel} between the wheel hub and body mass frame is used as the excitation signal for the test-rig with fixed body frame. In Figure 7, the normalized power spectral densities of the relative displacement h_{rel} and the measured road unevenness h are shown for a test road used in this study. As expected, the natural frequencies of the body and wheel are observable in the relative displacement frequency spectrum.

Free body mass, excitation via tire and wheel

Fixed body mass, excitation via wheel replacement adapter

Fig 5: Quarter-car test-rig of IAE with the mounted multi-link rear axle.

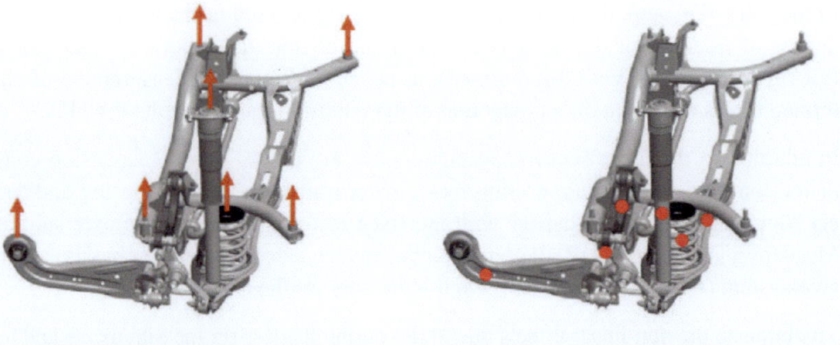

Fig 6: Sensor layouts for both investigated analysis methods.

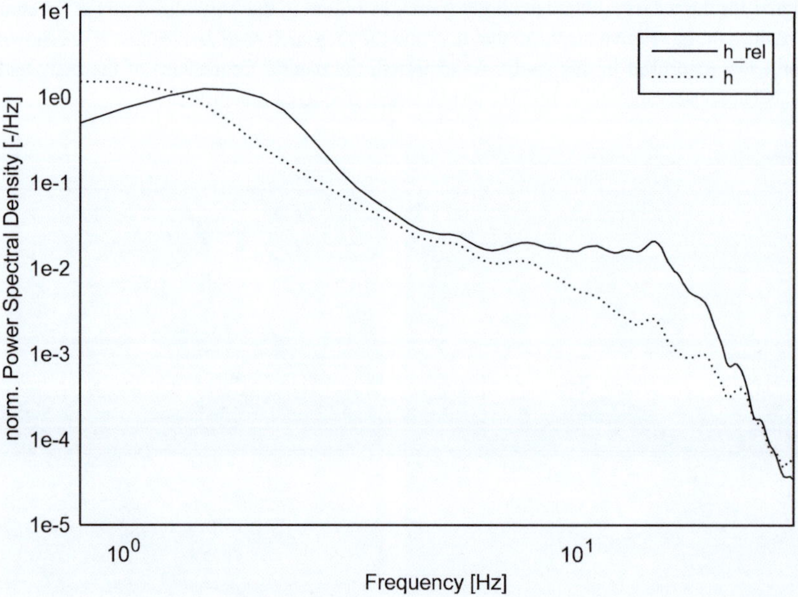

Fig 7: Normalized power spectral densities of the excitations for the quarter-car test-rig with free body mass (h, unevenness of driven road) and fixed body mass (h_{rel}, relative displacement between the wheel and body measured during excitation of the test-rig with free body using h).

In Figure 8, the transfer paths identified by the TPA are shown as amplitude spectra. The plotted curves in the figure correspond to:

- $Z_{bd,free}$ the body acceleration previously measured on the test-rig with free body as the target vibration, which is be broken down by TPA,
- $\Sigma\, Z_{i||bd}$: the sum of all single paths synthesizing $Z_{bd,free}$, which is for TPA similar to the (virtual) body acceleration of the test-rig with fixed body mass
- $Z_{S||bd}$: the spring (S) path,
- $Z_{D||bd}$: the damper (D) path,
- $Z_{LA||bd}$: the longitudinal control arm (LA) path,
- $Z_{Sub||bd}$: the subframe (Sub) path.

Comparing $Z_{bd,free}$ with $\Sigma\, Z_{i||bd}$, it can be stated that the experimental layout of the test-rig with fixed body mass and wheel replacement adapter excited by relative displacement achieves a good correlation with the body acceleration of the quarter-car with free body.

Furthermore, the spring and damper paths can be identified as the main dominating transfer paths at the body and wheel natural frequencies, respectively. The subframe path holds negative amplitudes, because the spring and damper are supported by the lower transverse link, which transfers the resulting counter-force via a bushing to the subframe. On the other hand, the longitudinal control arm holds the lowest contribution to the body acceleration.

In Figure 9, the transfer paths identified by the OTPA are shown as amplitude spectra. The plotted curves in the figure correspond to:

- $Z_{bd,free}$ the body acceleration previously measured on the test-rig with free body as the target vibration, which is analyzed by OTPA (same as in Figure 8),
- $\Sigma\, Z_{i||bd}$: the sum of all single paths synthesizing $Z_{bd,free}$,
- $Z_{S||bd}$: the spring (S) path,
- $Z_{D||bd}$: the damper (D) path,
- $Z_{TA||bd}$: the longitudinal control arm (LA) path,
- $Z_{LTL||bd}$: the lower transverse link (LTL) path,
- $Z_{UTL||bd}$: the upper transverse link (UTL) path
- $Z_{TR||bd}$: the tie rod transverse link (TR) path.

As it can be seen, the synthesized target vibration $\Sigma\, Z_{i||bd}$ achieves a good correlation with the actual target vibration $Z_{bd,free}$. However, the determined contributions do not match the actual path contributions determined by the TPA (see Figure 8). Although the damper path above the body natural frequency is correctly recognized as the dominant path, the qualitative participations of the remaining paths do not correspond to the actual path distribution.

Furthermore, the result of OTPA is very sensitive to the choice of the smallest singular values considered in the conditioning of the transmissibility matrix $\boldsymbol{H}$. It is not clear which value is most appropriate, because the match of the synthesized with the measured target response remains approximately the same. If, for example, very small singular values are considered in matrix conditioning, then the spring is identified as the dominant path over the entire frequency range. As a reason for false identification, the individual indicator accelerations have too high correlation with each other and thus do not allow a significant distinction for the OTPA.

For further investigation of the TPA and OTPA methods, simulations are performed using a validated multi-body model of the rear axle (see Figure 10). Simulations with the multi-body model show, that the OTPA can only detect qualitatively correct paths in the frequency ranges where the individual suspension linkages are excited in their own natural frequencies and therefore dominate the body acceleration in this frequency range. The indicator response measured on such a resonant excited suspension linkage shows a significant higher correlation with the target response than the other indicator responses do. As a consequence, the correlation of the indicator responses with each other and with the target response should be used for validation of the OTPA results. Only in case of sufficiently high differences in correlation, statements of the OTPA are credible. However, the simulations also show that there are no natural frequencies of suspension linkages in the frequency range up to 100 Hz, i.e. in the vibration and harshness frequency range.

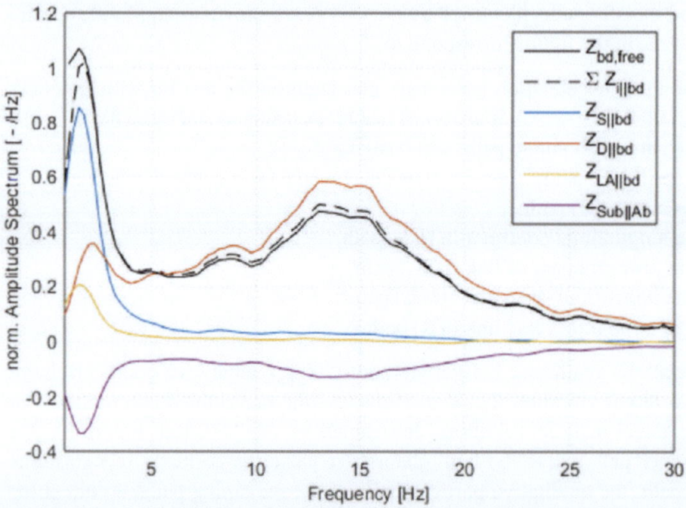

Fig 8: Vibration transfer paths identified through TPA.

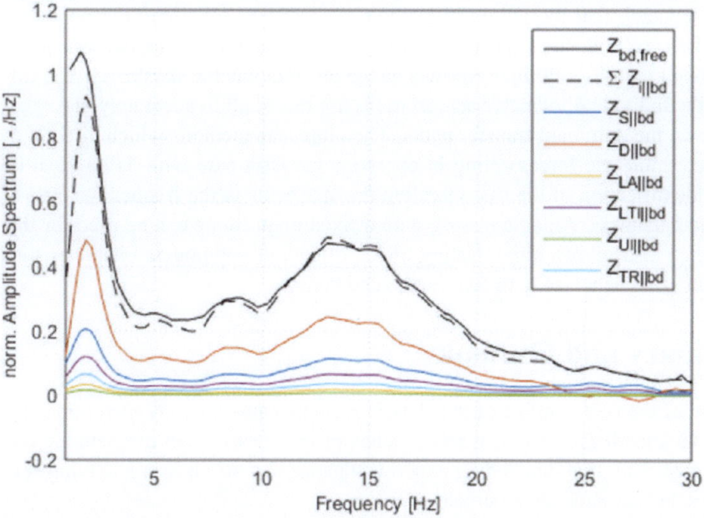

Fig 9: Vibration transfer paths identified through OTPA.

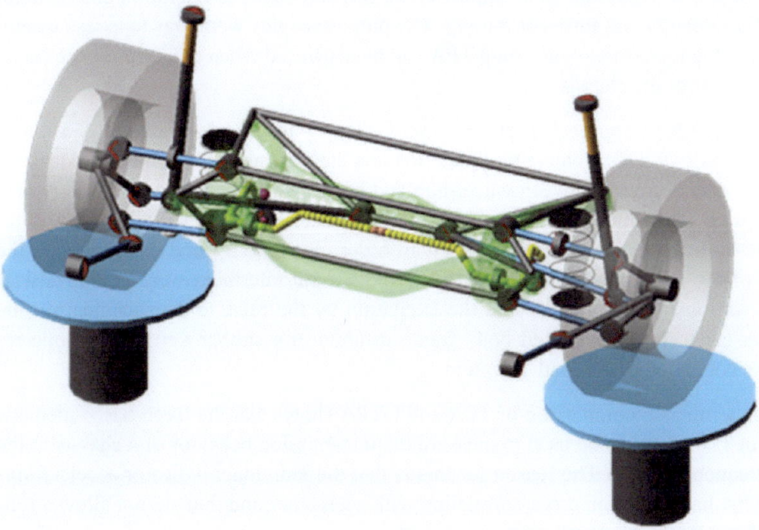

Fig 10: Multi body model of the multi-link rear axle.

It can be concluded, that the mere observation of vibrations measured during operation (as is the case with the OTPA) is not sufficient to achieve a breakdown of the transfer paths of a chassis in the frequency range of vibration and harshness. The differences in the individual indicator accelerations are too small to accurately and reliably detect even the dominant transfer paths. Therefore, the methods which directly or indirectly determine the forces acting in operation (as is in case with TPA) must be applied for identification of the true vibration transfer paths in the frequency range of vibration and harshness. An axle test-rig with a fixed main frame can be used for this purpose, which also allows quick testing of the effects of component characteristics on the transmission behavior between chassis and body.

4 Summary and Outlook

Optimization of the ride comfort of the chassis systems gains a steady increasing significance. The knowledge of transmission behavior in chassis when in operation can support the cost- and time-demanding expert evaluations of the chassis involving several road tests in late-stage development.

In this work, the basic principles of the transfer path analysis (TPA) were explained in context of an application to the vibration of chassis excited by road unevenness. Here the principles of the two different approaches of TPA and OTPA were described in more detail, whereby the latter requires only measurements in operation and is therefore significantly less time-consuming. The purpose of this work was to assess whether or not the more time-consuming TPA can be dispensed when determining the transfer behavior of the chassis.

To compare both approaches of TPA and OTPA in an experimental study, a multi-link rear axle of a test vehicle was mounted in a quarter-car test-rig and excited by the relative displacement between wheel-hub and body measured during a test on typical roads used for tuning processes. To avoid non-linear impacts of the tire, it was replaced by a wheel replacement adapter. In order to enable a force measurement, the body of the quarter-car test-rig was fixed and several interfaces were equipped with force sensors. It was shown, that the excitation by the relative displacement on the quarter-car test-rig with fixed body frame resulted in a similar vertical dynamic response compared to the road operation.

The experimental comparison of TPA and OTPA shows that the force-based methods (like in TPA) have to be used to measure the transmission behavior of a chassis in the low frequency range. The reason for this is that the individual indicator accelerations in OTPA have too high cross-correlation with each other and thus do not allow a reliable detection of the dominant transfer paths.

Simulations with multi-body models show, that in the frequency ranges in which individual suspension linkages are excited in their natural frequencies, vibration-based methods (like in OTPA) can probably be used to detect the dominant paths. The agreement of synthesized and measured target response should not be used for validation of the OTPA results. Rather, sufficiently high differences in correlation of the indicator responses with the target response should be taken into account for validation.

As a more practical alternative to the TPA with direct force measurement, more detailed studies on the validity of methods based on indirect force determination, in particular with mount stiffness, should be further investigated for application of TPA to chassis system in the low frequency range.

References

[1] Zeller, P.: Handbuch Fahrzeugakustik. Wiesbaden: Vieweg+Teubner Verlag 2012

[2] Genuit, K.: Sound-Engineering im Automobilbereich. Berlin, Heidelberg: Springer Berlin Heidelberg 2010

[3] Klerk, D. de u. Rixen, D. J.: Component transfer path analysis method with compensation for test bench dynamics. Mechanical Systems and Signal Processing 24 (2010) 6, S. 1693–1710

[4] van der Seijs, M. V., Klerk, D. de u. Rixen, D. J.: General framework for transfer path analysis. History, theory and classification of techniques. Mechanical Systems and Signal Processing 68-69 (2016), S. 217–244

[5] Janssens, K., Gajdatsy, P., Gielen, L., Mas, P., Britte, L., Desmet, W. u. van der Auweraer, H.: OPAX. A new transfer path analysis method based on parametric load models. Mechanical Systems and Signal Processing 25 (2011) 4, S. 1321–1338

[6] Klerk, D. de u. Ossipov, A.: Operational transfer path analysis. Theory, guidelines and tire noise application. Mechanical Systems and Signal Processing 24 (2010) 7, S. 1950–1962

[7] Bendat, J. S. u. Piersol, A. G.: Engineering applications of correlation and spectral analysis. A Wiley-Interscience publication. New York: Wiley 1993

[8] Gajdatsy, P., Janssens, K., Desmet, W. u. van der Auweraer, H.: Application of the transmissibility concept in transfer path analysis. Mechanical Systems and Signal Processing 24 (2010) 7, S. 1963–1976

Effect of friction reduction of magneto-rheological semi-active damper on ride comfort and vehicle dynamics

Kentaro Komori*, Gakuyo Fujimoto, Tatsuya Tsukamoto, Daijiro Endo

Honda R&D Co., Ltd. Automobile R&D Center

P.E. Pfeffer (Ed.), *8th International Munich Chassis Symposium 2017*, Proceedings, DOI 10.1007/978-3-658-18459-9_26

Abstract

In this decade, magneto-rheological damper (MR damper) has had several applications and widely spread to passenger vehicles. MR damper is a damper filled with magneto-rheological ferrofluid (MRF), which is controlled by a magnetic field using an electromagnet. This allows the damping characteristics of the damper to be continuously controlled by varying the power of the electromagnet. The wide variation and high response of damping force contribute to achieve an excellent performance in semi-active vehicle suspensions which can adapt to road conditions. However, MR damper shows the high friction characteristics, due to the unique sliding regime of internal parts of damper with MR fluid. This can be attributed to abrasive ferro-particles and chemical ingredients in MR fluid, causing obstacles for smooth sliding motion and improved ride comfort performance. Not only demand for handling stability and road-holding ability, but also demand for improved ride comfort has been increased. To respond to friction reduction of MR damper, we primarily investigate the potential use of MRF modification submitted to boundary lubrication for sliding contact of rod guide / piston rod, oil seal/piston rod, and piston / tube. Friction modifier chemical additives can be adsorbed and form low friction boundary film on sliding contact surfaces of the parts, leading friction reduction dramatically even in abrasive contact with ferro-particles. Thus, the modified MR damper shows sufficient low friction performance successfully. For effectiveness of the friction reduction on vehicle dynamics, vehicle vibration test result shows that the vehicle installed low friction MR damper can improve vibration transfer characteristic. Damper friction influences cut-off frequency characteristic effectively, in particular at smaller damper stroke in Bode diagram analysis. The gain in Bode diagram, which given by a vehicle body/vibration table acceleration, is decreased by friction reduction of MR damper dominantly at the region less than 3.0 mm in damper stroke. A similar tendency is found also in actual road input during vehicle running. Low friction MR damper can improve ride comport remarkably in terms of high-frequency vibration and harshness, while power spectrum density in frequency characteristic is scarcely different in large road input amplitude. These results suggest that the higher friction behaves as a steady damping force even if it might be unexpected damping force, so that low friction MR damper would enable the region hidden behind the friction force to control freely. Friction reduction of MR damper plays important roll to provide further variability of damping force and adaptability for innovative electrical control, utilizing notable high response damping. It is essential to realize the both of road-holding ability and ride comfort adapting to road condition with high accuracy and better quality, even for a prospective intelligent vehicle.

1 Introduction and motivation

In this decade, magneto-rheological damper (MR damper) has been widely spread to passenger vehicles. MR damper is a damper filled with magneto-rheological ferro-fluid (MRF), which is controlled by a magnetic field using an electromagnet. This allows the damping characteristics of the damper to be continuously controlled by varying the power of the electromagnet. The wide variation and high response of damping force contribute to achieve an excellent performance, most notably in semi-active vehicle suspensions which can adapt to road conditions [1]. In particular, the MR damper can control the damping force and generate a large damping force at a region of very low speed in damper motion, so that it can positively control transient characteristics of rolling motion and pitching motion for the vehicle. Sport utility vehicles (SUV) with high vehicle height and heavy weight tends to show relatively large roll angle and sports vehicles require posture control of the vehicle for high stability, so that MR damper has been mainly applied to SUVs and (super-) sports vehicles category from the viewpoint of road-holding ability and vehicle stability.

However, MR damper shows the high friction characteristics, due to the unique sliding contact regime of internal parts of damper. This can be attributed to abrasive ferro-particles and chemical ingredients in MR fluid. It is known that the component which has a high friction force contributes to suppress vibration, fluttering and shimmy of the steering or suspension systems to a certain extent during a vehicle running [2]. On the other hand, the excessive high friction force can be obstruction for the smooth motion of the chassis component. In particular, for control dampers, the excessive high friction can interfere with the damping force control resulting in diminishing controllable range, which is thought to have a great influence on vibration behavior of vehicle. Not only demand for handling stability and road-holding ability, but also demand for improved ride comfort has been increased directing toward expanding application window as sedan type vehicle, electric vehicle, and also future intelligent vehicles. It is essential to address the friction reduction of MR damper for achieving improved ride comfort and further controllability with high accuracy. This research approaches the possibility of improved ride comfort and advanced vehicle dynamics by friction reduction of MR damper.

2 Research concept

2.1 Friction in MR damper

A typical model of damping force curve for the MR damper is shown in Fig. 1. MR damper can generate the variable damping force; the maximum damping force is normally defined by the maximum current applied to an electromagnetic coil and the minimum damping force is obtained with no current applied. Therefore, the minimum damping force is primarily generated by a flow resistance of the MR fluid and a friction force of the MR damper. If the damper friction is excessive high, the friction force behaves as a constant damping force. The higher friction always generates a certain damping force even if it might be unexpected damping force. This can cause an unfavorable ride feeling particularly at a low piston speed region. It is suggested that reducing friction force of the MR damper would enable the region hidden behind the friction force to control freely. Friction characteristic plays important roles to provide further variability of damping force and expand controllability of the MR damper.

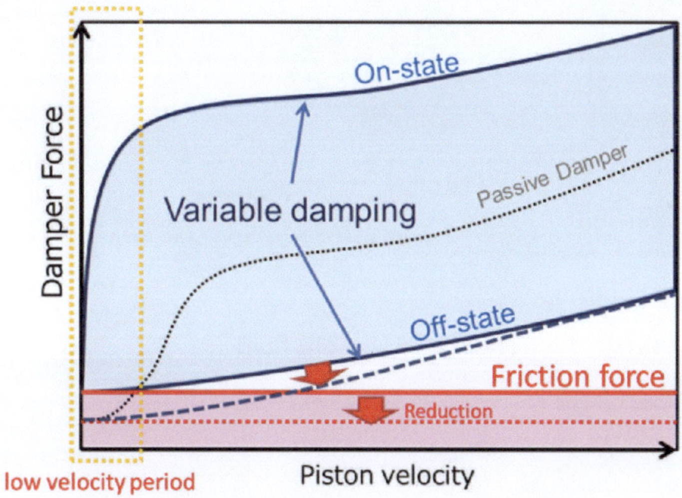

Fig. 1: Schematic image of damper force curve of MR semi-active damper.

2.2 Internal structure and damping mechanism of MR damper

A typical internal structure of the MR damper is shown in Fig 2. The inside of the damper cylinder is filled with MR fluid, and is sealed with an oil seal at the rod guide part and an O-ring seal at the free-piston part. The piston is provided with an annular gap flow path through which the MR fluid flows and an electromagnetic coil for applying a magnetic field to the MR fluid. The magnetic field is generated by the axial coil. A current is supplied to the electromagnetic coil though the lead wires which are conveniently brought out through the rod. The field strength that this will produce in the MR fluid depends on the entire magnetic circuit. The magnetic circuit of the example design is axisymmetric. The flux passes through the coil axially, expands radially outward through the disk at one end, through one MR fluid gap, back along the iron sleeve, radially inward through the disc at the other end and back into the core completing the circuit [1].

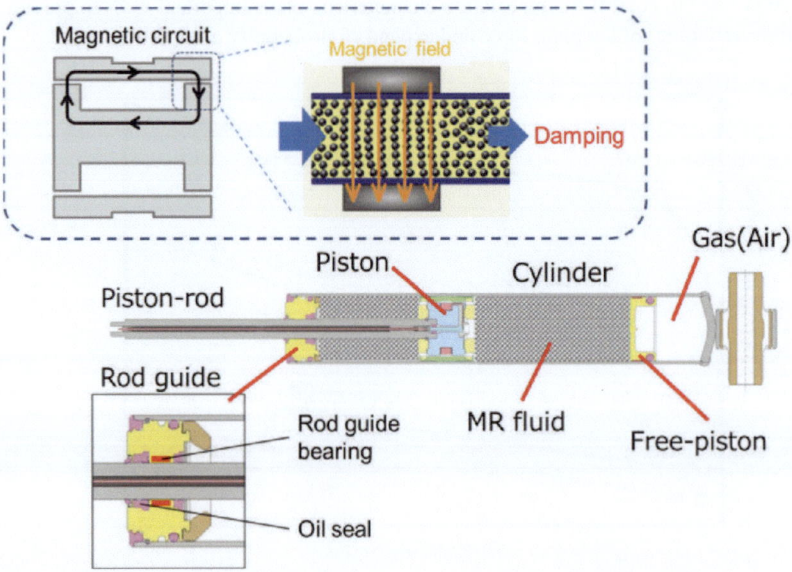

Fig. 2: Schematic illustration of typical internal structure of the MR damper.

MR fluid is formed by suspending numerous ferromagnetic iron particles in a low viscosity base-oil; typically poly-alpha-olefin (PAO) based oil is used for damper of vehicle suspension. The average diameter of the iron particles is a few micrometers with a range of 3–10 μm. A viscosity modifier and a surfactant are also added to keep stable dispersion of the iron particles without aggregation and minimize settling out. When a current is supplied to the electromagnetic coil, a magnetic field is applied to the MR fluid at the gap flow path. Simultaneously, the iron particles contained in the MR fluid form clusters so as to line up along the lines of magnetic force. Since the MR fluid flows through the gap flow path so as to inhibit the formation of clusters, the apparent viscosity of the MR fluid increases. This can cause a damping force, and the apparent viscosity can be varied with the strength of the applied magnetic field, so that the variable damping force can be continuously controlled. Thus, MR damper can obtain the high responsibility and the wide variable range of damping force, due to the immediate viscosity change of MR fluid itself and its variable flexibility. It is interesting to note that this can be achieved without a complicated mechanical structure.

On the other hand, the internal sliding parts of the MR damper work under lubricated condition with MR fluid containing the iron particles. The small solid iron particles affect the sliding interfaces during sliding contact between piston rod and oil seal, between piston rod and rod guide, and between piston and tube, causing a high friction. Furthermore, iron particles can cause an increasing abrasive wear of the sliding parts. Additionally, a mono-tube type structure has been typically applied to MR damper, so the internal pressure is relatively higher than the other semi-active control dampers. The higher internal pressure also influences the increase in the squeezing force of the oil seal. Thus, the MR damper has a unique friction characteristic attributed to the internal structure and its working mechanism.

2.3 Tribological approach to friction reduction of MR damper

A geometrical design has been refined for reducing friction force and improving wear resistance of the MR damper. The rod guide configuration, the rod guide bearing materials, or the sealing materials also have been improved so far. For example, in order to prevent iron particles in MR fluid from interaction with the sliding interface, a technique of supplying only the clear fluid of MR fluid to a rod guide sliding part isolated using a porous filter has been typically put to practical use for present products. This can provide a certain effect on reducing friction force; furthermore the wear of the rod guide bearing can be dramatically improved. However, it is still difficult to obtain a sufficient friction reduction, since the isolated rod guide configuration requires multiple oil seals which can increase seal friction. The effects on friction reduction appear to be limited in the mechanical method, since the friction coefficients cannot be suppressed fundamentally. The friction characteristics can depend on side force and vibration frequency of the damper. It is necessary to further reduce the friction force of the MR damper for achieving excellent ride comfort and improved dynamics of the vehicle.

It is generally known that friction characteristics, in particular friction coefficient, under lubrication conditions can be well explained by a model so-called the "Stribeck curve", as shown in Fig. 3. The variation of friction coefficient in this model can be interpreted by the difference in contact regime between two objects involving the state of the lubricant between the contact surfaces [3, 4]. The state of lubricant in the contact interface can change to depend on sliding speed typically. This gives the different distance and contact regime between two sliding surfaces. The properties of the adsorption film and the boundary film formed on the sliding surface play important roles, particularly in the boundary lubrication and the mixed lubrication region [5].

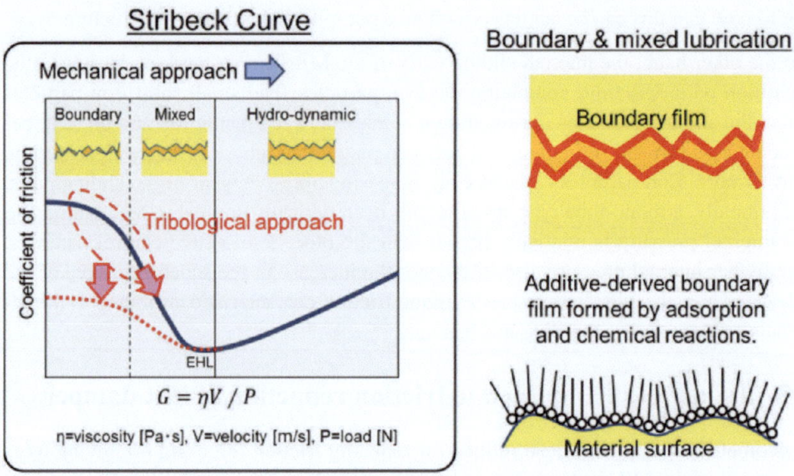

Fig. 3: Stribeck curve and typical model of boundary film formation on contact surface.

In this research, we focused on the fact that the chemical constituents of the conventional MR fluid are primarily specialized for dispersion of magnetic particles and do not contain chemical components to form a low friction boundary film effectively. It suggests that the expected performance can be realized by adding appropriate chemical components that induce a low friction to MRF. Furthermore, it also implies that the boundary film can suppress the abrasive attack of iron particles against sliding surfaces.

The important factors in MR fluid modification are as follows:

- Induced low friction boundary film formation with chemical additives.
- Respecting current MR fluid and keeping "magneto-rheological" function.
- Minimum interaction with existing ingredients.
- Simple adding process; easy for everyone.

We address the potential use of MR fluid modification submitted to improved boundary lubrication of sliding parts for the intrinsic improvement of friction reduction and wear resistance of the MR damper.

3 Numerical analysis of damper friction effect

It is essential to gain understanding of the influence of damper friction on the vibration behavior of vehicle. In order to analyze the effect of damper friction reduction, numerical analysis was conducted by means of a simple single-wheel model with two degrees of freedom.

The numerical analysis was performed in Matlab Simulink environment (The MathWorks, Inc.). The model has two springs, two dampers and two masses. Each element corresponds to a tire spring ratio: K_1, a suspension spring ratio: K_2, a tire damping coefficient: C_1, a suspension damping coefficient: C_2, unsprung mass: m_1, and sprung mass: m_2, respectively. The basic formulae are represented as the equation (1) and (2) shown below [6].

$$\ddot{X}_2 = -\frac{1}{m_2}\left(C_2(\dot{X}_2 - \dot{X}_1) + K_2(X_2 - X_1)\right) \quad (1)$$

$$\ddot{X}_1 = -\frac{1}{m_1}\left(C_2(\dot{X}_2 - \dot{X}_1) + K_2(X_2 - X_1) + K_1(X_0 - X_1)\right) \quad (2)$$

Where X_0, X_1, $\dot{X}_1$, $\ddot{X}_1$, X_2, $\dot{X}_2$, and $\ddot{X}_2$ indicates road input displacement, unsprung displacement, unsprung velocity, unsprung acceleration, sprung displacement, sprung velocity, and sprung acceleration, respectively. Since actual friction and damping force do not have linear properties against damper piston velocity, it is necessary to change the model that can look up a damping force table which had been collected from damping force measurements. Thus, the term $C_2(\dot{X}_2 - \dot{X}_1)$ must be substituted by DF. This substitution and the term DF makes it possible to use a "look up table" of damping force which includes friction force against piston velocity. Figure 4 shows the damping force look up table in range of -3 to 3 m/sec of piston velocity. A damping force and friction were measured by a test damper without applying electric current to MR damper, which is shown as standard line (STD No.4). The other 7 damping force curves were created by multiplying certain values to the standard line. Therefore, the all curves have exactly the same damping force but different friction force. Figure 4 shows detail in extremely slow piston velocity region. Peak to peak friction force varies from 20 to 300N.

The simulation was run under the condition as shown in Table 1. Road inputs were applied with the form of legalistic sine sweep wave in the frequency range of 0.1 to 30 Hz in 121 second. Road input Amplitude was set to vary in the range of 0.25 to 15mm with 0.25 mm steps. With each road amplitude condition, the simulation was run and power spectra density (PSD) was calculated as evaluation. The simulation time step condition was set to be 10 kHz which was necessary for observe friction effects in PSD spectrum, since the friction force changes in very short period of time when the piston stroke direction changes.

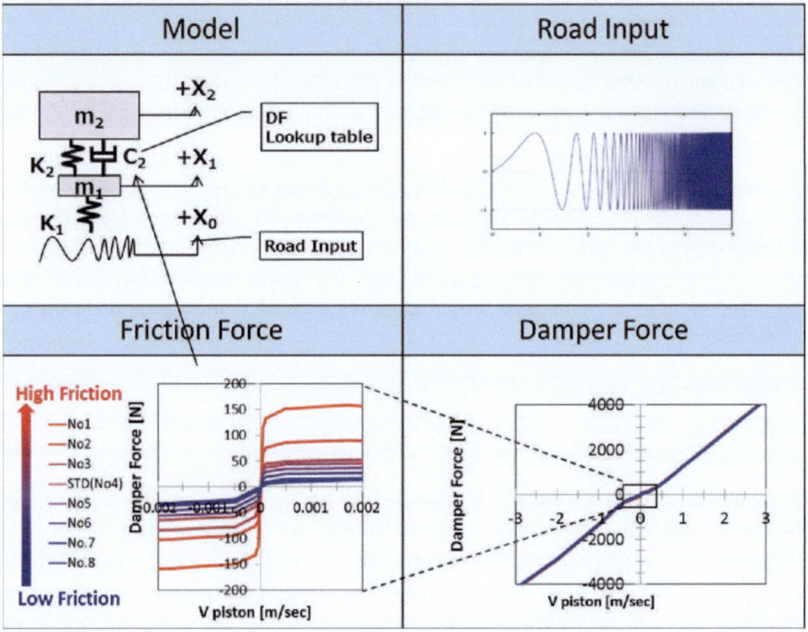

Fig. 4: Single wheel model and calculation regime for numerical analysis.

Table 1: Vehicle parameters for the single wheel simulation.

Parameters	m1	m2	K1	K2
Value	59 kg	710 kg	250000 N/m	30786 N/m

The vibration characteristics are shown in Fig. 5; it shows power spectrum density (PSD) as a function of vibration frequency for sprung body. When the amplitude of the road input is small, the PSD in the vibration characteristic diagram increases with increasing friction. On the other hand, when the amplitude of the road input is large, the variation of PSD with the change of the friction is small and the contribution of the friction appears to be small for PSD. According to the numerical analysis, the friction reduction of the damper is considered to be particularly effective for improving the ride comfort and controllability in a small input amplitude at a region of 3 mm or less. Furthermore, the variation of integral value of the PSD within the range of 0.25–3.25 mm in input amplitude with the damper friction force is shown in Fig. 6. The integral value of PSD decreases with decreasing damper friction force and becomes gradually constant at the region of the damper friction force of 30–40 N or less. This suggests that the effect of friction reduction until around 30–40 N at least will make sense significantly in the case of no side load applied on the damper.

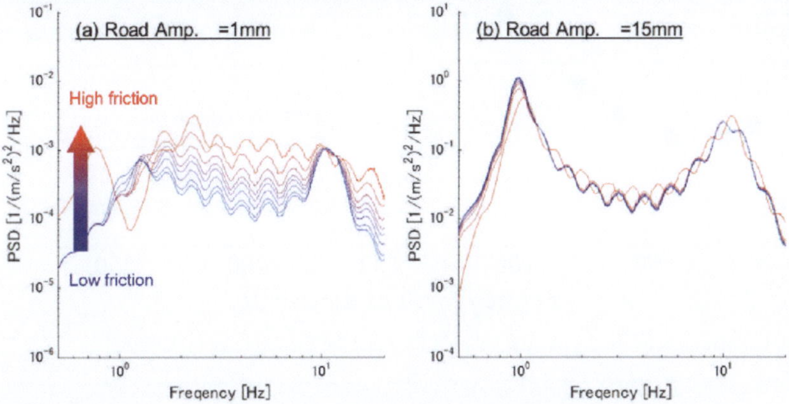

Fig. 5: Power spectrum density (PSD) as a function of vibration frequency for sprung body calculated by (a) small input amplitude; and (b) large input amplitude.

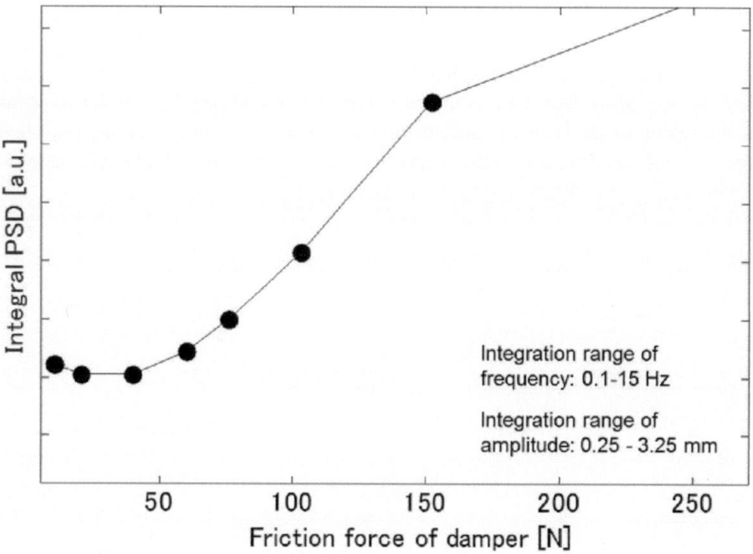

Fig. 6: Relation between the integral value of PSD and the friction force of damper.

4 Friction reduction of MR damper

Tribological approach to friction reduction of MR fluid

Basic friction model test was carried out for gaining an understanding of the effects of chemical additives. The modified MR fluids were prepared by simply adding the different types of additives to conventional MR fluid (MRF 126-BD, LORD Co.). In the friction model test, friction force of a rod (chromium plated steel) sliding against a rod guide bearing was examined under fully soaked condition with the modified MR fluid, as shown in Fig. 7. PTFE (polytetrafluoroethylene) -coated plain bearing (DDK86 bushing, Daido Metal Co., Ltd.) was inserted into the sliding part of the rod guide made of aluminum. The rod was set so as to pass through the PTFE-coated bearing inner cylinder part of rod guide. The rod guide was pressed against the rod under a load of 150 N. The test was performed at a sliding speed of 25 mm/s for 16 hours. The sliding stroke was of ±10 mm. Friction force measurements were conducted at a sliding speed of 0.15 mm/s under a load of 150 N at the time of 0, 1, 4, and 16 hours during the test. These were set so as close configuration to materials of actual damper parts and their working conditions. By way of comparison, the tests were also conducted using the clear fluids (base-fluids) from which iron particles were removed.

	MRF modification	Addition
Blanc	Standard MRF	-
HM-1	add EP agent: phosphate base	1 wt.%
HM-2	add Fatty acid	1 wt.%
HM-3	add FM agent-A: amine base	1 wt.%
HM-4	add FM agent-B: ester base	1 wt.%

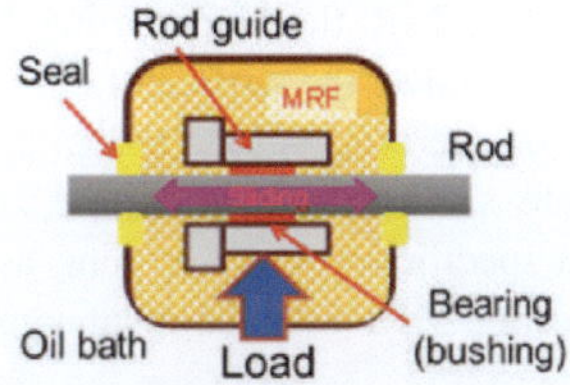

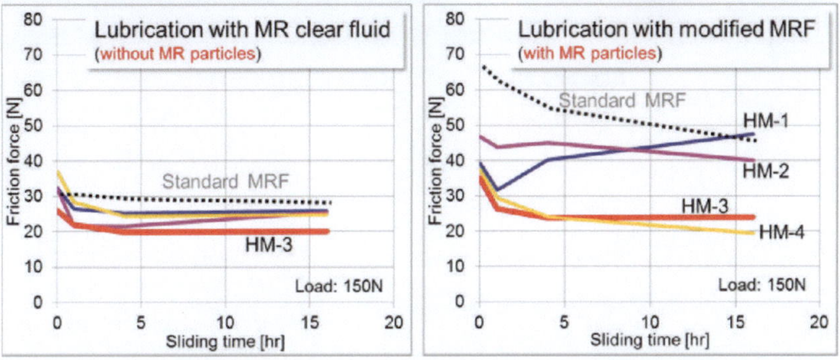

Fig. 7: Test condition and result of sliding test of rod guide lubricated with modified MR fluid.

Figure 7 shows the results of the friction model tests. Different friction characteristics are shown to depend on the types of additives. Friction force can be remarkably reduced in the amine-based or the ester-based FM (friction modifier) agent. It is considered that whether the additive can be adsorbed on the sliding surface and can form the boundary film is due to the chemical species of the additive. This implies that the friction force of the MR damper is largely dependent on the characteristics at the boundary or the mixed lubrication region, as expected. In particular, the modified MR fluid with amine-based additive shows the friction force almost equal to that in the clear fluid even in the presence of MR particles, and remarkably reduces the friction force regardless of the presence the MR particles. It suggests that the possibility of friction reduction can be shown not only for the rod guide part but also for the sliding between the piston and the inner cylinder of the damper, since the amine-based additive appears to act on both the metallic material and the resin material. Furthermore, it is possible to obtain a sufficiently low friction effect even when using an isolated type sliding structure such as a structure in which magnetic particles are eliminated by a filter, since it can act on the sliding surface as a chemical component in the base fluid.

In order to evaluate the durability of this low friction effect of the modified MR fluid with amine-based additive, the long-term sliding test was conducted under a high load condition (200 hours, 300 N). The test was carried out under severe condition with the modified MR fluid containing MR particles. The low friction modified MR fluid (MRF) shows the sufficient durability to maintain low friction characteristics even in the long-term/high load test in the presence of MR particles, as shown in Fig. 8. The results show that it is possible to obtain a remarkable low friction effect irrespective of a specific structure as long as it has a friction sliding structure lubricated by the modified MRF or a chemical component thereof.

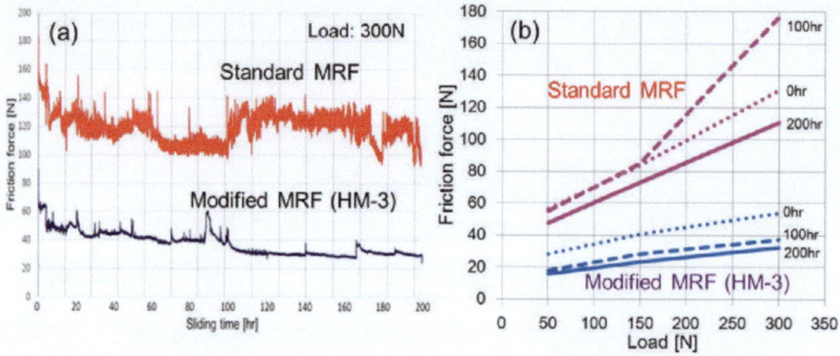

Fig. 8: Long-term durability test with high load; (a) friction force, and (b) load dependency.

Friction and damping performance of MR damper unit

The modified MR fluid with amine-based additive which shows a remarkable friction reduction in the friction model test were applied to prototype fabrication for a new "low friction" MR damper. The low friction MR damper was assembled only changing the MR fluid to modified one; the other components were remained same as conventional. Friction characteristic of the low friction MR damper unit was evaluated using hydraulic damping force measurement system (PKS-01303/ PP-80B, TOKYO KOKI Co., Ltd.) in comparison with a conventional MR damper. The friction force of damper was measured at very low speed so as not to generate the damping force. The axial force generated at piston-rod was measured in vibrating the cylinder of the damper unit in direction of rod axis. The vibration for the damper cylinder was input at frequency of 0.005 Hz (approx. 0.1 mm/s in stroke speed) with triangle wave form, and the vibration movement was of ±5 mm in stroke. The side load was applied to the damper cylinder at the position of rod guide part, in perpendicular direction of piston-rod axis. The damper friction measurements were performed with applying the side load of 0, 100, 200, and 300 N. The friction force of damper was defined as a total (peak-to-peak) of a friction force in compression motion and that in rebound motion; each of the friction force was measured at the position of the vibra tion center.

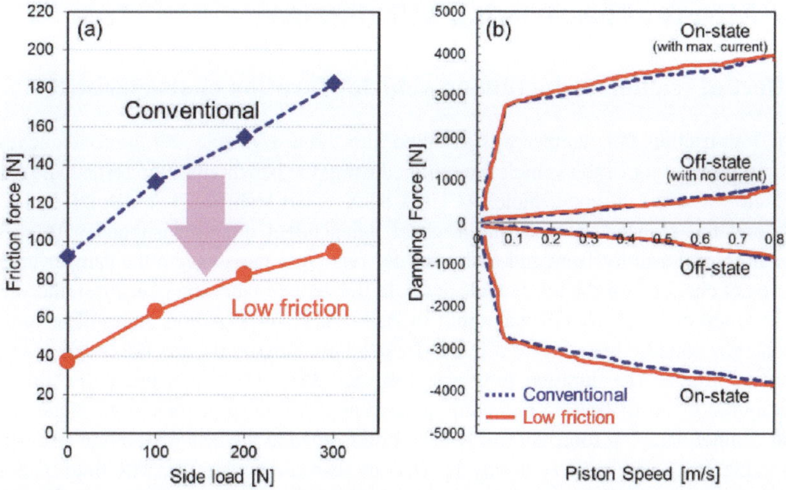

Fig. 9: (a) Friction force and (b) damping performance of low friction MR damper compared with conventional MR damper.

Figure 9 shows the relationship between the friction force of the MR dampers and the side load applied to the dampers. The low friction MR damper using the modified MR fluid remarkably reduces the friction force of the damper unit. The results also shows the low friction force of almost half of the conventional friction level can be achieved without depending on the load of the side force. The damper friction force at a side force load of 0 N is around 40 N, which is close to the upper limit of the saturation region of the PSD in the numerical analysis using simple single-wheel model. This implies the low friction MR damper may show sufficient effects for vehicle vibration behavior. Figure 9 also shows the damping force curve of the low friction MR damper unit compared with the conventional one. The damping characteristic inherent for MR damper is not impaired by the modified "low friction" MR fluid. The result allows us to say that the low friction MR fluid modification can work to reduce the friction force of actual MR damper unit effectively.

5 Effects of low friction MR damper

Effect of friction reduction on vehicle vibration characteristics

The low friction MR damper was installed into the test vehicle and the fundamental vibration behavior of the vehicle was studied using the bench vibration tester. The test vehicle was a sedan type vehicle (ACURA RLX, Sport Hybrid SH-AWD, 2016 model), and the suspension type was 4-wheel independent double-wishbone with split lower control arm for front and multi-link for rear. Any parts except the damper units were not changed for the test vehicle at all. In the vehicle vibration bench test, the vertical vibration (sine-sweep) was input to each wheel by vibration tables. The input frequency conditions were of 0.5-20 Hz for ±0.5 mm amplitude, and 0.5-30 Hz for ±5 mm amplitude. The duration time was 120 sec. We note that, in order to estimate purely the effect of friction reduction, current application to the electromagnetic coil and damper damping force control were not performed in the bench vibration test. The damping force was generated only by viscous resistance when the MR fluid passed through the flow path of the piston and the friction force of the damper unit. The vibration behavior of the vehicle was detected by the accelerometer set on the floor of the vehicle body.

The obtained vibration characteristic diagram is shown in Fig. 10. The low friction damper shows an effect on reducing the vehicle vibration gain which given as a ratio of vehicle body acceleration to vibration table acceleration. In the condition with small amplitude of 0.5mm, the gain for the low friction MR damper at the region of cut-off frequency range (3 to 8 Hz) decreases by 3–5 dB in comparison with the conventional higher friction MR damper. The effectiveness in friction reduction is also observed in the condition with larger amplitude of 5 mm; the gain decreases by about 1–2 dB at the region of 3 Hz or more, comparing the low friction MR damper with the conventional higher friction one. This is not so significant reduction as the case of smaller amplitude of 0.5 mm. The result of vehicle vibration bench test shows similar tendency of friction reduction effect to the numerical analysis result in single-wheel model.

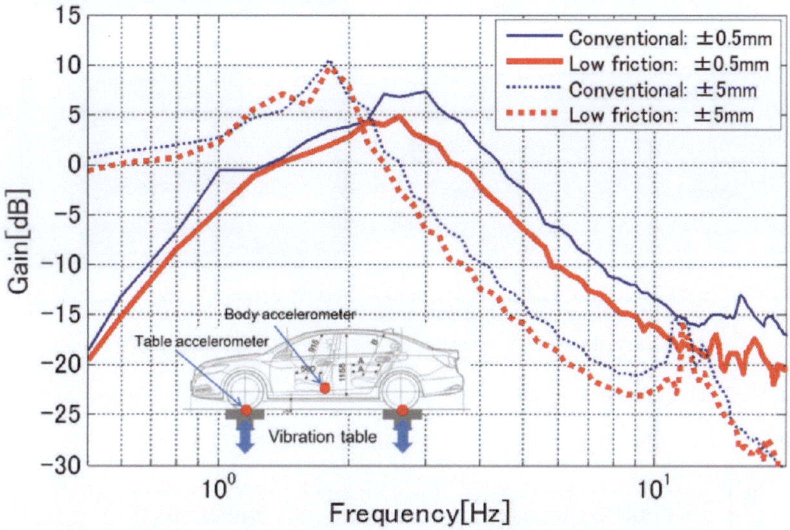

Fig. 10: Bode diagram of vehicle vibration characteristics measured by bench vibration tester.

Effectiveness in actual road input and actual driving environment

Vehicle ride tests were performed using the test vehicle installed with the low friction MR damper and vehicle vibration behavior against actual road input was investigated. The tests were carried out in closed test tracks (proving ground in Honda R&D Co., Ltd., Tochigi, Japan). The test vehicle has three accelerometers on its body, four accelerometers on its each knuckle, a low level electrical control unit (ECU) including power electric module for MR damper output, and a high level ECU (dSPACE Gmbh, Micro Auto Box II) in its trunk (in Fig. 11). The accelerometer signals were sent to high level ECU to feedback the current vehicle dynamical movement. The high level ECU output damping force instructions of each damper and pass them to the low level ECU. The low level ECU output each electrical current to four dampers coils, which made the dampers possible to change their damping force once in less than 20 msec.

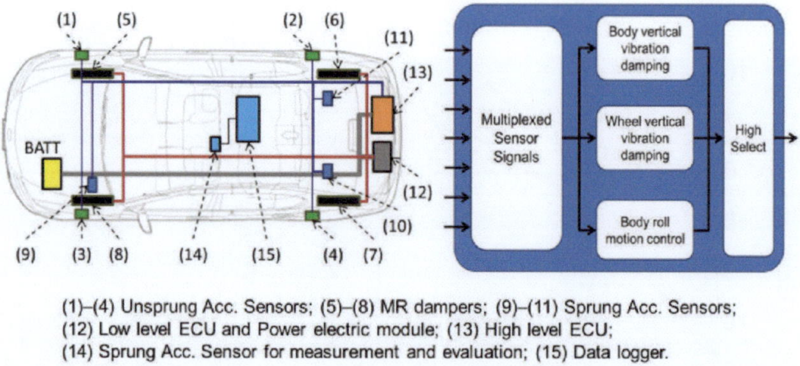

Fig. 11: Schematic illustration of test vehicle configuration with electrical control system.

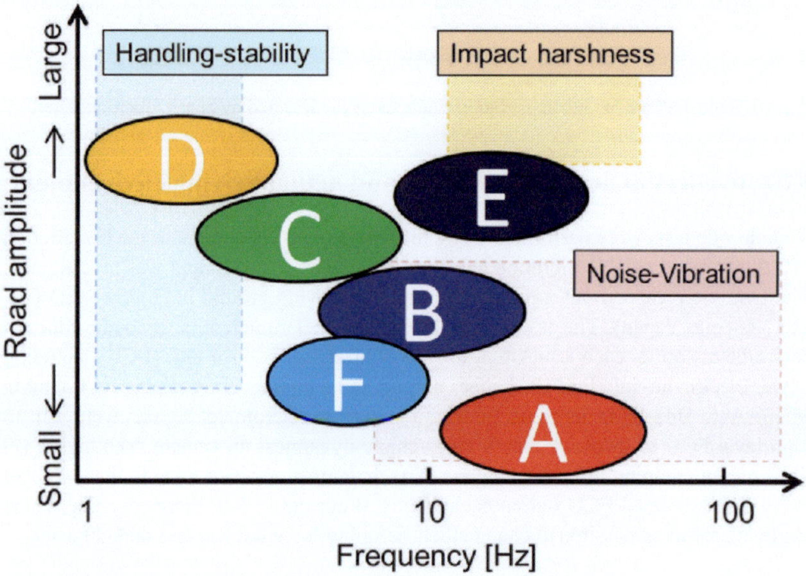

Fig. 12: Image of differences in road input condition between the test tracks.

Tests were carried out in six tracks which have different types of road surfaces in asperity and undulation, as shown in Fig. 12. Measurement of vehicle vibration was conducted using accelerometer set on the floor of the vehicle body.

Figure 13 shows the vehicle vibration characteristics measured over the all tracks except large undulation tracks (C and D). If the damping force control is not performed as in the case of the bench vibration test, the effect of friction reduction is shown in actual road input. The PSD at the region of cut-off frequency and higher frequency (unsprung region), for the low friction MR damper is reduced compared with that for the conventional higher friction one. On the other hand, the PSD at the region of lower sprung frequency rises in the low friction MR damper. Then, if the variable damping control is performed, the PSD at the region of the lower frequency (sprung region) also enable to be reduced; the PSD in the entire frequency range for the low friction MR damper is reduced. The results show the possibility that the vehicle with low friction MR damper can respond to the both of improving ride comfort and achieving appropriate damping force generation adapted to various road input conditions.

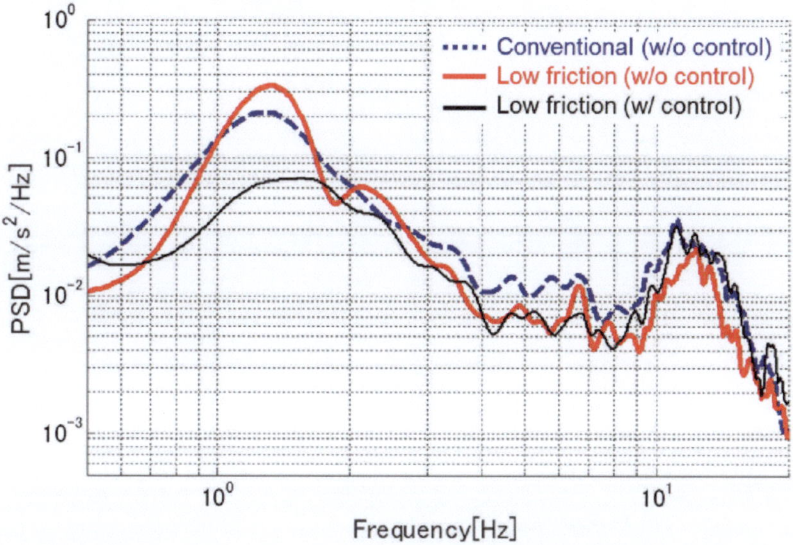

Fig. 13 Vehicle vibration characteristics measured over the all tracks except large undulation tracks -C and -D.

In addition, in order to clarify the contribution of the friction reduction effect under the actual driving environment, the measurement results without damping control are separately extracted and analyzed for each track with different characteristics of the road surface, as shown in Fig. 14. In particular, on the smooth road surface with small undulation, the effect of the friction reduction is shown remarkably, as shown in

Figs. 14(a) and (b). It is considered that inhibiting the generation of unfavorable damping force attributed to excessive high friction of damper can improve the ride comfort, high frequency vibration, and harshness, to achieve even better feeling and ride quality. On the other hand, for large road surface asperities and wavy road undulation(in Figs. 14(c) and (d)), the effect of friction reduction is smaller than that on a smooth road surface; so that, the contribution of the damping force variable control seems to be rather large. The results imply that the low frction MR damper pottentially can realize the both of improved ride comfort and road-holding ability in combination with appropriate and high-accuracy control for various road conditions.

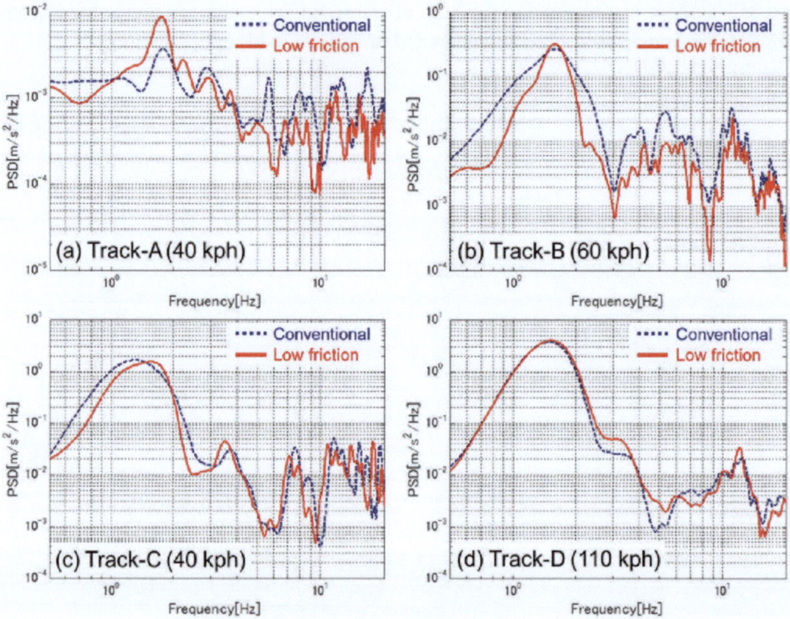

Fig. 14: Vehicle vibration characteristics (without damping control) separately extracted for the different road surface conditions: (a) track-A; (b) track-B; (c) track-C: (d) track-D.

6 Conclusions and future prospects

Friction reduction of MR damper, which has been difficult to realize so far, is achieved by MR fluid modification with adding simply "one drop" of chemical additive. Effect of friction reduction of the damper on vehicle dynamics is shown to depend on road input amplitude; the effectiveness is shown particularly for the smooth road surface with small undulation. The low friction MR damper can improve the ride comfort, high frequency vibration, and harshness, to achieve even better feeling and ride quality.

In this research, there is significance in that friction force appears to correspond to the role of damping force. The friction reduction can convert an unfavorable constant damping force to variable force to control freely. This suggests that it can be controlled with high accuracy adapting to the road conditions, and on the contrary, how to control this expanded window will play an essential role in the future. This also will be important to realize a favorable ride comfort with keeping road-holding ability of electric vehicle which has a heavy weight and a unique vehicle dynamics performance.

Furthermore, an up-and-down reciprocating motion of damper necessarily repeats stop-and-start, for example, at the return of compression into rebound. Since static friction is interposed during such motion, the transient friction behavior at motion initiation period becomes more important. This is a tiny second, but very significant not only for human perception but also for calculating in ECU. The friction reduction of damper unit might be able to improve the response lag corresponding to a few milliseconds at least. It is very difficult to control friction behavior itself by electric or electronic control; so that, friction behavior also will play important role for future intelligent vehicle and automated drive system in terms of whether vehicle can perform as controlled. It is very interesting that microscopic phenomena such as “molecular dynamics” will have a great possibility to contribute to “vehicle dynamics” control.

Acknowledgment

The authors are deeply grateful to Mr. J. Lopez, Dr. E. Vasiliu, Mr. J. Hart, and Ms. E. Ruble, from BWI Group, Dayton, USA, for co-operating with low friction MR damper research project, damper unit fabrication, and testing; Mr. G. Tessier, Mr. O. Raynauld, Mr. O. Robin, and Mr. C. Goergen, from BWI Group, Paris, France, for co-operating with damper control and vehicle ride test; Mr. T. Murayama, BWI Japan Company Ltd., Tokyo, Japan, for "every" technical support, worthy ride tuning, and valuable discussion. We also would like to thank Mr. S. Yamamoto and Mr. S. Sakanoue, from Idemitsu Kosan Co., Ltd., Japan, for co-operating with fundamental research of fluid modification. Finally, our heartfelt appreciation goes to Mr. M. Matsukuma for supporting project launch was of inestimable value for this research.

References

1. J. Dixon, "ER and MR damper", in "The Shock Absorber Handbook; Second Edition", WILEY, 2007, pp.303–326, ISBN: 978-0-470-51020-9.
2. K. Komori and T. Nagataki, "Friction Behavior of Diamond-Like Carbon Coated Ball Joint: Approach to Improving Vehicle Handling and Ride comfort", SAE International. J. Passeng. Cars – Mech. Syst., Volume 8, Issue 2, 2015, doi:10.4271/2015-01-1507.
3. H. Czichos, K.-H. Habig, "Tribologie-Handbuch: Tribometrie, Tribomaterialien, Tribotechnik" (Tribology Handbook), Vieweg+Teubner Verlag, Wiesbaden, 3. Auflage, 2010, ISBN: 978-3-8348-0017-6.
4. B. Jacobson, "The Stribeck memorial lecture", Tribology International, Volume 36, Issue 11, 2003, pp.781–789.
5. W. A. Zisman, "Friction, durability and wettability properties of monomolecular films on solids", in "Friction and Wear" R. Davis (Eds.), Elsevier, Amsterdam, 1959, pp. 110–148.
6. R. N. Jazar, "Quarter Car Model" in "Vehicle Dynamics: Theory and Application", 2nd Edition, Springer, 2014, pp.985–1030, ISBN: 978-1-4614-8543-8.

Development of an integrated chassis design analysis tool using Excel graphical user interface

Young Deuk Kim, Hong Kyu Jung, Moon Won Kang, Hyeon Seok Cho
Senior Research Engineer, Hyundai Motor Company, Korea

Sang Sup Kim
Professor, Kookmin university,
Graduate School of Automotive Engineering, Korea

P.E. Pfeffer (Ed.), *8th International Munich Chassis Symposium 2017*,
Proceedings, DOI 10.1007/978-3-658-18459-9_27

Abstract

Chassis performance is investigated analyzing geometry, kinematic & compliance (K&C) and ride & handling (R&H) to design chassis system of suspension, steering, tire etc. Commercial softwares as ADAMS/CAR are difficult for chassis design engineers to use because of complexity of modeling and analysis. So, chassis design engineers need simple tool to analyse geometry, K&C and R&H etc. to design chassis system. In this research, In-house Tool of geometry, K&C and R&H analysis based Excel graphical user interface(GUI) was developed. It is convenient and simple to use for chassis design engineers. Developed In-house Tool of geometry, K&C and R&H analysis could be used for target cascading using target setting and verification using chassis design data in V-cycle process. Chassis design engineers design chassis system according to target setting and test engineers verify performance using developed In-house Tool.

1 Introduction

Kinematic & compliance (K&C) and ride & handling (R&H) are analysed using multibody dynamics(MBD) commercial software as ADAMS/CAR[1] to verify chassis performance. Commercial software is difficult for chassis design engineers to use because of complexity of modeling method and analysis. It requires much time and cost for chassis design engineers to learn commercial software, make vehicle model and analyse results. Also the verification of analysis results is difficult. Test engineers expend much time and cost by trial and error of R&H performance tuning. Chassis control engineers use low DOF commercial software as CARSIM[2]. In-house Tool is required for chassis design engineers to investigate K&C characteristics and R&H performance with less time and cost also for control and test engineers to make good use.

R&H development process (V-cycle) was not well established and the process focuses on verification of chassis design and development improvement by trial and error. It is difficult to apply target cascading process using MBD commercial software. Some motor company and university use simple analysis code with low DOF for chassis performance verification [3-5].

In this research, the chassis design and R&H performance analysis tool was developed, optimized to chassis design and R&H development process. The reliability was investigated applying development vehicle data. It is useful for chassis design, analysis, test engineers in each phase of development. The tool could be used for target cascading using target setting and verification using chassis design data in V-cycle process. Chassis design engineers design chassis system according to target setting and validate chassis performance through geometry, K&C and R&H analysis. Chassis CAE engineer could perform sensitivity analysis and optimization of design parameters integrating optimization software. Therefore R&H development process of target cascading using quantified target could be established. Test engineers could predict performance according to design changes as spring, stabilizer bar, bush, damper etc. and improve R&H performance using the tool. It shorten development time and cost by decreasing trial and error. These activities are helpful to establish R&H development process (V-cycle).

2 Chassis design process and analysis tools

R&H performance development is dependent on subjective assessment. R&H development process of target cascading using quantified target was not well established. Prediction of R&H performance is difficult in the preceding phase. Therefore chassis design and test engineers expend much time and cost by trial and error in development process. The chassis design and R&H performance analysis tool was developed to improve chassis R&H development process including target cascading. Geometry, K&C and R&H analysis are possible in sequence and it is optimized to chassis R&H development process of our company. The tool maintains hard-point format of existing geometry analysis tool which chassis design engineers are using. Chassis design, CAE and test engineers can manage same design data format.

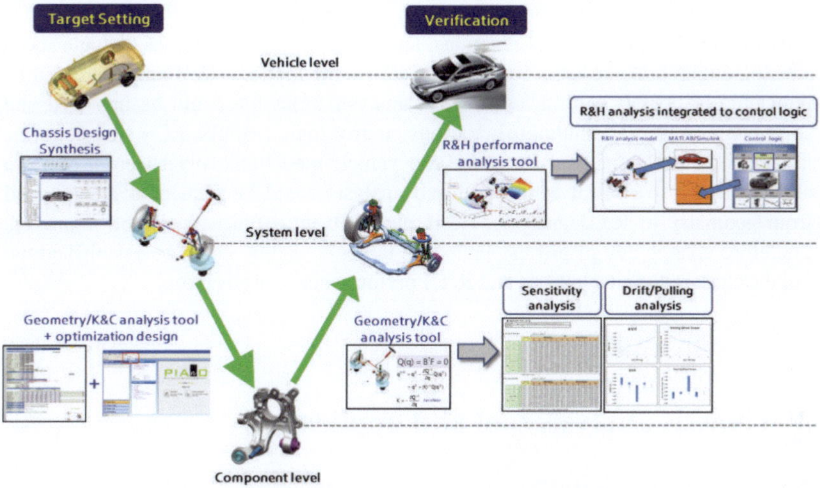

Fig. 1 Chassis design process and analysis tools

Fig. 1 shows chassis design process(V-cycle) and in-house analysis tools. In target setting phase, for satisfying vehicle performance target(using vehicle test/analysis DB including competing car), system properties(suspension, steering, tire etc.) could be determined by chassis design synthesis which was previously developed. System properties could be sequentially determined in each mode(ride, pitch, roll and lateral dynamics mode). Design variables in component level could be determined to satisfy system properties using the development tool connected with optimization design software. Objective functions are K&C characteristics and design variables are hard-point, bush stiffness, etc.

In verification phase, K&C characteristics in system level could be verified according to design variables setting in component level using the development tool. We could verify vertical, roll, longitudinal, lateral, aligning moment mode in K&C characteristics. Sensitivity analysis of design parameter could be performed for geometry and K&C characteristics. Drift analysis could be performed for K&C characteristics. Theoretical kingpin torque and drift distance were computed. Tire PRAT, which is appropriate to vehicle characteristics, could be determined. Also R&H performance in vehicle level could be verified according to K&C characteristics setting. The Functional

Suspension Model was used in R&H analysis. Suspension and steering characteristics were imported as polynominal functions. Driver and disturbance inputs (vehicle speed, steering angle, braking force, road profile etc.) were applied. R&H analysis could be integrated to chassis control logic. R&H analysis input file could be imported and solved in MATLAB/Simulink[6]. Various control logic (MDPS, ECS etc.) could be integrated to R&H analysis. Because same vehicle model and solver were used from chassis design to control application and analysis could be sequentially performed from geometry to R&H analysis, R&H development efficiency will be improved. Therefore chassis design process including target cascading could be established using the chassis design synthesis and R&H performance analysis tool.

3 Chassis analysis tool and application

3.1 Geometry analysis

Geometric characteristics of suspensions are analysed based on the hard-point information of suspension and steering system. Fig 2 shows several front and rear suspension types which are applicable for geometry analysis. The wheel alignment characteristics are analysed for wheel stroke(vertical geometry) and rack stroke(steering geometry). Suspensions are composed of a wheel and massless links modelled as complex joints.

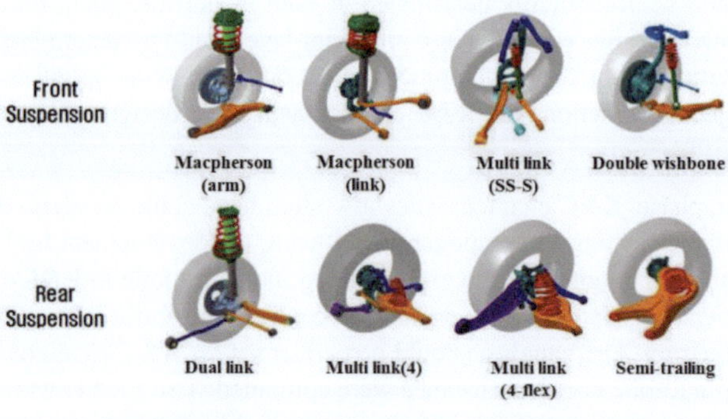

Fig.2 Suspension types of geometry analysis

Quarter-Car models are modelled using complex joints such as SS(Spherical-Spherical), RS(Revolute-Spherical) and Strut joints between wheel and car body. Equations (1) represents 4 kinematic constraint and 2 driving constraint equations in terms of 6 motion variables X of the wheel for front suspensions. The driving constraints are constraint equations for wheel vertical position and steering rack displacement. The motion variables X can be computed by Newton-Raphson method.

$$\Phi(X) = \begin{bmatrix} \Phi^K(X) \\ \Phi^D(X) \end{bmatrix} = 0 \tag{1}$$

Geometry analysis results are verified applying suspension hard point data. The results of front macpherson and multilink suspension are validated with the results of in-house tool[7] as shown in Fig.3 and Fig.4.

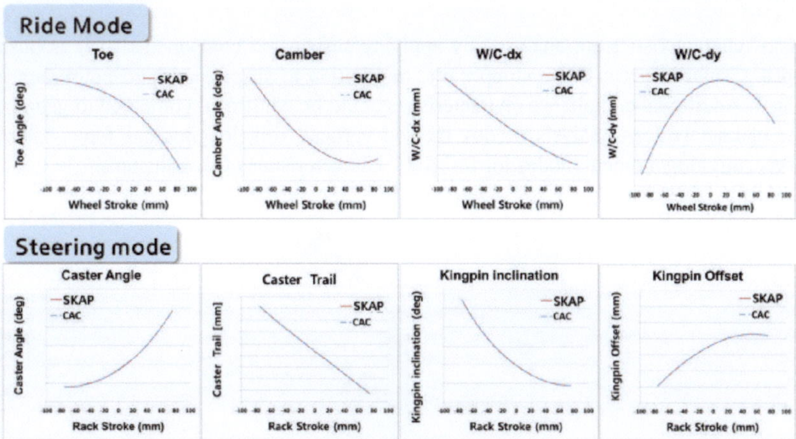

Fig.3 Geometry analysis of front macpherson suspension

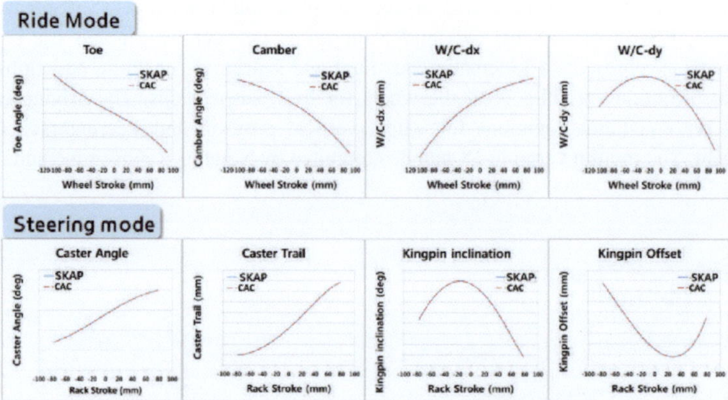

Fig.4 Geometry analysis of front 5 multilink suspension

3.2 K&C analysis

K&C characteristics are analysed by applying suspension components such as bush, spring, stabilizer bar, bump stopper etc. in addition to suspension hard-point data in Fig 5. Additional compliance characteristics could be confirmed compared to geometric characteristics. Analysis modes include vertical, roll, longitudinal force, lateral force, aligning moment modes for full K&C characteristics of suspensions.

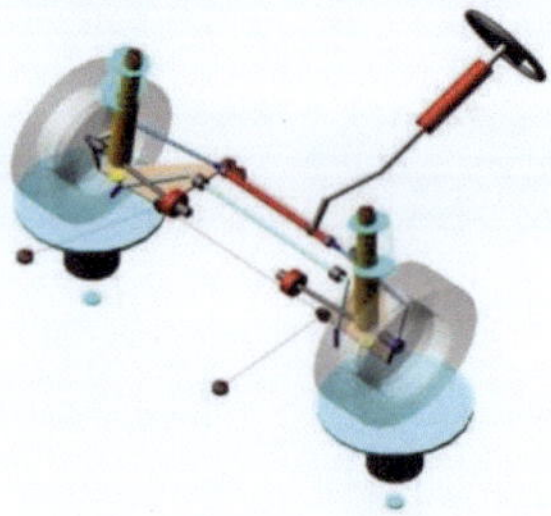

Fig.5 Half-Car suspension model

For computational efficiency multibody suspension systems are modelled using relative joint coordinates. Static equilibrium equations for K&C analysis could be expressed with no kinematic constraints in Equation (2),

$$Q(q) = B^T F = 0 \tag{2}$$

where q is the generalized coordinate of suspension model, Q is the generalized force corresponding to generalized coordinate and B is the velocity transformation matrix that relates generalized velocity to Cartesian velocity.

The generalized coordinate of suspension model could be computed by Newton-Raphson method from static equilibrium of the generalized force as Equation (3).

$$q^{(j+1)} = q^{(j)} - \frac{\partial Q}{\partial q}^{-1} Q(q^{(j)}) = q^{(j)} + (K)^{-1} Q(q^{(j)}) \tag{3}$$

$$j = 1,2,3..... \quad K = -\frac{\partial Q}{\partial q}$$

where K is the system stiffness matrix.

Jacobian matrix could be computed as Equation (4) by differentiation of Equation (2).

$$\frac{\partial Q}{\partial q} = B^T \frac{\partial F}{\partial q} + \frac{\partial}{\partial q}(B^T \hat{F}) = B^T \frac{\partial F}{\partial X} B + \frac{\partial}{\partial q}(B^T \hat{F}) = -K_1 - K_2 = -K \tag{4}$$

The system stiffness matrix is not computed by numerical differentiation, it could be computed by analytical method which can improve efficiency of computing time[8].

K&C analysis results were verified applying vehicle data. Suspension component data of bush, spring, stabilizer bar, bump stopper are used in addition to suspension hardpoint data. The results are validated with ADAMS/CAR in ride, roll and lateral force modes as shown in Fig.6. Design sensitivity analysis could be performed for parametric study of geometry and K&C characteristics.

Drift analysis could be performed for K&C characteristics. Theoretical kingpin torque and drift distance were computed. Tire PRAT, which is appropriate to vehicle characteristics, could be determined.

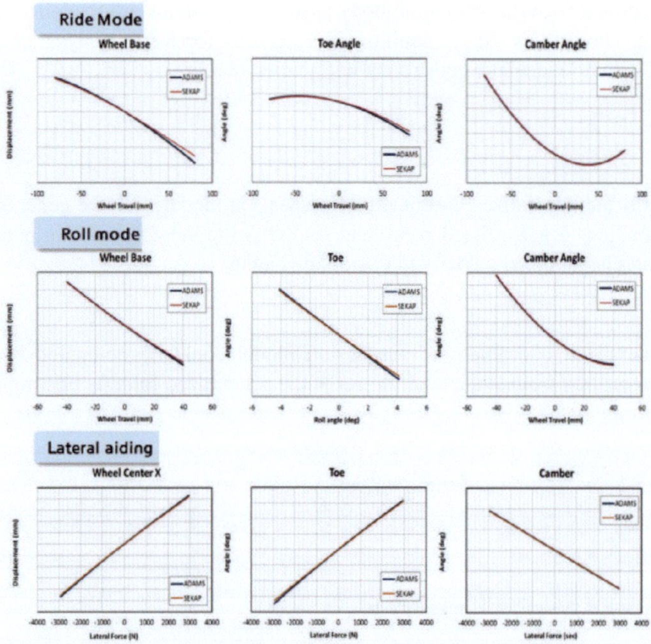

Fig.6 K&C analysis of front Macpherson

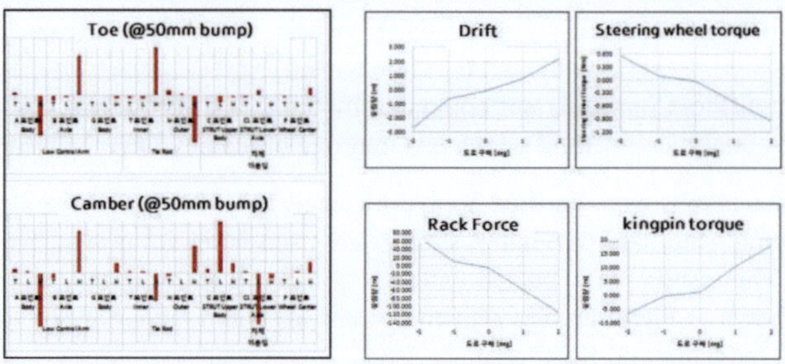

Fig.7 Sensitivity analysis of design parameter and drift analysis

3.3 R&H analysis

The Functional Suspension Model was used for R&H analysis. Suspension and steering characteristics are imported as polynominal functions from K&C analysis results[9,10].

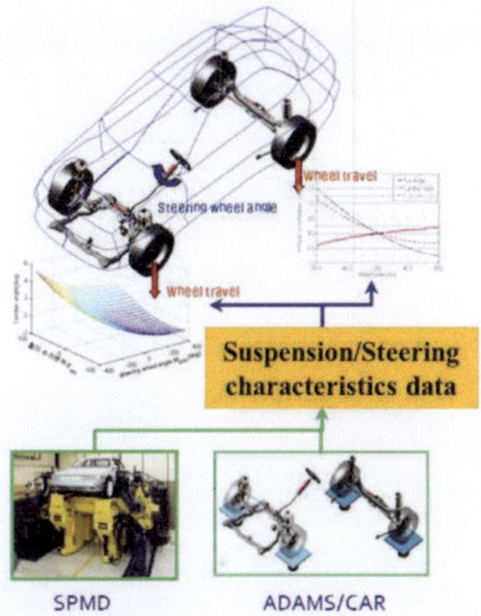

Fig.8 Vehicle model using Functional Suspension Model

The vehicle dynamics model could be modelled as 5 rigid bodies (vehicle body, 4 wheels). Each wheel moves relatively to vehicle body using the Functional Suspension Model. Each wheel is expressed as function of ride, roll and rack travel as equation (5). Ride and roll coupling model was used, which is applicable to both independent suspensions and axle suspensions as shown in Fig.9. Equation of motion could be expressed in Equation (6).

$$Z'_{w/b,front} = f(\delta'_r, \delta'_\varphi, s') \quad (5)$$

$$Z'_{w/b,rear} = f(\delta'_r, \delta'_\varphi)$$

Where δ'_r is ride travel, δ'_φ is roll travel and s' is rack travel.

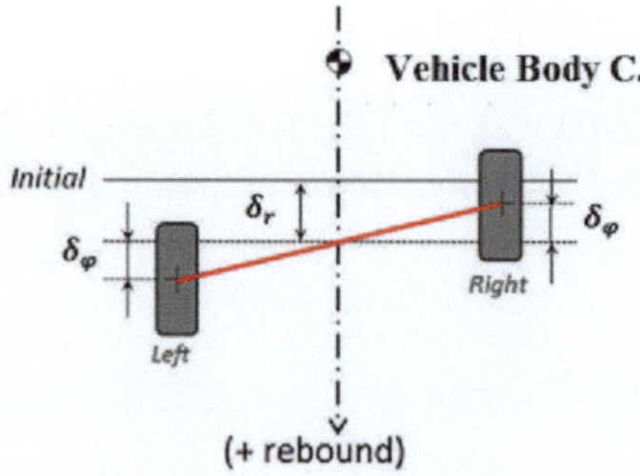

$$\delta_r = \frac{1}{2}(\delta_{right} + \delta_{left})$$
$$\delta_\varphi = \frac{1}{2}(\delta_{right} - \delta_{left})$$

$$\delta_{right} = (\delta_r + \delta_\varphi)$$
$$\delta_{left} = (\delta_r - \delta_\varphi)$$

Fig.9 Ride and roll coupling model

$$\begin{aligned} &\delta q^T(\bar{M}\ddot{q} - \bar{Q}) = 0 \\ &\bar{M} = B^T M B \\ &\bar{Q} = B^T(Q - MH) \end{aligned} \tag{6}$$

where q is the generalized coordinate, Q is the generalized force corresponding to generalized coordinate, M is mass matrix and B is the velocity transformation matrix that relates generalized velocity to Cartesian velocity.

R&H analysis results were verified applying vehicle data. Suspension and steering characteristics were imported as polynominal functions from K&C analysis results. The results are validated with ADAMS/CAR in Step Steer, Pulse Steer and Steady State Conering. Reliable R&H analysis tool could be used by chassis design engineers in the integrated chassis design environment.

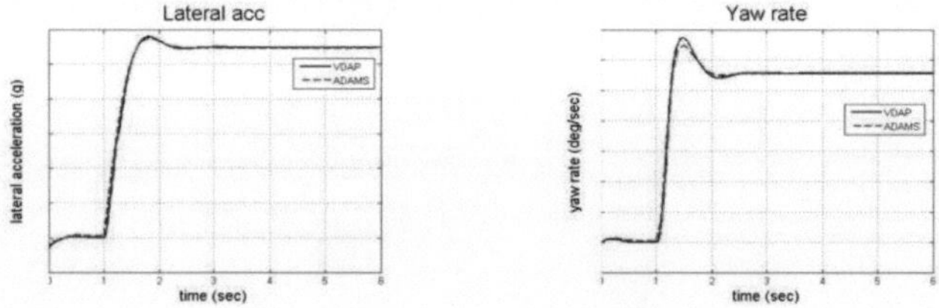

Fig.10 Step Steer analysis

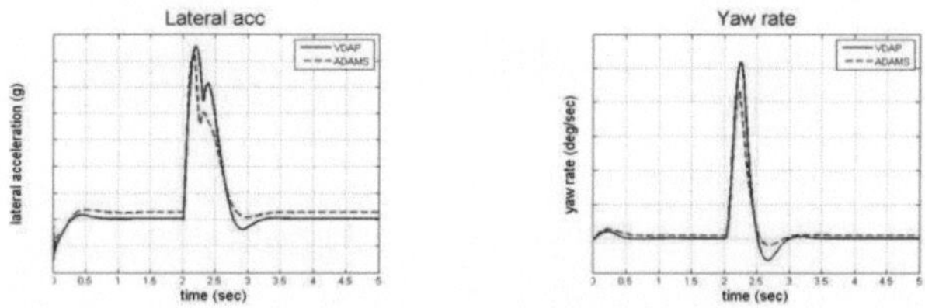

Fig.11 Pulse Steer analysis

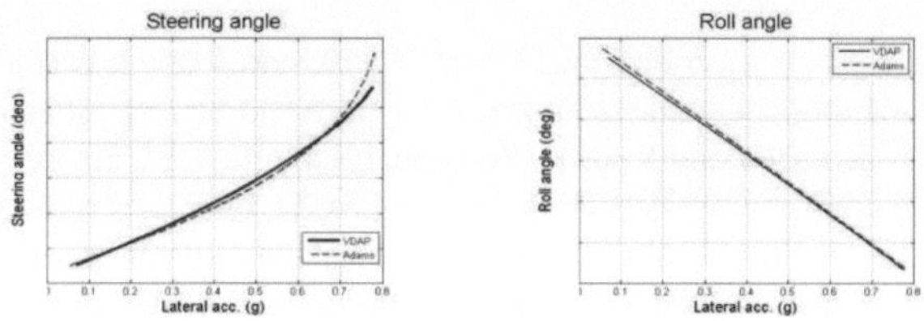

Fig.12 Steady State Cornering analysis

R&H analysis tool is integrated to chassis control models. R&H analysis input file could be imported and solved in MATLAB/Simulink. Therefore various control logic (MDPS, ECS etc.) could be integrated to R&H analysis. Because same vehicle model and solver were used from chassis design to control application and analysis could be sequentially performed from geometry to R&H analysis. R&H development efficiency could be drastically improved.

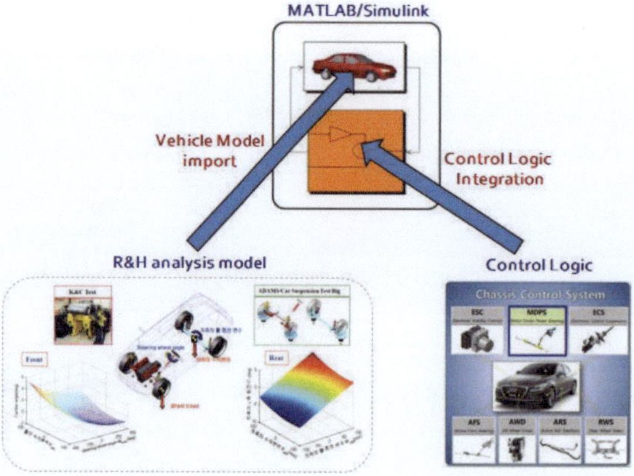

Fig.13 R&H analysis integrated to control logic

4 User Interface based EXCEL

Graphical user interface was developed using EXCEL and VBA. It is convenient and simple for chassis design engineers to use. Analysis could be sequentially performed from suspension geometry and K&C to R&H analysis in EXCEL environment. R&H design and development process is embedded as EXCEL templates form. Chassis design, CAE and test engineers can manage same design data format. Hard-point data in geometry analysis sheet is integrated to K&C analysis sheet. Therefore user can input additional information as suspension bush, spring, stabilizer bar, bump stopper etc. in K&C analysis sheet. Additional vehicle data for dynamic analysis can be input in R&H analysis sheet. Analysis results can be shown as format of summary sheet and plots. Therefore chassis design engineers can confirm comparison of results according to change of design.

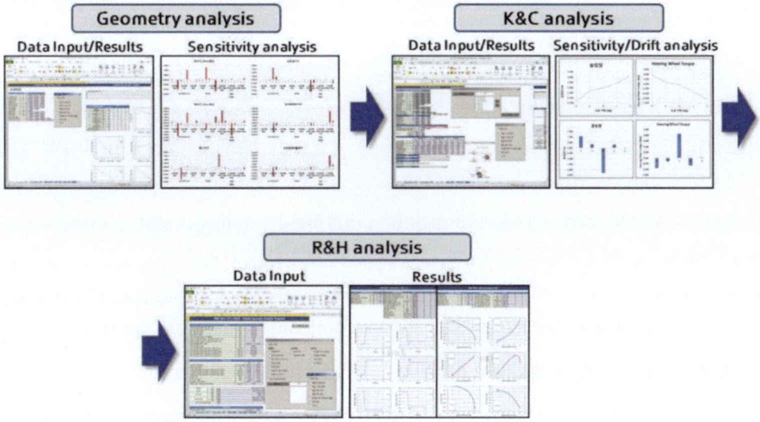

Fig.14 User Interface based EXCEL

5 Conclusion

A development of the chassis design and R&H performance analysis tool, optimized to chassis R&H development process, shows the following conclusions.

1) Chassis design engineers investigate design parameter and test engineers predict R&H performance using developed In-house Tool. Analysis could be sequentially performed from geometry to R&H analysis.

2) The tool could be used for target cascading using target setting and verification using chassis design data in R&H development process (V-cycle).

3) Chassis design engineers improve design completion and test engineers minimize performance tuning trial and error.

4) Integrating developed In-house Tool with optimization software, K&C characteristics and R&H performance could be optimized.

[Reference]

[1] http://www.mscsoftware.com/product/adams

[2] https://www.carsim.com/products/carsim/

[3] Ewald Fischer, "ADAMS/Car-AT in the Chassis Development at BMW", ADAMS Conference, 2001

[4] Karthik Vemireddy, "Development of a driving dynamics-oriented suspension design during the early concept phase", CHASSIS.TECH, 2015

[5] Young Deuk Kim, "A Development of the Chassis Design and R&H Performance Analysis Tool", HMC Industrial-educational cooperation research report, 2016

[6] https://www.mathworks.com/products/matlab.html

[7] UnKoo Lee, Byeonheui Ahn, "A Method to Analyze 'The Imaginary Kingpin Axis' in Multi-Link Type suspension Systems", SAE 930262 pp.11~26, 1993

[8] S. S. Kim, K. S. Choi, "Efficient Vehicle Dynamic Analysis Based on Quasi-Static Analysis of Suspension Model", International Journal of Automotive Technology, 1997

[9] Jung, H., Kim, S., Shim, J. and Kim, C., "Development of Vehicle Dynamics Model for Real-time ECU Evaluation System," AVEC, Aug. 2004

[10] Kim, S. and Jung, H., "Functional Suspension Modeling for Real-time Vehicle Dynamics Analysis Using K&C Test Data," 30th FISITA World Automotive Congress, 2004.

CHASSIS CONTROL SYSTEMS

SKYACTIV TECHNOLOGY to enhance 'Jinba-Ittai'

Tohru Yoshioka[1], Osamu Sunahara[1], Yasunori Takahara[1], Daisuke Umetsu[1], Hidenori Matsuoka[1], Tadanobu Yamamoto[1], Naokazu Kaneshina[1], Ryo Shimizu[1], Junya Takahashi[2], Makoto Yamakado[3]

[1] Mazda Motor Corporation, Japan,
[2] Hitachi, Ltd, Japan,
[3] Kanagawa Institute of Technology, Japan

P.E. Pfeffer (Ed.), *8th International Munich Chassis Symposium 2017*, Proceedings, DOI 10.1007/978-3-658-18459-9_28

1. Introduction

Mazda Motor Corporation has developed the technology architecture, SKYACTIV TECHNOLOGY, to deliver supreme driving pleasure to all the drivers while satisfying the requirements for the earth and societies. It improves both the vehicle performance and development efficiency and has been implemented onto all vehicle segments by pursuing the ideals of engineering principles. The architecture encompasses engine, transmission and platform as the basis of the building block strategy (Fig. 1).

Fig. 1: Building Block Strategy

In developing SKYACTIV TECHNOLOGY, we focused on "Jinba-Ittai," or oneness of driver and vehicle. Jinba-Ittai is based on the concept of human-centred design and requires optimization of all the design parameters of an entire vehicle to be adapted to human driver's characteristics. The primary challenge was to define the ideals of all the attributes and reallocate the functions onto all the systems and components of the vehicle.

In this paper, we first describe the outlines of SKYACTIV-CHASSIS, SKYACTIV-BODY and SKYACTIV engine. Then we explain the concept of SKYACTIV-VEHICLE DYNAMICS, which further enhances the vehicle dynamics by combining each system of SKYACTIV TECHNOLOGY. As its first application, we have introduced G-Vectoring Control (GVC) which enhances the dynamics performance with engine control and improves the closed-loop vehicle dynamics by modifying human driver's steering operation.

2. Targeted Dynamics Performance

Jinba-Ittai is the concept that combines vehicle performance precisely responding to a driver's operation, driving pleasure a human driver feels while driving, and perceived unity with the vehicle deepening inside a human driver through holistic driving experiences. SKYACIV TECHNOLOGY is to achieve supreme dynamics performance in all the vehicle segments by optimizing functional allocation based on the mechanisms of all the attributes. In addition to individual attribute targets, we paid great attention to the transient vehicle responses of longitudinal and lateral motions, body control, the relation between vehicle response and feedback to the driver as well.

Figure 2 describes an example of desired acceleration transition in a cornering maneuver with corresponding g-g diagram[1]. We also aimed to achieve the diagonal roll motion, or the combination of roll and pitch, so that a vehicle generates yawing motion effectively with the vertical load transfer from the rear axle to the front axle mainly induced by the downward motion at the front outer corner[2]. The feedback of steering wheel torque was also designed to correspond to the vehicle yaw rate (Fig.3). To achieve this consistent vehicle response, we pursued the ideals of each vehicle system such as chassis, body and engine, and integrated them to further improve performance.

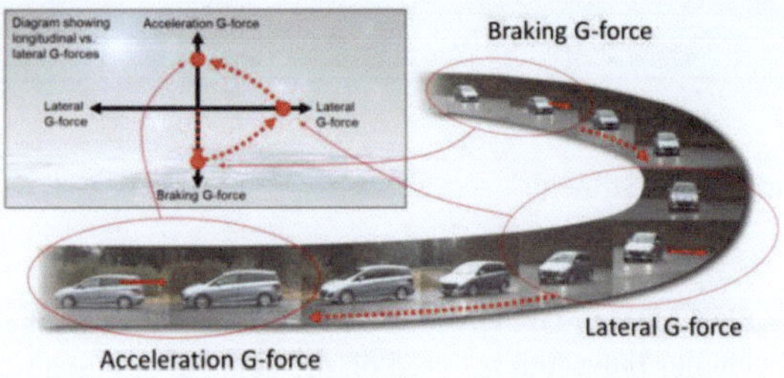

Fig. 2: Smooth Transition of Longitudinal and Lateral Acceleration in Cornering

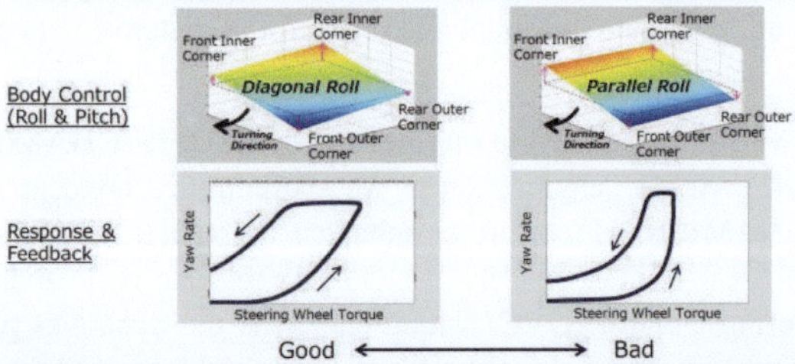

Fig. 3: Body Control in Cornering and Relation between Steering Wheel Torque and Yaw Rate

3. SKYACTIV-CHASSIS

For SKYACTIV-CHASSIS, we pursued breakthroughs in steering and handling, ride comfort and brake performance.

3.1 Suspension and Steering

Breakthroughs we aimed in designing the suspension and steering systems are between agility and stability, agility and ride comfort, and lightweight and stiffness.

We aimed for agility at a low- and mid-vehicle velocity and stability at a high-vehicle velocity. By applying a lower steering gear ratio to the steering system and enough under-steer K&C characteristic to the suspension system, we raised the yaw gain at a lower vehicle velocity while suppressing it at a higher vehicle velocity. The electric power-assist steering was also designed to keep the steering effort adequate for the yaw rate.

We also improved ride comfort while keeping agility at a low- and mid-vehicle velocity. By reallocating all the K&C characteristics among the suspension components, we achieved the targeted agility with reduced suspension springs, top-mount rubber springs and damping forces (Fig.4). The hard points of the suspension were also altered. Figure 5 shows the example of the rear suspension where the attachment point of the trailing arm is moved upward. This makes the force coming from the ground be transmitted more in the vertical direction than in the longitudinal direction so that ride comfort is improved due to the reduced longitudinal force at the trailing arm bush.

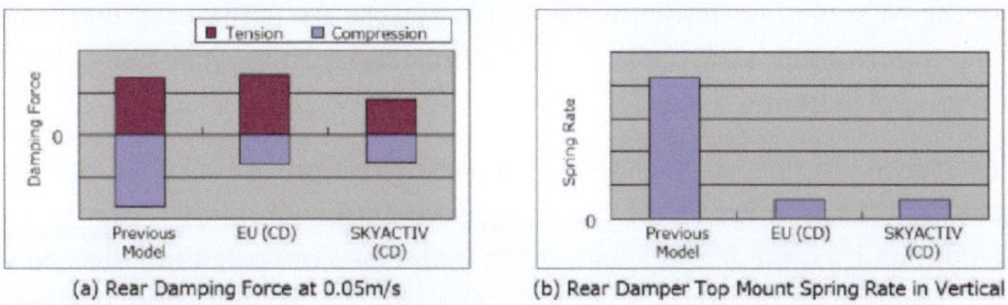

Fig. 4: Reduced Damping Force and Mount Rubber Spring

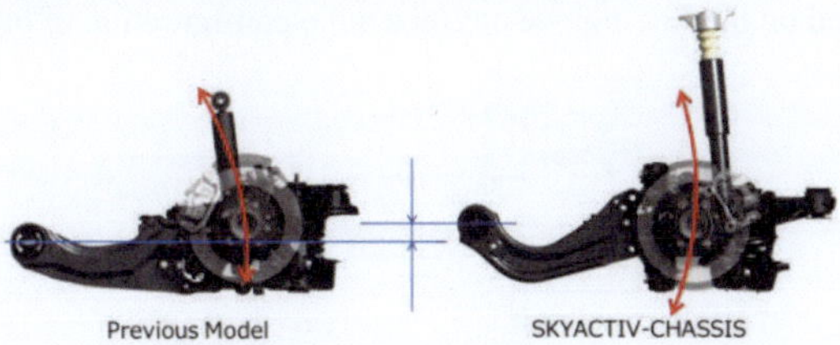

Fig. 5: Height of Attachment Point of Trailing Arm

We pursued lightweight and stiff components by optimizing the structure of each member and improving the welding technology. In designing sub-frames, we shortened the offset between the points of force input and its support, enlarged the sections of the members which sustain the load and increased the distance between the parts from the stage of layout study. In designing frames, we maximized the weight efficiency by optimizing the welding location and welding method (Fig.6).

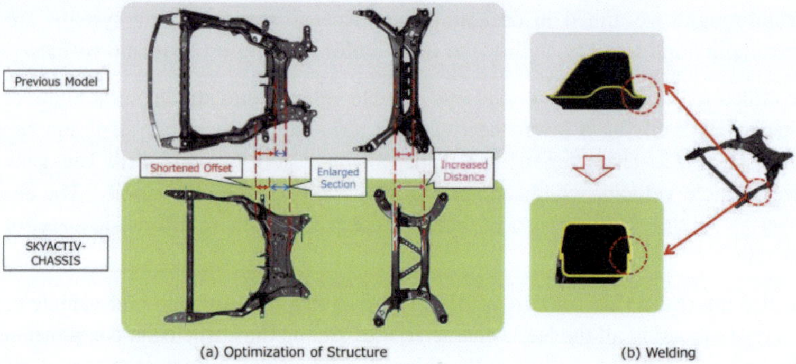

Fig. 6: Lightweight and Stiff Sub-frames

3.2 Brake

Breakthroughs we aimed in designing the brake system are between effectiveness and brake noise, lightweight and brake temperature, and rolling resistance coefficient (RRC) of tires and stopping distance.

We applied quality engineering to define the configuration of brake caliper in a way that ensures the effectiveness as well as suppresses brake vibration. After installing a motor onto the brake system and rotating the motor, we observed the electric current of the motor while the brake fluid pressure is applied. We attributed the current deviation of the motor's power supply to the vibration energy which may induce brake noise, and based on the finding, we optimized the configuration of brake caliper.

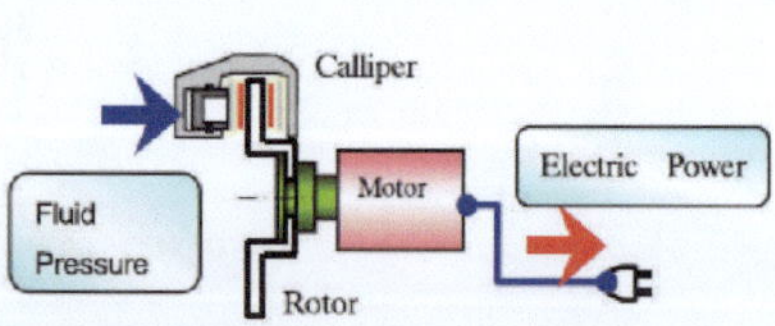

Fig.7: Illustration of Brake Energy Measurement System

We also applied quality engineering to renew the configuration of brake rotor in a way that ensures lightweight as well as improves cooling capability. We defined the dimensions of the brake rotor as control factors and evaluated temperature rise, deformations due to heat, and weight. As the result, we realized the lightweight brake rotor relative to necessary braking force as shown in Fig.8.

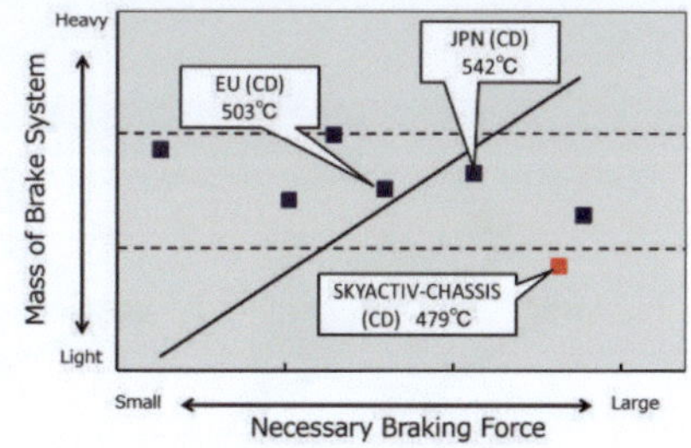

Fig. 8: Mass of Brake System and Cooling Capability

While the low tire RRC contributes to fuel efficiency of the vehicle, it often deteriorates the stopping distance. We improved the stopping distance by suppressing the pitching motion with the increased anti-lifting force at braking through the suspension configuration as shown in Fig. 5.

3.4 Body Control in Cornering

Figure 9 shows how each corner of the vehicle moves during transient in right cornering at an initial vehicle velocity of 60 kph. In the very early stage of cornering, the front outer corner moves downward so that the induced vertical load transfer from the rear axle to the front axle yields yawing moment effectively. The pitch motion of nose-dive is kept until the steady state cornering in a way that ensures the diagonal roll defined as the target shown in Fig.3.

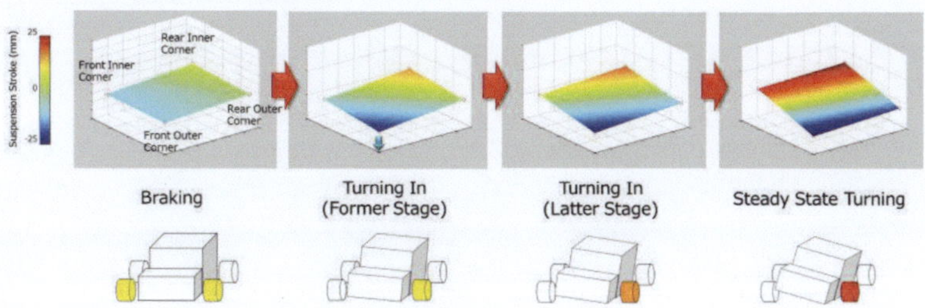

Fig. 9: Body Control in Transient Cornering

3.5 Application of SKYACTIV-CHASSIS

We have applied SKYACTIV-CHASSIS to all the vehicle segments. Instead of applying some specific parts in common, we developed the consistent chassis characteristics by adapting the detailed suspension components to vehicle dimension as shown in Figs. 10 and 11.

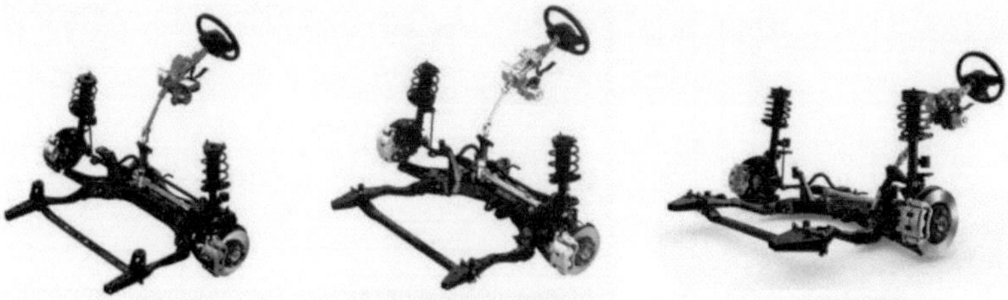

(a) CX-3 (b) Mazda3 (c) CX-9

Fig. 10: Outlook of Front Suspension

(a) CX-3 (b) Mazda3 (c) CX-9

Fig.11: Outlook of Rear Suspension

4. SKYACTIV-BODY

Since the body performance influences all the vehicle attributes, it is important to optimize the body design in various aspects such as stiffness, crash safety and lightweight. A perspective for breakthroughs we took is optimization of body structure and material replacement.

4.1 Optimization of Body Structure

We optimized the body structure from the viewpoints of straight frames, continuous structure and multi-load path.

4.1.1 Straight Frames

We renewed the base structure based on the principles of structural design. In the past, there were bends and dents on the frames due to the layout restrictions with existing powertrain and drive-shafts. By designing the top-hat, platform and powertrain at the same time, we realized the straight frames without bend and dent as shown in Fig. 12.

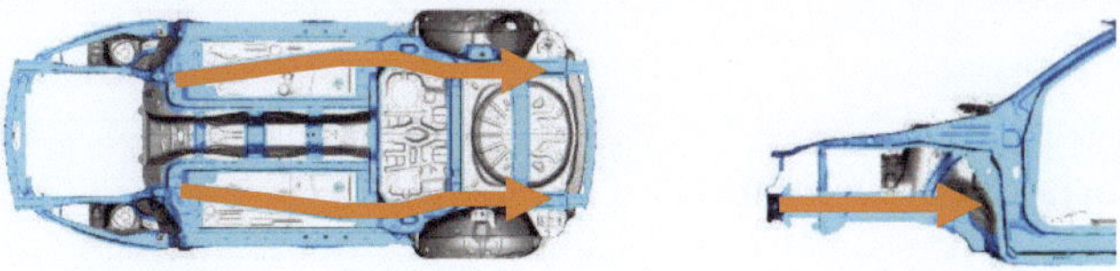

Fig. 12: Straight Frames

4.1.2 Continuous Structure

The top-hat should also function as the structural members. The body stiffness has increased with three ringed structures which comprise of the reinforcements of the platform and the top-hat structures such as the roof rail and B-pillar (Fig. 13). The suspension attachment points are also optimized.

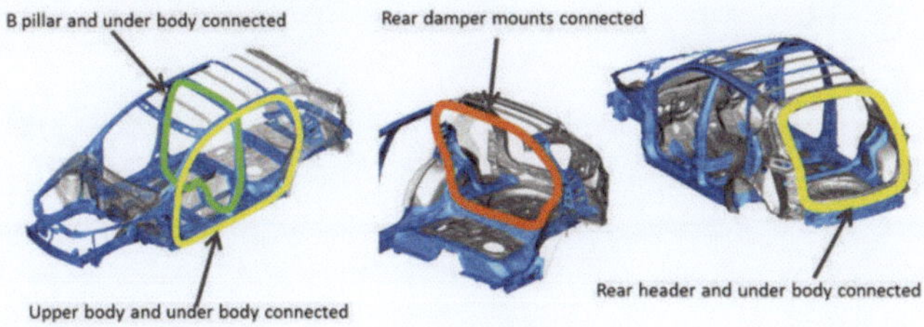

Fig. 13: Continuous Structure

4.1.3 Multi-Load Path

We realized the lightweight and strong body structure by diversifying the load input to various components as shown in Fig.14.

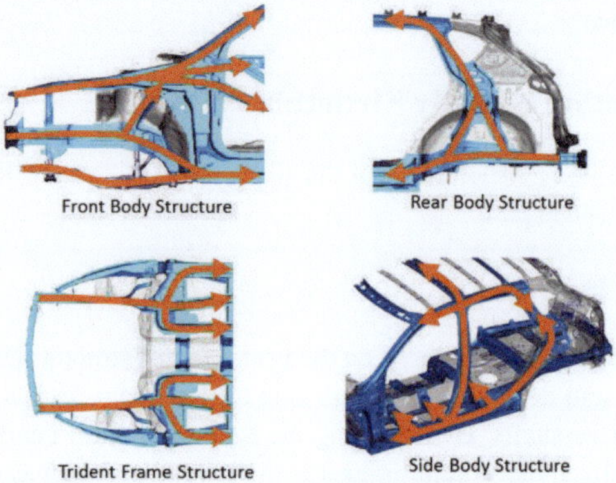

Fig. 14: Multi-Load Path

4.2 Material Replacement

We applied the high-strength steel which absorbs larger crash energy to the necessary components for crash safety after keeping their thickness so that they ensure enough body stiffness based on CAE analysis. The example of high-strength steel application of Mazda 2 is shown in Fig. 15.

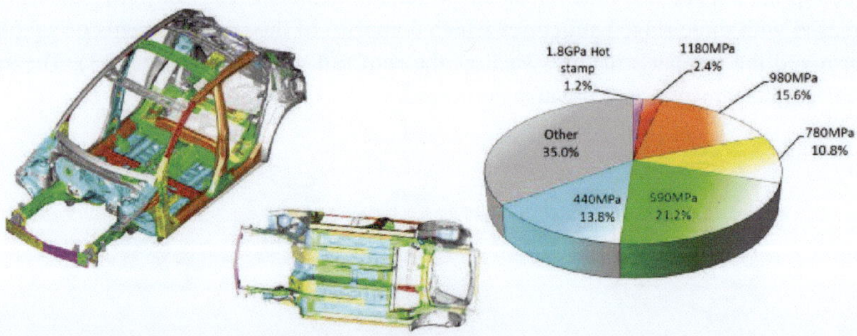

Fig. 15: Application of High-Strength Steel (Mazda 2)

4.4 Application of SKYACTIV-BODY

We have applied SKYACTIV-BODY to all the vehicle segments. Instead of applying some specific parts in common, we developed the consistent body characteristics by adapting the detailed frame arrangement to vehicle dimension, body type and the trend of material development as shown in Fig. 16.

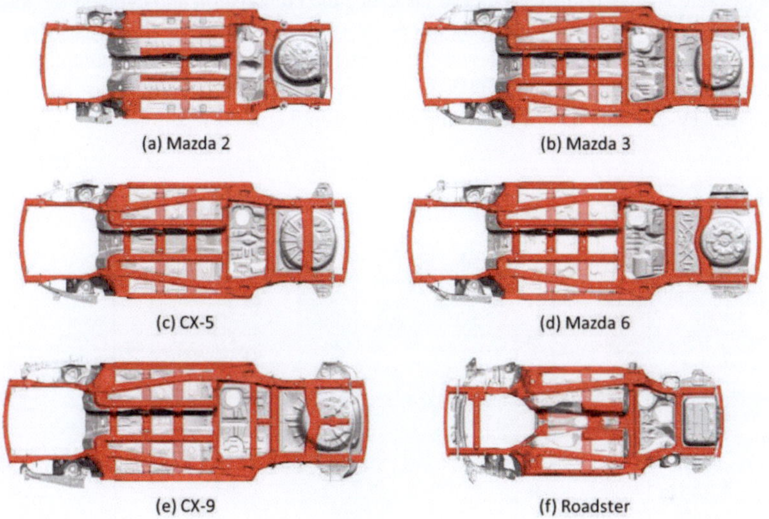

Fig. 16: Body Frame Structure of Each Segment

5. SKYACTIV Engine

We pursued the ideals of combustion regardless of the fuel type in SKYACTIV engine.

5.1 Pursuit of Ideal Combustion

We extracted seven control factors to influence the combustion and evolved the internal combustion engines by bringing the control factors near to the ideals as shown in Fig. 17. As the result, we achieved the overwhelmingly low fuel consumption and quick response with SKYACTIV-G of the record-high compression ratio and SKYACTIV-D of the record-low compression ratio.

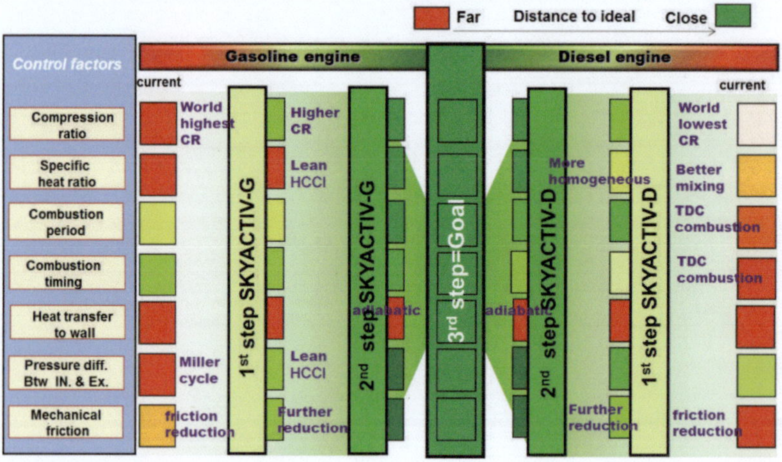

Fig. 17: Evolution of Engine

5.2 Application of SKYACTIV Engine

We have applied SKYACTIV engine to all the powertrains. In the application, we did not take an ordinary short-sighted approch such as the simple commonization of engine dimensions but reallocated the functions to each of the control factor so that the combustion characteristics are the same throughout the engine types and displacements (Fig. 18). This resulted in superior performance regardless of engine displacement.

Fig. 18: SKYACTIV-G and SKYACTIV-D

6. SKYACTIV-VEHICLE DYNAMICS

Vehicle dynamics is the closed loop performance of a human driver and the vehicle, and we have to ensure Jinba-Ittai in the driving environments which change every moment due to road undulations and so on. SKYACTIV-VEHICLE DYNAMICS is the concept of control technology to further enhance vehicle dynamics by integrating all the systems of SKYACTIV TECHNOLOGY rather than operating each system independently (Fig. 19).

The first application of SKYACTIV-VEHICLE DYNAMICS is GVC[3]. GVC enhances vehicle dynamics by controlling engine torque under the idea to generate the maximum cornering motion with the minimum energy.

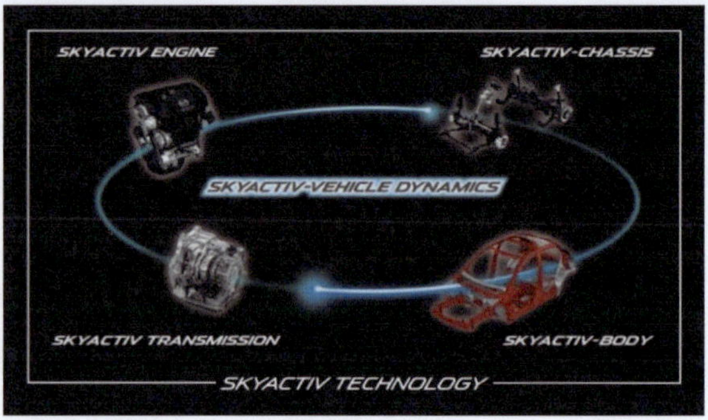

Fig. 19: Concept of SKYACTIV-VEHICLE DYNAMICS

6.1 Concept of GVC and System Configuration

Experienced drivers achieve harmonious transitions in lateral and longitudinal accelerations shown in Fig. 2. They realize the smooth g-g transition by decelerating and accelerating the vehicle in response to the lateral jerk shown in Fig. 20.

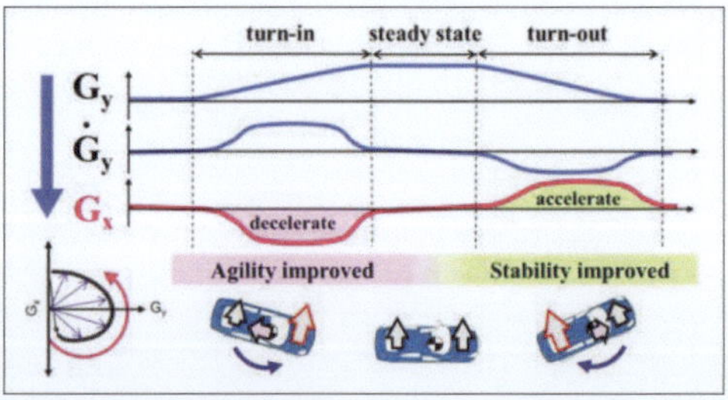

Fig. 20: Concept of G-Vectoring Control

In GVC, the longitudinal acceleration can be expressed in Eq. (1)[4][5],

$$G_{xc} = -sgn(G_y\,\dot{G}_y)\,C_{xy}|\dot{G}_y| \tag{1}$$

where G_{xc}, C_{xy}, G_y and $G_{y\_}dot$ are targeted deceleration, control gain, lateral acceleration and jerk, respectively. We used SKYACTIV engine to realize the targeted deceleration thanks to its quick response. The system configuration is outlined in Fig. 21. The system in Powertrain Control Module predicts the jerk from steering wheel angle and vehicle velocity, defines the targeted deceleration, derives necessary engine torque reduction and reduces the engine torque by considering current engine status.

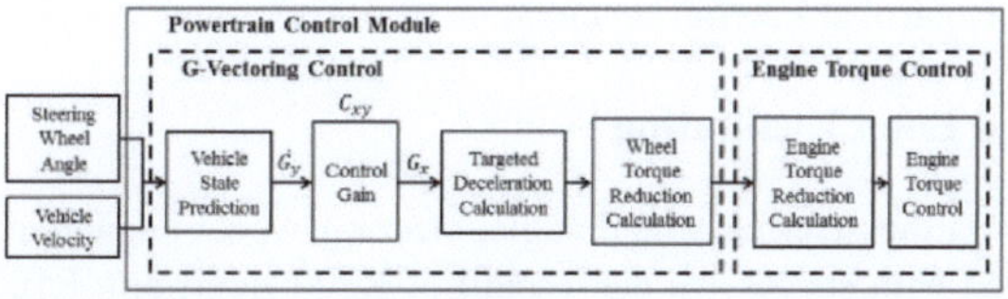

Fig. 21: System Configuration of GVC

6.2 Improvement of Vehicle Dynamics due to GVC

6.2.1 Open Loop Evaluation

The g-g diagram in J-turn is shown in Fig. 22. The vehicle velocity is 40 kph, and the steering wheel angle increases from 0 deg to110 deg in about 3 seconds. The longitudinal deceleration rises in harmony with lateral acceleration from the beginning of turning with GVC.

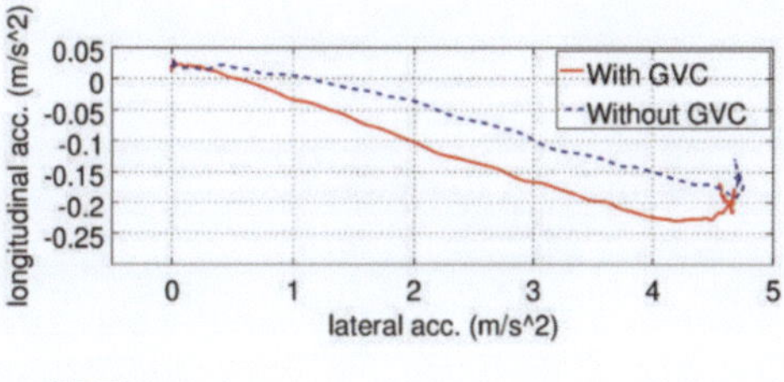

Fig. 22: g-g diagram of J-Turn

The harmonious longitudinal acceleration also induces the diagonal roll mode as shown in Fig. 23, which enhances smooth turning of the vehicle as explained in Fig. 3.

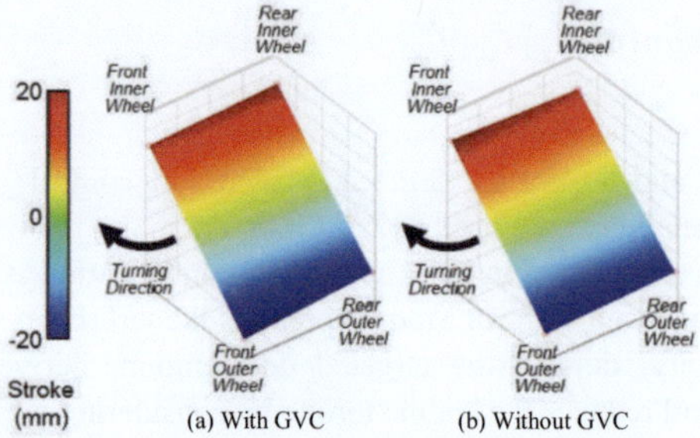

Fig. 23: Body Control at Lateral Acceleration of 0.25g

6.2.2 Closed Loop Evaluation

Figure 24 shows the steering wheel angle, steering wheel angular velocity and yaw rate on the road with successive curve radii of R200 rightward, R100 rightward and R100 leftward. The overall profile of steering wheel angle and yaw rate with and without GVC resembles to each other, because the driver traces the designated track, but the deviations of the steering wheel angular velocity is clearly suppressed with GVC which implies that the driver controls the vehicle with more confidence.

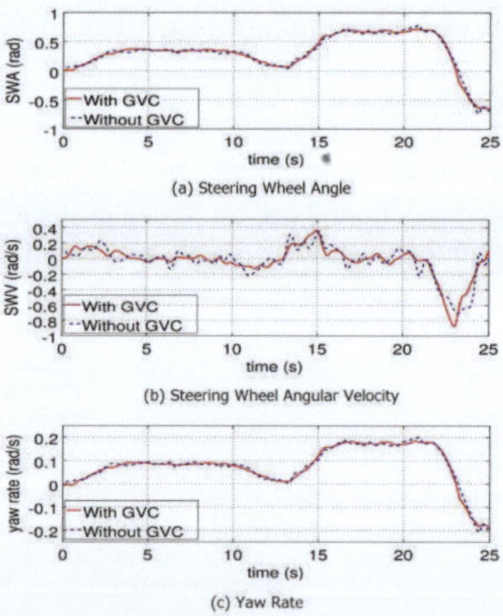

Fig. 24: Driver's Operation in Lane Trace

6.3 Influence of GVC on Driver's Operation and Behavior

The GVC improves the performance of human-vehicle closed loop system. Figure 25 shows the comparison of lane change maneuver with a human driver and steering robot which emulates the driver's steering operation with GVC. Though there is a significant difference between the cases with and without GVC when a human driver drives, the difference is very small with the steering robot operation. This infers that the GVC improves a human driver's operation itself better than a vehicle's mechanical responses.

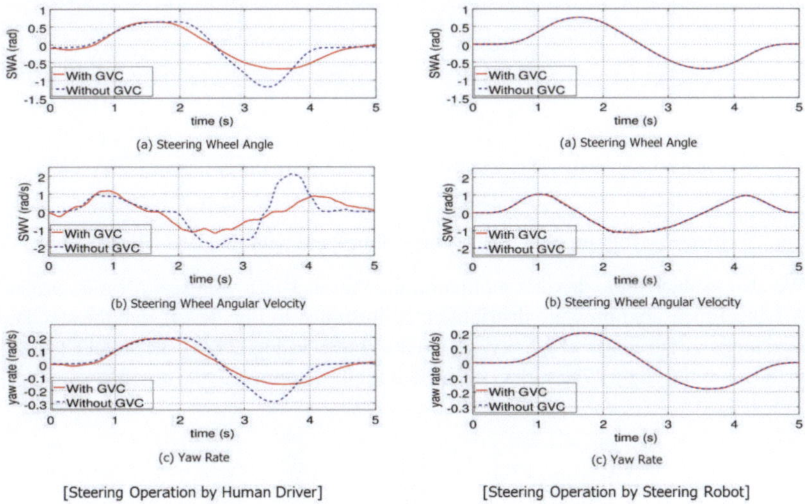

Fig. 25: Comparisons of Cases with/without GVC under Steering Operation by Human Driver (Left Colum) and Steering Robot (Right Colum)

To observe how a human driver changes, we evaluated his muscle activities. Figure 26 shows the muscles we measured. They are sternocleidomastoid muscles and upper trapezius muscles which sustain the driver's head against the external forces such as inertial force during driving. The electromyogram with and without GVC in a left J-turn is shown in Fig. 27. In each chart, two samples of electromyogram of each muscle are plotted, and the vertical dotted line represents the time of the maximum lateral jerk. In a left turn, the left and right sternocleidomastoid muscles and the left upper trapezius muscle should sustain the driver's head against the inertial force to the right.

In Fig. 27, the electromyograms of these muscles reach the peak value prior to the maximum lateral jerk with GVC (dotted red circles), but they don't without GVC. The muscles to sustain the driver's head seem to work more effectively with GVC.

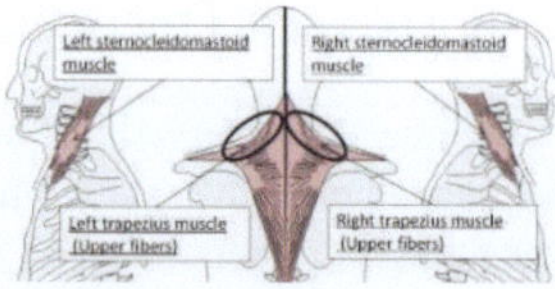

Fig. 26: Location of Muscles for Measurement

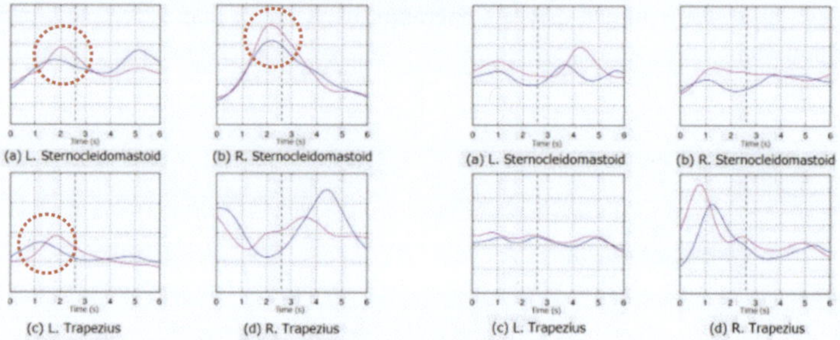

Fig. 27: Electromyogram with GVC (Left Colum) and without GVC (Right Colum)

We also evaluated the driver's motion on the seat in J-turn with seat pressure sensor. A typical measured pressure distribution is illustrated in Fig. 28 (a), and the standard deviations of the center of gravity (COG) are shown in Fig. 28 (b). The total COG deviation is reduced by 27% with GVC, which implies improved ride comfort.

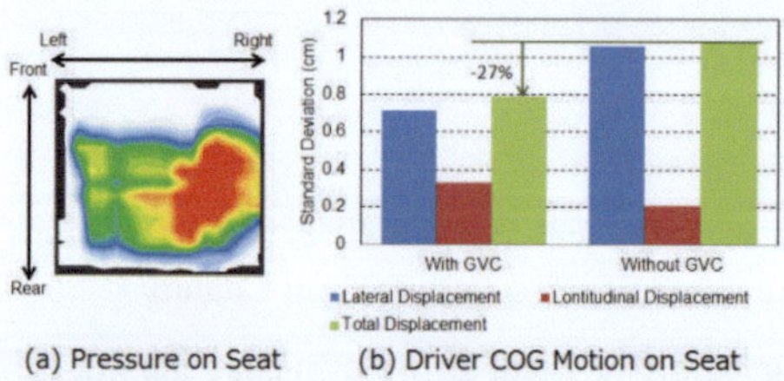

Fig. 28: COG of Driver on Seat

Figures 27 and 28 infer that GVC improves the human driver's behavior during maneuvers and that he or she can respond to the vehicle behavior change more precisely and appropriately. We can expect this benefit in daily driving since GVC functions even at very subtle driving situations such as straight-ahead driving with small steering modulation. GVC contributes to the total quality of driving for human drivers.

As the mechanism of the change in a human driver's behaviors due to GVC should be explored more in the future, we will further develop the concept of SKYACTIV-VEHICLE DYNAMICS from the viewpoint of human-centered development.

7. Conclusion

Mazda Motor Corporation has developed SKYACTIV TECHNOLOGY to deliver supreme driving pleasure, sustainable environmental performance and uncompromising safety performance to all the drivers. We pursued the ideals of each system and applied the developed concept and characteristics to the vehicles of all the segments after achieving excellent performance as well as high development efficiency. We further developed SKYACTIV-VEHICLE DYNAMICS by integrating each system of SKYACTIV TECHNOLOGY to enhance vehicle dynamics which is energy efficient and adapt itself to human drivers. We introduced GVC into the market as its first application.

To realize the future where the earth, society and the vehicles will exist in harmony, we are committed to develop supreme vehicle performance while satisfying all the requirements for the earth and society without compromise. We will develop SKYACTIV TECHNOLOGY successively and enhance Jinba-Ittai based on the philosophy of human-centered development.

8. References

[1] Milliken, W & Milliken, D., “Race Car Vehicle Dynamics”, SAE International, 1995.

[2] Kodaira, T. et al., “Improvement of Vehicle Dynamics Based on Human Sensitivity (Second Report) -A Study of Cornering Feel-,” SAE Technical Paper, 2007-01-0447.

[3] Yoshioka, T. et al.,” Development of G-Vectoring Control System Based on Engine Torque Control,” AVEC’16, 2016.

[4] Yamakado, M. et al., “An experimentally confirmed driver longitudinal acceleration control model combined with vehicle lateral motion,” Vehicle System Dynamics, Vol.46, 2008, pp. 129-149.

[5] Yamakado, M. et al., “Improvement in vehicle agility and stability by G-Vectoring control,” Vehicle System Dynamics, Vol. 48, 2010, pp231-254.

New torque vectoring functions for modular electrical propulsion

Rob Kraaijeveld, Senior Chassis Specialist
Martin Dorn, Vehicle dynamics engineer
Klaus Wolff, Vice president Vehicle
FEV GmbH

Michael Schubert, Group Leader electrical propulsion
ISEA-RWTH Aachen

P.E. Pfeffer (Ed.), *8th International Munich Chassis Symposium 2017*,
Proceedings, DOI 10.1007/978-3-658-18459-9_29

Abstract

With the increasing electrification of passenger cars, more and more vehicles will be equipped with multiple E-motors. Having 2 motors on one axle offers the opportunity to implement functionality without additional hardware.

By applying different torques on the left and right wheel, the vehicle driving direction can be influenced by the resulting yaw moment. The function is known as torque vectoring. Torque vectoring is not new and is also implemented in mechanical systems.

The main objective of such systems is to achieve better traction during cornering and to increase the steering response of the vehicle. FEV has investigated additional possibilities for torque vectoring systems based on dual E-motors. These functions focus more on the daily use for typical customers.

A simulation model has been built to develop and investigate the new functions. As next step a development platform was used to demonstrate the results on an actual vehicle.

In this paper the focus is on the development of the function "perturbing force compensation".

This function has the aim to support the driver in circumstances where there is a side force on the vehicle that would require the driver to use the steering wheel to keep driving straight ahead.

Examples of forces are: side wind and road banking, both situations where the correction is needed for longer periods >1s. The function can also serve as a compensation for E-motor tolerances that would give an off-set to the steering wheel effort and angle.

The base of the function is an observer that compares the steering wheel angle and torque to the driving direction in relation to vehicle speed. Simulation results show the effectiveness of the function. Finally, the potential risks of torque vectoring have been analysed to secure a functional safety compliant development.

1 Background and introduction

Electric power trains will become more common in the near future. Reasons are the global environmental awareness, urbanisation effects and the diminishing oil related resources, giving impulses to develop new propulsion technologies.

Where the vehicle stability of traditional combustion engine driven vehicles is mainly controlled by Electronic Stability Controllers acting mainly on the brake system, electric propulsion can offer all new possibilities. The vehicle controller that can improve the efficiency by intelligent braking, can even increase the stability of the vehicle by intelligently shift traction torque to the individual wheels to avoid critical situations.

Vehicle active safety and therewith the controllers will play an increasing role in the near future. Since it is not known yet what topology will be mainstream for future propulsion systems, researchers and industry are looking into possibilities to build modular and scalable systems. This offers the possibility to incorporate one, two, three or four modules in a vehicle, depending its requirements and purpose. This approach however, requires the definition of control systems that can cope with the new requirements set for these single wheel controlled propulsion.

1.1 Masstab

MAs:Stab is the name of the research project that was the basis for this work. The overall goal of the project was to support future developments of scalable modular electric propulsion components. The project was sponsored by the German government under number: 01MY12008C. Special focus was on the requirements for the dynamic controller behaviour. An independently two wheel driven vehicle was built to investigate the vehicle dynamic effects of the system.

Gefördert durch:

Bundesministerium
für Wirtschaft
und Energie

aufgrund eines Beschlusses
des Deutschen Bundestages

MAs:Stab
Modularer Antriebsstrang: Stabilität bei Mehrfachantrieben

In his paper the validation and optimisation of the controllers and algorithms was in focus during simulations and later during evaluations of the electric components on a test rig and of the vehicle on a test track.

1.2 Torque vectoring

Torque vectoring is not new in vehicles and is researched since the 1930's. The goal of torque vectoring is pointing (vectoring) drive torques to specified wheels. Since the beginning, starting with the development of the first limited slip differential by Ferdinand Porsche, the main focus was on optimal traction during acceleration.

After the introduction of mechatronics, more intelligent systems for controlling the differential were developed that can adjust the torque distribution depending on the actual need and conditions and have the ability to support more agile steering and increase the stability.

Modern ESC systems achieve similar results by creating a yaw moment by means of applying the brake of a chosen wheel or wheels.

Multiple, independently controlled electric motors create new freedom to generate wheel individual torques, both positive and negative. This allows the development of functions, not possible with current torque vectoring systems. Moreover, is does not result in increased costs for additional components.

1.3 Controller

1.3.1 Requirements

Since the controller development was started at the same time as the function development, requirements needed to be set for the controller and other hardware components to perform the tasks it has to perform.

To define these requirements, a generic vehicle model in dSpace ASM was used to perform vehicle dynamic simulations with a simple torque vectoring control model. Based on the targeted functions, the agility increase function was defined as the one that sets the highest requirement on the system. The ultimate performance of this function was seen during a step steer input.

In this function the aim is to increase the yaw rate build up in the first phase of the maneuver where a steering wheel angle is present and the yaw rate generated by the tyre lateral slip is lagging behind. It also reduces the typical yaw rate overshoot in a step steer maneuver with a negative yaw moment in the overshoot phase.

The target used for defining the requirements was set to increase the yaw rate response by 20%. The simulations showed that based on this, the following dynamic targets for the control system can be defined:

- Requirement for torque delay: < 10 ms
- Requirement for torque accuracy: < 2,5 Nm (2,5% of the nominal torque)
- Requirement torque build up (0 … 100% of the nominal torque): < 100 ms

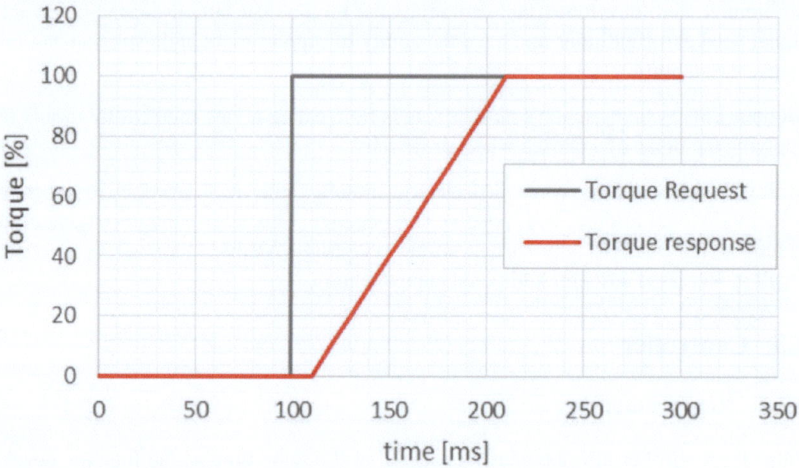

Figure 1: Control requirements regarding torque build up, based on first simulation results

1.3.2 Implementation details of the electrical drive system

It is well known, that electrical drive systems offer very high dynamic torque capabilities and good torque accuracy. However, depending on the motor type and control technique certain limitations exist. Especially when additional optimisation goals, such as improved efficiency, control robustness and functional safety must be achieved.

Torque production in electrical machines relies on two different effects, the reluctance force and the Lorenz force. The latter is dominant in permanent magnet synchronous machines (PMSM) and induction machines (IM) which are widely used in traction applications. The torque produced by these machines is given by the product of the magnetic flux-linkage in the machine and the torque-producing current component through the stator coils. In PMSM the flux-linkage is inherently present due to the permanent magnets but in IM drives like the one considered in this paper an additional flux-producing current component must be applied to the machine. This additional current increases ohmic losses and degrades efficiency of the IM drive system.

However, due to advantages in the field weakening region at high speeds and lower material cost, the IM is still a very attractive traction machine, especially when field weakening is also applied at light loads in the base speed region. The flux linkage is then reduced to an optimal value where only as much flux is present as needed for the desired torque production. This leads to a significant current reduction and a competitive efficiency.

Both, PMSM and IM drives, easily meet the aforementioned dynamic requirements in terms of torque development when flux-linkage is present in the machine. The torque dynamic is only limited by the dynamic of the controller and usually chosen at several hundred hertz. On the left of Figure 2, the step response of the IM drive has been simulated in Matlab/Simulink with constant nominal flux-linkage. The desired torque is reached after less than 10 ms. If flux-linkage is reduced to achieve maximum efficiency, the flux-linkage has to be increased before the desired torque can be produced. Contrary to the torque producing current, flux-linkage does not instantaneously change with the flux-producing current, but responses like a low-pass with the rotor time-constant, which is defined by the machine design and usually in the range of 100 ms to 1 s. Because torque is the product of flux-linkage and torque producing current this significantly reduces the torque dynamic of the IM drive.

To overcome this drawback of the IM drive, in this work a flux-linkage boosting current trajectory optimizer has been implemented that accelerates flux-linkage production by dynamic utilization of the available inverter current and optimized current sharing of both current components. The simulation results in the right of Figure 2 demonstrate that both current components are immediately increased without exceeding the maximum inverter current, whereas priority is given to flux-production. As soon as the desired flux-linkage is reached, flux-producing current is reduced and more current is available for torque production. The desired torque is then reached after only 60 ms.

Experimental results for the achieved torque dynamic are given on the left of Figure 3. The data has been recorded by logging of the CAN messages for torque request and the estimated torque feedback of the inverter which are updated every 20 ms. For validation of the estimated torque feedback, the torque was measured by a shaft torque transducer connected to the same CAN bus. The measurements show, that the efficiency optimized IM drive meets the torque dynamic requirement. Furthermore, only torque increase is affected by the limited flux-linkage dynamic. The response to a torque step down can be considered as instantaneously. The measurements also show that the CAN communication has a significant influence. To meet the delay requirement of 10 ms, the update rate of the torque request message must be set appropriately.

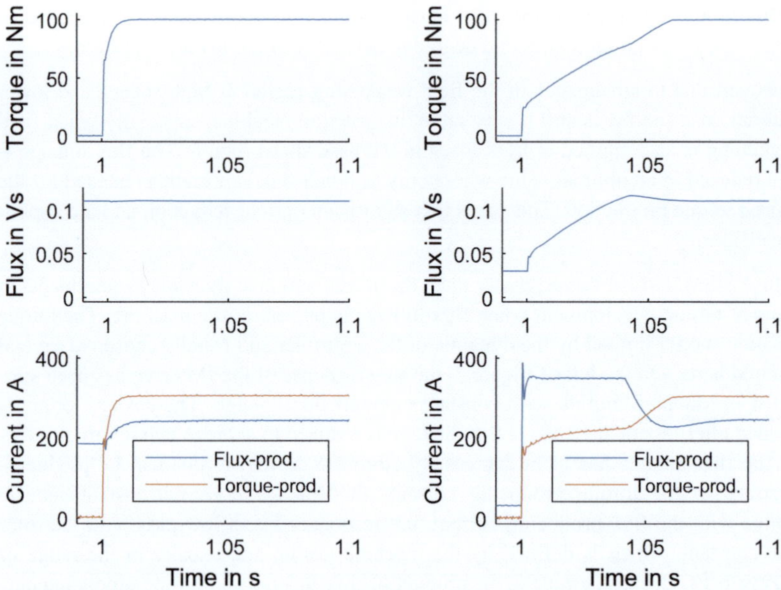

Figure 2: Simulation results for torque step reference at 1 s and 2000 rpm. Left: constant nominal flux linkage; Right: Dynamic efficiency optimized flux linkage

To achieve the torque accuracy requirement, a direct field orientated control algorithm was used. By stator-flux alignment and utilization of a back-EMF based stator flux observer, the torque controller becomes insensitivity to IM parameter variations [3]. Only at very low speeds, where back-EMF is low, significant torque deviation occurs. Experimental results in the right of Figure 3 prove that the accuracy requirement can be met for a wide operating range. Only at low speeds above 50 Nm the torque deviation is significant. Further improvements can be achieved by parameter adaption utilizing a thermal model.

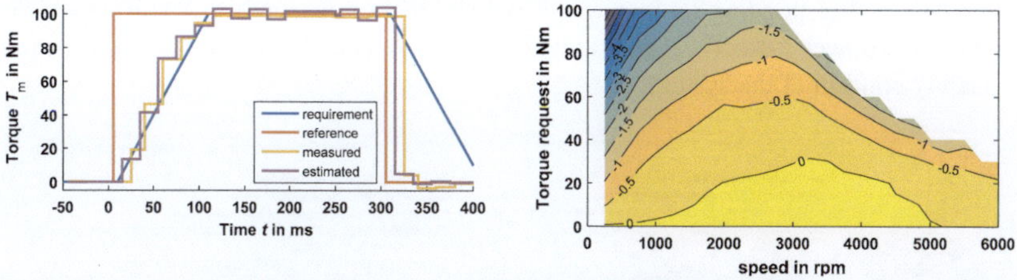

Figure 3: Left: Recorded test bench CAN-data of the drive system for a torque step request from 0 to nominal torque at 2000 rpm. Logger sampling rate: 5 ms. Message update: 20 ms. Right: Resulting torque deviation of the drive system

1.4 Functional safety

Functional safety is an important aspect in the development and series production of electronically controlled vehicle systems. Electric propulsion falls into this category and controllers for electric vehicles must be developed in such way that the functional safety is considered. Here the main focus is on unintended driving or braking torques deviating from the driver request.

When multiple motors with individual controllers are used, a new risk arises, which is the unintended yaw moment that can occur because of the different drive torques left and right.

A risk analysis was performed according to the automotive safety standard ISO 26262 regarding the stability of the vehicle with independently controlled rear wheels.

For this, risks regarding controller failures are defined, test cases generated and by means of simulation these risks are rated. This rating has been done resulting in an ASIL (Automotive Safety Integrity Level). The resulting level defines the requirements for the following series development of the controller.

A risk analysis according to ISO 26262 consists of 3 steps:

- In step 1 the functional hazard analysis will show what the functional failures are and their impact on the vehicle behavior. For instance: A failure in the torque vectoring function agility can give a wrong torque distribution, leading to a too high yaw moment which subsequently leads to an undesired direction change.
- In step 2 driving situations are chosen where this failure can lead to a danger of creating an accident.
- In the 3rd and last step the resulting risk matrix (danger in each driving situation) is rated according normalized factors. If risks rated with ASIL A until ASIL D are

found, a corresponding safety target is defined that corresponds to a safe status (for instance power cut-off). These safety targets (and safe states) are the basis for the safety concept of the system.

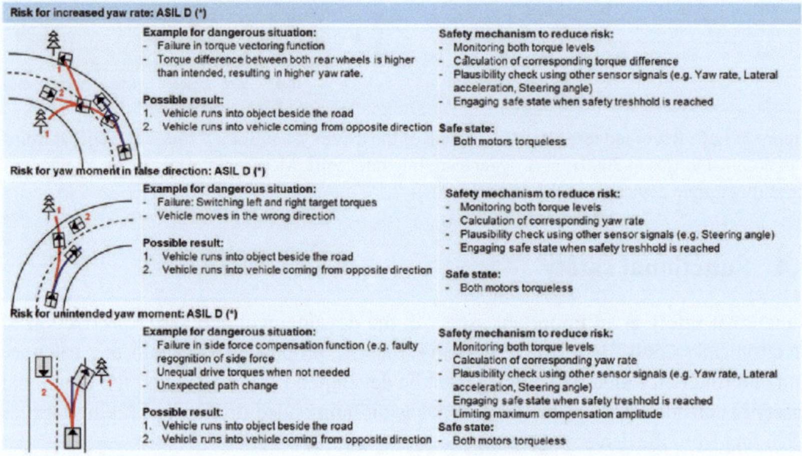

Figure 4: Summary of the possible failures and their connected rated risks

2 Functions

2.1 Investigated functions

At the start of the project many possible functions were discussed and evaluated. For the further work 5 functions were chosen for further implementation, which appeared to be the most relevant under the consideration of the modular scalable approach:

- Differential functions
- Agility increase function
- Side force compensation
- Motor characteristic
- Traction control

In the overview below the functions are described in detail

Differential function (including limited slip)

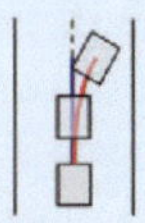

Aim of the function:
Open differential:
- Duplicate open differential.
- Allow wheel speed differences between left and right under equal traction torques (not necessary in case of dual engine system but added for completeness)

Limited slip differential:
- Duplicate limited slip differential.
- Unequal speed but different torques at the wheels is possible.

Function description:
Open differential: An open differential allows unequal wheel speeds, as needed for cornering. The traction torque is always the same on both wheels and limited to the available traction on the wheel with the lowest grip. The vehicle does not pull because of unequal wheel toques.

Limited slip differential: Such differential allows different wheelspeeds but also different traction torques. This is of benefit to increase traction during cornering or unequal road adhesion. This system can cause (strong) pulling under traction conditions.

Agility increase

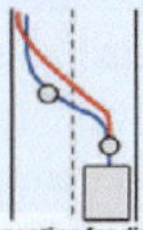

Aim of the function:
- To increase the direction change (yaw rate) during steering meneuvers.
- Also the decrease of yaw rate can be used to improve the stability during steering maneuvers.

Function description:
A yaw moment, generated by different traction torques on the rear wheels, increases the yaw rate build up and (optionally) the steady state level during cornering.

Compensation for disturbing forces as side wind and road banking

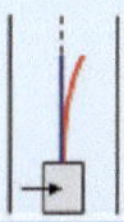

Aim of the function:
To have vehicle drive straight ahead with reduced driver input in case of disturbing side forces or vehicle asymmetries.

Function description:
A yaw moment, generated by unequal traction forces on the rear wheels, corrects the driving direction to reduce or delete the need for steering wheel corrections

Traction control

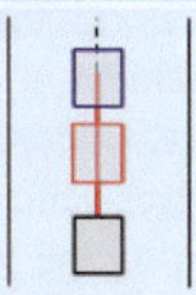

Aim of the function:
Reduces the traction on each wheel to optimise the wheel slip.

Function description:
On each wheel the traction torque is limited to the values that produces maximum traction torque.

Compensation function for uneven motor characteristics

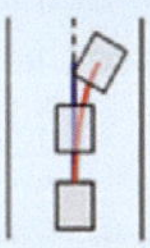

Aim of the function:
To reduce vehicle pulling or path deviation caused by asymmetric motor characteristics.

Function description:
The propulsion system consists of two electric motors. In this case the actual torques can divert from the requested torque, causes vehicle pull. This effect is monitored and correction factors, adjusting the output torque are determined for long term deviations.

Increased regenerative braking

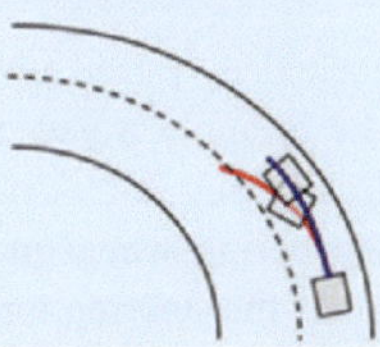

Aim of the function:
To control the stability that might reduce from high regenerative (electric) braking torques. This allows higher regeneration rates.

Function description:
During high regeneration torques, the rear axle can oversteer. By reduction of the torque to the inner wheel this effect can be reduced to keep the desired stability.

Figure 5: Summary of the possible failures and their connected rated risks

All the above mentioned functions have been developed and tested in the simulation model. Focus of the development was to prove the functionality, that is, to show that the desired effect is achieved and to validate the controller performance with the specified requirements.

Most functions are not new and also well described in other publications. Two functions however were of more interest from a vehicle dynamics point of view. Firstly the differential functions. Replicating a standard open differential was not so complicated but it brings up the question whether this function is of any use for this application. The same is valid for the limited slip differential. It was possible to achieve the desired functionality but also this function does not seem to be the optimal solution for a single wheel driven vehicle. The ideal differential function for this vehicle will be a combination of aspects offering good traction and stability under all circumstances. Deeper investigation to define this function were not performed but will be part of a separate new study. The second function is the side force compensation. This function was studied more in depth and is described in the next chapter.

2.2 Side force compensation function

Drivers of modern vehicles expect that if they want to drive straight ahead, the effort needed for this task is as low as possible. For this, the first requirement is a good chassis and steering set up that delivers a good straight ahead stability. Under normal conditions this is satisfying but there are circumstances where the driver-vehicle control system is disturbed and that is when there is a side force acting on the vehicle, caused by side wind or a road banking, both pushing the vehicle in the direction away of the acting force. Driver assist systems are developed that are used in EPS steering systems. These assist systems can use their steering torque sensor to determine a high steering effort during straight ahead driving and adjust the assist torque to reduce the torque needed by the driver.

In this work it was investigated if it is possible to achieve the same effect by means of torque vectoring, without the information of steering torque.

The first question to be answered is what the target to be reached is. Based on the standard torque vectoring functions a simple solution would be to apply a yaw moment to keep the yaw rate related to the steering wheel angle, assuming both sensors are available in the vehicle. The driver will always have the steering wheel in the "correct" position but he will not be satisfied by this if he still feels the steering torque needed to counteract the side force.

The target should therefore be to reduce the steering torque itself. Simulations were used to define the relation between the three parameters: steering wheel angle, yaw rate and steering torque. The following graphs show the relation in the simulated vehicle for the condition without side force and a situation with 40 km/h side wind.

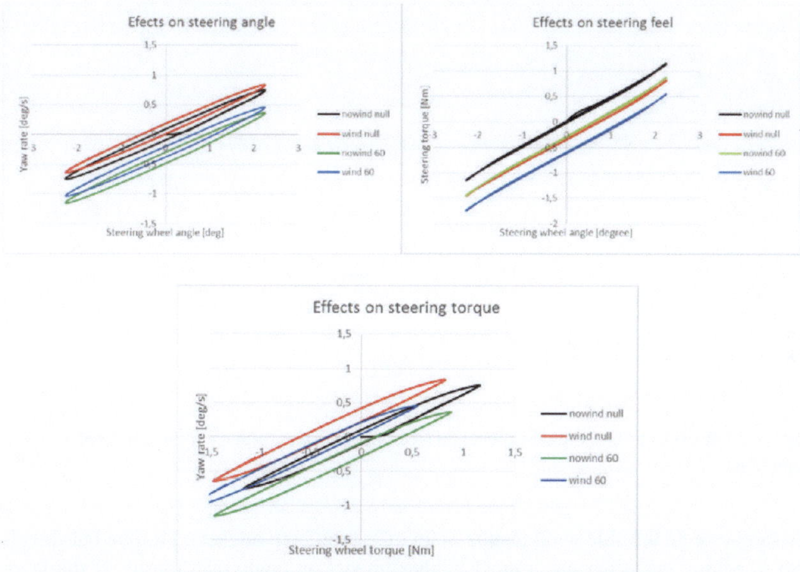

Figure 6: Effects of side wind and torque vectoring fixed rate (60 Nm) on the relations between steering angle, yaw rate and steering torque. The vehicle was driven at 80 km/h and a sinus input with 4 s period time was given to the steering wheel

What can be seen in the plots is that side wind has a small effect on the steering wheel angle (from zero to 0.3 degrees). The impact on steering torque is much higher. Here we reach 0.5 Nm, which is quite noticeable to the driver and already a level that may be reduced.

In this example a torque vectoring level of 60 Nm was applied. This was chosen as the best compromise for steering torque reduction. Higher values would even reduce the steering torque further but eventually lead to negative driving torques on one of the wheels. Although this is physically possible and also tested in the vehicle, it has the risk of unwanted side effects such as NVH issues. In the picture below the effect of the torque vectoring level on steering angle and torque are given, all based on the situation driving straight ahead with 40 km/h side wind where 115 Nm traction is needed to keep the vehicle at the constant speed.

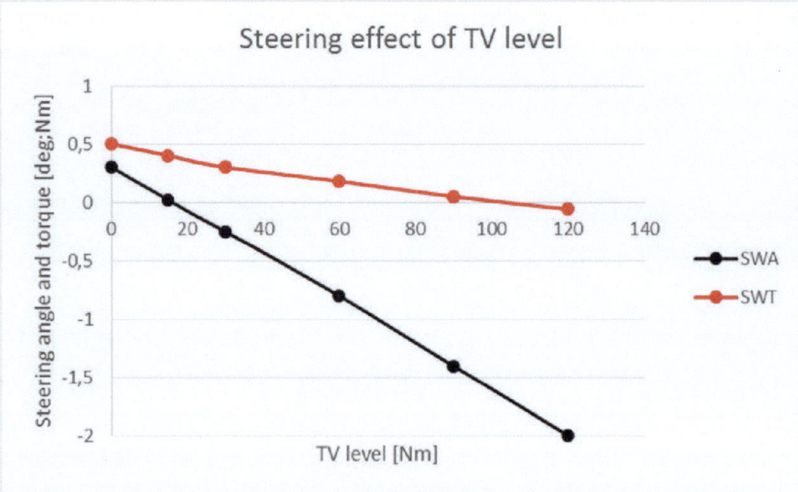

Figure 7: Effects of torque vectoring level on steering angle and torque during side wind condition

As can be seen, the side wind results in an off set on both angle and torque but the effect of torque vectoring has a much higher impact on angle than torque. If the level would be adjusted to compensate only the steering wheel angle, 15 Nm would be sufficient but the effect on torque is minimal. To reach an acceptable level 60 Nm is needed. As can be seen the effect of this is that the steering angle becomes negative. This may seem unnatural (the driver is steering away from the direction of the side force) but it is known that for small steering actions the torque is the main observed parameter and the small negative angle will not be noticed.

With this information the control strategy was developed. At first the deviation of the steering angle in relation to the yaw rate is determined. This done by monitoring both signals and determine the average deviation from the nominal characteristic, which is one that is depending on vehicle speed. If a threshold is reached and is observed for longer than 1s, the function becomes active and generates an estimated torque vectoring level to reach the set target. From this time, the actual deviation is calculated not by comparing it with the base characteristic but with the characteristic fitting the current torque vectoring level. Based on this the level is adjusted until it falls under the threshold, from which time the function becomes inactive.

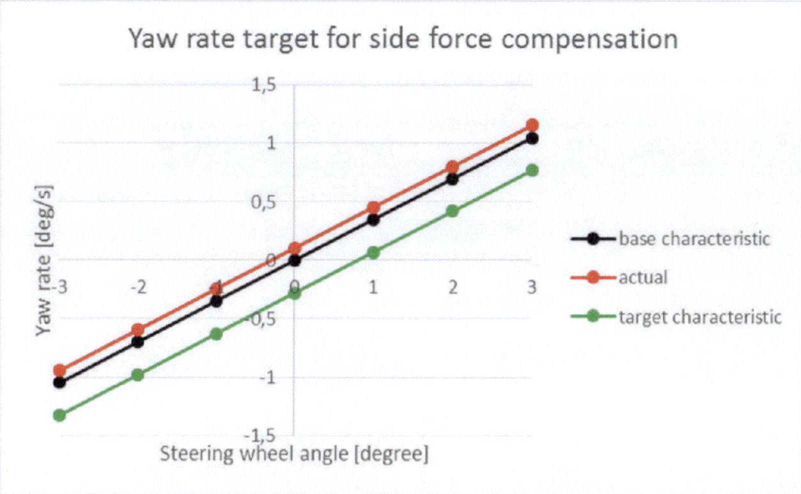

Figure 8: Moving target for yaw rate characteristic due to side forces

The function has been implemented and tested in the simulation and delivered the results as targeted. It is however clear that because of the small off set in the characteristic the performance in realistic conditions may become difficult. The feasibility of the function will depend heavily on the actual behaviour of the vehicle. The vehicle used in the simulations was not very sensitive to side forces and the investigated disturbing levels were also low. This means that for other vehicles and higher levels of disturbance better results can be expected. A limitation however will remain in the requirement to avoid negative torques on one of the wheels.

3 Simulations

3.1 Simulation model

The vehicle model was changed during the research project because of a change in development platform. The final model however was correlated to the final vehicle by means of main vehicle dynamics tests: Steady state circle tests, Step response tests and frequency response tests.

The model was build, using the dSpace ASM tool which has an advantage over other tools that it is completely build in Simulink and no co-simulation is needed to incorperate the foreseen Simulink sub models: Torque vectoring controller, motor models and motor controllers. Figure 9 gives an overview of the final model. The combustion engine of the base model as well as the differential were removed and replaced by the 2 electric motors, driving the wheels directly, using a reducer on each wheel.

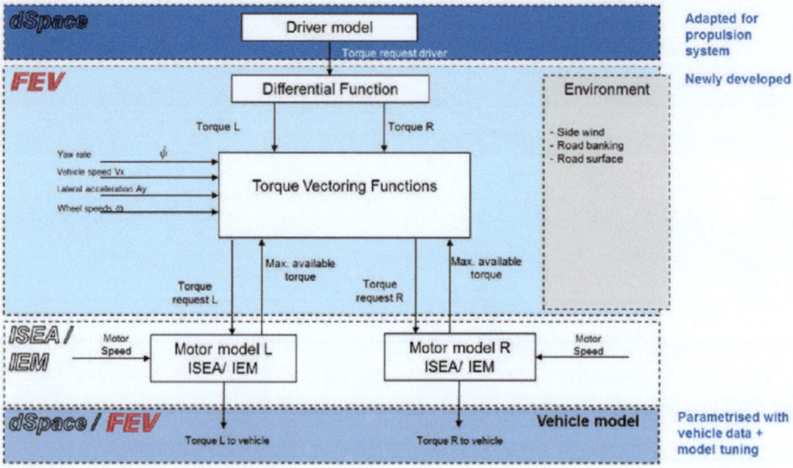

Figure 9: Overview of the simulation model

3.2 Model validation and function development

Because of the nature of the project (parallel development), the vehicle model was developed before the development platform was built. The function base development therefore has been done with a status of the model with estimated parameters and limited component characteristics. In the last phase of the project, the vehicle became available and validation of the model could be performed. The vehicle dynamic properties of the vehicle were tuned based on standard vehicle dynamic tests: steady state circle, step steer and frequency response.

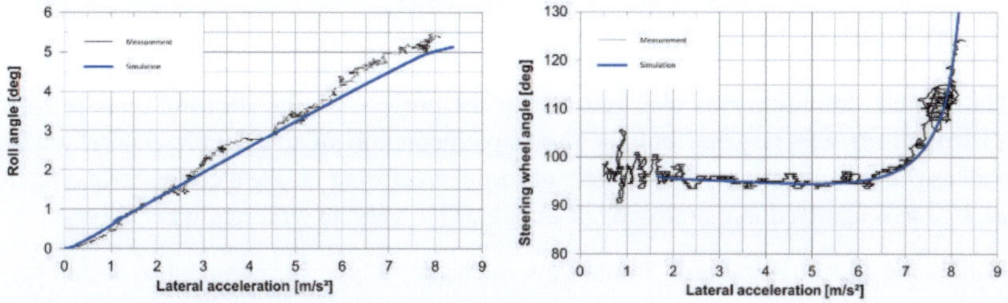

Figure 10: Model optimisation results for steady state test

Next step was to determine the torque vectoring characteristic. For this the vehicle was driven straight ahead at different speeds with fixed –straight ahead– steering wheel. From that status, vectoring torques, defined as positive on one and negative on the other wheel, were applied and the effect on both yaw rate and steering torque was determined. The results of these tests are used for parameter tables for all the torque vectoring functions and their development.

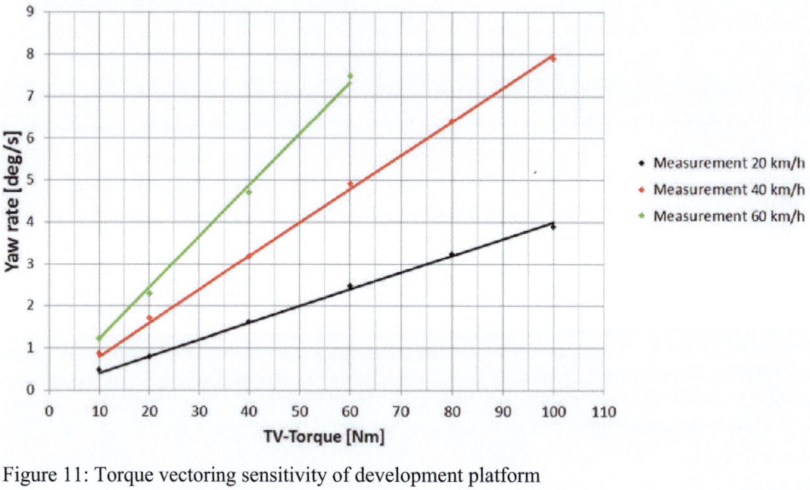

Figure 11: Torque vectoring sensitivity of development platform

What is clear from the graph is that the yaw rate effect is linear in the whole range up to the maximum torque (Note: the mentioned torque is motor torque and the wheel torque is 10 x higher related to the reductions)

4 Testing

The vehicle definition was changed during the project. At the start the Streetscooter B14 platform was targeted but for the implementation of the double powertrain as well as the complication of the effects of torque vectoring on the steering torque feedback, it was changed to a prototype of the new C16 vehicle with rear wheel drive.

During the test phase several limitations of the prototype were encountered for performing a full validation of all torque vectoring functions. Regarding the side force compensation function the main issue was that the chassis set up in combination with the steering system resulted in very little returning toque and steering torque feedback. This made it impossible to give a final verdict on the performance of the function. Future research must show the effectiveness and comfort of the function.

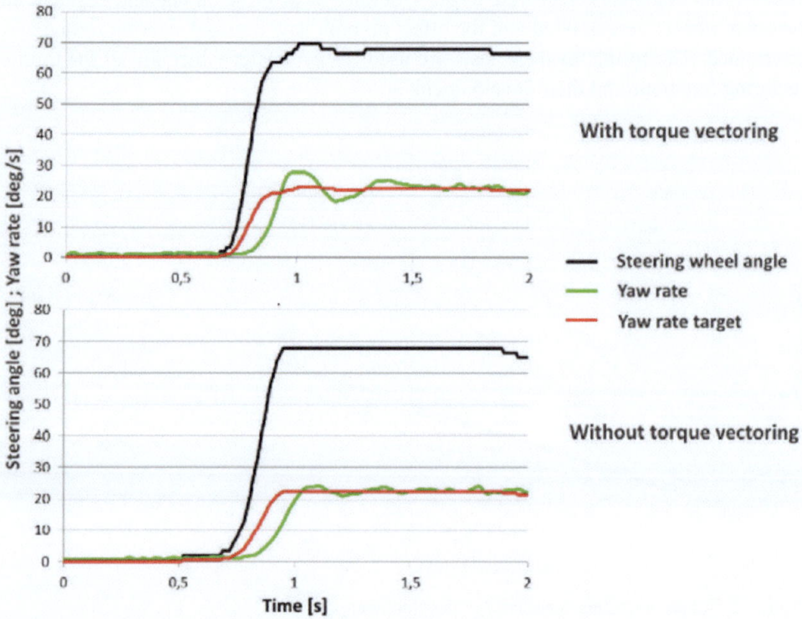

Figure 12: Yaw response increase of agility function

The general effects of the torque vectoring functions however are tested and confirmed their potential, although the control system in the vehicle appeared to have time delays that could not be improved during the research project. The graph below for instance shows that the 20% increase in response is reached but the target stability increase (less overshoot) not. It was investigated however that it is not related to the motor controller but to the mentioned vehicle controller and bus system.

Figure 13: The C16 development platform

[1] *How torque vectoring has the potential to significantly reduce the conflict between stability and response yet enhance ride and fuel economy in electric vehicles.*
Malcolm Burgess
Vehicle Dynamics International

[2] *Torque Vectoring for improved vehicle dynamics.*
Peter van Vliet, Robert Bosch GmbH
Vehicle Testing Expo, 2010.06

[3] *On the torque accuracy of stator flux observer based induction machine control.*
Schubert, Michael, Scharfenstein, Daniel und De Doncker, Rik W. Nadi, Fiji: IEEE, 2016. 2016 IEEE Symposium on Sensorless Control for Electrical Drives (SLED). S. 1--8.

[4] *ISO 26262-3:2011 – Road vehicles – Functional safety – Concept phase*

Controller concept for automated lateral control

Felix Tigges, M.Sc., Research assistant

Florian Krauns, M.Sc., Research assistant

Alexander Hafner, M.Sc., Research assistant

Dr.-Ing. Roman Henze, Chief engineer

Prof. Dr.-Ing. Ferit Küçükay,
Director Institute of Automotive Engineering (IAE)

Technical University Brunswick

P.E. Pfeffer (Ed.), *8th International Munich Chassis Symposium 2017*,
Proceedings, DOI 10.1007/978-3-658-18459-9_30

1 Introduction

Driving safety as well as improving efficiency or enhanced comfort are crucial objectives in the automotive development. Latest lateral control functions like lane keeping assist already improved the active safety in the last few years, while the lane change assist is still waiting for its breakthrough. Thereby ADAS (**A**dvanced **D**river **A**ssistance **S**ystems) release the driver more and more from his tasks during driving on the motorway. By doing and extending this relief ADAS are smoothing the way towards highly automated driving. For ADAS as well as automated driving a robust and high precise lateral controller is required in the near future.

Current lateral control functions may use different control algorithms for different task like lane keeping or lane change, by no later than switching to high automated driving a single controller is needed to cover all desired lateral control tasks. This work focuses on a controller architecture that is capable of steering the vehicle in every relevant speed and lateral acceleration range; this includes driving on the edge of lateral limits to ensure maneuverability during high dynamic evasive maneuvers.

A well-known problem of current lane keeping assists in terms of a subjective desirable behavior is an oscillating within the lane markings. The developed controller uses multiple inputs besides the measured track offset that work in different cascades. By working within inner and thereby faster control loops a track offset is already reduced before it really occurs. In addition to the cascade control a pilot control based on the single-track-model is used to eliminate lag of the system response.

The developed lateral control reaches remarkable results in the simulation of different freeway-scenarios including miscellaneous disturbances like squalls as well as in high dynamic driving on the “Nürburgring Nordschleife”. Another advantage of the controller is its model-based architecture which allows a very fast parameterization on other vehicles. Most required parameters are also needed in the ESC and thereby known for every new vehicle.

The system architecture for autonomous vehicles is subdivided by the Institute of Automotive Engineering into the superior groups “environment perception”, “action strategy” and “longitudinal and lateral control”. Therefore this paper and so the following explained controller will receive the information of a planned trajectory from the “action strategy”.

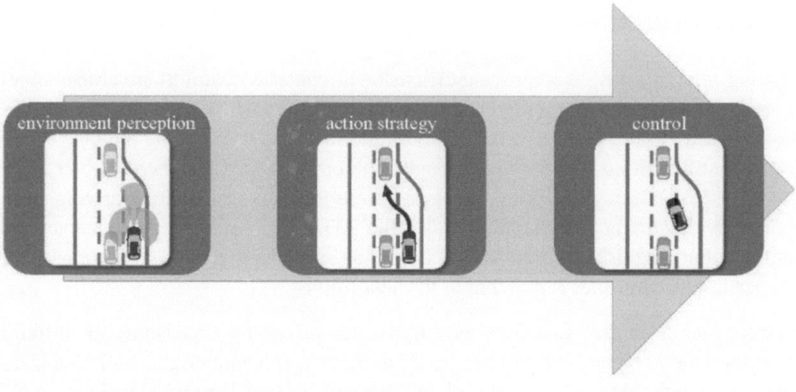

Picture 1: System architecture for autonomous vehicles

To design the controller in advance an exemplary trajectory for a lane change will be presented. Subsequent the control algorithm will be explained in detail. Afterwards the controller performance under interference is shown to prove the robustness. The Institute of Automotive Engineering uses a high-quality tool chain, consisting of offline-simulation, DVRS (**D**ynamic **V**ehicle **R**oad **S**imulator) and multiple test vehicles. Thus the designed lateral controller is used in the DVRS to evaluate how a passenger wants to be driven by an advanced driver assistance system or automated driving systems and finally installed in the test vehicles to reach another step towards automated driving.

2 Trajectory for lane change

In the future system architecture, as shown in picture 1, the longitudinal and lateral controllers will receive a planned trajectory and velocity profile from the action strategy based on the environment perception. To design the lateral controller parallel or even prior to the action strategy and the environment perception an exemplary trajectory is needed.

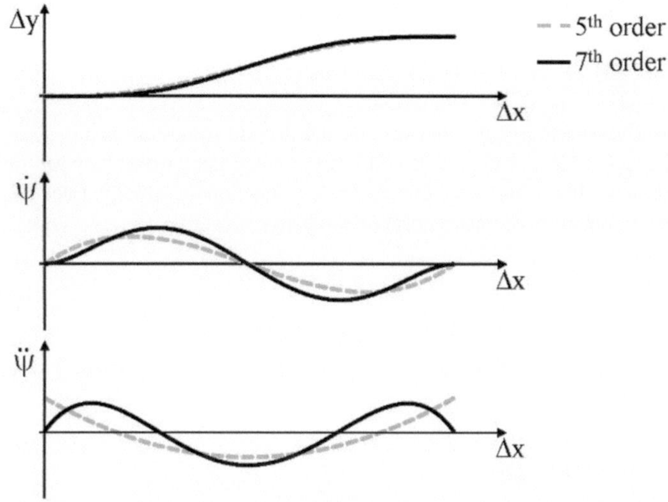

Picture 2: Comparison of polynomial of fifth order and polynomial of seventh order

A previous lane change controller at the IAE used a fifth order polynomial, as shown in picture 2. The trajectory is displayed in the upper diagram as lateral offset versus longitudinal offset under the constraints of a constant speed and a given maximum lateral acceleration. Initially the polynomial of fifth order seems to be a good way to connect the current and the desired lane with a smooth transition. A more detailed analysis of the trajectory in terms of yaw velocity shows an unsteady transition between driving in a straight line and the beginning or ending of the polynomial of fifth order. The yaw acceleration of the polynomial of fifth order in the lower diagram confirms the previous statement with its not equal to zero values at the beginning and the ending of the polynomial.

An oscillating behavior of the controller output was often observed in former applications in case of a difference between the actual velocity and the velocity used to parameterize the controller. To avoid such an unintentional behavior the polynomial of seventh order will be used to design the controller and also will be applied in the future action strategy for normal lane changes. The lateral offsets versus longitudinal offsets of both polynomials are almost the same as the upper diagram in picture 2 shows, but the yaw velocity of the polynomial of seventh grade shows a flatter slope at the beginning and ending than the previous viewed polynomial of fifth grade. Also the polynomial of seventh grade in the lower diagram starts and ends with a yaw acceleration equal to zero and therefore doesn't request any steps in the vehicles behavior.

The lateral offset value y is calculated by the following equation:

$$y(x) = y_e \left[a_7 \left(\frac{x}{x_e}\right)^7 + a_6 \left(\frac{x}{x_e}\right)^6 + a_5 \left(\frac{x}{x_e}\right)^5 + a_4 \left(\frac{x}{x_e}\right)^4 \right] \tag{1}$$

in dependency of the current longitudinal value x and the end values of the polynomial x_e and y_e. The end value y_e can be adjusted to the needed track offset between the actual and the desired lane, while the end value x_e is a function of velocity, track offset y_e and the maximum allowed lateral acceleration $\ddot{y}_{max}$:

$$x_e = b\,v\,\sqrt{\frac{y_e}{\ddot{y}_{max}}} \tag{2}$$

The constant b can easily be defined by an exemplary parameter set consisting of x_e, v, y_e and $\ddot{y}_{max}$.

3 Controller concept

Picture 3 shows the overall controller concept with the nominal trajectory as input value in the form of lateral value y, yaw velocity $\dot{\Psi}$ and yaw acceleration $\ddot{\Psi}$,as described in the previous chapter. The controller itself consists of three parts, the feed forward control, the **i**nternal **m**odel **c**ontrol (IMC) and a PID-Controller. Every part of the controller has a share of the overall steering angle which is applied on the steering wheel. In the offline simulation a multi body simulation model is used to simulate the car. If there are interferences respected in the simulation, they will be added to the model reaction caused by the steering angle.

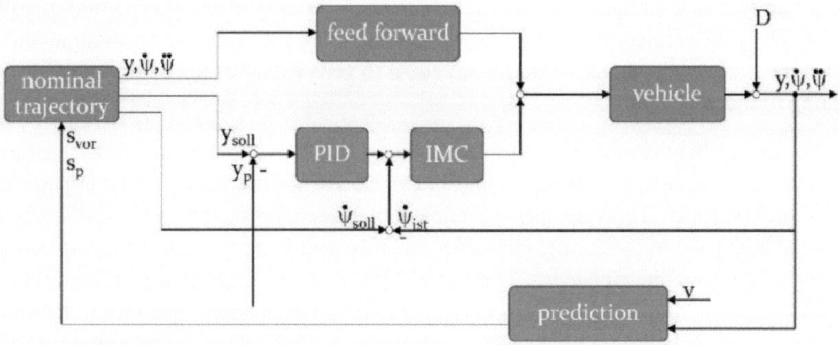

Picture 3: Controller concept

The reaction of the car is led back to a prediction algorithm, which calculates the vehicles position with respect of a prediction time. The actual yaw velocity is directly led back and compared to the nominal yaw velocity; the discrepancy between these two yaw velocities is the input value of the internal model control. The IMC will be explained in chapter 3.2 in detail, its output value is a steering angle, which is added to the steering angle provided by the feed forward control. The PID controller uses the discrepancy between the nominal and predicted lateral position as input to calculate a needed yaw velocity to close the gap between these two lateral positions. This desired yaw velocity is passed on to the IMC which converts it into a steering angle.

3.1 Feed forward control

As mentioned in the previous chapter a feed forward control is used in the control algorithm. The idea behind the use of a feed forward control is to shorten the response time of the vehicle in terms of a lateral change in the nominal trajectory. While a feedback controller needs a discrepancy between a nominal and actual 470ehaviour, the feed forward controller calculates the needed steering angle to follow the nominal trajectory before discrepancies occur.

By providing the calculated steering angle on the vchicle at any point of the planned trajectory the car will show no discrepancy to its nominal 470ehaviour on condition that the model used in the feed forward control is faultless. This shows the second advantage besides a shorter response time. The feed forward controller reduces the control fault and thereby the input of the feedback controller. So the feedback controller doesn't have to deal with the already planned trajectory and can be more efficiently adjusted on eliminating interferences like crosswind.

The feed forward control is based on the idea of inverting the vehicles transmission behavior of steering angle to yaw velocity. So the yaw velocity profile planned by the "action strategy" can be used as input signal to calculate the needed steering angle to follow the nominal trajectory. Since the transmission behavior of the vehicle is very complex and non-linear it's not usable in a control algorithm. Instead the transmission behavior of a single track model is used:

$$\frac{\dot{\Psi}}{\delta_L} = \frac{1}{i_L l} \frac{v}{\left(1 + \frac{v^2}{v_{ch}^2}\right)} \tag{3}$$

With:

$$v_{ch}^2 = \frac{l^2\, c_{\alpha VC}\, c_{\alpha H}}{m\,(l_h c_{\alpha H} - l_v c_{\alpha VC})} \tag{4}$$

This basic equation can be inverted and used in the control algorithm. So the feed forward control can determine the needed steering angle on condition that the single track model is properly parameterized.

Picture 4 shows the simulation results of a double lane change at a velocity of 130km/h without any kind of feedback control. The upper diagram shows the nominal and actual trajectories, below the discrepancy between nominal and actual lateral position is displayed. The third diagram shows the nominal and actual yaw velocities. The steering wheel angle and its proportions of feed forward and feedback control are presented in the lower diagram. Arising from the lower diagram the feedback control is switched off in this simulation to show the performance of the feed forward control. The feed forward control is very accurate in terms of following a planned trajectory if there are no interferences. As the second diagram shows the maximum lateral offset in this maneuver accounts less than two centimeters.

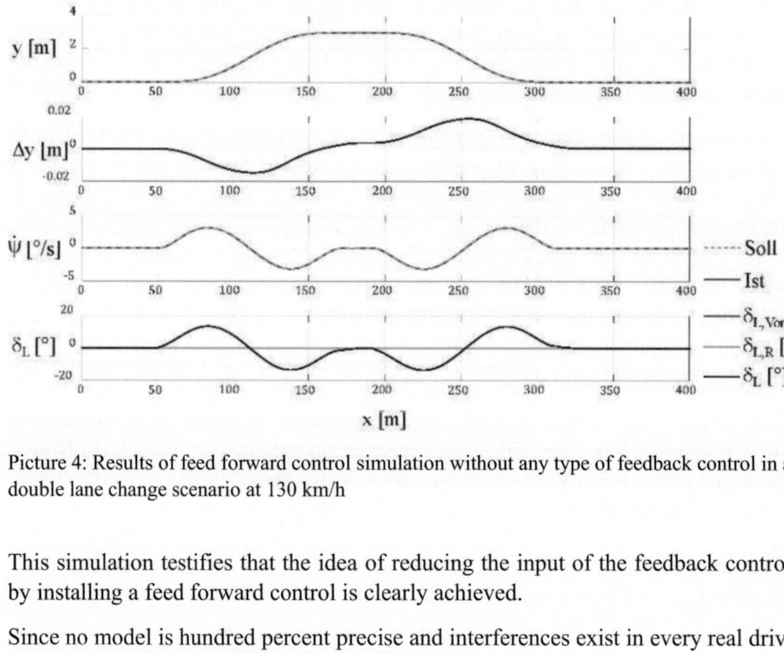

Picture 4: Results of feed forward control simulation without any type of feedback control in a double lane change scenario at 130 km/h

This simulation testifies that the idea of reducing the input of the feedback controller by installing a feed forward control is clearly achieved.

Since no model is hundred percent precise and interferences exist in every real driving situation and additional feedback controller will be introduced in the next chapter.

3.2 Yaw rate control

The yaw rate controller is designed as an IMC (**i**nternal **m**odel **c**ontrol) which is an model based controller. Picture 5 shows the part of the control algorithm relevant for the IMC. On the left side is the input of the IMC; the nominal yaw rate and the yaw rate desired by the PID-Controller. On the right side are the vehicles transmission behavior and the steering angle of the feed forward control as an additional input. The IMC itself is located between the input and the vehicle block.

The internal model control is based on the idea of linearizing the vehicles transmission behavior in a model with the transmission behavior $\hat{G}(s)$ and thereby using a model of the vehicles behavior in the control algorithm. It works pretty similar to the feed forward control explained in the previous chapter. The controller gain consists of the inverted transmission behavior of the model and a filter $F(s)$, as picture 5 shows. [1]

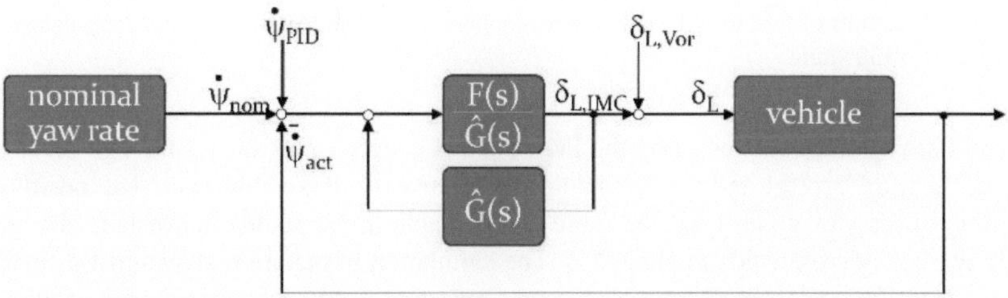

Picture 5: Block diagram of the IMC yaw rate controller with nominal yaw rate and an additional yaw rate desired by the PID-Controller as input

The effect of the controller gain on the vehicle is led back and compared to the effect on the model of the vehicle $\hat{G}(s)$. Thereby the controller can on the one hand react on the interferences affecting the vehicles behavior and causing a deviation between the actual and nominal yaw rate and on the other hand cut out model uncertainties.

Picture 6 shows a simulation of the same maneuver as in picture 4, this time with the feed forward control and the IMC yaw controller as feedback control. The second diagram displays the lateral deviation between the nominal and actual trajectories. In comparison to the simulation without the feedback loop, the discrepancy is reduced from 2 centimeters to approx. 2 millimeters.

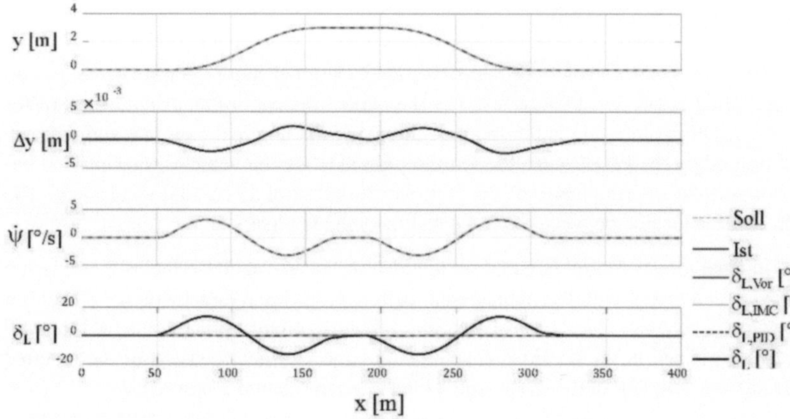

Picture 6: Results of feed forward and internal model control simulation in a double lane change scenario at 130 km/h

The steering angle requested by the IMC-control is very small due to the high precise feed forward control and no occurring interferences in this simulation. The positive effect of the yaw control will be more recognizable in the simulation of lane change during heavy crosswinds in chapter 4. The simulation in picture 6 shows just a small positive effect on accuracy and even more important a still very smooth, not oscillating steering angle, which gives the passenger a comfortable and secure feeling.

The IMC-Controller is a very effective and high precise tool to calculate the needed steering angle to solve discrepancies in yaw velocity. Its similarity to the feed forward control is another benefit in terms of controller application and parameterization. Except of the filter constant T_f used in the filter $F(s)$ every needed parameter is also used in the feed forward control. Actually all the parameters of the single track model are already known from the ESC-parameterization.

3.3 Track offset control

As the previous chapters have shown, the combination of the feed forward and yaw velocity control is a high accurate system to follow planned trajectories. The feed forward controller is a very good tool to reduce the yaw controllers' effort and so the yaw controller can react really fast on occurring discrepancies. However there is one weakness of this combination, its ability to reduce a deviation just in the moment it happens. If the car doesn't behave like it's supposed to, for example during crosswinds, the yaw controller reduces the deviation in yaw velocity but doesn't take into account the already arose lateral offset. Thereby a car without a track offset control would follow a trajectory parallel to the nominal trajectory.

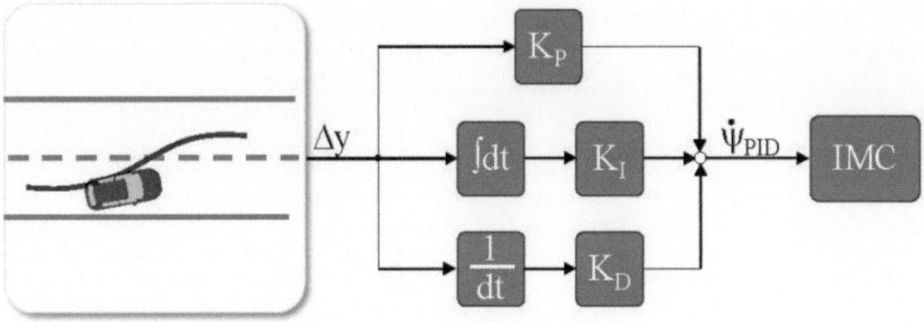

Picture 7: Block diagram of the PID controller with lateral offset as input and desired yaw rate as output

For this reason a PID controller is installed in the lateral control which will be supplied by cameras and laser scanners.As already mentioned the PID controller is designed with a desired yaw rate $\dot{\Psi}_{PID}$ as output instead of a steering angle. The desired yaw rate is then passed on to the IMC controller, which translates it into a steering angle.

Picture 8 shows again the same maneuver as the last two diagrams, a double lane change at a speed of 130 km/h. This time all three parts of the lateral controller are activated. The second diagram shows the lateral deviation between nominal and actual trajectory. In comparison to picture 6 is evident that the offset is reduced from 2 millimetres to 0.3 millimetres. Still the second development goal, a smooth steering angle, is reached while improving the accuracy.

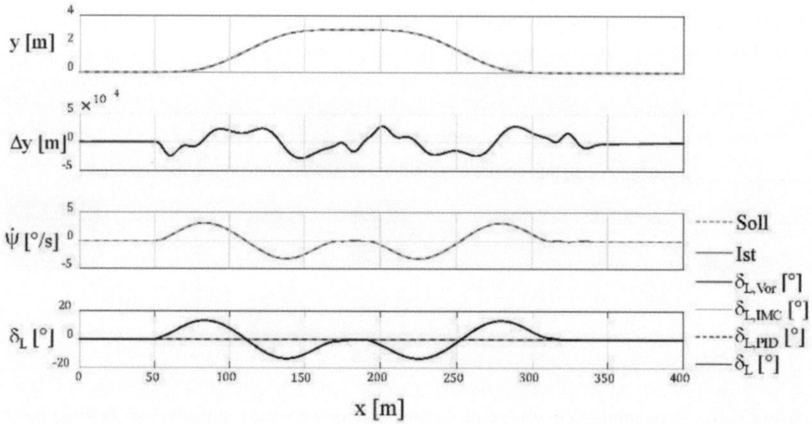

Picture 8: Results of combined feed forward, IMC and PID controller simulation in a double lane change scenario at 130 km/h

4 Controller performance under interferences

The previous tests were used to design and develop the controllers' architecture and its parameters. The simulations have shown that the chosen combination of control algorithms is a high-precise lateral controller for double lane changes on highways.

To test the controllers' ability of cutting out interferences during driving on highways several test with various interferences were made. Picture 9 shows the known simulation of a double lane change at a speed of 130 km/h, this time with a heavy crosswind with a wind speed of 100 km/h. The crosswind creates an exemplary lateral force of $F_{Ly} = 2000$ N and is applied in gusts as shown in the first diagram.

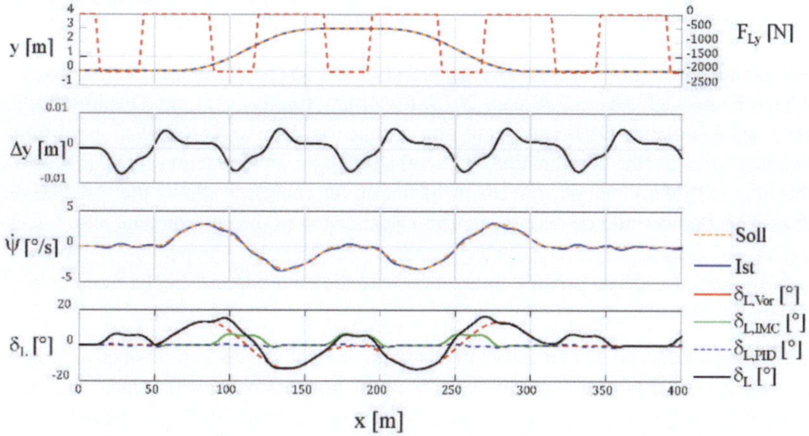

Picture 9: Results of combined feed forward, IMC and PID controller simulation in a double lane change scenario at 130 km/h during crosswind speed of 100 km/h

The lower diagram shows again all different parts of the steering angle. The red dotted line of the feed forward control still executes the same movement as in the other simulations. The summed up steering angle that is actually applied to the car deviates a lot more from the feed forward control than in the previous simulations. This is mainly because of the intervention of the IMC-controller which is plotted with a green solid line and has its swings when crosswind occurs. The blue dotted line indicates the steering angle as result of the lateral deviation, as visible these steering angles are very small, that means the yaw controller works really good and is able to avoid lateral deviation before it occurs. Even under these really tough conditions the controller output shows no oscillating behavior and thereby gives the passengers a safe and comfortable feeling. The third diagram shows the nominal and actual yaw velocities with just small deviations between those two. The lateral offset in the second diagram presents also just small deviations under these circumstances.

5 Lateral Controller in DVRS

The following chapter will show the integration of the presented lateral controller into the driving simulator. Therefore, the simulator itself will be shown, including its hardware and software. Afterwards the possibilities of using self-made models, the integration and its results will be demonstrated.

5.1 The DVRS

The grand opening of the NFF (Niedersächsisches Forschungszentrum Fahrzeugtechnik) in Brunswick was in February 2015. According to this event, the Dynamic Vehicle Road Simulator (DVRS) was put into service too. Due to its technical innovative, unique character (the combination of lateral and longitudinal reaction force created by the tires with the dynamic movement of the driver cabin), it allows the research on chassis and powertrain development. The visualization of the environment allows testing of Advanced Driver Assistance Systems (ADAS) and Cooperative Automated Driving Functions. The driver's interaction with the environment can be monitored in different scenarios.

Any type of scenarios can be shown to a large number of drivers in reproductive scenarios and a save driving environment. A uniform programming environment and programming methodology enables the scenario exchange between different simulators like the DVRS and single-user workstations. Functions and scenarios can be built up and tested by the engineer at his computer while a survey takes place at the simulator.

The DVRS, as shown in picture 10, is based on a Volkswagen Golf VII GTD. To save mass and volume, the drivers' cabin was cut off after the B-pillar. Removing the combustion engine and replacing it by an electromotor, the climate compressor, the alternator and the vacuum pump of the braking system are powered. The rest of the driver cabin is based on the original car so that the driver feels like in a real vehicle driving down a real road. To provide the feeling, based on the acceleration of the driving car, a hydraulic hexapod is installed. The high range of movement (30-degree pitch and roll angle, 40-degree heading angle) supplies manifold opportunities of dynamic driving.

Getting the visualization, five 48" curved monitors are installed in front of the driver cabin which allows a 180-degree round view. Replacing the side mirror glasses with 7" widescreen monitors, the rearward traffic can be focused. The observation of the driver is realized by five installed cameras. These focus on the drivers face, his body, the feed and the pedals, the steering wheel and the drivers' field of view. Additionally, an intercom system and a physiological monitoring system are installed.

Besides the mentioned hardware, visualization software is required. Therefore a package is installed, which enables the visualization of the scenarios, the calculation of the motion system and the integration of self-made Driver Assistance Systems. Scenarios can be built up by a fictional route you have in mind or by a special event. Next you can get a scenario from a real world route. Therefore a tool is built up, which needs a high quality image of the trajectory from laser scanners as an input. Afterwards it converts these data into a scenario which can be used at the Dynamic Vehicle Road Simulator. [2]

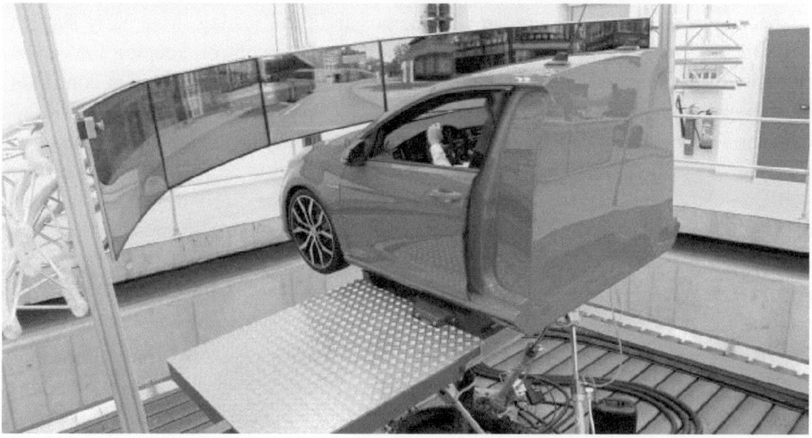

Picture 10: DVRS – Dynamic Vehicle Road Simulator of the IfF in the NFF

5.2 Using self-made models

To focus the research on the drivers' behavior within the automated driving, accorded systems need to be integrated in the simulation. Models, for example based on Simulink, can be converted into c-code and transferred into dynamic link libraries (dll). These libraries can be added into the simulation software. Afterwards the driver can use for example Heading Control (HC) and doesn't have to steer the vehicle on his own.

Other systems were invented at the Institute of Automotive Engineering (IAE) too. To get a high number of different lateral and longitudinal control systems a modular lateral and a modular longitudinal control system was invented. With this opportunity different steps of automated driving can be shown in the DVRS. Besides the mentioned systems, sensors and environment models were invented and integrated. To get a higher flexibility of the vehicle management and the hexapod control a self-designed vehicle dynamic model is integrated. This enables the full parameterization of the vehicle, including the motor management, the powertrain, the chassis and the force-feedback of the steering wheel.

The models are needed to do a comparison between the drivers' behavior in a real car and in the simulator. Therefore the models can be integrated into the test vehicle, as mentioned in the following chapter and in the DVRS too. Different types of drivers can do a test drive in a real car, testing the mentioned systems, and afterwards do a survey in the simulator. This is needed to compare the drivers' reaction and to verify the quality of the simulator.

A special advantage of the simulator is the possibility of testing in an early stage of development. The lateral control is able to do a heading control, keeping the car between two lanes, or to do a lane change. First, each function had its own controller, managing the path following of the calculated trajectory. Different settings and a redundant adjustment provides a high engineering effort. With the new mentioned controller, two improvements can be shown. First, you need to adjust just one controller at a time when you need to get a different system behavior. Second, you can compare the same system in two different environments, the simulator and the test vehicle.

6 Lateral Controller in test vehicle

Lateral control algorithms are implemented on test vehicles of the IAE for multiple purposes. In general, the lateral (and longitudinal) chassis control is a fundamental module in the architecture of an automated car which executes trajectories that are provided by a maneuver and trajectory planning module. One step earlier, the algorithms are also used for the prototypical development and customer-oriented optimization of ADAS functions like the lane keeping assist or lane change assist ("hands on" or "hands off") as well as for the realization of specific test conditions, e.g. for driving defined maneuvers at automotive proving grounds as a substitute for a steering robot.

Picture 11: TIAMO – Testing of Integrated Automation and Monitoring Systems

The model based controller which is presented in this paper is partly integrated in the IAE test car TIAMO (Testing of Integrated Automation and Monitoring Systems, picture 11) by using a "Rapid Control Prototyping" (RCP) platform (dSPACE Autobox). The vehicle is a series production Volkswagen Golf VII which is modified so that bus signals of existing ADAS functions can be manipulated within certain limits. In this way lateral control for highway driving can be realized by sending a target steering wheel torque as an input to the Electric Power Steering (EPS). For using this actuator interface it is necessary to add another control cascade which converts the steering angle, which is the initial control output, into a steering torque. There are also several sensor inputs that are used for prototype functions. A mono camera transmits an estimation of the lane position and course, the ESC provides driving dynamics data like speed, accelerations and yaw rate. As a reference to the series sensors the test vehicle is also equipped with an additional IMU and an environment measurement system based on laser scanners.

The primal application of the controller concept in the described test car is a "hands off" lane keeping control which is aided by a feed forward control based on a quasi-stationary single track model of the car, which is estimated in real time. The parallel feedback controller contains a comparison of the actual driving trajectory to the target lane which involves an estimation of the ego motion and a preview of the road course. In the first place, the feedback controller was developed as a softly interfering function which supports the driver by staying within the lane. Therefore, a slight oscillation between the lane markings is a sufficient control behavior. For developing the function further with the focus on low-error automated lateral control without steering intervention of a driver it is necessary to enhance the structure to reach better precision and responsiveness.

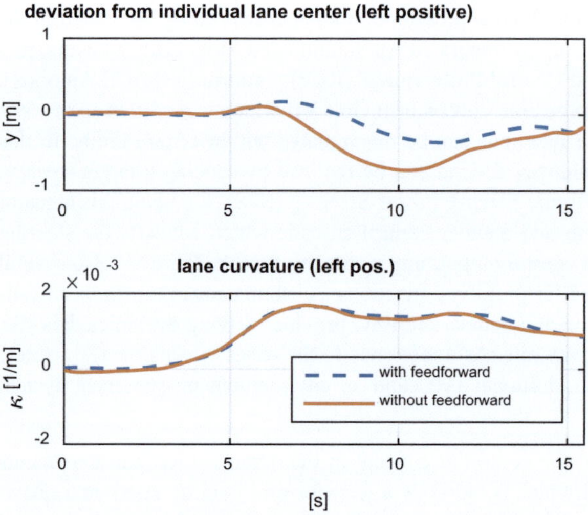

Picture 12: Comparison of a controller with feed forward and without feed forward control

In picture 12, a qualitative result of a comparative test of the two control variants (with and without feed forward) is shown. Initially, the test vehicle was kept within a straight road section by the lateral control at a speed of 50 km/h without greater differences between the variants. After 3.5 s a left turn begins which is visible through the rising curvature (measured by camera). After this point, it is obvious that the controller with the feed forward module steers more suitable and stays closer to the middle of the curved lane while the other controller heads straight ahead for a longer period of time and drifts closer to the right edge of the lane. It is expected that the improved timing behavior will show even more advantages if it will be used for following more complex trajectories like lane changes or slip roads. As a consequence, the next step is to transfer the complete concept for automated lateral control to a real time model that is executable on RCP hardware with the available actuator interfaces.

[1] [Lunze, Jan: Regelungstechnik 1 (8. Auflage), Springer]

[2] [HAFNER, A., SONKA, A., HENZE, R., KÜÇÜKAY, F. (2016). Integration of real measurement data into the DVRS. In: Proceedings of chassis.tech plus 2016, 7th International Munich Chassis Symposium.]

STEERING.TECH SECTION

STEERING SYSTEMS AND FUNCTIONS

Dynamic All-wheel steering (DAS)

Dr. Anton Obermüller | AUDI AG

P.E. Pfeffer (Ed.), *8th International Munich Chassis Symposium 2017*,
Proceedings, DOI 10.1007/978-3-658-18459-9_31

Dynamic-all-wheel steering | DAS

chassis.tech$_{plus}$ 21 June 2017

Dr. Anton Obermüller | AUDI AG

2 Dynamic-all-wheel steering (DAS) - ChassisTech 2017 | 21 June 2017 | Dr. Anton Obermüller | AUDI AG

Agenda

01. Motivation

02. Analogy

03. Architecture

04. Tuning

05. Performance

06. Animated graphics

3 Dynamic-all-wheel steering (DAS) - ChassisTech 2017 | 21 June 2017 | Dr. Anton Obermüller | AUDI AG

Motivation
start of development

$1 + 1 \overset{!}{=} 3$

- improve vehicle dynamics and horizontal comfort
- flexibility, modularity and quick changes
- use of existing components

4 Dynamic-all-wheel steering (DAS) - ChassisTech 2017 | 21 June 2017 | Dr. Anton Obermüller | AUDI AG

Motivation
potential

conventional steering

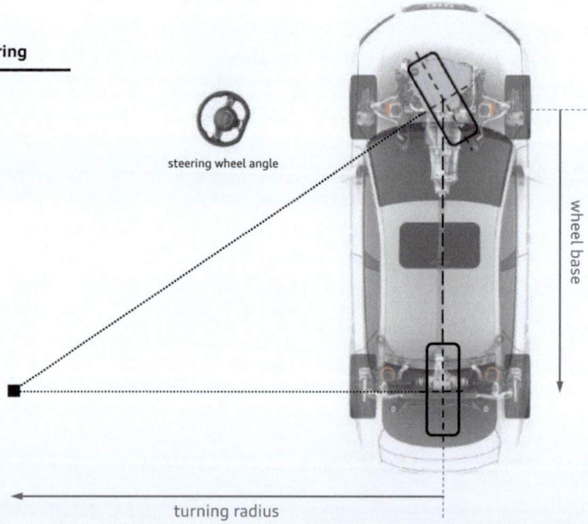

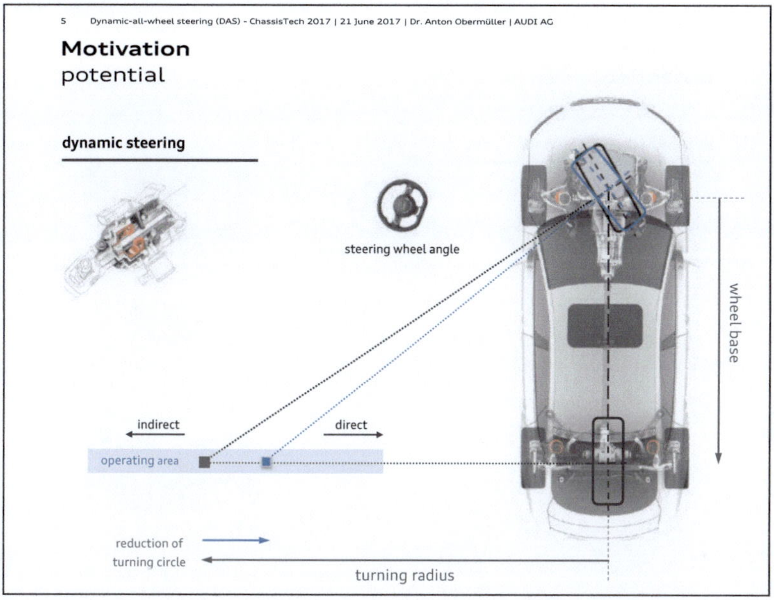
5 Dynamic-all-wheel steering (DAS) - ChassisTech 2017 | 21 June 2017 | Dr. Anton Obermüller | AUDI AG
Motivation
potential
dynamic steering
steering wheel angle
wheel base
indirect
direct
operating area
reduction of
turning circle
turning radius

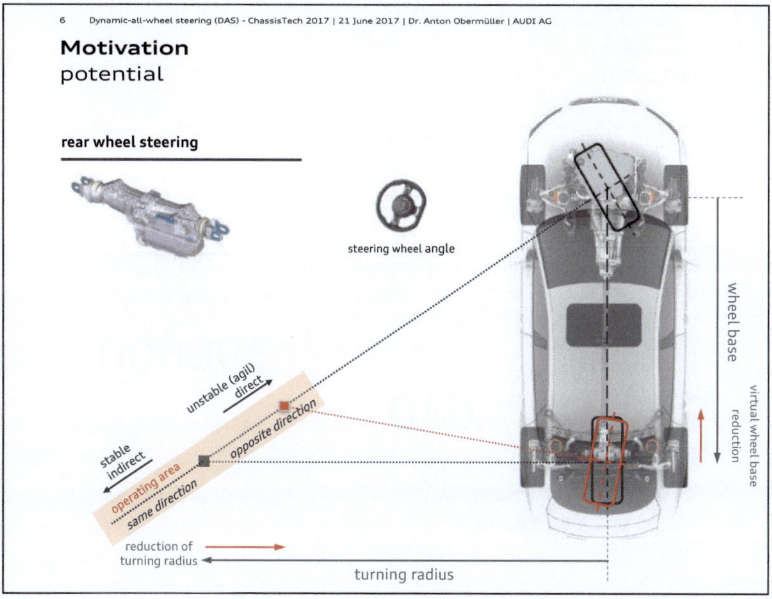
6 Dynamic-all-wheel steering (DAS) - ChassisTech 2017 | 21 June 2017 | Dr. Anton Obermüller | AUDI AG
Motivation
potential
rear wheel steering
steering wheel angle
wheel base
virtual wheel base
reduction
unstable (agil)
direct
stable
indirect
opposite direction
operating area
same direction
reduction of
turning radius
turning radius

7 Dynamic-all-wheel steering (DAS) - ChassisTech 2017 | 21 June 2017 | Dr. Anton Obermüller | AUDI AG

Motivation
potential

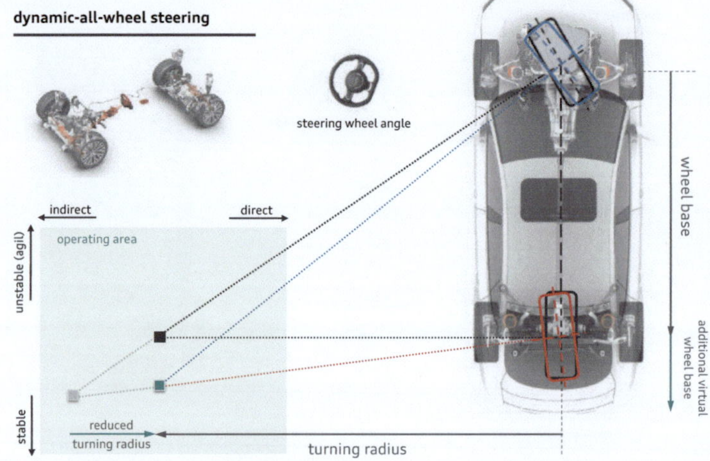

8 Dynamic-all-wheel steering (DAS) – chassis.tech$_{plus}$ 2017 | 21 June 2017 | Dr. Anton Obermüller | AUDI AG

Motivation
complexity

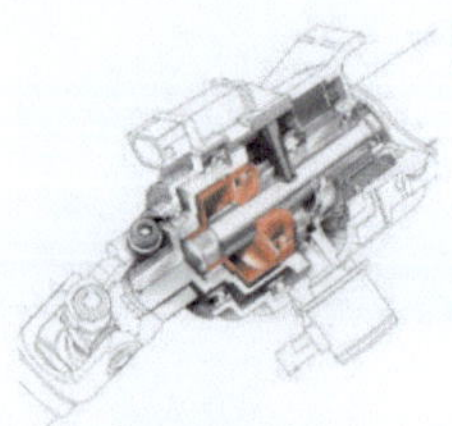

620

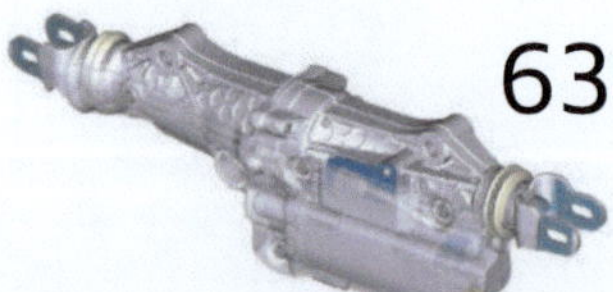

630

= 1250

or

390600

9 Dynamic-all-wheel steering (DAS) - ChassisTech 2017 | 21 June 2017 | Dr. Anton Obermüller | AUDI AG

Analogy
complexity

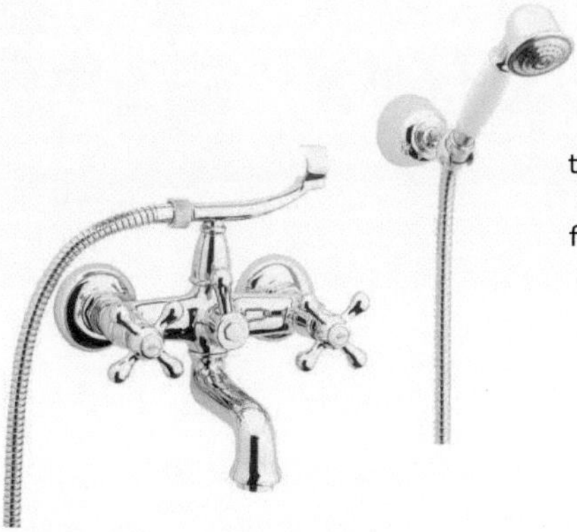

temperature

flow volume

1 1

[Quelle Bild: http://fishzero.com/2017/03/24/dusche-unterputz-armatur-austauschen]

10 Dynamic-all-wheel steering (DAS) - ChassisTech 2017 | 21 June 2017 | Dr. Anton Obermüller | AUDI AG

Analogy
complexity

starting point:

- architecture
- tuning

[Quelle Bild: http://www.sopomarkt24.de/grohe-bad-armaturen-sets-armatur-thermostat-brause-fuer-dusche-o-badewanne/kueche-bad-wc/a-3064]

Agenda

01. Motivation

02. Analogy

03. Architecture

04. Tuning

05. Performance

06. Animated graphics

Architecture
complexity

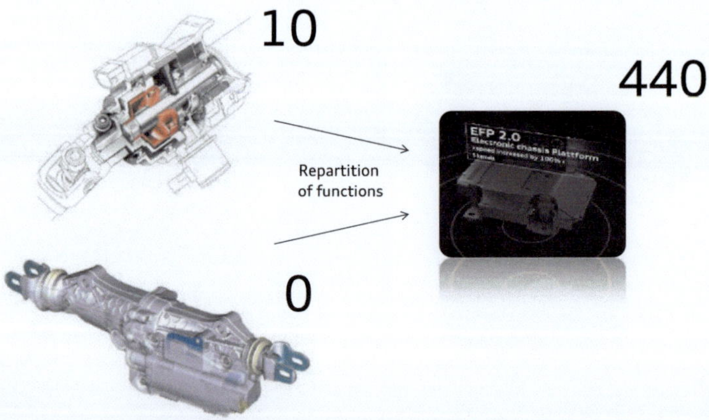

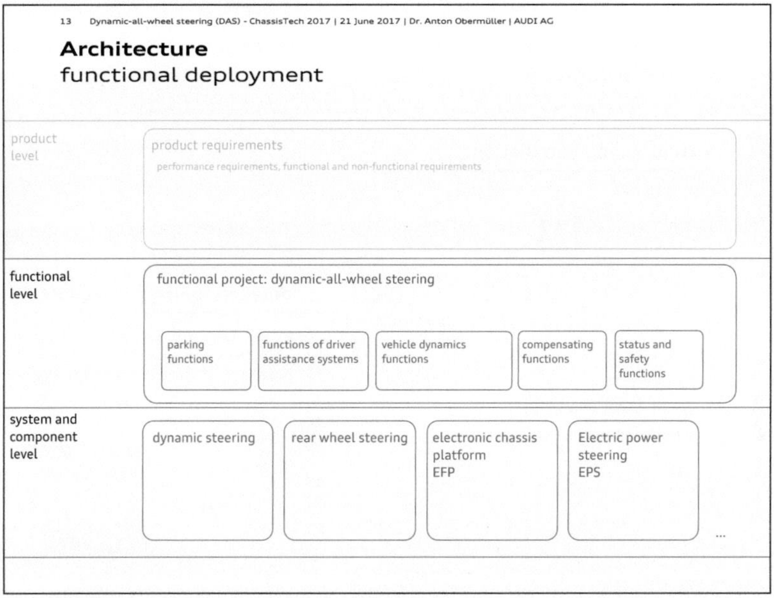
13 Dynamic-all-wheel steering (DAS) - ChassisTech 2017 | 21 June 2017 | Dr. Anton Obermüller | AUDI AG
Architecture
functional deployment
product level
product requirements
performance requirements, functional and non-functional requirements
functional level
functional project: dynamic-all-wheel steering
parking functions
functions of driver assistance systems
vehicle dynamics functions
compensating functions
status and safety functions
system and component level
dynamic steering
rear wheel steering
electronic chassis platform EFP
Electric power steering EPS
...

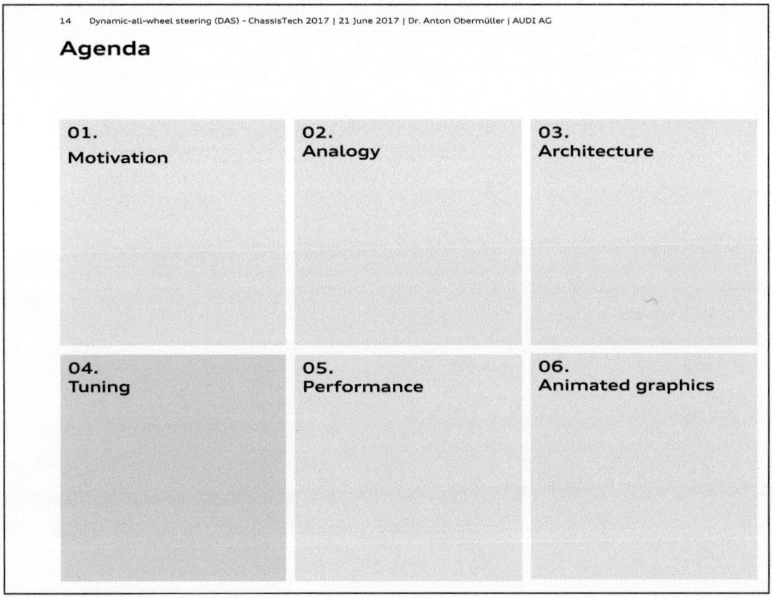
14 Dynamic-all-wheel steering (DAS) - ChassisTech 2017 | 21 June 2017 | Dr. Anton Obermüller | AUDI AG
Agenda
01. Motivation
02. Analogy
03. Architecture
04. Tuning
05. Performance
06. Animated graphics

15 Dynamic-all-wheel steering (DAS) - ChassisTech 2017 | 21 June 2017 | Dr. Anton Obermüller | AUDI AG

Tuning
virtual tuning

virtual tuning (simulation)

70% virtual tuning

detailed tuning in road tests

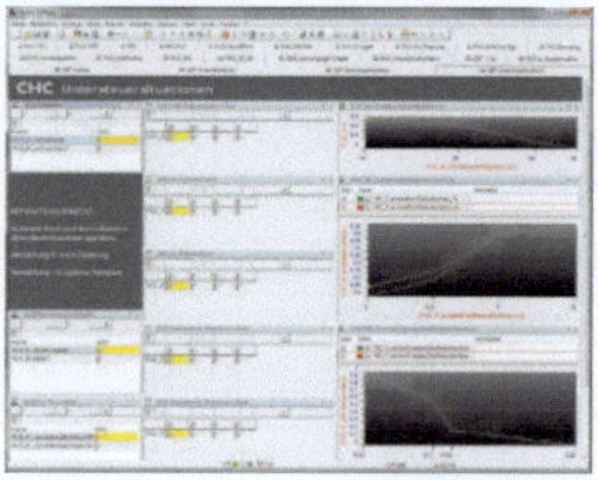

- parameter studies
- system behavior analysis
- vehicle performance simulations

16 Dynamic-all-wheel steering (DAS) - ChassisTech 2017 | 21 June 2017 | Dr. Anton Obermüller | AUDI AG

Agenda

01. Motivation	02. Analogy	03. Architecture
04. Tuning	05. Performance	06. Animated graphics

17 Dynamic-all-wheel steering (DAS) - ChassisTech 2017 | 21 June 2017 | Dr. Anton Obermüller | AUDI AG

Performance

Air suspension, damper control

Active body control

Dynamic-all-wheel steering **DAS**

Torque Vectoring **quattro Sportdifferenzial**

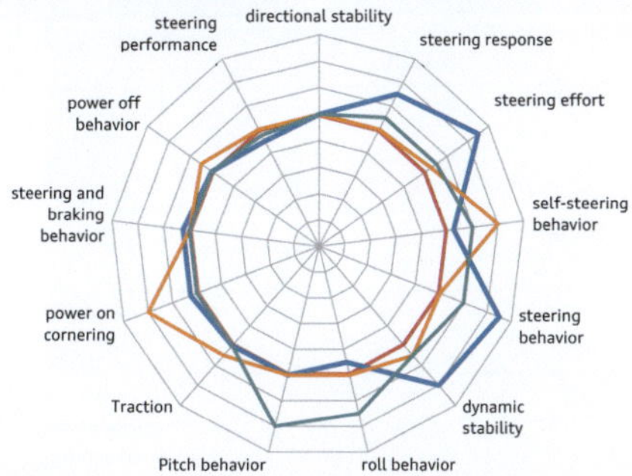

[Quelle: Carsten Jablonowski, AUDI AG]

18 Dynamic-all-wheel steering (DAS) - ChassisTech 2017 | 21 June 2017 | Dr. Anton Obermüller | AUDI AG

Performance
benefits

Benefits for the customer:

In the city
- easy parking
- reduced turning radius and required space
- maneuverability
- reduced motion sickness

On ordinary highways
- agility
- sportiness
- driving like running on rails

On autobahn
- stability
- comfortable driving

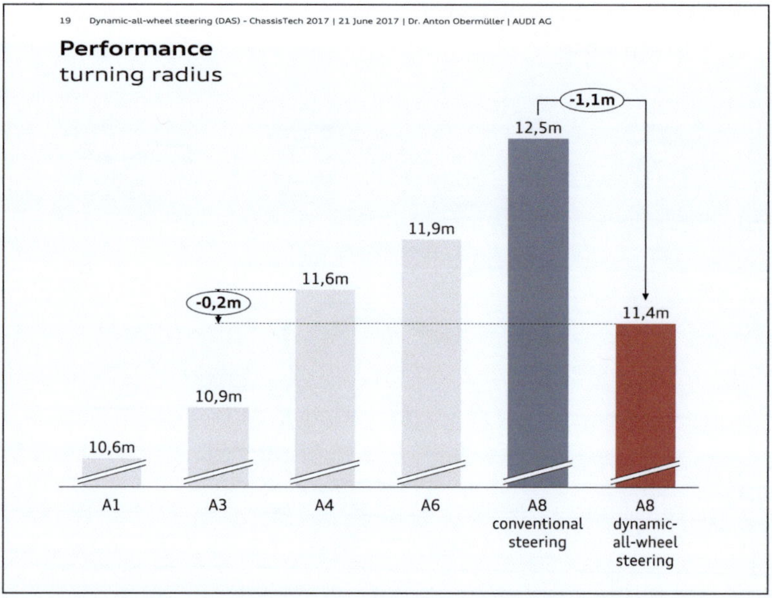
19 Dynamic-all-wheel steering (DAS) - ChassisTech 2017 | 21 June 2017 | Dr. Anton Obermüller | AUDI AG
Performance
turning radius
-1,1m
12,5m
11,9m
11,6m
-0,2m
11,4m
10,9m
10,6m
A1
A3
A4
A6
A8 conventional steering
A8 dynamic-all-wheel steering

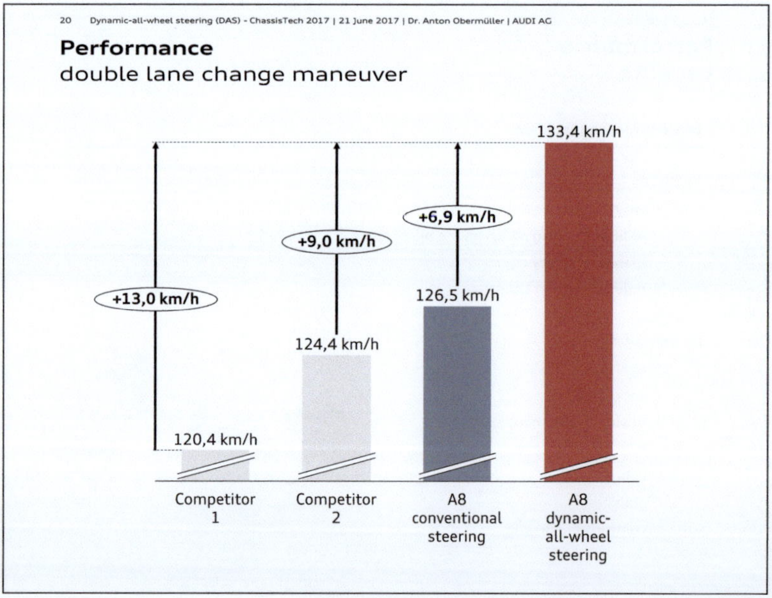
20 Dynamic-all-wheel steering (DAS) - ChassisTech 2017 | 21 June 2017 | Dr. Anton Obermüller | AUDI AG
Performance
double lane change maneuver
133,4 km/h
+6,9 km/h
+9,0 km/h
+13,0 km/h
126,5 km/h
124,4 km/h
120,4 km/h
Competitor 1
Competitor 2
A8 conventional steering
A8 dynamic-all-wheel steering

Animated graphics
dynamic-all-wheel steering

Thank you!

Development of Electric Power Assisted Steering (EPAS) for global emerging markets such as India

Anand Mule, S. Matsagar, J. K. Sinha,
Tata Technologies Limited, India;
S. Salunkhe, Tata Motors Limited, India

This manuscript is not available according to publishing restriction.
Thank you for your understanding.

P.E. Pfeffer (Ed.), *8th International Munich Chassis Symposium 2017*,
Proceedings, DOI 10.1007/978-3-658-18459-9_32

The choice of hand positions on the steering wheel

Christian Strümpler, TAKATA AG

This manuscript is not available according to publishing restriction.
Thank you for your understanding.

P.E. Pfeffer (Ed.), *8th International Munich Chassis Symposium 2017*, Proceedings, DOI 10.1007/978-3-658-18459-9_33

VIRTUAL DEVELOPMENT METHODS

Systematic model-based vibration analysis of a controlled electric power steering system

Prof. Dr.-Ing. Gerd Wittler, Hochschule Esslingen;

B.Eng. Michael Haßenberg, Prof. Dr.-Ing. Hermann Henrichfreise,

CLM – Cologne Laboratory of Mechatronics, TH Köln;

Dipl.-Ing. Hermann Briese, M.Sc. Thomas Schubert,

DMecS GmbH & Co. KG, Köln

P.E. Pfeffer (Ed.), *8th International Munich Chassis Symposium 2017*,
Proceedings, DOI 10.1007/978-3-658-18459-9_34

Abstract

In [1, 2] a LQG/LTR-design is presented to control the steering wheel hand torque of electric power steering systems. Undesirable oscillations (parasitic eigenmodes) are caused by elasticities, which are not considered in the controller design. This paper shows the impact of these parasitic eigenmodes on the behaviour of the control system. Based on a detailed model analysis the root causes are shown. The results of this model analysis are used for a new control approach.

1 Introduction

To control electric power steering systems (EPS systems) different control algorithms are used. In [1, 2] a LQG/LTR-design is presented to control the steering torque of EPS systems. One advantage of this approach is that the steering feel can be adjusted independently from the design of the controller. Precondition for the application of the steering feel is a robust control system. For this reason the robustness of the control system will be regarded in this paper. The existing controller design considers only the elasticity of the torsion bar. An EPS system contains many more elasticities. These elasticities are necessary among others for noise reduction, compensation of fabrication tolerances and decrease of shocks on the steering mechanism. Because these elasticities are not considered in the controller design, they can lead to undesired oscillations (residual eigenmodes) and also stability problems (eigenvalue spillover) [3]. The residual modes take adverse effect on the control system therefore, they are called parasitic modes.

This paper presents a systematic and model based analysis of these vibration phenomena. For this reason a detailed model will be derived in chapter 2 which contains all occurring concentrated elasticities of the steering mechanism. To outline the impact of these elasticities on the control system, the existing LQG/LTR-controller will be analysed in combination with this detailed model in chapter 3. The root causes of the problems presented in chapter 3 will be shown in chapter 4 by a detailed model analysis. Finally a first counter measure to improve the controller will be presented in chapter 5. All investigations are done using an axially parallel power steering system. The presented method can also be applied to other configurations.

2 Modelling of the Steering Mechanism

Fig. 1 shows a physical model of the steering mechanism with eight degrees of freedom (8 DOF). The individual rigid bodies are marked up with the indices S (steering), P (pinion), R (rack), N (nut), M (motor), C (casing), V (vehicle), WL (wheel left) and

WR (wheel right). Elasticities in the model are considered by the torsion bar stiffness c_{tb} (torque sensor), recirculating ball screw stiffness c_{NR}, belt drive stiffness c_{MN}, stiffness of the axial nut bearing c_{NC}, stiffness of the casing attachment c_{CV} and the stiffness of the wheel attachments c_{RW}. The mass moment of inertia and the mass of the rigid bodies are named J_x and m_x, while the index is representing the body. Gear ratios are labelled with i_{xy}, where the index x symbolizes the driving and the index y the driven body. The equations of motion of this model have been derived, linearized and transformed into state space representation. Based on the model with eight degrees of freedom a simplified model with three degrees of freedom (3 DOF) has been determined by a model reduction.

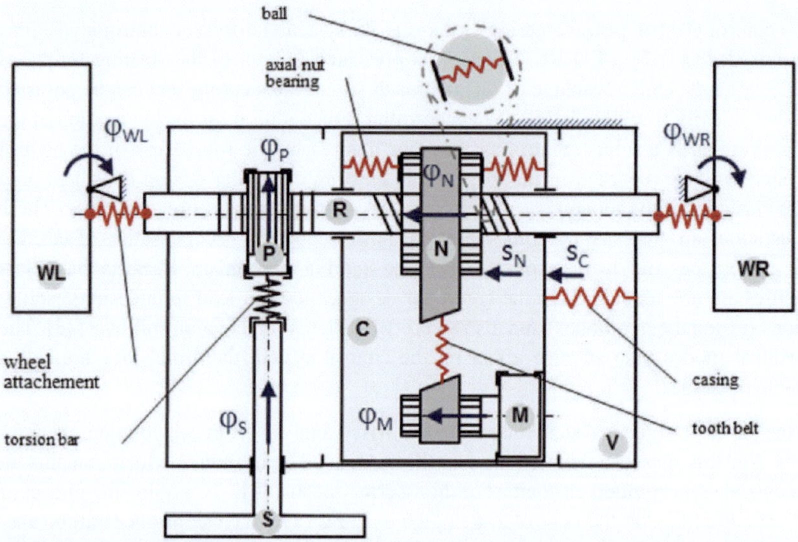

Fig. 1: Physical Model of the Steering Mechanism with eight Degrees of Freedoms

If only the torsion spring is considered and all the other connections are assumed to be rigid, the mass properties of the bodies M, N, R, P, WL and WR can be combined. From this step a model with two degrees of freedom (2DOF) results from [1, 2]. All these different models are extended by a model of the current controlled EPS-motor. These models describe the behaviour of the control plant.

3 Control System with different Models for the Steering Mechanism

In Fig. 2 the control system is shown which consists of a dynamic LQG-compensator and a model of the control plant. The LQG-compensator consists of an optimal state space controller (LQR) and an optimal state space observer (LQE). The torsion bar torque T_{tb} of the drivers steering torque control in [1, 2] can be seen as the controlled variable. The optimal state space controller (LQR) tunes the controlled variable T_{tb} on the externally generated reference steering torque T_{req}. A disturbance variable compensation eliminates internal friction forces and external forces. Furthermore this controller enables an active vibration damping caused by the elasticities of the torsion bar. Because some state space variables could not be measured directly, an optimal state space observer (LQE) is necessary for the reconstruction of the missing state space variables and for the estimation of the disturbance variables. The controller is designed so that the system saturations are not reached during normal operation. Because of this the bandwidth of the controller is approximately 80 rad/s. To demonstrate the differences in the behaviour of the controller, the two degrees of freedom model (2 DOF) is compared with the eight degrees of freedom model (8DOF).

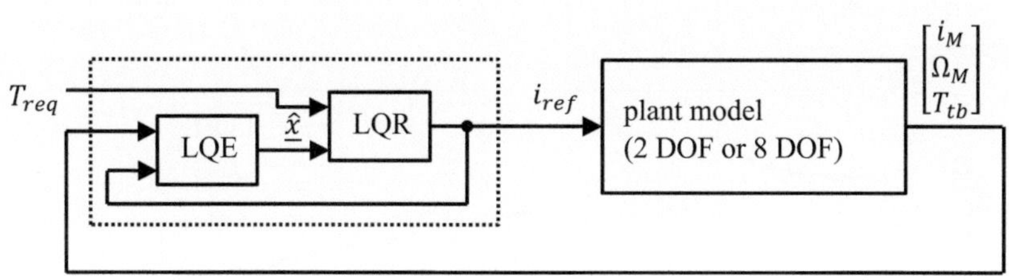

Fig. 2: Control System with dynamic LQG-Compensator

In terms of the controller design model the control system shows good results as expected. The reference step response in Fig. 3 shows a fast and well damped transient response of the torsion bar torque. The plots for LQR and LQG are identical.

Based on the frequency response of the open loop control, the robustness characteristics can be evaluated. Therefore the control loop is cut open at the input of the control plant. The transfer path of this input to the output of the controller will be considered. Fig. 4 shows that the control loop has a phase margin of 71° and an infinite magnitude margin.

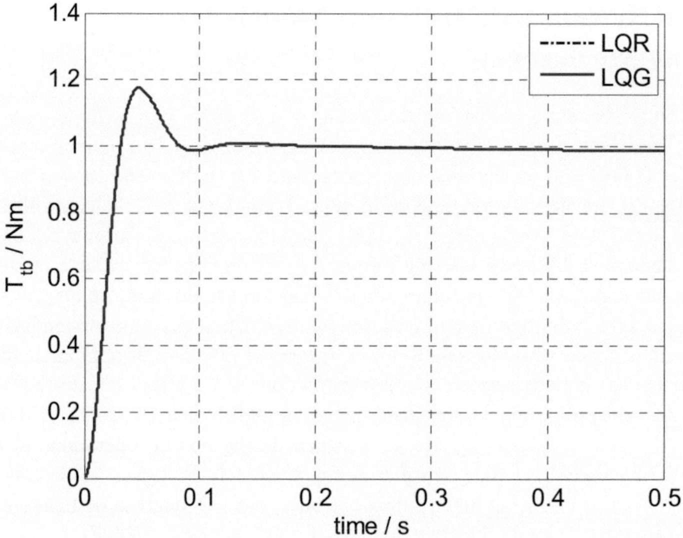

Fig. 3: Reference Unity Step Response, Plant Model with Two Degrees of Freedom

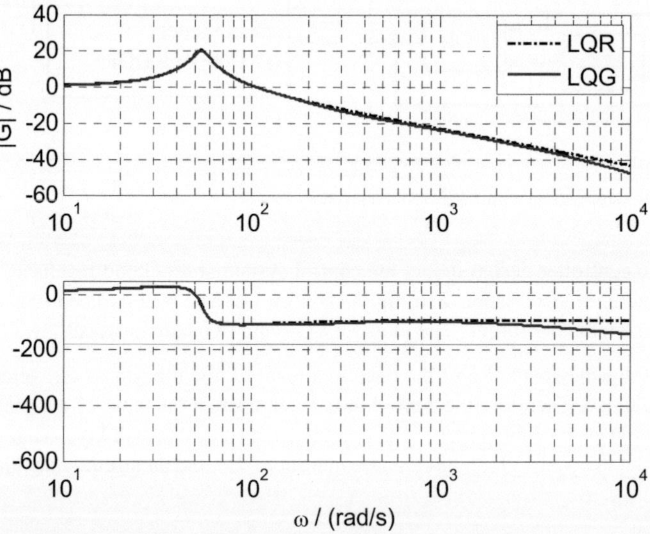

Fig: 4: Open-Loop Bode plot, Plant with two Degrees of Freedom

If the same controller will be connected with the eight degrees of freedom plant model, a low damped oscillation can be seen in the reference unity step response of Fig. 5. This leads to an undesired steering haptic and indicates a low stability margin. The frequency response of the open loop in Fig. 6 confirms this observation. A phase margin of 64° and a minimal magnitude margin of 1.5 dB can be perceived. By small changes of some stiffnesses in the steering mechanism model, the control system can become instable. In a real system these changes may be caused by nonlinearities, production tolerances or aging.

A comparison of the frequency response of Fig. 4 and Fig. 6 indicates differences at low frequencies. The resonance frequency is shifted from 54 rad/s to 48 rad/s. Above 100 rad/s the frequency responses differ significantly. The root causes for that are the differences in the used models of the steering mechanism. They will be analysed more in detail in the next chapter.

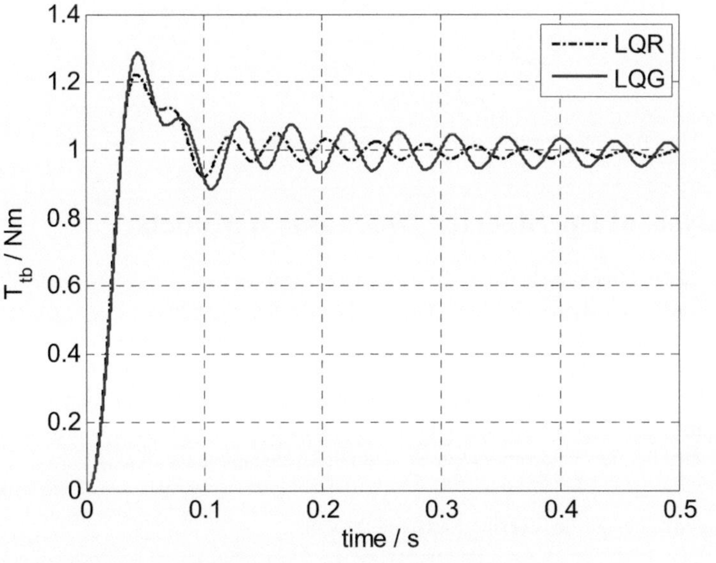

Fig. 5: Reference Unity Step Response, Plant with Eight Degrees of Freedom

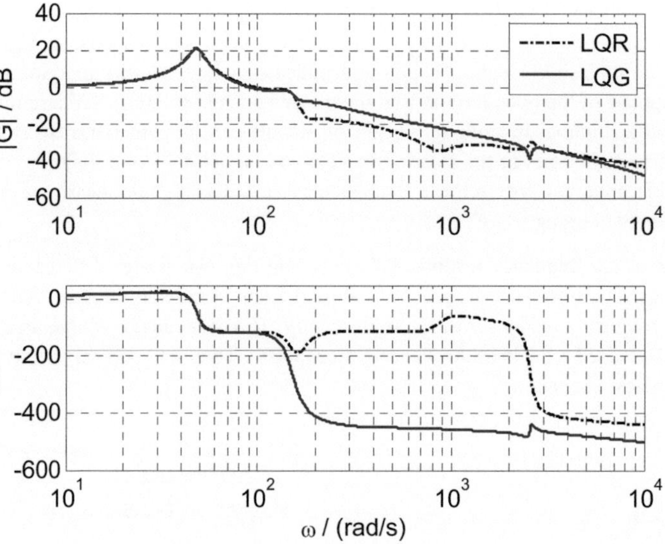

Fig: 6: Open-Loop Bode plot, Plant with eight Degrees of Freedom

4 Analysis of the Steering Mechanism Models

In this chapter the differences between the two and the eight degrees of freedom model (2 DOF and 8 DOF) are shown. For this purpose the relevant eigenvalues, frequency responses and eigenmodes for the control behaviour where examined. In Table 1 the magnitudes of the eigenvalues of the models are presented. The eight degrees of freedom model has two real eigenvalues for the rigid body movement and seven conjugate complex eigenvalue pairs, which can be assigned to the elastic oscillations. The aforementioned shift of the resonance frequency becomes clear at the different eigenfrequencies ω_{01}.

Table 1: Eigenvalues of the Steering Mechanic Models

	1/T	1/T	ω_{01}	ω_{02}	ω_{03}	ω_{04}	ω_{05}	ω_{06}	ω_{07}
8 DOF: $\lvert\lambda_k\rvert$ / (rad/s)	0	3,7	48,43	148,4	174,4	951,5	1949	2569	7340
2 DOF: $\lvert\lambda_k\rvert$ / (rad/s)	0	3,7	54,43						

Considering an equivalent spring-mass system, the eigenfrequency ω_{01} of the eight-degrees of freedom model can be approximated very well by equation

$$\omega_{01} \approx \sqrt{\frac{c_{res}}{J_{res}}} \tag{1}$$

with the expressions of in series-connected springs

$$c_{MP} = \frac{c_{NR} \cdot c_{NC} \cdot c_{CV}}{c_{NC} \cdot c_{CV} + c_{NR} \cdot c_{CV} + c_{NR} \cdot c_{NC}} \cdot \frac{1}{i_{PR}^2} \tag{2}$$

and

$$c_{res} = \frac{c_{tb} \cdot c_{MP}}{c_{tb} + c_{MP}} \tag{3}$$

and the resulting inertia

$$J_{res} = \frac{J_S \cdot \left(J_M + \frac{J_N}{i_{MN}^2}\right) \cdot i_{MP}^2}{J_S + \left(J_M + \frac{J_N}{i_{MN}^2}\right) \cdot i_{MP}^2} \tag{4}$$

By increasing the values for the ball screw stiffness c_{NR}, for the axial nut bearing stiffness c_{NC} and for the casing stiffness c_{CV}, the eigenfrequency ω_{01} of the eight degrees of freedom model moves towards the eigenfrequency ω_{01} of the two degrees of freedom model. The controller is improved in the direction of the original control behaviour (Fig. 3 and Fig. 4) by increasing these stiffness values.

The abovementioned differences between the two models are also visible in the frequency response from the driving torque T_M of the EPS-motor to the torsion bar torque T_{tb} shown in Fig. 7.

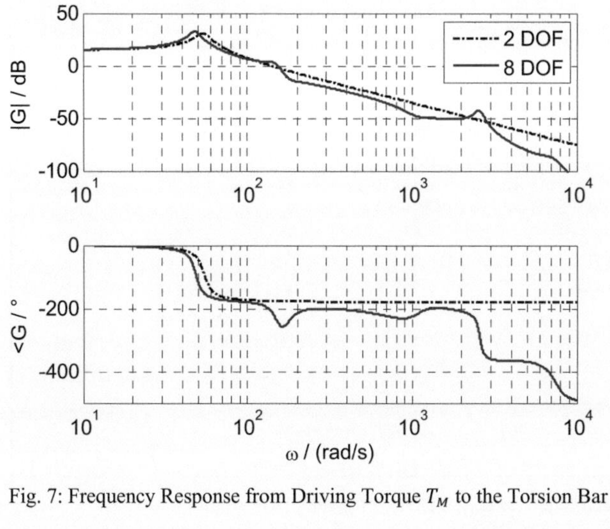

Fig. 7: Frequency Response from Driving Torque T_M to the Torsion Bar Torque T_{tb}

As can be seen in Fig. 7, significant differences already occur in the frequency range up to about 200 rad / s, which is relevant to the controller. The model with eight degrees of freedom has a further resonance frequency at 148 rad/s, which can be assigned to the eigenfrequency ω_{02}. A notch at 174 rad/s can be assigned to the elastic connection of the wheels and can be reproduced with

$$\omega_T \approx \sqrt{\frac{c_{RW}}{\frac{J_W}{i_{RW}^2}}}. \tag{5}$$

In the following, the eigenmodes are analysed for a further insight into the dynamics behaviour of the steering mechanism. Through representing the magnitudes and the phase angles of the position coordinates, it is shown how the individual bodies are involved in the eigenmodes. Hereby, the magnitudes are transformed to a common level by use of the gear ratios. For the control system only the eigenmodes to the eigenfrequencies ω_{01} and ω_{02} are relevant since the remaining eigenfrequencies lie above the bandwidth of the control system. In the model with two degrees of freedom, the first eigenmode (Fig. 8, left) is determined by the torsion bar stiffness. The other stiffness values in this model are assumed to be infinite and the connections between the corresponding bodies are assumed to be rigid. Because of this, these bodies have the same deflections. For a better comparison with the model with eight degrees of freedom the amplitudes and phase angles of these bodies are shown in the eigenmodes. They are identical for the bodies because of the rigid connections.

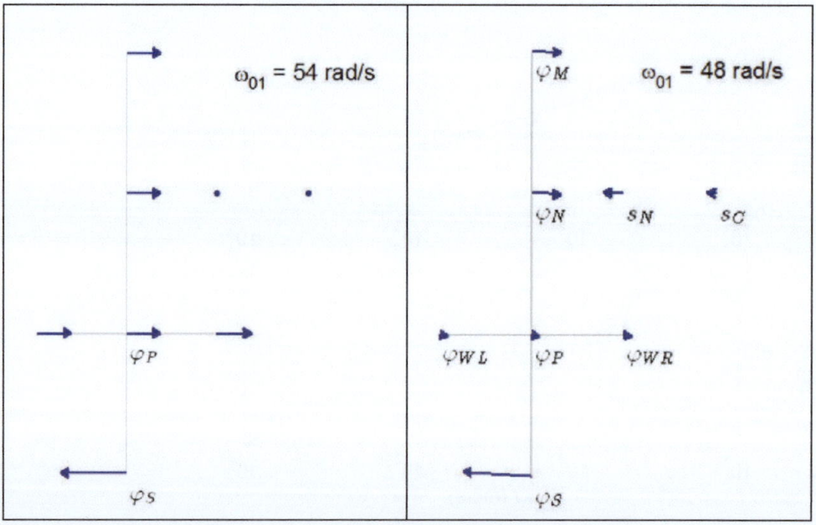

Fig. 8: Eigenmodes of Eigenfrequency ω_{01}, 2 DOF (left) and 8 DOF (right)

The comparison of the two eigenmodes in Fig. 8 shows that the models have a similar behaviour, but some amplitudes differ in their size. Based on the figure for the model with eight degrees of freedom it can be seen that the elasticities of the ball screw c_{NR}, the axial nut bearing c_{NC} and the casing c_{CV}, are deformed and so they have an impact on the eigenmode. The above described differences in the eigenfrequency ω_{01} are caused by these stiffnesses.

The deflection of the wheels is dominant in the second eigenmode (Fig. 9, left). The aforementioned stiffness values have also an impact on the vibration behaviour. This eigenmode has furthermore an impact on the torsion bar torque

$$T_{tb} = c_{tb}(\varphi_S - \varphi_P) \tag{6}$$

that is used for the controller as a measurement variable. In the frequency response of the open loop control in Fig. 6 a significant reduction of the phase margin can be seen, which is caused by this eigenmode.

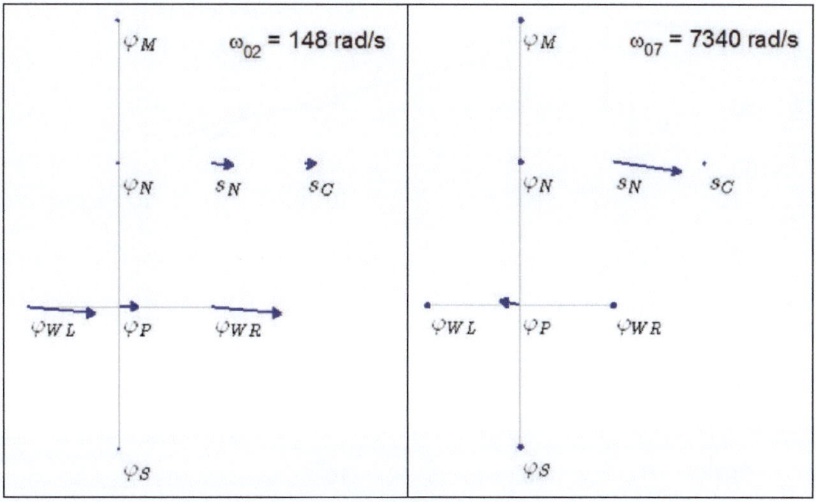

Fig. 9: Eigenmode of Eigenfrequency ω_{02} (left) and ω_{07} (right)

The eigenmodes of the eigenfrequencies ω_{03} up to ω_{07} have a small impact on the continuous controller. However, depending on the sampling frequency and the observability, this can lead to aliasing effects in the discrete controller realization. The observability can be estimated by means of the measured variables. For example, the eigenmode to the eigenfrequency ω_{07} is shown in the figure (Figure 9, right), which is represented by the amplitude for φ_P in the measurement of the torsion bar torque.

5 Improved controller design

In this section, a controller design is presented, which is more robust against parasitic eigenmodes in the control plant. The structure of the controller and the measuring values used are retained (Fig. 2).

With the findings from the model analysis a model with three degrees of freedom (3 DOF) was derived by physical model reduction. It contains all relevant eigenmodes for the controller design. The frequency response in Fig. 10 shows, that the detailed and the reduced model have almost identical behaviour in the relevant frequency range. With this model a new controller design was carried out. Fig. 11 shows the frequency response of the open loop with the newly designed controller and the model with eight degrees of freedom for the control loop.

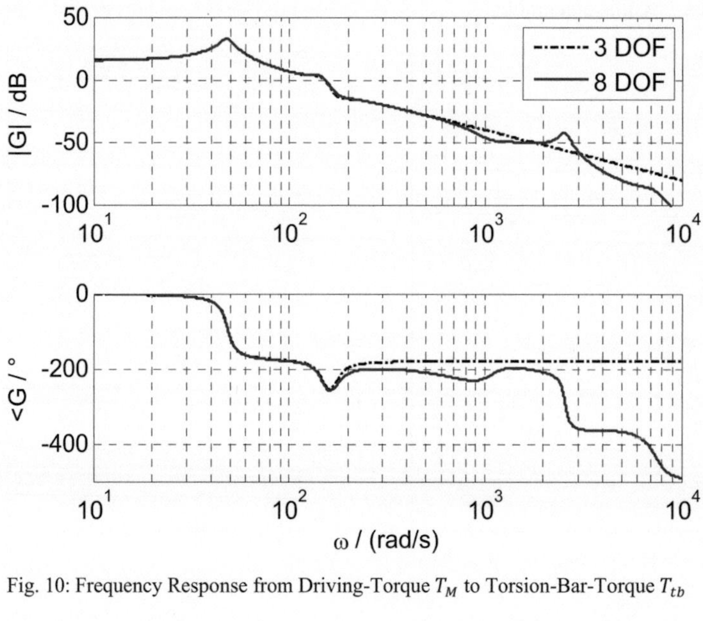

Fig. 10: Frequency Response from Driving-Torque T_M to Torsion-Bar-Torque T_{tb}

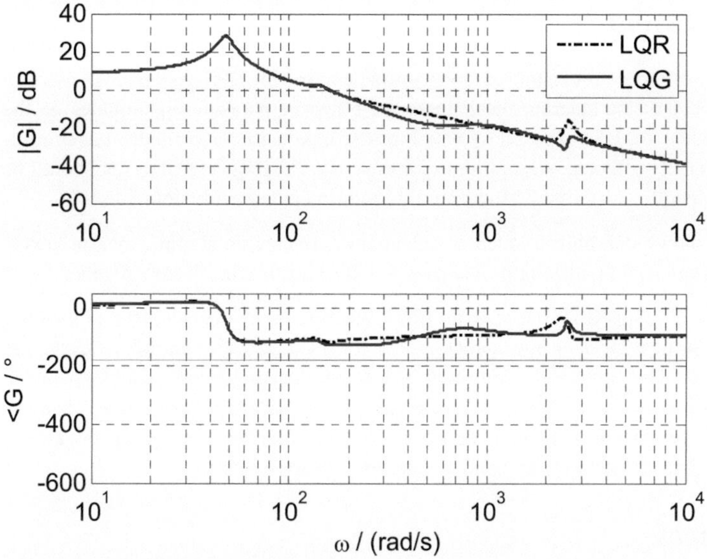

Fig. 11: Open-Loop Bode plot, Control Plant with Eight Degrees of Freedoms

Compared to the original controller design from section 3, significantly greater stability margin can be shown. With the same bandwidth, the control has a phase margin of 52 ° and an infinite magnitude margin (Fig. 11). In the case of a variation of the stiffness values in eight degrees of freedom plant model, however, robustness problems for the controller turned out. The analysis and a further improvement of the presented control concept is the subject of the current work.

6 Conclusion

This paper examines an existing steering torque control for EPS-systems. By use of a detailed model of the steering mechanism the effect of parasitic eigenmodes on the control was shown. It is exposed by which of these parasitic eigenmodes the controller was adversely affected. An option to improve the existing control system could be to increase specific stiffnesses. However, this solution is generally not realisable.

It was also shown that high-frequency eigenmodes are contained in the measured variables. This can lead to aliasing problems when the controller is discrete realized.

Based on the findings from the analysis of the steering mechanics model and the control system, a new steering mechanics model for an improved controller design has been derived. This improved controller has a greater robustness against parasitic eigenmodes but still does not have sufficient robustness against parameter changes. The aim of the current work is to improve this controller in terms of its robustness, nonlinearities in the steering mechanism and discretisation effects.

References

[1] H. Henrichfreise, J. Jusseit, H. Niessen, “Optimale Regelung einer elektromechanischen Servolenkung”, VDI-Tagung Mechatronik 2003, Fulda

[2] O. Graßmann, H. Henrichfreise, H. Niessen, K. v. Hammel, “Variable Lenkunterstützung für eine elektromechanische Servolenkung”, Elektronik im KFZ, Haus der Technik, Stuttgart: 2003

[3] M. J. Balas, “Feedback Control of Flexible Systems”, IEEE Transactions on Automatic Control 1978

Virtual verification of automotive steering systems

Matthijs Klomp, Marcus Ljungberg, Ramadan Salif
Volvo Cars, Gothenburg, Sweden

Michael Attinger, Holger Bleicher, Steven Hoesli, Tim Kratzer
Robert Bosch Automotive Steering GmbH,
Schwäbisch Gmümd, Germany

P.E. Pfeffer (Ed.), *8th International Munich Chassis Symposium 2017*,
Proceedings, DOI 10.1007/978-3-658-18459-9_35

Abstract

The vehicle industry is in a transformation where software and electronics are revolutionizing the way we engineer the cars of the future. This is particularly true for steering systems, which have developed from passive mechanical systems to now enabling advanced driver support systems and the evolution toward fully autonomous driving. With this ever increasing complexity, relying only on physical testing is no longer practical due to slow feedback loops from testing back to development and the lack of repeatability. The question addressed in this paper is how computational methods can help to increase test coverage, shorten development cycles and enable continuous integration of software for steering systems. In particular the development, validation and application of methods to virtually release steering systems for passenger vehicles is presented.

1 Motivation

Demands on steering systems have steadily grown as cars have become faster and heavier and safety requirements have increased. In the past, these requirements primarily impacted mechanical and hydraulic systems, but in recent years an increasing emphasis is placed on electronics and software functionality [1].

As the rate of functional growth continues to increase, it is becoming necessary for more complex products to reach maturity earlier in development cycles [2]. In addition, testing that was previously possible only through subjective methods must be performed using objective criteria as time and prototypes become increasingly scarce [3]. Also repeatability of testing, required for regression testing, where the revisions of the software are re-tested and analysed for differences in results [4], is a challenge in physical prototypes.

In the current product development cycle, the road release presents a significant obstacle to addressing these challenges. This development step represents the conclusion from the engineering and testing departments that the product is mature enough to be operated in the field (for example on public roads) and currently necessitates subjective evaluation in a physical prototype. Thus the development of new methodologies for this application are critical to maintaining a competitive technical advantage.

To address these challenges, virtual methods are developed to enable testing of solutions before hardware is built and to find issues early – front-loading, as well as fully automatic, repeatable and safe testing of real driving situations. This paper will present how virtual methods have been developed specifically to address the road release of steering systems which is believed to be a critical step in enabling continued functional growth and innovation with steering system development in particular.

2 Organization

The use of vehicle simulation for a road release has mutual advantages for both vehicle manufacturers (OEM) and suppliers, and the development of these methodologies must be correspondingly collaborative. In addition to the rapidly evolving context of current and future automotive steering systems, this changes the typical division of work for a release.

2.1 Collaboration

By developing the model evaluation criteria and methodology together, the team is able to identify and address problems quickly, improving the quality of results using the shared expertise of a vehicle manufacturer and steering supplier. Key aspects of this teamwork are:

- Organizational buy-in and commitment from both sides.
- Agreed standard maneuver definitions (e.g. ISO standards) for validation of the virtual vehicle environment – proprietary test methods are not shared.
- A dedicated team with project managers from both vehicle manufacturer and supplier with the method development as a separate project outside the ordinary project work. Regular project meetings both face-to-face and online.
- Use of an existing vehicle for the method development and validation.

The division of responsibilities for the release itself remain as in the past, as shown in Figure 1. Rather than making a physical prototype available, an OEM provides the supplier with a virtual prototype and environment that is used in parallel or as a replacement to a traditional release. The supplier's process and release content remain its own.

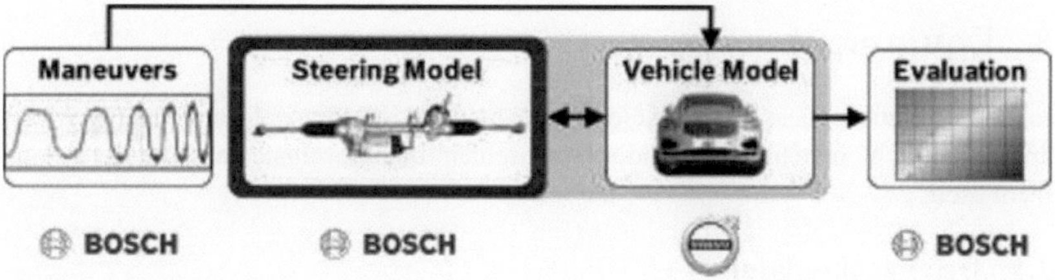

Figure 1: Breakdown of responsibilities with a release using virtual methods. As before, OEM remains responsible for vehicle-level content.

2.2 Prioritization

Although the release process remains the responsibility of the supplier, both the OEM and supplier can achieve benefits through the use of vehicle models for release activities A breakdown of typical product development cycle with respective priorities of team members is shown in Figure 2, where it is shown that the primary OEM benefit is in the early stages of the development when physical prototypes are scarce and expensive. For the system supplier, the primary benefit is during cyclical regression testing where short feedback loops and repetitive testing are required. All of these are key strengths of the virtual methods developed in this project.

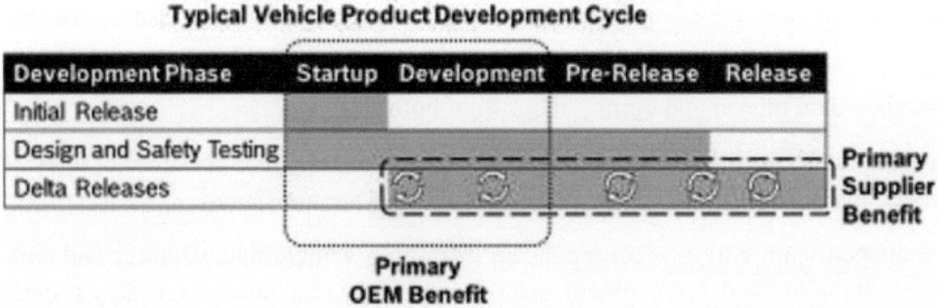

Figure 2: Breakdown of typical product development cycle with respective priorities of team members.

By focusing on implementing virtual methods in phases of the release where there is a common benefit, we can ensure that both parties achieve increased test coverage and shorter development cycles at the same time

3 Equipment

In order to achieve these goals efficiently, the method development should be organized appropriately by matching the product requirement being evaluated and the test equipment used.

3.1 Driving simulator

The Volvo Cars driving simulator, or DiM (Driver in Motion) platform, was developed to subjectively assess full vehicles and vehicle subsystems using virtual models instead of physical prototypes [5]. It consists of a full size cockpit, resting on a hexapod that is placed on a platform capable of planar motion. In total the cockpit has 9 degrees of freedom (full 2D-motion in the base platform of a hexapod and full 3D-motion of freedom between the top and base platform of the hexapod) which enables exceptional perception fidelity related to steering feedback in particular.

3.2 Steering system test bench

Steering system test benches offer physical testing of a complete steering gear and steering column. The construction is largely common between Volvo and Bosch:

- Two linear drives or hydraulic cylinders connected to the tie rods to represent the road loads. They can be force or position controlled.

- A steering robot capable of representing the driver either with torque or angle controller in open or closed loop control.

The inputs to the steering robot and tie rod actuators can be taken either from measurements in a physical vehicle, from full vehicle simulations (e.g. in Adams/Car or IPG CarMaker) or by having a vehicle model in the loop. Simple models can also be implemented in Simulink, for example if it is enough to represent the road forces with a spring load.

At the Robert Bosch Automotive Steering location in Schwäbisch Gmünd, there are a number of similar test benches developed in-house, however the System-HiL in particular offers advantages for virtual verification (see Figure 3).

Figure 3: System HiL at Robert Bosch Automotive Steering

In addition to the typical construction described above, the entire system under test is installed inside of a climate-controlled chamber capable of maintaining temperatures between -40°C and 120°C. There is also a possibility to test in a "driver closed loop" environment via a driver-seat mock-up with a real steering system and virtual vehicle. This mock-up can be installed in place of a steering robot, allowing a driver to "feel" steering behaviour during a simulation. In this way the System HiL is useful for evaluating simpler subjective phenomena in addition to system-level temperature effects.

4 Vehicle and Steering System Models

In this section, the development and validation of vehicle model systems and the steering system model shown in Figure 1 are described in more detail. The validation of the complete environment (steering and vehicle system) is described in Section 5.

4.1 Vehicle model

The basis of the virtual prototype used for the release (IPG CarMaker) are high fidelity vehicle models (Adams/Car) parameterized from component specifications and/or measurements. The positions, weights and component properties are hence modelled in the Adams/Car model. The conversion process to obtain a Carmaker model is divided in different sections for each subsystem of the vehicle.

The body is defined to have the same weight, center of gravity, inertia, chassis weights and position of rotating/translating parts.

The suspension properties are parameterized according to the Adams/Car model whereas the kinematics and compliance (K&C) are defined in look-up tables. In order to obtain the kinematics data, parallel and anti-parallel wheel motion is simulated in Adams including the steering sweep from end to end lock. The compliance requires force and torque load on wheel simulations simultaneously with parallel wheel travel. Additionally the K&C look-up tables may be based on measurement data (if available) to increase model fidelity. The standard steering system model from CarMaker is replaced with a specific model described below, combined with a steering wheel and column model specific for this project.

The tires models are the standard Magic Formula (MF) tire models parameterized from flat-track measurements based on the TYDEX standard tire interface and data format. In addition to the standard MF-Tire model [6] in CarMaker (version 5.1 and 6.1), the 3rd party MF-Swift tire model [7] with additional low-speed parking features was evaluated.

Finally, the powertrain is replaced with vehicle specific model supplied from the powertrain department and the aerodynamics are parameterized on data from either CFD simulations or wind tunnel measurements.

4.1.1 Sub-system Validation Methods

A validation of the kinematics and compliance is presented in Figure 4 showing of the lateral wheel centre displacement and the longitudinal toe-in compliance of the vehicle model compared to the Adams/Car model and measurements in a K&C rig.

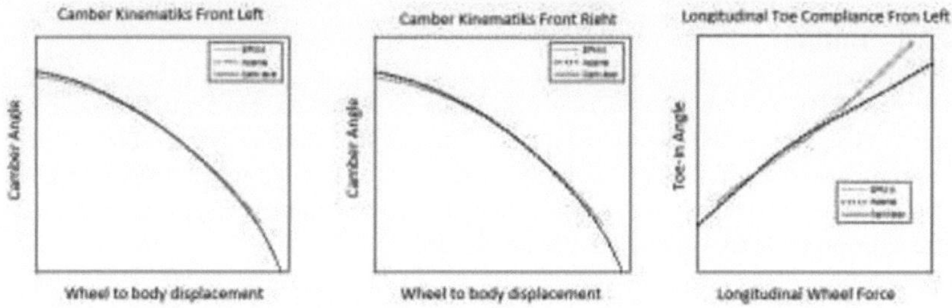

Figure 4: Lateral wheel centre displacement & longitudinal toe-in compliance.

Due to the varying fidelity requirements across the vehicle dynamics range, it is necessary to evaluate models with various complexities. Where conventional tire models have been showing potential in vehicle dynamics simulations above 10 km/h speeds, the need to simulate at lower speeds (ex: parking maneuvers) increase the model fidelity requirements. Figure 5 shows an example of steering wheel torque at standstill for various tire models compared to a real measurement.

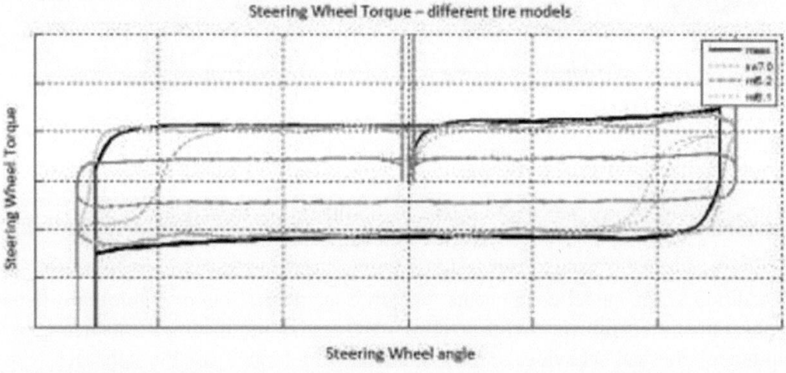

Figure 5: Steering wheel torque versus steering wheel angle of different tire models compared to measurement.

From Figure 5 it can be seen that the best correlation to measurement is obtained with the 3rd party tire model specifically designed to deal with the challenges of low-speed and stand-still tire force generation.

4.2 Steering model

For release activities where a steering model is preferable to a real steering system, a system model is used together with a software-in-the-loop (SiL) model of the software. In this application, a high importance is placed on model precision and validation.

4.2.1 Build

The model for the complete steering system can be separated into controller and plant model. The basis of the controller model is the complete steering system software in C code, compiled for Windows applications with a make-based process. By using the original software rather than a simplification, it can be guaranteed that as many software effects as possible are captured in the simulation.

There are, however, differences in implementation, hence hardware specific code sequences have to be replaced (e.g. direct memory access) and SiL-specific additional features like sensor hardware as well as vehicle bus emulation are necessary.

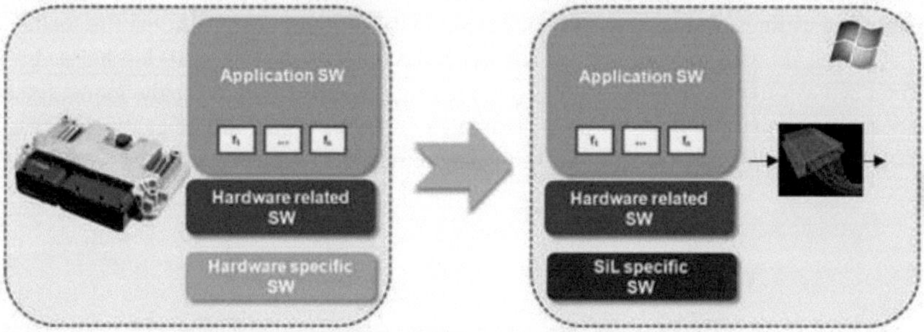

Figure 6: Visualization of SiL controller model build

The plant model covers effects of the mechanical and electrical parts of the steering system.

In addition to the model components, two different solvers are available for different applications: a simplified, real-time (1ms step) model and a complex model, able to capture all physical behaviour of the system, but with larger, variable step sizes.

4.2.2 Parameterization

Parameterisation of the SiL model is based upon the release parameters as defined by the software and application team. Access to these parameters at the controller model is similar to what is available in real prototypes (via the a2l file), and DCM tunings are transferable automatically.

Mechanical parameters such as inertia or geometries are derived from drawings, whereas functional characteristics such as friction or stiffness are derived from specific measurements performed with a nominal steering system on a test bench.

4.2.3 Validation

Before the steering system is integrated into the final vehicle simulation environment, a subsystem validation is performed. The first step in this process is the correlation of the controller model in order to confirm that the model has been built and parameterized correctly. Then a correlation of the complete steering model takes place, after which select fine tuning of the model parameters may be necessary to achieve an optimal correlation. Finally, a subset of maneuvers are used for rating the model quality in three frequency ranges: static response, command response (< 5 Hz), and reaction to disturbances (> 5 Hz). The model correlation quality across these frequency ranges can be seen below in Figure 7.

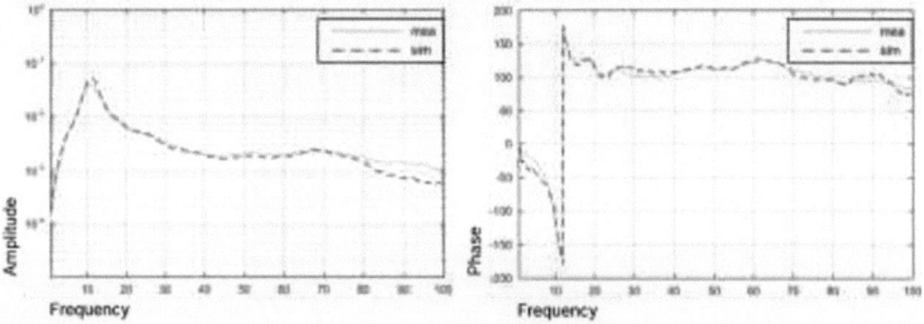

Figure 7: Transfer behaviour of SiL steering model (torsion bar torque – rack force)

5 Complete Simulation Environment Evaluation

Once the individual models have been validated and integrated into the complete simulation environment, the next step is to correlate the complete environment to a real vehicle to ensure that simulation results align closely with real-world maneuvers.

5.1 Data Collection

The following categories of ISO standards were used for the correlation: transient response (ISO 7401:2011), free steer (ISO 17288-1/2:2011), and on-centre handling (ISO 13674-1:2010).

In addition to the ISO maneuvers, a complete road release was logged to assess the model correlation using the supplier's own internal maneuvers and requirements.

Throughout these maneuvers it is essential to ensure not only that the vehicle dynamics (ex: lateral acceleration, roll angle, yaw rate, etc.) are recorded with an inertial measurement unit (IMU), but also that the interfaces to the steering system (ex: input shaft torque, rack force, etc.) and the internal steering system signals (ex: assist torque, rack position, etc.) are collected. An overview of these measured quantities is shown in Figure 8 below. This enables a much more thorough analysis at the steering system level.

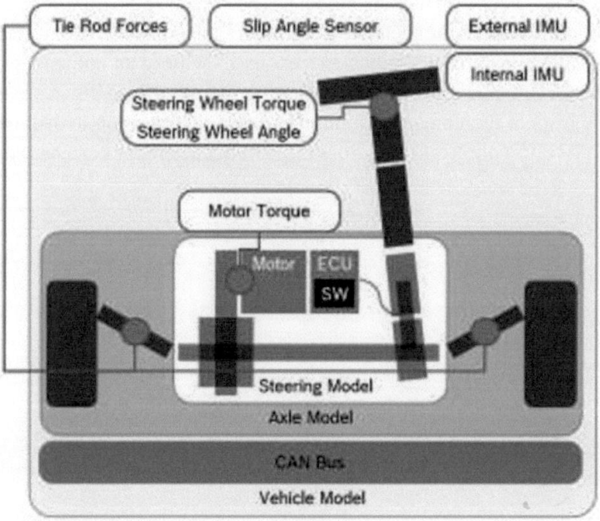

Figure 8: Diagram of select signal sources

5.2 Post Processing

After collection, the data must be interpolated and post-processed at Bosch to align the simulation results to the real vehicle measurements. After this step, the correlation is performed with an automated evaluation script in Matlab to facilitate quick comparisons. Notable deviations are communicated openly and discussed in the team.

5.3 Results

Overall the simulation results with focus on the steering system compare favourably to reality. As anticipated, low speed maneuvers (under 10 km/h) and highly dynamic maneuvers (steering speeds greater than 600 deg/s) show some deviations. Two sets of maneuvers were selected for presentation that represent the present successes (ISO standards) and the challenges (parking).

5.3.1 ISO Validation Standards

To validate the simulation tool on a complete vehicle level, the procedure described in ISO 19364 and ISO 19365 are a valuable baseline. Figure 9 presents the body sideslip angle at rear axle versus lateral acceleration at constant radius with tolerance boundaries according to the standard. Figure 10 presents a sine with dwell time series with the last amplitude of steering wheel angle before stability controller intervention.

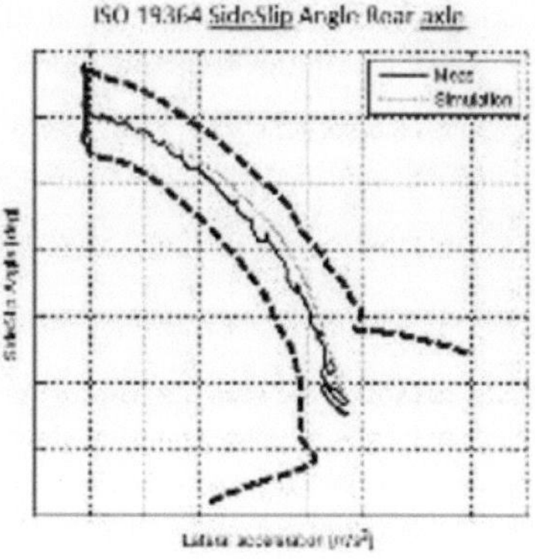

Figure 9: ISO 19364 Body sideslip angle at rear axle, validation with boundaries.

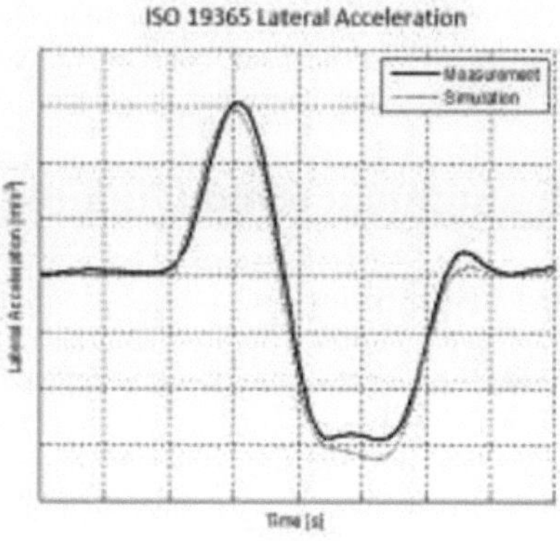

Figure 10: ISO 19365 Sine with Dwell on last steering wheel angle amplitude before stability control intervention.

Table 1 presents the metrics and tolerances stated in the ISO. The results are confirmed to be within the tolerances.

Table 1: ISO 19365 metrics validation.

ISO 19365 – Metric	Difference	Tolerance
First Peak	Pass	±15%
Second Peak	Pass	±20%
Time of Zero Crossing	Pass	±0.1s

5.3.2 Parking

In order to validate the tire model for standstill maneuvers a set of parking maneuvers were designed. As the previous MF5.2 tire model does not include turn slip equations that are essential for correlation to reality these maneuvers have been used to validate the MF-Swift 7.0 tire model with enhanced modelling of standstill dynamics.

The maneuvers are performed with a constant steering wheel velocity at both standstill and at low vehicle speed, consisting of multiple cycles of lock-to-lock steering inputs.

6 Summary and Outlook

6.1 Summary

Using simulations and test rigs to replace full vehicle testing in the area of road release is an ambitious but necessary goal to achieve the new customer requirements and shortened lead times of the future. Through close collaboration between Volvo Cars and Robert Bosch Automotive Steering a first step has been taken in achieving this.

The focus up until now has been towards achieving a satisfactory correlation between physical and virtual prototypes by using well-known ISO standard maneuvers. Parking is also being evaluated but since there is no ISO standard for parking custom maneuvers were defined.

Since the steering gear is the focus of the release, the important signals relate to the steering gear's interface to the vehicle. Those interface points are the input shaft and the tie-rod ends and that is where the external forces and torques need to be close enough to reality, as specified by requirements from the supplier.

Adequate correlation was achieved for all quasi static maneuvers, i.e. steady state and on center steering. The challenge remains highly dynamic maneuvers (ex: frequency response), particularly those with inputs greater than 3 Hz. Parking maneuvers are also a challenge due to tire model restrictions but solving this is ongoing, one possible solution being the implementation of MF-Swift 7.0 in the simulation environment.

6.2 Achieved goals

Apart from strengthening the collaboration between vehicle manufacturer and the system supplier, key to meeting the challenges in today's rapidly evolving automotive environment, the discussed methodology has achieved the following:

- Test coverage has been expanded. The amount of maneuvers that are performed during a road release has been expanded and is reproducible. This enables better control when performing regression analysis.
- The benefits of simulating multiple variants simultaneously have been proven. When it comes to the breadth of vehicle variants that can be included in the road release, improvements have been made. Since creating a new vehicle variant is a matter of exporting a MBS model, the limiting factor becomes the post-processing of the data for all simulated variants.
- Virtual methods has been tested in an ongoing vehicle project. Though physical prototypes were used in parallel while virtual testing was performed, continuously analyzing and improving the process is key to ensuring sustainable innovations.

6.3 Next steps

The next steps within the scope of this project are to extend the testing capabilities so that parking situations can be assessed and to continue to improve the correlation to reality.

Improving this correlation is continuous work – a better correlation leads to more test cases can be moved from a physical to a virtual prototype vehicle. In addition the analysis of further subjective acceptance criteria in further detail will be necessary. As such maintaining and improving the collaborative infrastructure developed through this project will be essential.

Moving forward these new capabilities will be put to use in the next great challenge of the automotive industry which is fully autonomous driving. The development and verification of these systems require the capabilities of repeatability and scalability, and the virtual methods developed and presented in this paper are key contributions to meet these challenges.

References

[1] P. Pfeffer and M. Harrer, Lenkungshandbuch, Springer Vieweg Verlag, 2013.

[2] P. Milbaier, "Evaluation of the Robustness of an EPS Control System in an Early Stage of the Product Development Life Cycle," in *ATZ chassis.tech plus*, Munich, 2017.

[3] S. Grüner, "Objektive Bewertung von Lenkgefühl und Anwendung im Rahmen einer virtuellen Lenkgefühlsabstimmung," in *ATZ chassis.tech plus*, Munich, 2017.

[4] G. J. Myers, The Art of Software Testing, John Wiley & Sons, 2004.

[5] "Safe Simulation at Volvo Cars," *Agenda Magazine,* pp. 5-6, 3 2015.

[6] I. J. Besselink, A. J. Schmeitz and H. B. Pacejka, "An improved Magic Formula/Swift tyre model that can handle inflation pressure changes," *Vehilce System Dynamics,* vol. 48, no. sup1: IAVSD Supplement, pp. 337-352, 2010.

[7] A. J. C. Schmeitz, I. J. M. Besselink and S. T. H. Jansen, "TNO MF-SWIFT," *Vehicle System Dynamics,* vol. 45, no. sup1, pp. 121-137, 2007.

Steering System Models - An Efficient Approach for Parameter Identification and Steering System Optimization

J. Lin, T. Kloos, P. E. Pfeffer

The original version of this chapter was revised.
The correction to this chapter can be found at
DOI 10.1007/978-3-658-18459-9_58.

P.E. Pfeffer (Ed.), *8th International Munich Chassis Symposium 2017*,
Proceedings, DOI 10.1007/978-3-658-18459-9_36

Introduction

In the product development of the automotive industry, it is very common nowadays to use an X-in-the-Loop development method through the V-model process. Within the procedure, a lot of components are required to present as a model in the virtual environment at the early development stage. Therefore a huge effort is made to set up physical simulation models. However, as complicated as a model can be, a valid set of parameter which could reproduce the real physical behaviour is also needed in the end.

Lin et al. has developed advanced steering models (Pfeffer MXsteering models) to represent most of the steering systems on the market (1). With the motivation stated above, an efficient method to derive parameter values for the model was developed, so that the models can finally be used in the development process.

Methodology

In order to develop a systematic methodology for the steering model parameter identification, the following steps have been undertaken:

Firstly, the identification process must be designed with the knowledge of the steering model structure. To identify the parameters and design the test procedure accordingly, it is important to conduct the sensitivity study of the model parameters and establish the corresponding relationship between the steering response and the model parameters.

Secondly, it is necessary to set up an appropriate test environment, suitable for the parameter identification. In Munich University of Applied Sciences, two steering test benches have been set up with advanced control techniques. They can apply a demanded load and measure the response of a steering system accurately. The steering test benches and the steering models share the same interface to the vehicle model and the driver. Therefore the influence of the vehicle model is minimized. It is only required that the rack force is in the same range as the target vehicle.

Thirdly, the test procedure must be designed according to the requirements. Owing to the advanced control algorithm of the steering test bench, the test procedure design is much more convenient and effective. The test procedure can be divided into two catalogues: steering characteristics tests and driving manoeuvre tests (HiL tests). The former tests identify the physical characteristics of the system while the latter tests provide a measurement for the iterative optimization of the steering model parameters.

Finally, appropriate post processing algorithms and an optimization method are defined, so that the facility is established completely. Typically the optimization method which is used, is a multi-objective optimization in order to minimize the error between the measurement and simulation in multi-dimensions.

The Structure of the Advanced Steering System

The Advanced Steering System Model (ASM) or Pfeffer MXsteering model has been developed in order to represent and simulate accurate characteristics of steering systems (1). The model comprises a mechanical module and a power assistance module. The required input signals for the model is either steering wheel angle or steering wheel torque. A dynamic vehicle model should provide the rack force input to the steering model, according to its rack position.

Figure 1 shows the steering model layout with possibilities to switch from a steering wheel angle input (top) or a steering wheel torque input (bottom). The power assistance module refers to the assistance of the hydraulic powered steering system (HPS) or electric powered steering system (EPS) for the respective steering systems. The mechanical module provides the steering and rack velocity to the power assistance module, while the power assistance module feeds back the assisting force and the friction caused by the power assistance module such as the sealing friction caused by the pressure in a HPS system.

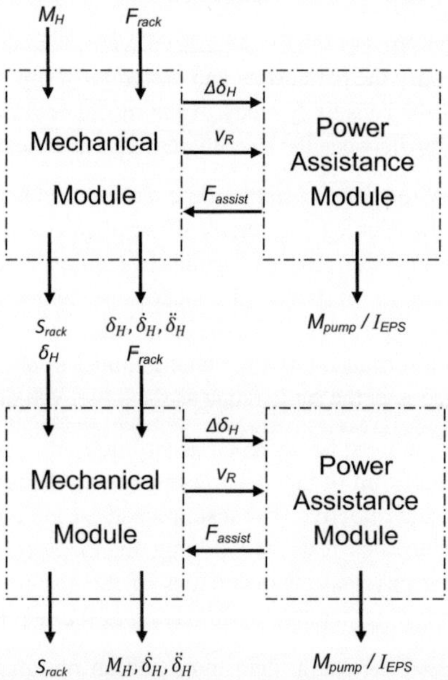

Figure 1: Steering model layout (δ_H – steering wheel angle; V_R – steering rack velocity; $\Delta\delta_H$ – torsion bar twist angle; F_{assist} – assistant force of the power assistant module; M_H – steering wheel torque; M_{pump} – pump output torque of the hydraulic power assistance; I_{EPS} – EPS current)

The mechanical module is shown schematically in Figure 2. It contains all mechanical components which transfer the torque from the steering wheel to the tie rods, i.e. the steering wheel, steering column, hardy disc, torsion bar, pinion, steering rack and the occurring damping and friction effects. The components are modelled with spring-damper elements and the friction elements are applied to the column and the rack. The power assistance is applied to the column, close to the steering wheel in the case of a column assist type while the assistance to the rack otherwise. Figure 2 shows one example draft of an EPS column assist type.

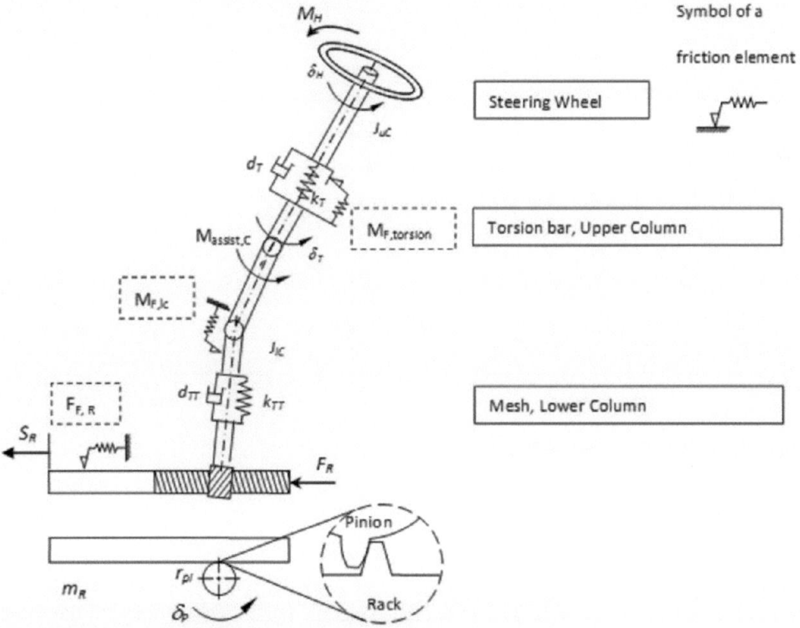

Figure 2: The mechanical steering module (EPS column assist type)

The friction elements play a significate role in the steering behaviour. The ASM is using an "Exponential-Spring-Friction-Element" with a parallel "Hyperbolic-Tangent-Element" (ESFH) (1). Other common friction models such as Dahl models with or without Stribeck effect and LuGre models are also considered in the optimization routine (2). They share a similar behaviour as the ESFH (see below). The identification method regarding the friction model can also be used for other friction models with small adjustments.

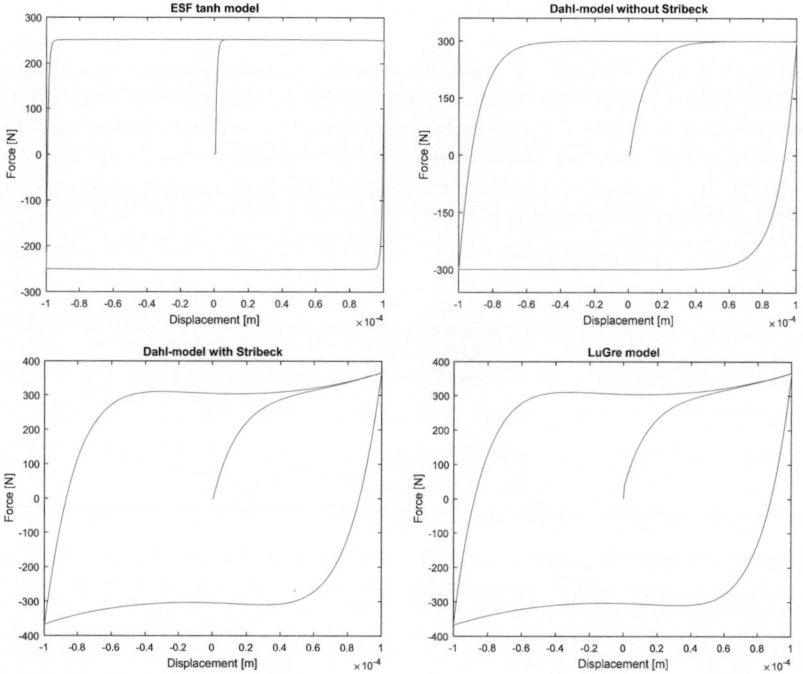

Figure 3: Friction models comparison

Sensitivity study

A series of sensitivity tests were conducted to identify the sensitive parameters in relation to the steering behaviour. The virtual steering test bench was set up with the same interfacing as shown in Figure 1. The input data remains the same throughout the whole sensitivity study. Different driving manoeuvres were applied, which generate the sensitivity matrix as shown in Figure 4. The identified sensitive parameters are the boost curve, the friction elements and the stiffness of system. Different test cases were conducted in order to identify the relationship between the steering responses to the steering model parameters.

The used manoeuvres are transition tests focusing on the quasi-static behaviour of the system, weave tests, which relate to the on-centre behaviour and sinus sweep tests, which exhibit the dynamic behaviour.

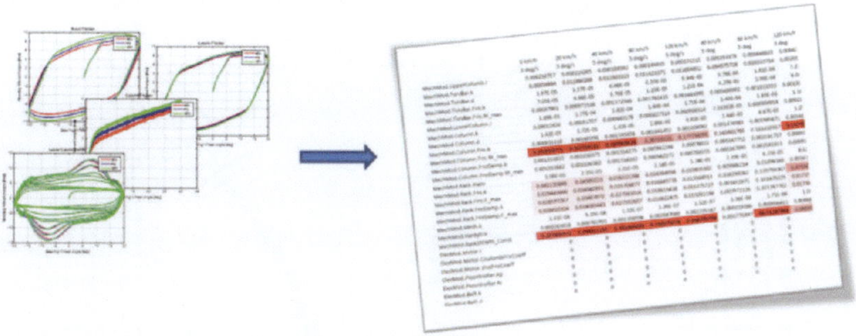

Figure 4: Steering model parameter sensitivity study

The Steering System Test Bench

The system design of the steering system test bench at the Munich University of Applied Sciences consists of two linear motors and a steering wheel actuator. The components with the associated control units and the support for the test subjects are connected to a solid steel platform. The actuators can be arranged individually according to the test configuration. The steering system test bench concept introduces a two-sided force, providing a realistic tracking force and is characterized by very high dynamics during the force application.

With the integration of a vehicle model into the simulation environment, the hardware-in-the-loop capability of the steering system test bench is established. This means that tests can be carried out on the whole vehicle in the virtual environment. Any vehicle dynamics simulation software such as CarMaker, dSPACE ASM or VI-CARREALTIME can be used at the steering system test bench. The software is integrated as a library module into the existing Simulink model at the steering system test bench and is connected to the required interfaces. The vehicle model exchanges information with the steering system test bench and the steering system in bidirectional direction. The interface is the same as that of the Pfeffer MXsteering models, which facilitates the ideal optimization environment independent of the vehicle models.

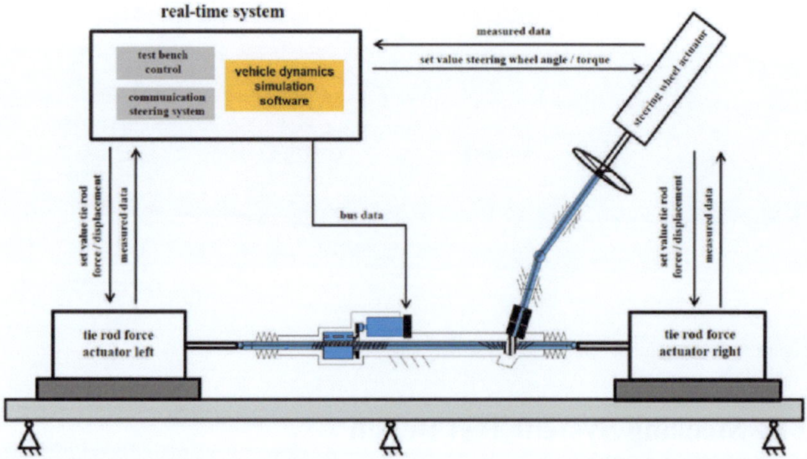

Figure 5: Steering system test bench (3)

The Test Procedure and the Post Processing

The test procedures were designed to extract useful information for the steering parameters. They are divided into two catalogues: the system characteristic tests and the driving manoeuvre tests.

System Characteristic Tests

Friction tests for column friction, rack friction and overall friction

For the column friction tests, the cardan joint is disconnected, and the steering wheel is turned from one end to another, to evaluate the friction level of the steering column at different steering wheel velocities. This test is done with a detached electric motor to exclude the influence of the electric-magnetic field. The same configuration applies to the overall friction test, but with a connected cardan joint. For the rack friction tests, the cardan joint is also disconnected and the steering rack is pushed from one end to the other, to evaluate the friction level of the steering rack at different rack velocities. Figure 6 and Figure 7 show a measurement and the results after the post processing.

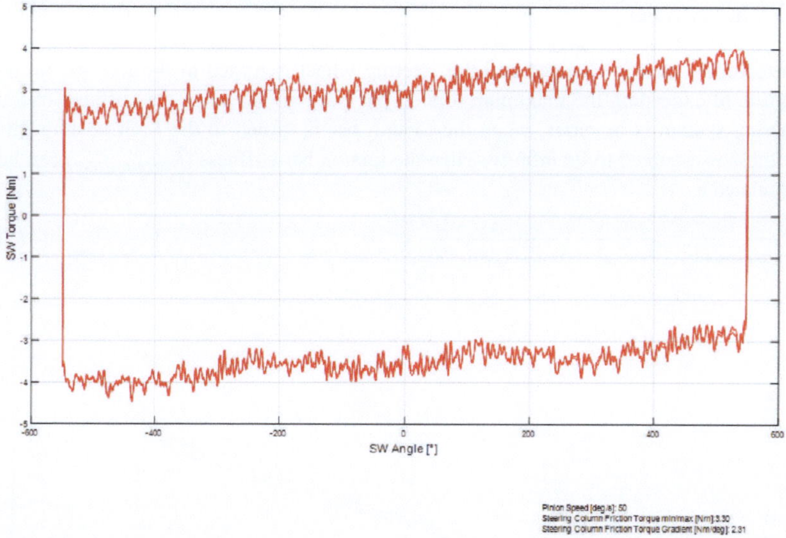

Figure 6: Overall friction test at 50 deg/s

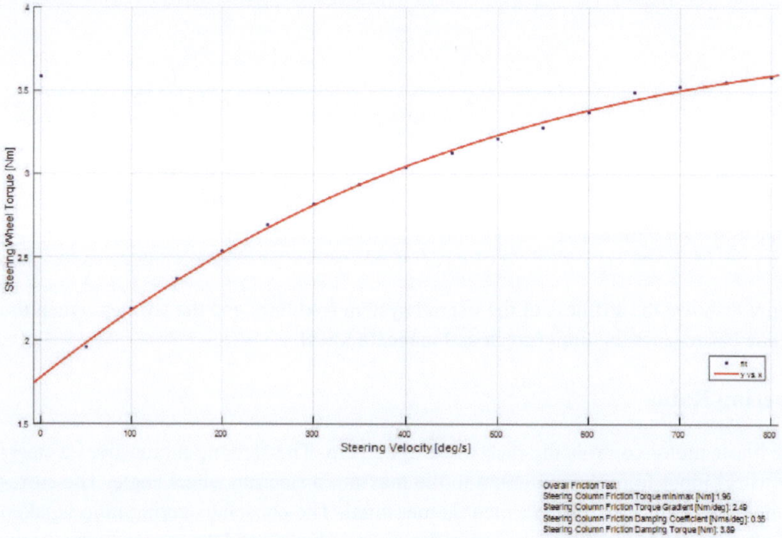

Figure 7: Overall friction test at different steering velocities

System Stiffness

The steering rack is clamped and the steering wheel is turned to the left and to the right until exceeding the maximum torsion bar twist angle. Since the stiffness of the steering system is in series, when the torsion bar is turned to the limit position, its stiffness is assumed to be infinity. Then the torsion bar stiffness ($k_{torsion\_bar}$) can be calculated as:

$$\frac{1}{k_{torsion\_bar}} = \frac{1}{k_{overall}} - \frac{1}{k_{without\_torsion\_bar}} \tag{1}$$

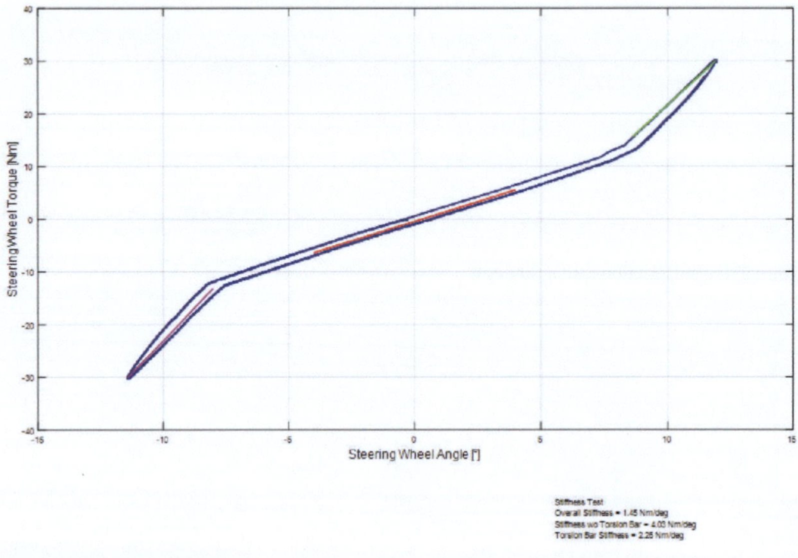

Figure 8: System stiffness test

Figure 8 shows the stiffness of the overall system (red line) and the stiffness when the torsion bar reaches the limit (green and magenta lines).

Steering Ratio

The linear motor controls the rack force to be zero. The steering motor gives a steering wheel input from the minimum to the maximum steering wheel angle. The corresponding steering rack displacement is measured. The constant steering ratio is taken out of the measurement by calculating the mean value around the steering wheel centre. When it comes to the variable steering ratio, a lookup table is generated. The measurement and the calculated ratio are shown in Figure 9 as an example for the constant steering ratio.

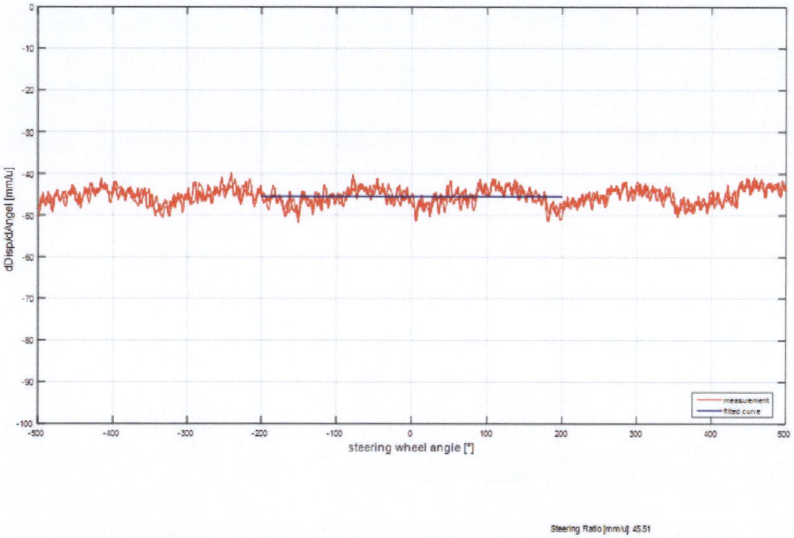

Figure 9: Steering ratio test

Boost Curve Test

The steering rack is clamped by one linear motor while a steering wheel torque is applied continuously (from minimum to maximum torque e.g. ±10 Nm) to the steering wheel at characteristic vehicle speeds (e.g. 0, 10, 20 … 100, 120, 140 km/h).

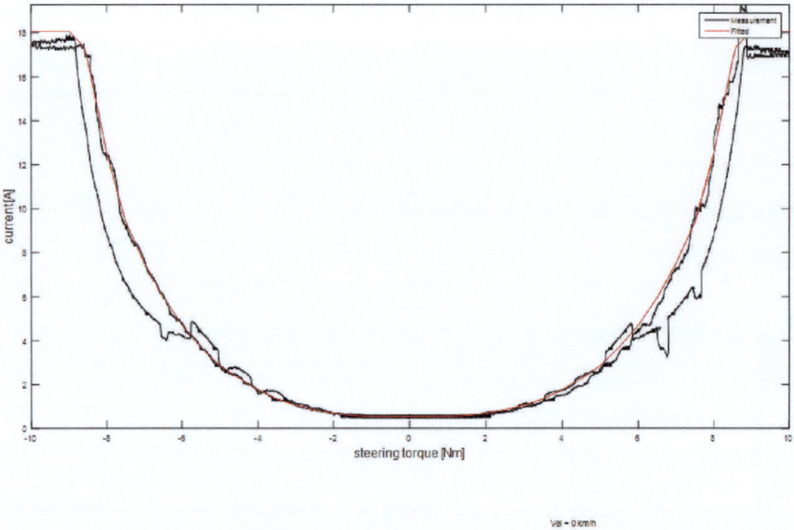

Figure 10: Boost curve measurement at 0 km/h

The measured boost curves (Figure 10) are used to generate a look-up table for the steering model at different vehicle speeds. They are fitted by piecewise cubic interpolation with a constant step torque input. The offset at 0 Nm is removed accordingly.

Driving manoeuvre tests

The driving manoeuvre tests include transition tests, weave tests and dynamic tests such as sinus sweep or step input tests.

The transition tests were used to evaluate the static steering response and the assistance level. The tests were done with a slowly increasing steering wheel angle at a vehicle speed of 0, 10 … 140 km/h with a steering wheel velocity of 5 deg/s.

Weave tests can be used to investigate the distribution of the friction level within the system. It was done by controlling the steering wheel with sinusoidal input with different amplitudes and vehicle speeds. The area in the steering wheel torque vs. steering wheel angle hysteresis depicts the centre feel of the system.

Dynamic tests such as sinus sweep tests are used to assess the dynamic response of the steering model. For example the sinus sweep tests were typically done up to 2.5 Hz at different vehicle speeds.

The identification process is summarized in Figure 11. First, the target steering model and its initial parameter values must be set up. The second step is to calculate the physical values from the characteristic test results, which provide a realistic initial value to the whole process. Afterwards the driving manoeuvre test results are used to optimize the sensitive parameters, which are difficult or impossible to measure, by using an iterative optimization routine (Figure 12). During this step, it is possible to identify the measurement error in the characteristic test results, which the red lines show. The last step is to combine the steering model with a target vehicle model and verify the simulation with a vehicle measurement. Since the steering model parameters are identified independently from the vehicle model, the verification at this step is mainly there to validate the vehicles suspension model. Figure 13 to Figure 15 show examples of fitted results of the different manoeuvres. The process is automated and therefore very efficient.

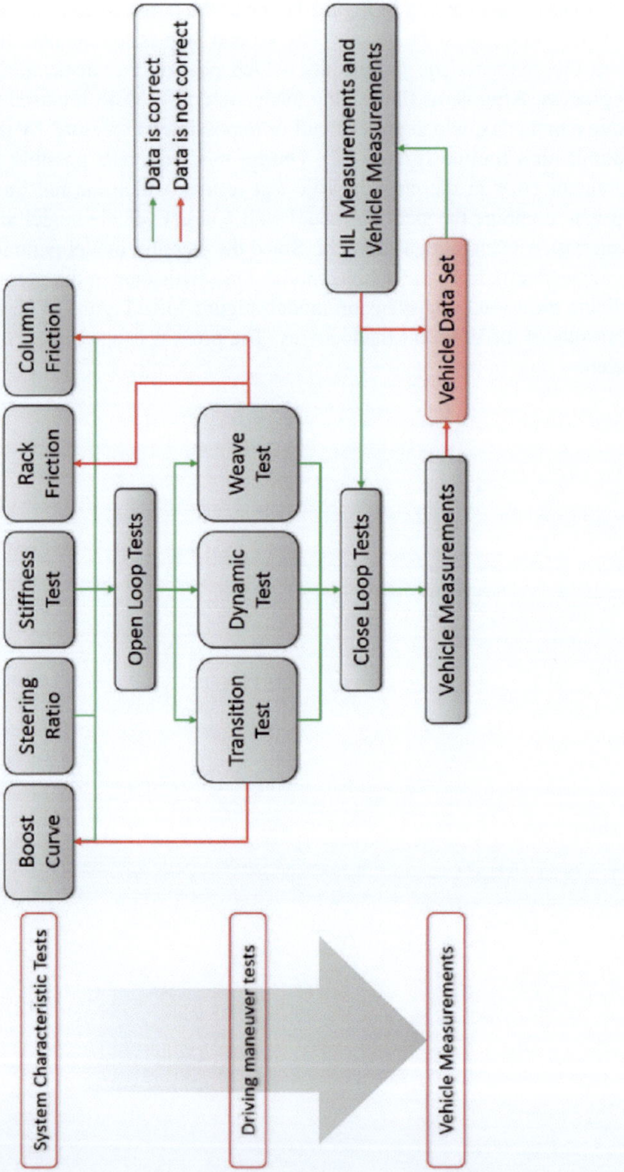

Figure 11: Steering model parameter identification process

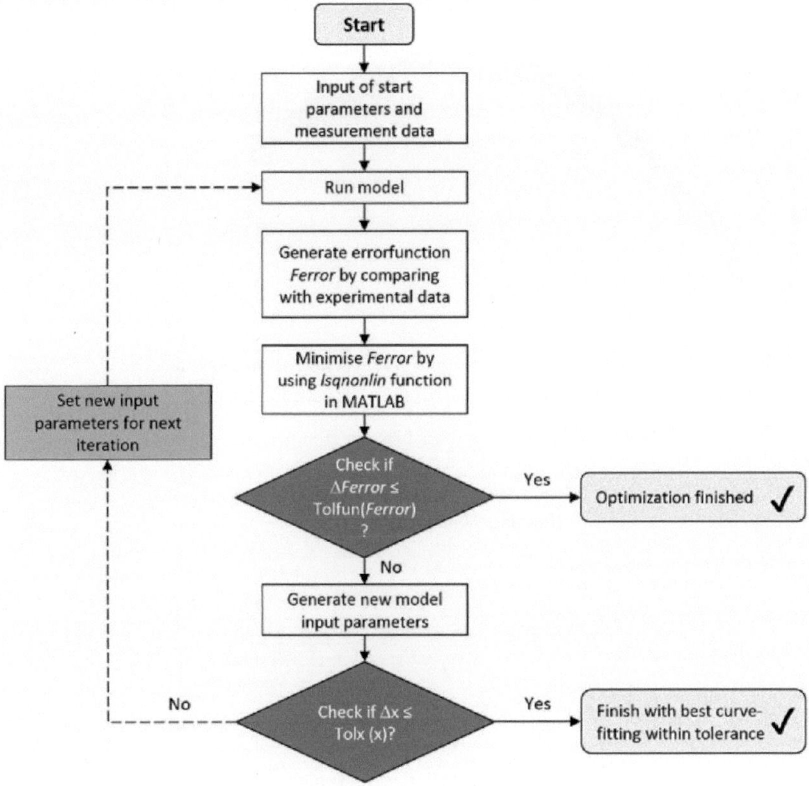

Figure 12: Iterative optimization routine

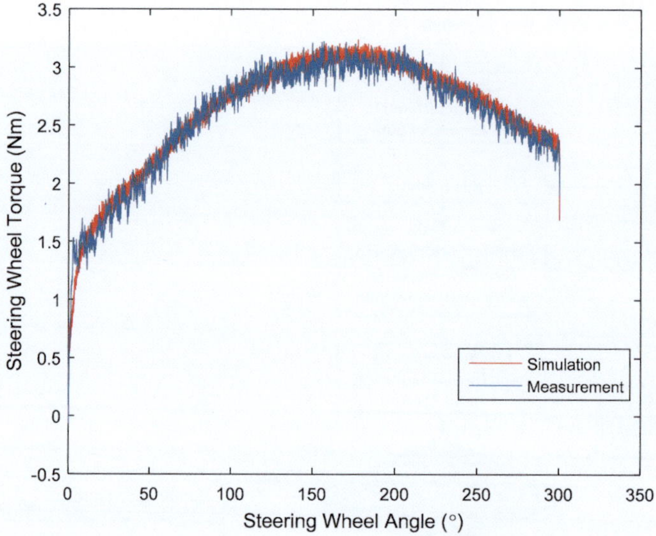

Figure 13: Transition test result

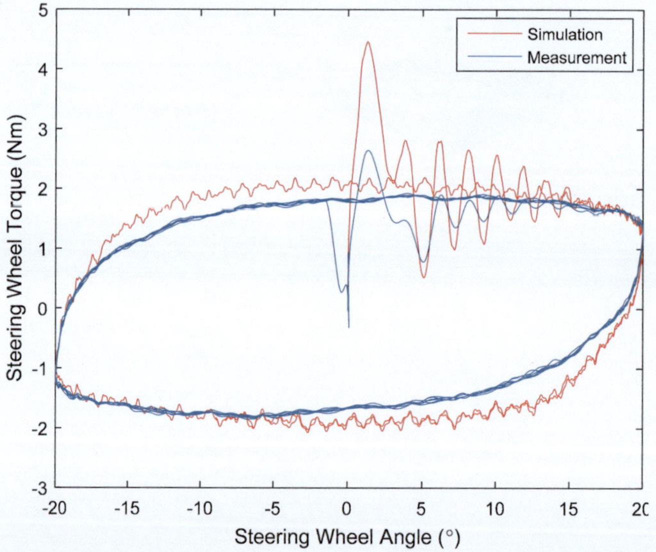

Figure 14: Weave test results

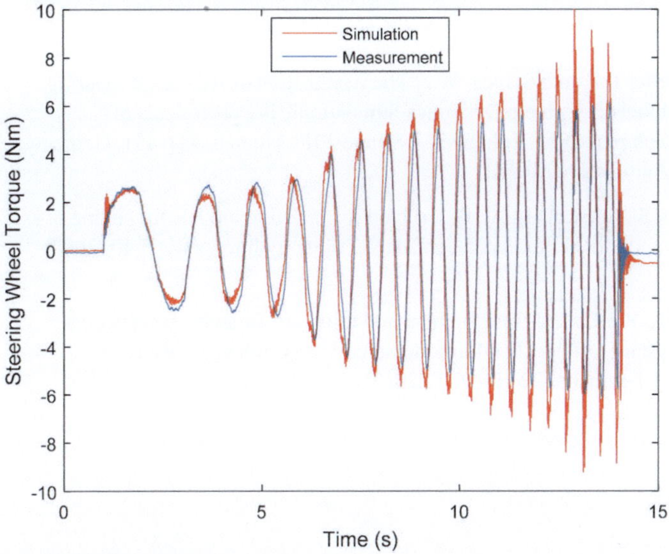

Figure 15: Sinus sweep test results

Conclusion

The methodology introduced in this paper demonstrates an efficient approach to identify parameters for steering models. A valid set of steering parameters is derived and is ready to use for the development process in a much shorter time compared to the conventional methods. The application of the advanced steering test bench and the optimization algorithm facilitates the automation of the whole process and eventually leads to a cost effective product development.

Reference

[1] Lin J., Pfeffer P.E. and Schick, W., "The Application of Advanced Steering System Models in Vehicle Dynamics Simulation", in Proceedings of Chassis.Tech plus 2011 Volume 2, 7-8 June 2011,Munich, 441-451, Germany, Springer Automotive Media

[2] Andersson S., Soederberg A., Bjoerklund S., "Friction models for sliding dry, boundary and mixed lubricated contacts", Tribology International 40 (2007) 580–587

[3] Schimpf R., Pfeffer, P.: Development and testing of steering systems on a steering system test bench, 14. International VDI-Conference Reifen – Fahrwerk – Fahrbahn, 22.10.-23.10.2013

STEERING FEEL

The new Hyundai i30's steering feel – integrated performance development by HiL tests and DoE

A. Contini, T. Schöning

Hyundai Motor Europe Technical Center GmbH,
Rüsselsheim, Germany

P.E. Pfeffer (Ed.), *8th International Munich Chassis Symposium 2017*,
Proceedings, DOI 10.1007/978-3-658-18459-9_37

Abstract

One of the main focuses during the new i30's development was laid on further enhancements of steering feel to match European customer's taste reaching high steering precision through defined on-center, natural feel by smooth torque build-up and a good connection by linear feedback.

Translating these subjective characteristics into objective targets allowed an early entrance in the project with simultaneous engineering combining simulation, laboratory tests and real road testing.

The key tool for bringing these areas together is hardware-in-the-loop (HiL) testing that is used for parameter tuning to match the desired steering feel performance.

By overcoming the challenge of a smooth integration into the common tuning work, HiL-testing allows testing without the real vehicle available, covering of all possible variants concerning tires, weight, powertrains etc. and thus to finally reduce real-road testing.

The new Hyundai i30's motor driven power steering systems (MDPS) is tuned by a high number of parameters that influence the steering feel. The number of combinations and the possible counter effects must be controlled for each tuning parameter set in all driving areas to make sure that the performance is on target. HMETC introduced Design-of-Experiments (DoE) to automate this process and to make full use of the possibilities of simulation based performance tuning.

Based on above approach technologies like HiL and DoE have become standard tools for HMETC in creating an efficient process finally making sure the desired steering feel is reached for all vehicles.

Keywords: Model-based calibration · Design of Experiments (DoE) · Steering-in-the-Loop · Steering feel tuning · Steering system test bench · Behavior models

1 Targets for the new Hyundai i30's steering feel

Hyundai developed for the new i30 a completely reworked chassis and a more direct steering to deliver dynamic driving experience while keeping high levels of ride comfort. The main target for the steering feel was to match European customers taste reaching high steering precision through defined on-center, natural feel by smooth torque build-up and a good connection by linear feedback. A multi-link rear axis with a new geometry allowed developing a better chassis suspension and providing a dynamic and compliant ride. Laboratory testing on the one hand and real road testing on the other hand where used to improve the new steering. The Hardware-in-the-Loop (HiL) testing is used for tuning the steering parameters and it allows testing without the real vehicle available. HiL has become a standard tool for Hyundai's steering and suspension development to finally reduce real-road testing.

An important plot (Figure 1) used frequently during the steering tuning process is the steering hysteresis generated with a sinusoidal steering wheel angle input [1]. This open-loop test is suitable to describe the steering characteristics and will be used below for further explanations:

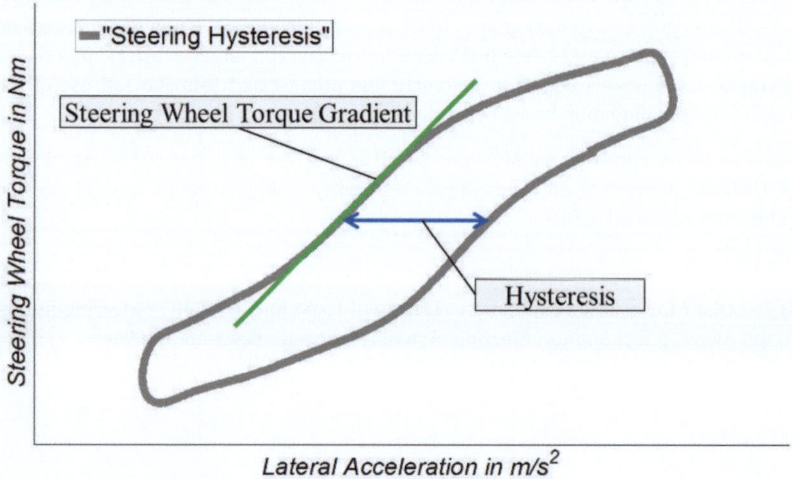

Figure 1: Steering hysteresis with objective parameter (ISO 13674-1)

2 Overview of steering tuning process at Hyundai Motor Europe Technical Center

The classical approaches of tuning steering behavior consist in doing changes at the mechanical system as well as the ECU logic and evaluate the changed steering behavior subjectively on the proving grounds and on public roads. Due to the increase of variants and so the increase of work load a powerful tool chain was designed to support the tuning engineers in the car and to widen out the classical tuning approach [2, 3, 4, 5 and 6].

At the beginning it is necessary to set the targets for the wished steering DNA fitting to the brand (step 1, in Figure 2). A special attention has to be turned to the translation of subjective assessment criteria into objective parameter, which can be directly used later during the tuning process. A virtual vehicle model needs to be build up to generate trustable simulation data from the vehicle, e.g. lateral acceleration, yaw rate, etc. (step 2). The validation process (step 3) reveals the trustability of the simulation model and its limits. The amount of work for the right model quality depends on the target setting in step 1. Once the validation is finished the automated HiL-steering-tuning (step 4) can be started. Correspondent to the actual vehicle development state, changes conducted in the real car need also to be updated in the vehicle model, e.g. when changes at chassis are done (shock absorber, etc.) to improve the vehicle behavior, this has to be conducted in the vehicle model also to keep the model quality ensured in the validation step (3). Step 5 is reasonable to be done early, to get an early feedback of the current tuning state from a test driver. This corresponds to an intermediate validation step conducted subjectively. The arrow from step 5 to step 4 corresponds to a communication loop. When the set targets are reached by using the automated HiL-steering-tuning (step 4) the suggested ECU map can be transferred to step 6: the fine tuning in the real car.

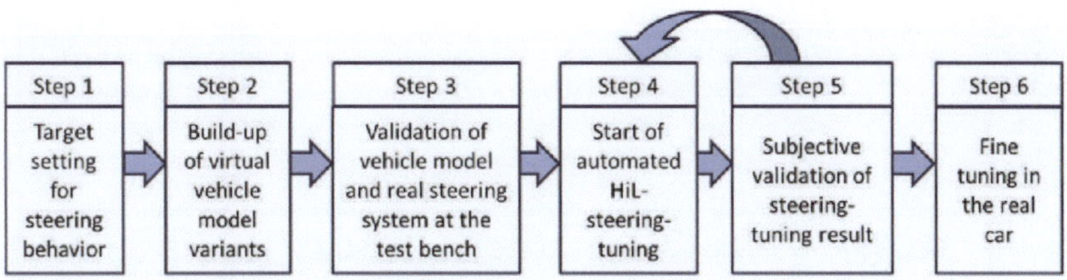

Figure 2: Steps necessary for steering tuning process at HMETC

The steering tuning process is supported by subjective and objective engineers. The experience has taught that a very good collaboration between both is very beneficial for the whole process. To quote an example the utilization of the automated HiL-steering-tuning helps to analyze the influence of specific ECU parameters on the steering feel. In Figure 3 it is shown that by changing the ECU parameter "ECU 5" the steering hysteresis can be kept (red horizontal line) and the steering wheel torque gradient can still be changed (red raising line) to the needed direction. With this knowledge increase also the subjective tuner can finalize his task faster.

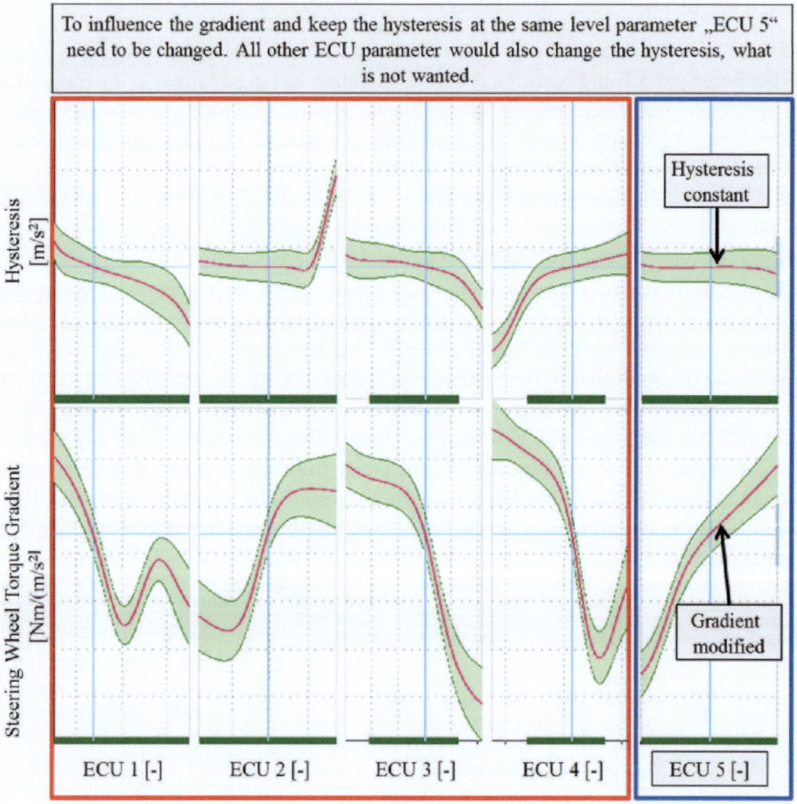

Figure 3: Intersection Plots showing the influence of ECU parameters

3 Tuning parameters and their influence on steering feel

Several ECU parameters of the steering ECU where varied up to +/-70% of their initial values. By generating a DoE with a test list consisting of about 200 tests the results fitting best can be found through the corresponding behavior model. In following example tow results with enough differentiation should be generated – the same on-center but a different off-center effort was wished.

The Figure 4 shows the lower and the upper limits for the steering wheel torque over the lateral acceleration. The strong variances of the ECU parameters lead to a strong variance of the steering behavior. At the end two results fitting the set target curves where found. Both results are behaving same in the on-center area and in the off-center area the wished differentiation was generated. The corresponding ECU values for the two results where put out through the behavior model.

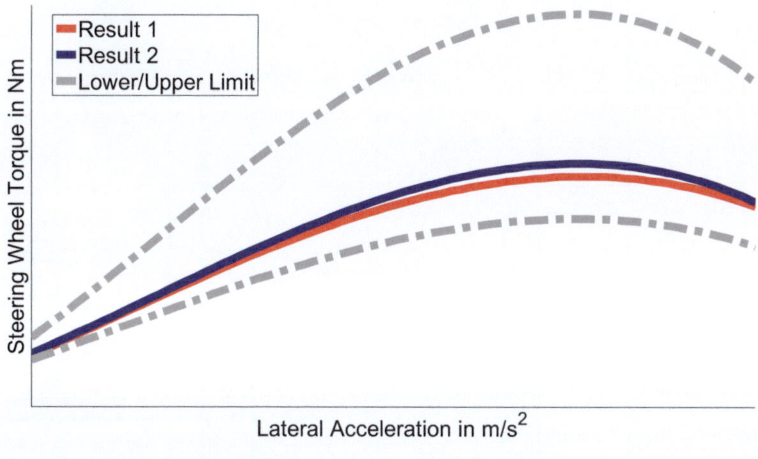

Figure 4: Test area by using DoE

One plot of the behavior model is shown in the following Figure 5 where an intersection plot is presented. To reach the steering behavior with the higher off-center effort (s. a. Result 2 in Figure 4) especially changes of the ECU parameter "ECU 4" and "ECU 5" where required compared to the parameters for Result 1. By changing this ECU parameter the desired off-center behavior was reached. Changes of about -30% of the initial values had to be made to reach a 5% higher *off-center parameter*. Simultaneously the *on-center parameter* was kept.

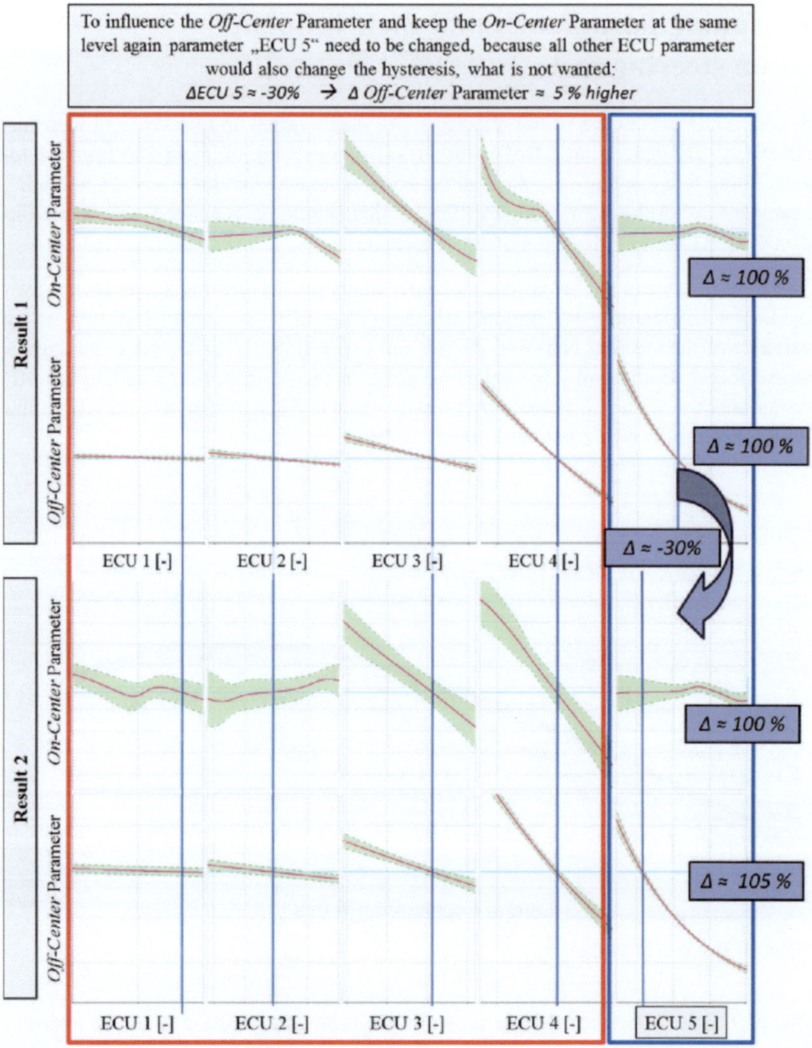

Figure 5: Intersection plots for Result 1 and Result 2

The intersection plots are based on the build mathematical behavior models. These models offer the user fast and reliable results.

The model quality for the *off-center parameter* is shown exemplary in the following Figure 6 where the adjusted coefficient of determination r^2adj to evaluate the quality of the model reaches a very high fitting (nearly 1). These statistical values are used to know how trustable the behavior models are.

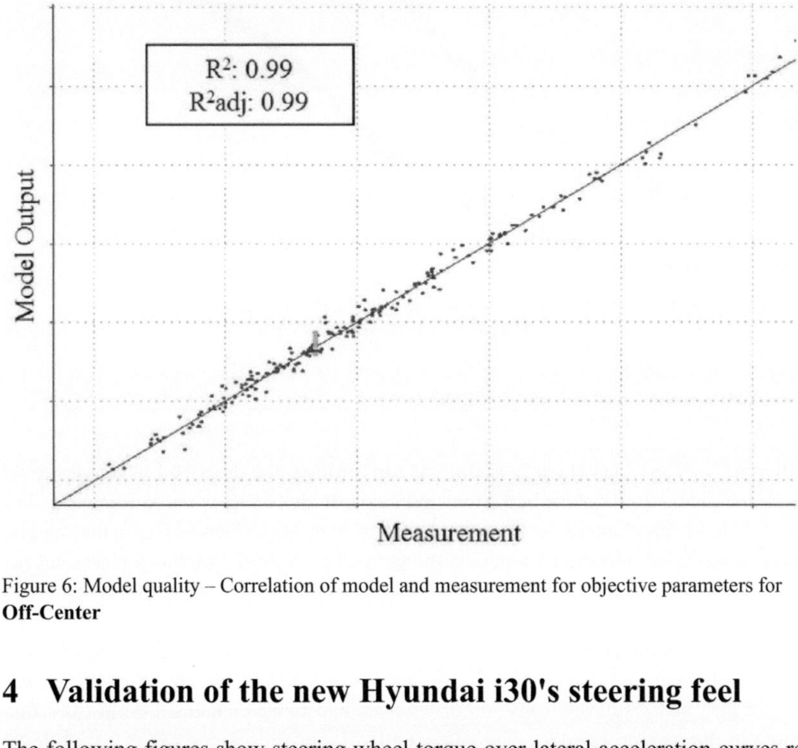

Figure 6: Model quality – Correlation of model and measurement for objective parameters for **Off-Center**

4 Validation of the new Hyundai i30's steering feel

The following figures show steering wheel torque over lateral acceleration curves related to the steering behavior and so to the steering feel.

Figure 7 shows the hysteresis generated by doing the weave test. The red dashed curve shows the measurement at the HiL-steering-test-bench after the tuning process at the test bench was finished. With the same vehicle variant and the identical steering tuning logic a weave test was also conducted on the proving ground. After executing a test plan of several dynamic tests a comparison between proving ground and steering test bench measurements can be conducted.

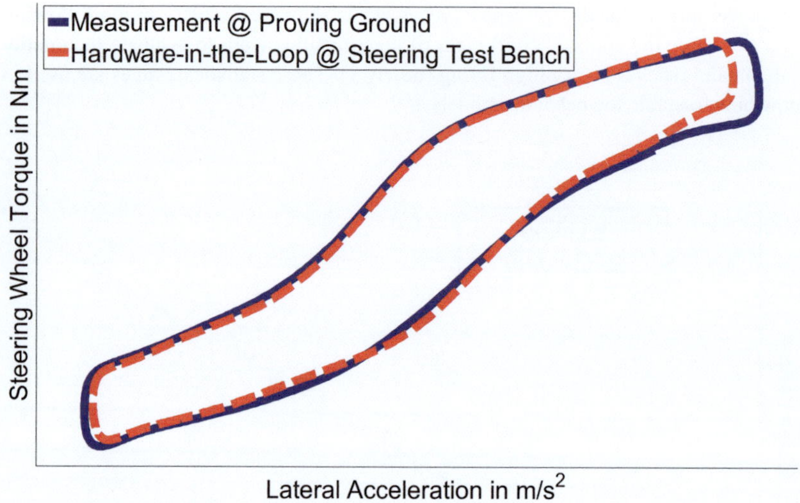

Figure 7: Measurement of the same vehicle variant with the identical steering tuning logic on the proving ground and on the HiL-steering-test-bench

After the DoE test (described in section 3) run through, the described behavior model was generated. By using the behavior model proper ECU values could easily be chosen to reach a target curve for the correspondent maneuver. These ECU values can be implemented also into the ECU inside the car and be used directly as a base map for further subjective tuning. In Figure 8 two steering behaviors effectuated by two different ECU maps are compared. The dashed red curve is the result of the automated HiL-steering-tuning (step 4) and the solid blue curve the result of the subsequent fine tuning in the real car (step 6). The figure shows that the HiL-steering-tuning is already near to the final result and only minor changes had to be conducted during the fine tuning in the real car (step 6). The automated HiL-steering-tuning offers a high level for a basic ECU map where to start from in the fine tuning phase.

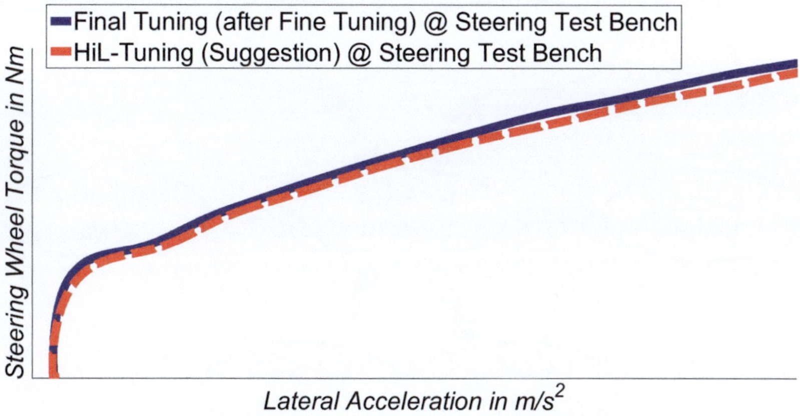

Figure 8: Comparison of automated HiL-steering-tuning result with the fine tuning result

5 Challenges of multilateral cooperation and their benefits

To establish a proper process a multilateral collaboration was necessary. The collaboration has to be done together with different departments, with the parent company Hyundai Motor Company in South Korea and with the suppliers.

The design department supports the preparation of adapter parts, with design information about the steering parts and the preparation of the vehicle model for the simulation. The objective team allocates all necessary real car measurements needed for the vehicle model build up as well as the validation step. The steering supplier provides the tuning software. Together with the test bench developer improvements and adaptations are conducted to increase the measurement quality or to extend test scenarios.

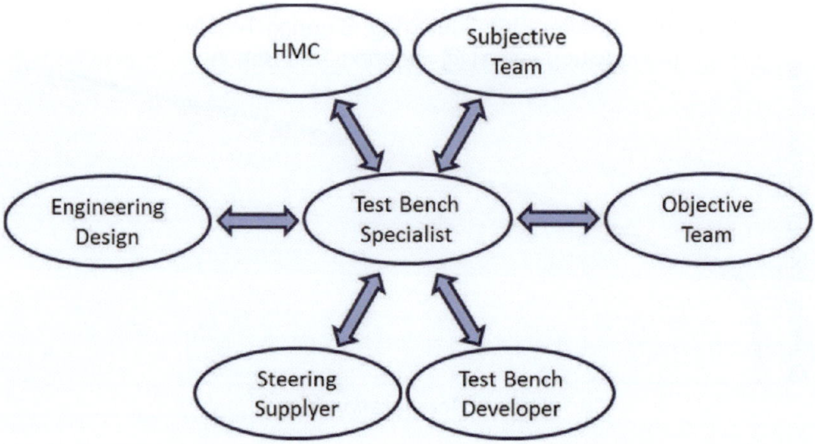

Figure 9: Overview of usual collaboration to set up a HiL-steering-tuning-process at HMETC

The communication with the subjective team guarantees that the HiL-tuning map comes up to the expectations and can later on be used for the subsequent fine tuning in the real car. Like described in

Figure 9 this communication loop is also leading to a second validation step. By seeing the influence of ECU parameters at objective measurements the fine tuning step can still be supported.

6 Media Feedback

The steering performance was a key point during development strongly supported by state-of-the-art HiL-testing and was also confirmed in initial drive tests with selected journalist during vehicle development showing a broad acceptance of the steering feel being 'natural with clear feedback' and having 'good precision' .

The first comparison tests after market launch in 2017 showed the competiveness of the new i30 with the best-in-class receiving positive feedback especially about its ride and handling character and finishing on top spots in various magazines:

- "Good steering tuning with good center feel offers a lot of emotion" [7]
- "Precise and well-balanced steering feedback" [8]
- "Good steering: harmonic effort build-up, smooth on-center, good response" [9]

Steering is matching European taste.

7 Outlook

HMETC is still working hard to further improve i30's performance for upcoming variants with different body styles and powertrain variants to also fulfil the high expectations concerning suspension performance of an even broader group of customers.

Additionally to i30, other Hyundai and Kia models' steering tuning process will be supported by HiL and DoE to deliver a matching steering feel over all segments.

References

1. ISO 13674-1 Road vehicles – Test method for the quantification of on-centre handling – Part 1: Weave Test
2. T. Schöning (HMETC GmbH): Improvement of Steering Feel – Virtual Approach with HiL, IPG apply & innovate 2014
3. A. Honisch, M. Lugert, T. Schöning (HMETC GmbH), S. Hakuli (IPG Automotive GmbH): Improvement of Steering Feel – Virtual Approach with HiL, ATZ 06, 2015, Volume 117
4. Specka, F.; Contini, A.: Bewertung und Optimierung des Lenkverhaltens im Gesamtfahrzeug am „Steering-in-the-Loop“ Prüfstand, VDI Wissensforum Reifen – Fahrwerk – Fahrbahn 2013
5. M. Lugert, T. Schöning, P. E. Pfeffer, A. Contini, H.-M. Koegeler, B. Schick: Model Based Steering ECU Calibration on a Steering-in-the-Loop Test Bench, Chassis tech. 2015
6. Nippold, C.; Kücükay, F; Henze, R.: Analysis and application of steering systems on a steering test bench, SAE Automotive and Engine Technology, 2016
7. Englert, P.; Kühler M.; Auto Zeitung comparison test; 08/2017
8. Renz, S.; Auto Motor Sport, 5/2017
9. Pfannkuchen, T.; Blick; 08/2017

Model-based online optimization of EPS controller using HiL test benches

Modellbasierte Online-Optimierung von EPS-Applikationen an HiL-Prüfständen

Christian Wagner, Maximilian Flormann, Thorsten Meister,

Dr.-Ing. Roman Henze, Prof. Dr.-Ing. Ferrit Küçükay

P.E. Pfeffer (Ed.), *8th International Munich Chassis Symposium 2017*, Proceedings, DOI 10.1007/978-3-658-18459-9_38

Abstract

Achieving the best driving-feel is the main purpose of the vehicle development process. Particularly active and passive chassis control systems are applied with elaborate real test runs in order to achieve this.

Electric Power Steerings (EPS) offer a variety of adjustments to achieve a brand-specific or type-specific driving- and steering-feel. The base for this is the subjective expert driver´s impressions. In the past, studies have shown that the method of objectification is an adequate approach in order to describe subjective impressions by using a combination of characteristic values. During the lifetime of a production vehicle, some constructive changes are made. These changes have an impact on characteristic values as well as on driving- and steering-feel.

In order to identify suitable characteristic values, the Hardware in the loop (HiL) test bench at the Institute of Automotive Engineering (IAE) from TU Braunschweig is used. This test bench is equipped with additional sensors to identify smallest differences between characteristic values caused by different software applications or hardware changes.

This work focuses on developing a method for model based online optimisation using HiL test benches to optimise the behaviour of EPSs caused by constructive changes in the lifetime of a vehicle. First, a sensitivity analysis is used to identify the effects of application parameters on the EPS-systems behaviour. Depending on the influence (linear or nonlinear relations between application parameters and characteristic values) it is possible to choose an optimisation algorithm for the next steps. Additionally, a loss function is needed to identify the best application setup. This is either based on characteristic values specified by the OEMs or characteristic values derived from the objectification.

The online optimisation analyses the results of measurements during a testplan. This leads to an adaptive testplan which improves the model in the requested area in regard to the characteristic values. First measurements show precise results for this method in case of optimising single application parameters towards a loss function. Furthermore the method will be expanded to optimise different application parameters to a loss function.

1 IfF Hardware in the Loop test bench

Main request for model based online optimisation are the precision of actuators and the resolution of defined state variables. To fulfil the requirements, the steering system test bench of the Institute of automotive engineering is equipped with additional sensors and precise actuators. These components allow, to measure minimal changes in the application.

Figure 1 shows the schematic setup and the signal flow of the steering test bench. In this setup the steering gear and the intermediate steering shaft are the hardware components to be tested.

First on the list are the actuators to simulate the required input and counterforce for the investigation. The servo motor brings the input for the single measurements, based on specified parameters and normalized driving manoeuvres. The second actuator is a linear motor to provide the rack forces via a modified track rod. This track rod is equipped with two track rod joints to compensate a potential coaxial behaviour of the rack and the linear motor's axle. The choice of a linear motor has been made deliberately to achieve the highest possible accuracy of the force. In this investigation the focus is on on-centre steering, so the maximum possible force of the linear motor (2500 N) is sufficient. In addition to this, the linear motor is equipped with a transducer measuring the current position of the stator. This measured displacement act as the Input for the vehicle model computing the required rack force in real time.

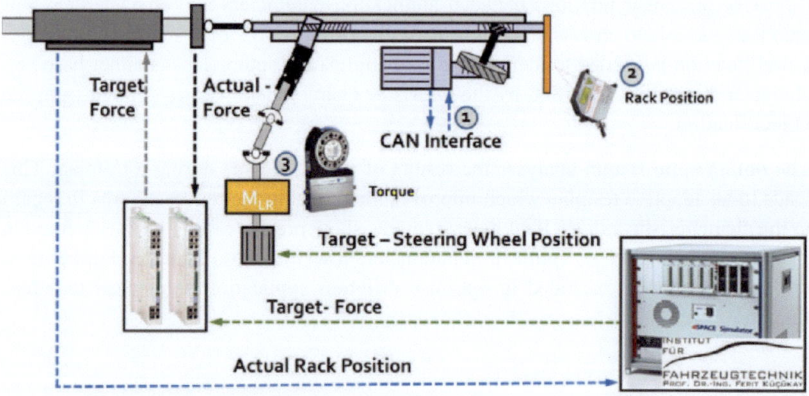

Figure 1: Schematic setup and signal flow of the steering test bench

Figure 1 shows the measuring point on the HiL steering test bench. With the CAN interface (1) it is possible to measure internal steering parameters. In particular, the focus is on the steering torque measured by the torque rod and the steering angle as input of the steering system. Furthermore, this Hardware in the Loop test bench is equipped with additional sensors. First to mention is the optoNCDT laser sensor (2) to quantify the rack position. This sensor has a measurement accuracy of 0,15 µm to determine smallest differences from different steering applications. The second additional instrument is a T12 Torque sensor (3) to determine smallest differences in the steering torque.

1.1 Vehicle model on the Hardware in the Loop test bench

One of the key parts of the HiL-test-bench is the vehicle model. For this purpose, a vehicle model developed by the IAE is used. The first steps to validate the model are real driving tests to record specific vehicle state variables used to describe the vehicle behaviour in the simulation. Afterwards it is possible to parameterize the model. Finally, the validate model is ready for usage at the Hil-test-bench.

The following Figure 2 shows the comparison between the real driving test and the simulation for a sinus manoeuvre. Cause of the small amount of deviations, this model is considered as valid.

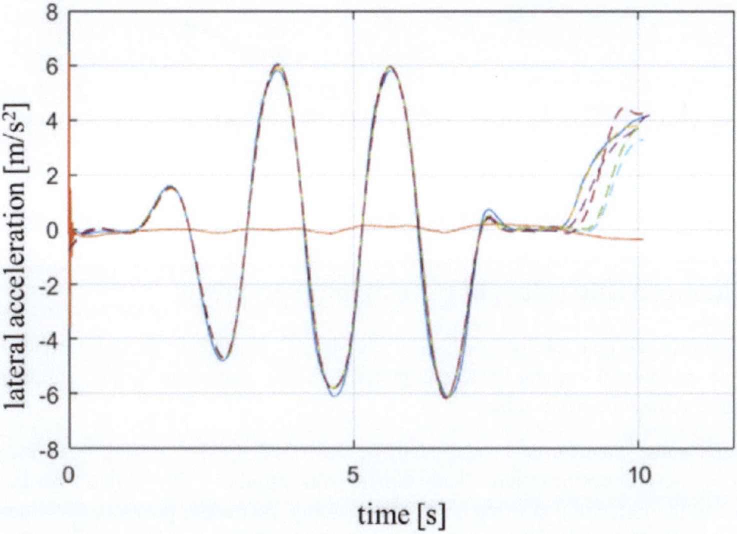

Figure 2: Simulation of a Sinus manoeuvre

1.2 Repeatability

This part shows the repeatability and resolution accuracy of the Hardware in the Loop test bench at the IAE. Here, a sinus weave test at 100 km/h simulated speed serves as an example.

The figures 3 a) and b) show the measurements mean normalized steering torque on the steering wheel angle. Each measurement consists of 7 periods and is repeated 4 times. The first and last period of each repetition is excluded. This results in 20 periods for the evaluation. In addition to the mean data the standard deviation with a confidence interval of 99% for all 20 periods is shown in figure 3 b). The largest amount of the deviation is as big as the measurement resolution of the described external torque sensor. Summarized the repeatability of the IAE's HiL test bench is very good with little to no variation within of the measurements.

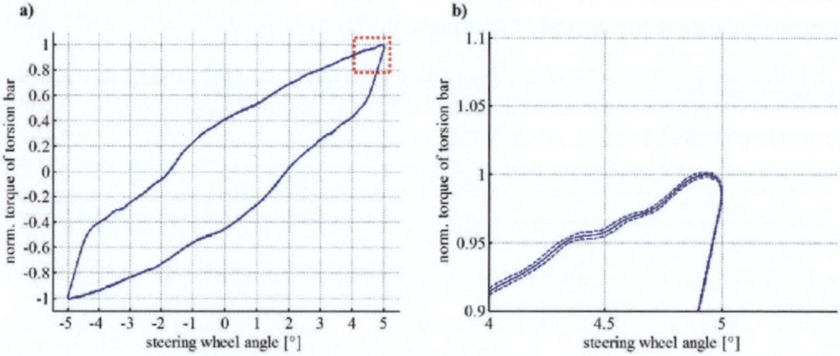

Figure 3: Investigation of measurement repeatability
a) steering hysteresis; b) detailed section of hysteresis

1.2 Characteristic values of steering measurements

Figure 4 shows the characteristic values of the steering hysteresis for the described weave test. In general, typical hysteresis parameters like maximum torque, abscissa and ordinate define the basic values.

Hereinafter, some characteristic values will be described for the steering hysteresis and later on used for optimization. Characteristic value number 1 (CV 1) describes the abscissa and the remaining steering angle after passing the middle position. CV 2 describes the ordinate and the required steering torque out of the on-center position. The next characteristic value (CV 3) designate the steering torque by the maximum steering angle during this manoeuvre. CV 4 indicates the specific area of the hysteresis.

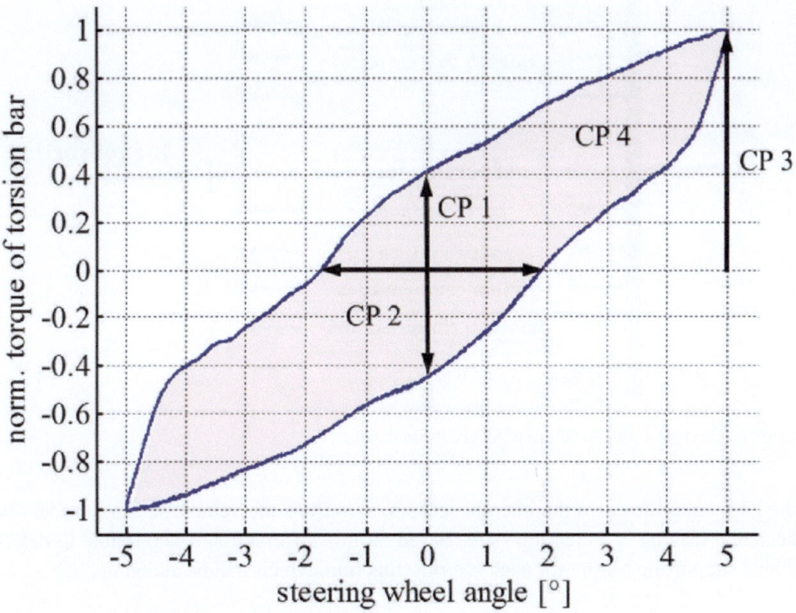

Figure 4: characteristic values of steering hysteresis

2 EPS Controller

Mainly the investigation at the steering bench focus on steering feel functions, as well as designing the current controller which has a significant impact on the steering feel. Electric Power Steerings (EPS) offer a variety of adjustments to achieve a brand-specific or type-specific driving- and steering-feel. The bases for this are subjective expert driver´s impressions. The EPS Controller enables a supporting torque which is regardless of driver. Based on a brand-specific steering feel it is possible to define the supporting steering torque with corresponding functions. However, the steering functions contain of a large amount of parameters which influence the steering feel. To develop the brand specific characteristic on a new steering system a lot of time is needed. Figure 5 shows a schematic overview of the steering functions of an EPS. This function can be divided in “advanced functions” and “steering feel function.

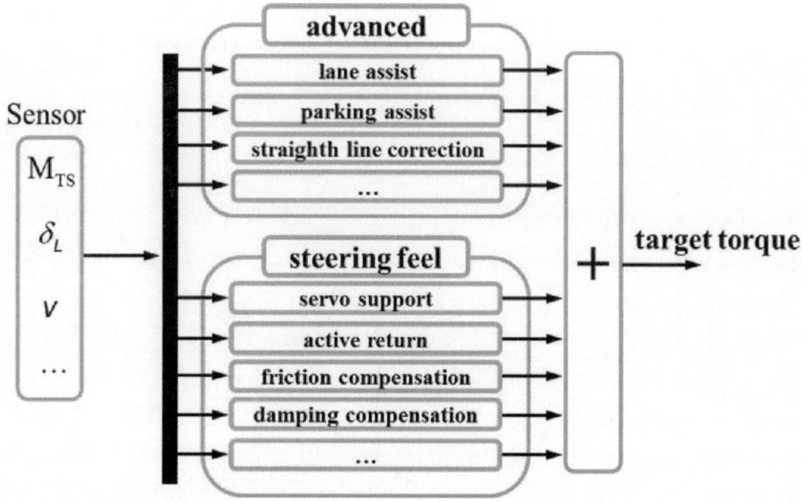

Figure 5: Overview and structure of steering functions

The EPS controller uses the internal sensors, as well as the vehicle sensors to identify the actual driving and steering state. Based on this information it is possible to determine a supporting torque for each steering function, which will be added up.

2.1 steering functions

This part introduces 2 steering functions with their operating principle and a sensitivity analysis.

Servo Support

Servo Support is described as the important basic function of an EPS Controller. Their purpose is to deploy a force to support the driver within his driving task. Based on the servo support parametrisation the balance of rack forces can be designed. This is expressed on a noticeable steering torque for the driver

Regarding to various literature, the servo support is an approximation of a quasi-steady-state balance of rack forces. This balance is described by the following equation:

RackForce = *isteering* * TorsionbarTorque+ *iservo* * ServoTorque

Parameter *isteering* depicts the transmission of steering torque to rack force. Also parameter *iservo* depicts the transmission of Servo torque to the rack force. In case of a quasi-static driving situation the rack force is in balance with the described forces above. Therefore, the steering torque has to be supported by the driver and the servo torque is the amount of torque/force summed out of the steering functions. Based on this equation there is a variety of combinations of servo and steering torque to supporting the lateral forces. After all, this combination influences the sensed torque by the driver and the related steering feel.

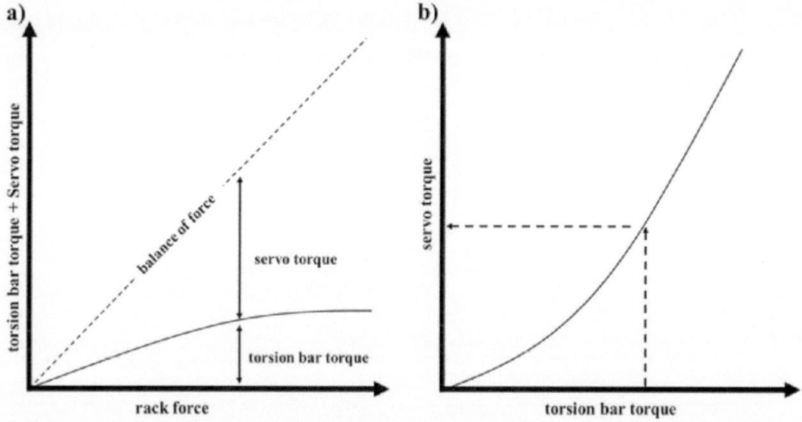

Figure 6: a) balance of force on steering rack; b) relation between servo and torsion bar torque

Furthermore, this function is depending on the actual vehicle speed. Compared to other driving situations, the rack forces are the highest at standstill and slow vehicle speeds. Therefore, a higher supporting torque can be added to enable a comfortable steering, having no differentiable response from the street. At higher speeds this function has to be applied to connect the lateral acceleration and forces to the actual steering torque.

Active Return

Today's axle constructions of modern vehicles have insufficient rewind behaviour. This issue is addressed by the active return function. This function supports the rewind behaviour in the straight driving. Therefore, the supporting torque depends on the steering angle, which in general leads the steering wheel to centre position. Therefore, this function has influence on the steering feel, because it has an effect on the force balance mentioned above.

2.2 Sensitivity of steering functions

First we will take a look at the sensitivity of the steering functions. Thus, the application parameters will be introduced and the effect on the characteristic values will be described.

The first function is the servo support. This application parameter resizes the supporting torque, i.e. the parameter compresses the characteristic curve of figure 6 b). As follows, the characteristic value will be defined as SERVO-GAIN. The default parameterisation is equal to the current steering version of the vehicle.

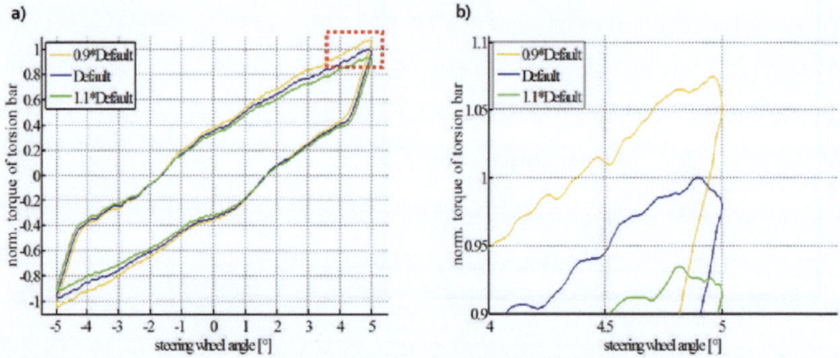

Figure 7: Investigation of servo support sensitivity on steering hysteresis
a) steering hysteresis; b) detailed section of hysteresis

The sensitivity analysis is performed for a SERVO-GAIN variation of +-10%. Afterwards it is possible to identify a linear or non-linear behaviour of the characteristic values based on the application variation. Figure 7 a) shows the hysteresis of the normalised steering torque on the steering angle for the default setting and the variation of +- 10%. Figure 7 b) shows a section of deviation points. It is obvious that the SERVO-GAIN has a great influence on the maximum steering torque.

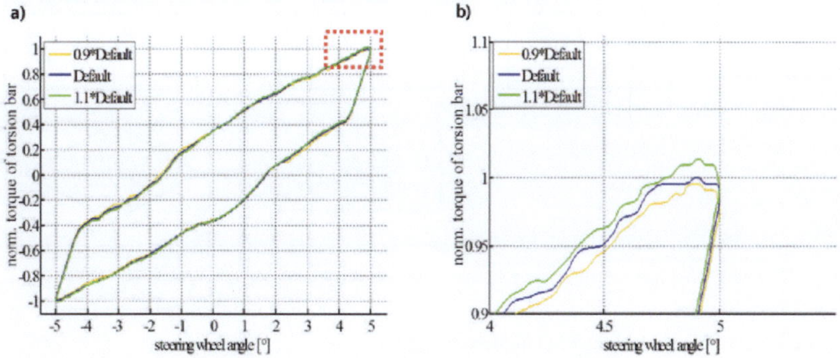

Figure 8: Investigation of active return sensitivity on steering hysteresis
a) steering hysteresis; b) detailed section of hysteresis

For the active return the procedure is identically. This function scales the support while decreasing the steering angle. The application parameter defined as AR-GAIN in Figure 8 a) and b) shows the hysteresis for the default setting and the variation with +- 10% to the default setting.

Online Optimisation

Fitting the behaviour of a specific steering system to a reference behaviour by using an automated test bench changing the steering systems application parameters, requires an algorithm that systematically computes the next parameter set to be tested. The approach utilised in this paper uses an optimisation algorithm implemented on the HiL test bench to do this with regard to a so called loss function. This function characterises whether a tested application parameter set is an improvement or not.

In order to find the optimum of this given loss function, various algorithms are compared. The algorithms are tested regarding their robustness and rate of convergence. As a result of this evaluation, the *Particle Swarm Optimisation* provides very robust results and a very high rate of convergence. Therefore, the PSO algorithm is implemented for the online optimisation on the test bench.

The *Particle Swarm Optimisation* Algorithm as introduced by (Eberhart & Kennedy, 1995) is a methaheuristic swarm-based algorithm mimicking the movement patterns of a swarm of birds. The Algorithm can be divided into two main stages.

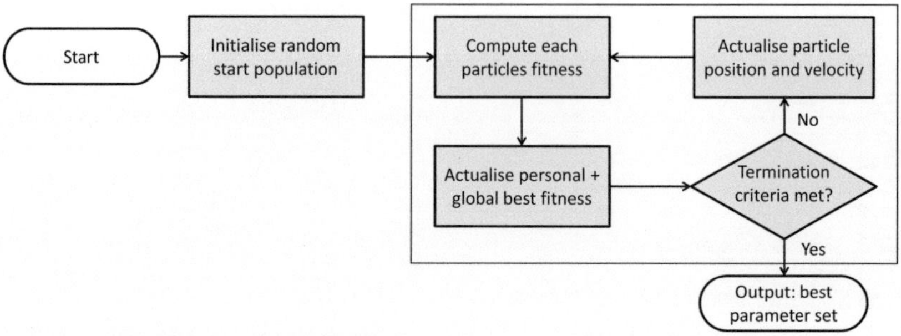

Figure 9: Flow Diagram of the PSO implementation

In the first stage, a finite set of particles are initialised randomly. These particles have a position $\overline{x_i} = (x^1, \dots, x^n)$ and a velocity $\overline{v_i} = (v^1, \dots, v^n)$ in the $\overline{x_i}$ domain. The fitness-value of the particles is their function value $f(\overline{x}_i)$

The second stage is iterated until the stopping criterion is met. Here, the fitness value of the particles is updated. Each particles personal best position/fitness combination as well as the global best are stored. Hereafter, the stopping criterion is checked. As a last step of this iteration, the particles' velocities are updated according to the following formula.

$$v_i^d(t+1) = \chi \cdot v_i^d(t) + \varphi_1 \cdot \hat{r}_i^d \cdot \left(p_{Best}{}_i^d(t) - x_i^d(t)\right) + \varphi_2 \cdot \tilde{r}_i^d \cdot \left(g_{Best}{}^d(t) - x_i^d(t)\right)$$

As a last step within the iteration loop, the particles' positions are shifted according to their new speed $v_i^d(t+1)$.

$$x_i^d(t+1) = x_i^d + v_i^d(t+1)$$

The inertia weightχ, according to (Clark & Kennedy, 2002), has been set to

$$\chi = 2/\left|2 - \varphi - \sqrt{\varphi^2 - 4\varphi}\right|$$

With $\varphi = \varphi_1 + \varphi_2$ and $\varphi_1 = 2.8$ and $\varphi_2 = 1.3$.

In the case of this test bench based optimization of a control parameter set, a population of 20 particles has been proven to be sufficient.

The implementation of the particle swarm optimisation algorithm shown in this paper uses the steering systems application parameters described in chapter 2 as the positioning space of the particles. Thus, the particles are moved within the application parameters' domain within specific boundaries, defined by the sensitivity study described in chapter **2.2**. The optimization algorithm's loss function, defining the parameters fitness, consists of the characteristic values introduced on chapter **1.2**. The purpose of the online optimisation is to minimise the difference between the reference steering system's characteristic values and the characteristic values of the steering system with implemented changes.

Results of the Online Optimisation

This part shows results of the online optimization in order to confirm this method. Figure 10 shows the default and optimised hysteresis. The optimised hysteresis is a result of 20 randomly initialised particles and 20 populations. In comparison to the default hysteresis, the optimised hysteresis has less mismatches. The used characteristic values achieve the demanded value.

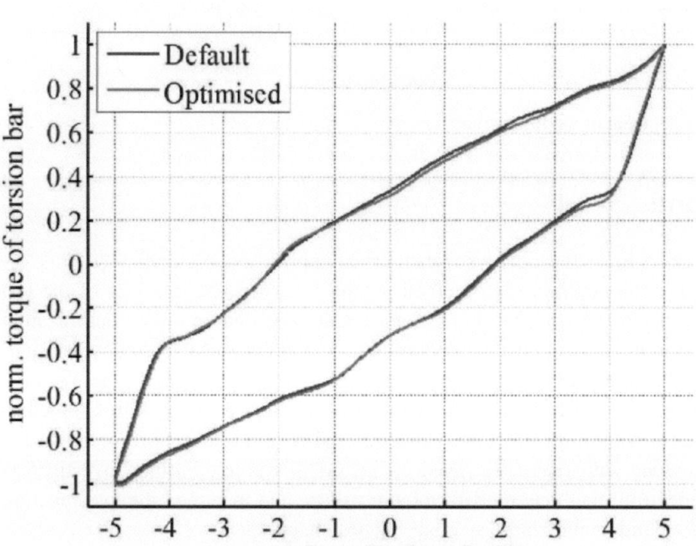

Figure 10: Validation measurements

The loss function for this optimisation consist of all 4 characteristic values CV1, CV2, CV3 and CV4. Figure 11 shows the result of the loss function for each particle in each population. This convergence yields the completion of an optimised state.

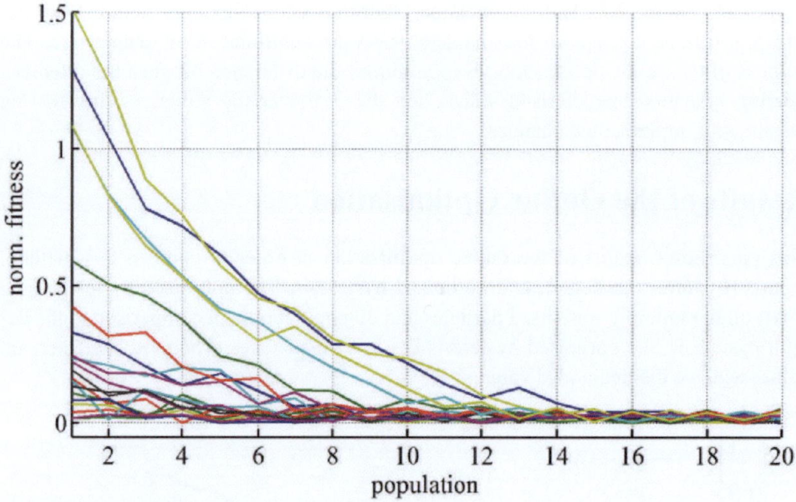

Figure 11: Development of the fitness function

Conclusion

The present paper focuses on the test bench based application of steering systems, especially on an online optimisation method. In the first step, the qualities of the test bench are evaluated. This forms the basis for a test bench based online optimization for steering system applications. To determine a loss function for optimisation algorithms, some characteristic values are introduced. Finally, the comparison between the default and optimised state shows less mismatches. These are based on missing characteristic values in the loss function. However, the method enables to optimise the steering behaviour with regard to given characteristic values. Further investigation will lead to identify brand based characteristic values, and to extend the loss function with new values and include more steering function parameters.

Literature

Eberhart, R. & Kennedy, J. (1995): "A *new optimizer using particle swarm theory"* Micro Machine and Human Science, 1995. Proceedings of the Sixth International Symposium on, Pages 39-43.

Clerc, M. & Kennedy, J. (2002): "The Particle Swarm – Explosion, Stability, and Convergence in a Multidimensional Complex Space" Trans. Evol. Comp 6.1, Pages 58-73.

Bröcker, M.: New control algorithms for steering feel improvements of an electric powered steering system with belt drive. Veh Syst Dyn Int J Veh Mech Mobil 44(Suppl 1), 759–769 (2006)

Brand, W., Brenner, P.: *Functional development under conditions similar to series production to optimize the steering feel*: 21st Aachen Colloquium Automobile and Engine Technology, pp 649–660 (2012)

Karch, G., Grüner, S.*: Mechatronische Lenksysteme (Mechatronic Steering Systems)*: Band 55, Heft 6: pp 281–289: at— Automatisierungstechnik (2009)

Peter Pfeffer, Manfred Harrer Hrsg.: „Lenkungshandbuch – Lenksysteme, Lenkgefühl, Fahrdynamik von Kraftfahrzeugen" 2. Auflage, Springer Vieweg

Evaluation of the robustness of an EPS control system in an early stage of the product development life cycle

Paul Milbaier, Stefan Grüner, Markus Heger, Alexander Gaedke

Robert Bosch Automotive Steering GmbH

P.E. Pfeffer (Ed.), *8th International Munich Chassis Symposium 2017*,
Proceedings, DOI 10.1007/978-3-658-18459-9_39

Introduction

On one hand, today's Electric Power Steering (EPS) steering systems have demanding steering feel requirements [1] that in many cases requires the EPS controller to be tuned at stability limits. On the other hand, EPS systems, especially those with axially parallel steering system (EPSapa) have a very complex design with divergent requirements regarding system stability and system durability. Furthermore, these systems have a relative wide range of bearing clearances that result in a wide range of servo drive stiffness tolerances.

In order to achieve these requirements a mechanical system needs to be designed that allows for a highly dynamic control performance with suitable disturbance compensation. In addition, this design should take into account varying boundary conditions such as different vehicles with different wheels, road surfaces or loads. Especially the unknown dynamic behaviour of the vehicle axle makes it difficult to forecast the final steering feel in the vehicle since the axle dynamics have a strong influence on the steering system dynamics (see Figure 1).

In contrast, the time to test and validate the applied steering feel in the vehicle is decreasing due to the availability of prototypes [2] and decreasing developing time.

For these reasons, an initial design of the EPS system needs to be found with an adequate robustness level. This allows a successful application of a steering feel in the presence of these varying boundary conditions and eliminates the risk of expensive design changes in a later phase.

This article will present a method that evaluates the robustness of EPS systems, especially EPSapa systems, by calculating boundaries for the dynamic front axle behaviour and gives an insight what design parameter can be changed in order to improve this robustness. This method expands upon other common stability analysis methods since it provides helpful results without requiring the knowledge of the final steering controller tuning.

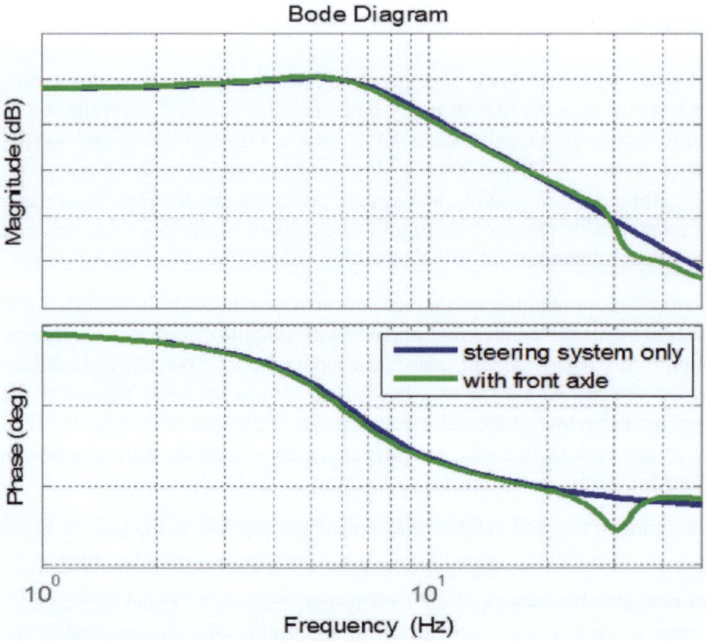

Figure 1: Open Loop of the steering system transfer function (motor torque→torsion bar torque) to illustrate the influence of the front axle dynamics [4].

Analysis method

The control loop for the steering system (Figure 2) consists of three parts, the controller, the EPS-System and the effective dynamics of the front axle that are considered as a multiplicative disturbance.

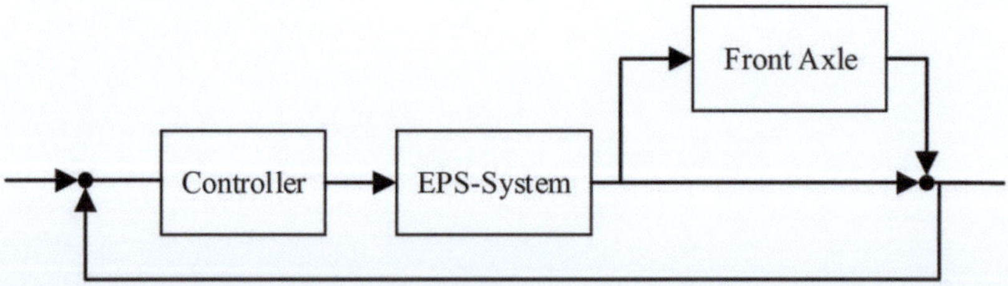

Figure 2: Control loop for the steering system [4].

Representation of the EPS-System (EPSapa)

The mechanical EPS part is represented by a linear mass-spring-damper model (Figure 3). The model represents the mechanical EPSapa system with a fixed steering wheel. This mass-spring-damper model can be parameterized with only few parameters, which are taken from the design data or from suitable measurement data. Nevertheless, there are some simplifications, such as the neglected friction and lumped stiffness, damping and masses.

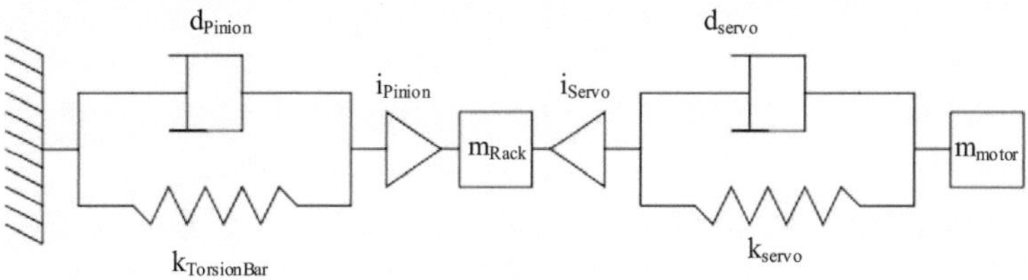

Figure 3: Linear mass-spring-damper model of an EPSapa system [4].

Representation of the Steering Feel Controller Dynamics

The steering feel controller is represented by the absolute value of the complementary sensitivity of the closed loop as given in Figure 2 without the steering axle dynamics. This complementary sensitivity was calculated in SIL simulations over various projects. Enveloping these complementary sensitivities, least upper bounds for maximal, average and basic control dynamics was identified (Figure 4).

High controller performance represents a steering controller, which allows steering feel tuning without any compromises. This includes, among others, the potential to realize a fully disturbance free steering feel.

An average controller performance represents a steering controller, which allows steering feel tuning with a limited tuning range. Hence, the potential to realize disturbance free steering feel is reduced.

Finally, the basic controller performance represents a steering controller, which allows a steering feel tuning with a strongly limited tuning range. This also affects the potential to realize a disturbance free steering feel to a large extend.

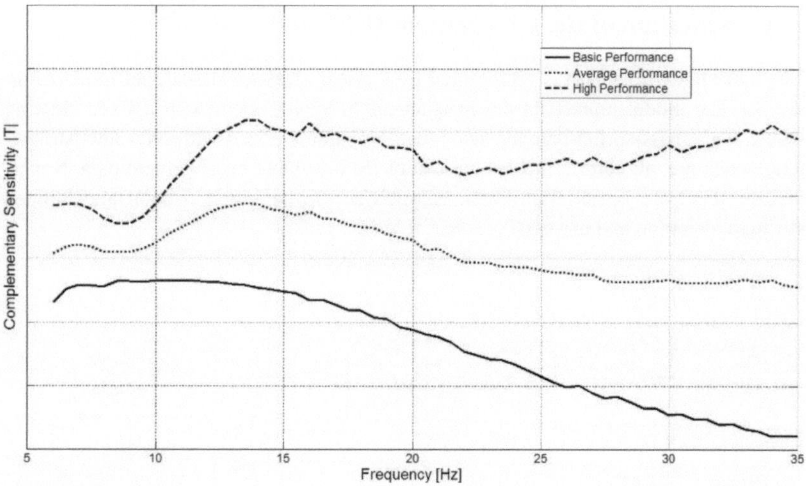

Figure 4: Resulting least upper bounds of the complementary sensitivities for various EPSapa projects [4].

Representation of the Front Axle Dynamics

The front axle dynamics describe the response of the rack force to a rack displacement. These dynamics can be obtained using a simulation or by a vehicle test. For a simulation-based approach, a multi body model of a chassis can be used. In this simulation model, the excitation of the rack displacement can e.g. be done by applying a motor torque on the servo unit. To get this transfer function, the signals of the rack position and the rack force are required (Figure 5).

If a vehicle is already available, the measurement of the axle dynamics can be done in the vehicle. Therefore, the vehicle is placed on different surfaces and is excited by a piece-wise constant sine with increasing frequencies. This allows an accurate system identification with a good coherence.

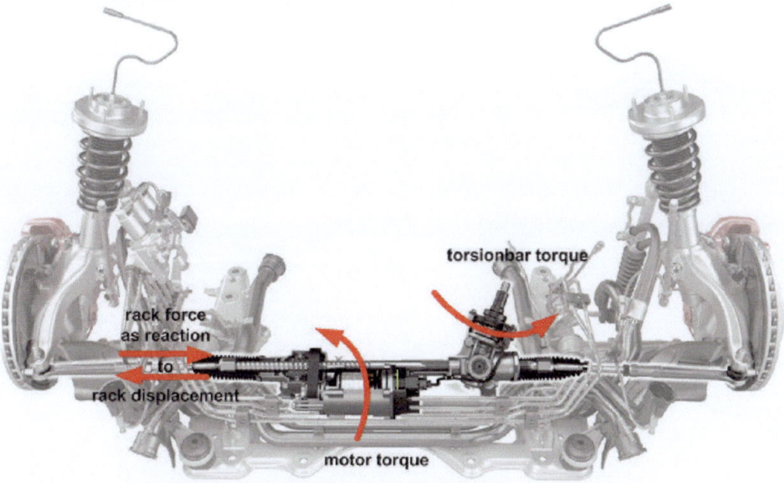

Figure 5: Illustration of the relevant measurement values [3].

To ensure an adequate measurement accuracy of the rack position, an acceleration sensor is used since common rack position sensors do not have the needed bandwidth, especially in higher frequencies. The transfer function (frequency response of the force to a rack displacement) can then be built by integrating the acceleration signal in frequency domain.

A typical run of the axle dynamics on different surfaces with varying friction values and varying loads, is illustrated in Figure 6.

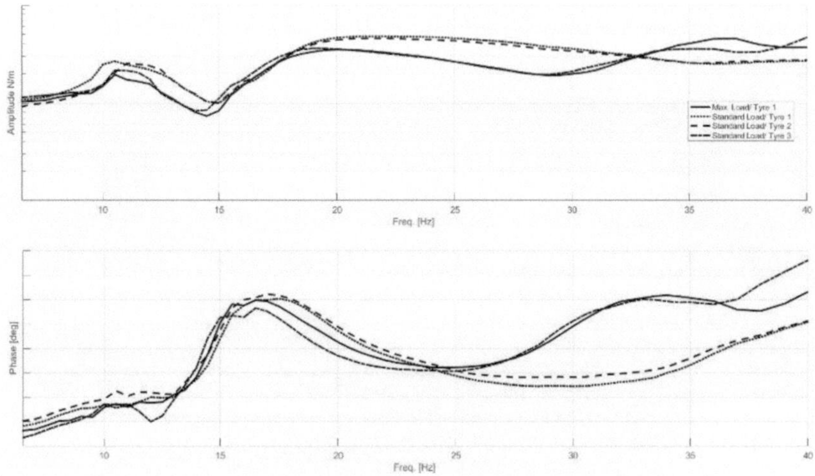

Figure 6: Example of the axle dynamics with different loads and different tyres [4].

Robustness Criteria

The robustness criteria is based on the assumption of a multiplicative perturbation of the steering system by the axle dynamics. If certain limits of the investigated controlled system are assumed (i.e. complementary sensitivity), one can determine the area of those axle frequency responses which result in a non-critical system (Figure 7). Following this, the current front axle dynamics, that are measured in vehicle or simulated with CAE techniques, are compared to this robustness boundary. Through the comparison, it can be concluded if the controlled EPS system considering the vehicle front axle, has sufficient robustness reserve to be tuned stable with an appropriate steering feel controller. The criteria itself is defined as the relative distance to the admissible axle area.

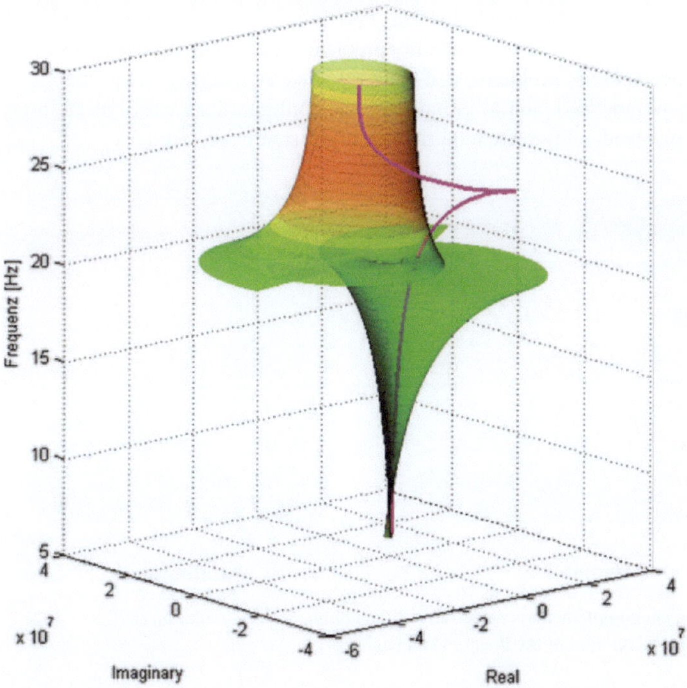

Figure 7: The area within the tube is the rage of the admissible axle dynamics. The current axis (pink line) violates this boundary for the frequencies of 20-23 Hz. Instabilities are expected in this frequency range [4].

Validation of the Method

The validation of this method was twofold. First, a comparison of the linear model to an extended nonlinear SIL-Modell was done. An equivalent for the complementary sensitivity was calculated using the nonlinear SIL-model at specific operating points and then compared to the one of the linear model. The nonlinear simulation and the calculation with the linear model showed comparable results (Figure 8). Second, a questionnaire was completed by the in-house vehicle test engineers to evaluate the experience from past projects. The questionnaire contained questions related to the kind of instabilities, at which frequencies they occurred and what kind of steps were needed to resolve these instabilities. Furthermore, the engineers divided the project or the steering feel application into three performance groups. In total, we were able to evaluate 27 configurations.

In another step, we divided the configurations into three groups. Group one represents all the configurations which have no stability problems at all, in group two all the configurations which stability problems, that were resolved by modification of the application or by new functions, and in group three the configurations where instabilities could only be resolved with severe modification of the steering system.

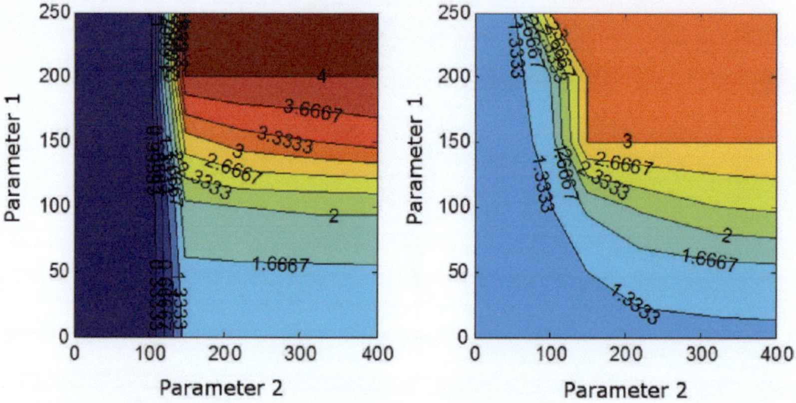

Figure 8: Equivalent complementary sensitivity of the nonlinear SIL simulation (left) in comparison to the calculation of the linear model (right) [4].

To compare this evaluation with the calculation result, the robustness value (relative distance to the admissible axle tube) for each configuration was calculated.

The comparison is illustrated in Figure 9. In this illustration, the evaluation of the questionnaire is mapped to the horizontal axis and the calculation result on the vertical axis. It shows a high correlation of the in vehicle observation and the calculation result. Moreover, there is no configuration, which was considered challenging in vehicle and had an uncritical robustness value and vice versa.

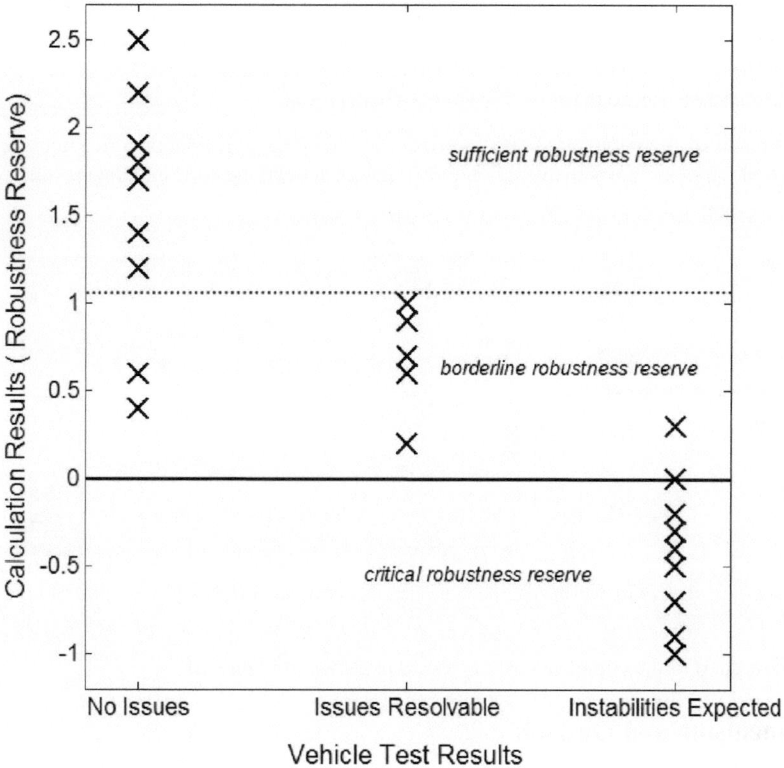

Figure 9: Comparison of calculation results and the in vehicle observations. This plot shows that the results of the vehicle test and the calculation result has a very good correlation [4].

Integration into development process

The method is implemented in the standard EPS motor performance design tool ELENA. This in-house tool is used in the early development process to design an EPS system that meets all OEM performance requirements. In this design process, the tool already uses most of the parameters, which are needed to parameterize the linear model described above. Thus, the implementation focuses on the visualization of results and a functionality, which allows the user to identify the sensitive parameter that improve the overall system robustness. The tool itself is organized in tree GUI tabs, the input tab, the parameter study tab and the results tab. The input tab contains the selection of some additional parameters, which are not part of the EPS system design, and the selection of the front axle and the controller dynamics. After the calculation, which is finished after just a couple seconds, the user is able to see the robustness results in a simple representation in the result tab. If the result is not satisfying, a parameter study can be done to identify the parameters needed to be modified to get a better result.

The result of the parameter study is printed in a contour plot as shown in Figure 8. Each of the plots shows the variation of two parameter and the corresponding result.

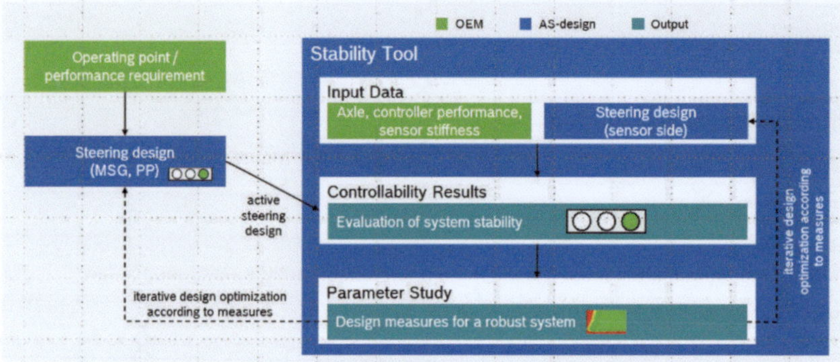

Figure 10: A flow chart, which describes how the robustness tool is used [4].

A flow chart of the complete design process in shown in Figure 10.

Conclusion and Outlook

This tool is implemented into the Bosch AS product development process and is utilized in early project phases. Nevertheless, it is not intended nor possible to replace some of the stability calculation processes like the SIL stability process or the stability tests in the vehicle itself. To a greater degree, this method helps Bosch AS to avoid potential setbacks and delays in started projects. So, if in a project some challenges or stability risks have been identified at the start of the project, it is supported with detailed stability simulation using common CEA methods. Currently, this method is used only for EPSapa systems. An adaptation to EPSdp (Dual Pinion Steering System) is currently in development and will be released by the end of 2017.

Furthermore, this method enables the cooperation between the OEM's chassis development team and the steering system development team at Bosch AS. In this way, it aids in the development of complementary products, improving product quality on both sides by detecting potential design risks at an early phase. This has the added benefit of accelerating developing cycles and reducing development costs.

Literature

[1] Klomp M.; Ljungberg M.; Attinger M., Hoesli S., Kratzer T. (2017): Virtual Verification of Passenger Vehicle Steering Systems. In Chassis.tech plus 2017

[2] Grüner S.; Werner T.; Käpernick B. (2017): Objectification of Steering Feel and Application in the Context of Virtual Steering Feel Tuning. In Chassis.tech plus 2017

[3] Grüner S.; Gaedke A.; Hsu, H.; Harrer, M. (2012): The New EPSapa in the Porsche 911- Innovative Control Concept for a Sports Car Typical Steering Feel In Chassis.tech plus 2012

[4] Robert Bosch Automotive Steering GmbH

BRAKE.TECH SECTION

HIGH-PERFORMANCE BRAKES AND TEST METHODS

Regenerative braking during high-performance driving in super sports vehicles

Jorge A. Jáuregui,

ESP-Application Expert for Super sport vehicles
Engineering customer projects ESP, Bosch Engineering GmbH

P.E. Pfeffer (Ed.), *8th International Munich Chassis Symposium 2017*,
Proceedings, DOI 10.1007/978-3-658-18459-9_40

Introduction

In current times, where the emission regulations push car manufacturers to their limits of design, the electric motors have shown to be a very effective help to decrease emissions without compromising driveability. Such vehicles driven by the combination of an internal combustion engine and an electric motor are called Hybrid.

However, not all of the Hybrid cars are meant to consume less fuel and be silenced by the usage of the electric motors; the super-sport vehicle manufacturers see another beneficial aspect of the electric motors, the great amount of instant torque that can be used to boost the acceleration of the car.

One of the biggest challenges for hybrid super sport vehicle's design is to find an optimal battery size, which allows boosting during many laps (to have a constant performance) but without adding too much weight in order to keep high lateral and longitudinal dynamics.

In order to be able to reduce the battery size, it is very important to be able to recharge the battery during driving as often as possible and this is done through regenerative braking. This regenerative braking should then be done in every braking manoeuvre and the amount of braking energy that has to be done with the electric motor should be as big as possible.

One of Bosch Engineering's customers is one of the most respected sport cars manufacturer in the world and they aimed to build a hyper car, which should show Formula 1 technology and still be a legal road car. This car would be a Hybrid car with a very powerful internal combustion Engine and electric boost, both delivering their power only to the rear axle. Their specification was very strict regarding how many laps the car should be able to boost before the battery was empty and which safety standards if should fulfil. This lead to a very interesting balancing act between regenerating as much energy as possible with the electric motor but not too much to avoid instable situations.

This was the challenge that Bosch Engineering GmbH had to solve for this project; we had to adapt our ESP® hev-II in order to reach the extreme performance standards of our customer…daily business!

1 The challenge

1.1 Customer Requirements

The target of our customer for this project was to be able to drive a defined number of Laps (10 Laps) on a defined reference track in race driving style (at the limit) with full Hybrid boost potential. This would mean that the powertrain performance has to remain constant during these 10 laps. Of course, tire and brakes temperatures (and other temperature depending factors) influence the overall performance of the vehicle over this amount of laps.

1.1.1 The technical challenge

The target of keeping boosting performance constant on a great amount of laps and the aim for a small battery (due to weight reduction), leads to a conflict of interests regarding the energy balance. The car needs a lot of energy during a long period of time, but the energy storage is very small. The only solution for this conflict is to recharge the battery as much as possible (and as often as possible) in order to delay the emptying of it.

This solution leads again to a conflict with the original ESP® hev-II strategy. This strategy avoids (as explained below in this paper) regenerative braking in extreme situations in order to avoid instability and safety issues. All of this leads to the technical challenge; the software of the ESP had to be changed in order to allow regenerative braking also while driven at the limit (entering EBD or even ABS control in many situations) without compromising the stability of the car.

1.2 Reference Track

The racetrack defined as reference track for this project is a private track, which is about 13 km long and has 12 curves with different radiuses. This track is mainly used to test Formula one cars and super-sport street vehicles.

1.3 Vehicle Characteristics

The vehicle is a Hyper car showing Formula 1 technology transfer (Hybrid similar like KERS) that can perform nearly like a Formula 1 on the race track, but also be comfortable and ecological on the streets. In order to achieve this, the car has not only a big internal combustion engine, but also an electric motor, which boosts when the driver requests it to reach excellent lap times.

The Powertrain of this vehicle (Fig. 1) contains a high-Rev V12 Engine with about 800HP, a dual clutch transmission an Electronic differential (it is only RWD) and an electric motor with more than 150HP. Due to the fact, that the electric motor is directly coupled to the rear axle, there is also a second electric motor used as a generator to charge the high voltage Battery in standstill situations. During driving, the big electric motor connected to the rear axle is used for boosting during acceleration and as a high voltage generator during braking.

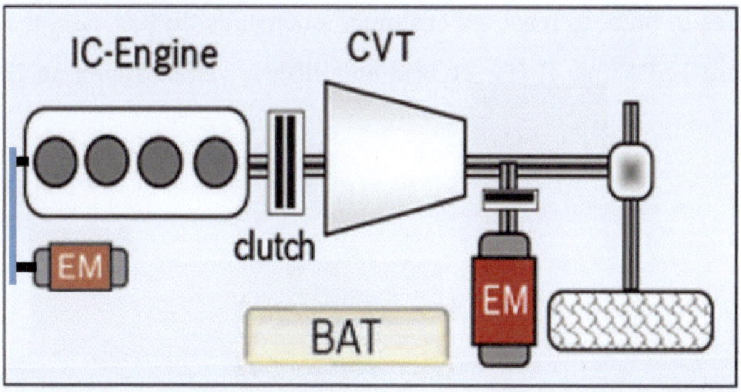

Fig.1 Powertrain Configuration

The Vehicle is a high performance vehicle with a weight Distribution from over 60% on the rear axle, an extremely low centre of gravity, High performance tires, ceramic brakes and aerodynamic load for higher speeds. It has a very neutral behaviour specially during braking, which leads to a reduced available braking force on the rear axle in order to maintain stability.

Due to the fact, that this vehicle has to be a State of the art exhibitor and a brand statement of a sport cars manufacturer, it has to be the fastest production car of all times around the defined track. This leads to a very tight compromise for heavy components like the high voltage Battery and hybrid components. This is why, the HV-Battery has a reduced size in order to reduce the overall mass of the vehicle; and this was very important to be able to keep a very high cornering speed due to nonlinear behaviour of the tires.

1.4 Used ESP® System

1.4.1 Hardware

The Development done by Bosch Engineering GmbH bases always on available Bosch products and adds features in order to reach all customer's demands. In this case, the base product is a standard ESP® hev-II (for Hybrid and electric vehicles with an II configuration) as shown in Fig.2.

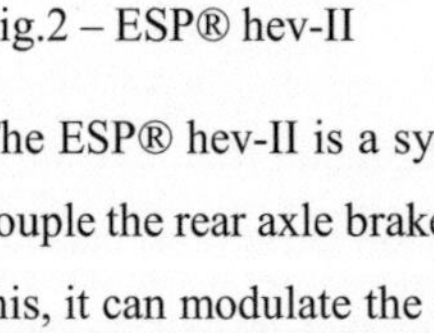

Fig.2 – ESP® hev-II

The ESP® hev-II is a system which is designed in such a way, which allows it to decouple the rear axle brakes from the brake pedal every time the driver brakes. By doing this, it can modulate the fluid pressure going to the rear brakes with the pump without the driver feeling anything on the pedal (Fig.3). This independent control of the brakes allows the free distribution of braking force between the electric motor and the hydraulic brakes.

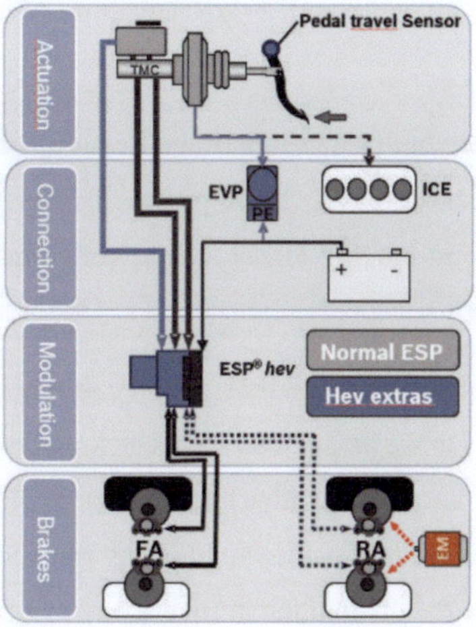

Fig.3 Simplified Schema of the ESP® hev-II System

1.4.2 Software

The original Software of the ESP® hev-II has all the features of the normal ESP® and in addition includes a very complex CRBS functionality (Cooperative Regenerative Braking Systems). This feature covers solutions where the variations of the additional regenerative brake torque done by the electric motor are compensated by a hydraulic pressure modulation (Fig. 4)

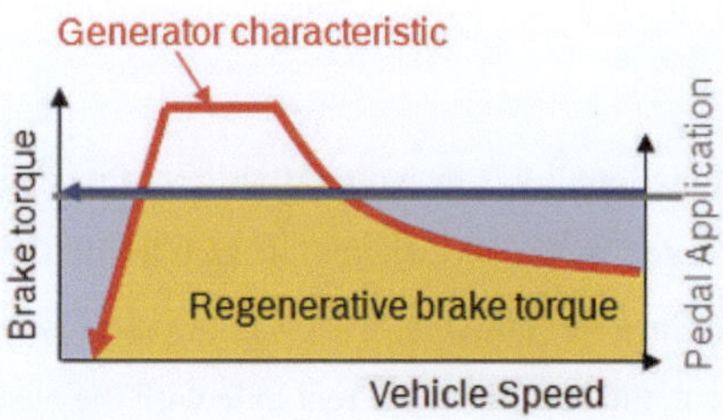

Fig. 4 CRBS Principle: intelligent system coordinates both brake actuators /brake torque sources

Depending on the tuned parameters, the software builds a Driver Brake Request from the pressure of the master cylinder and the push rod travel. These parameters can be tuned to reach different braking feeling (different braking forces for the same brake pedal inputs) depending on the customer's wish.

This Driver Brake Request is then used to calculate how much braking force should go to the rear axle (the front axle is directly coupled to the brake pedal). This braking force on the rear axle is then distributed by the CRBS into Regenerative torque of the electric motor and braking fluid pressure on the rear brakes (which is build up by the ESP-Hydraulic). Usually, the CRBS tries to optimize the distribution in such a way that the electric motor can regenerate as much energy as possible in this situation and the rest is done with the brakes.

Another feature of the CRBS is the Stability margin. This function calculates how close a potential instability might be during a braking manoeuvre or how dynamic the braking manoeuvre is. It considers many factors, the most relevant are lateral and longitudinal acceleration and the tire slip (how close the actual tire slip is to the ABS slip) in all four tires. Once these indexes for each wheel are calculated, the wheel that is the closest to ABS is taken into account to make the system more conservative.

With this calculated Stability Margin, the function shuts down energy regeneration if the car is close to the limit.

The main Purpose of the Stability margin functionality is to avoid instability of the rear axle due to a too high braking torque generated by the electric motor. Due to the fact, that the Electric motor brakes the complete rear axle (both rear wheels in the same way) and the ABS regulation is wheel independent, the braking of the rear axle with the electric motor in an ABS situation might lead to an unwanted situation. These unwanted situations are mainly two:

- The electric motor is reducing the braking torque to reduce the braking slip on one wheel and thus reducing the braking torque on both sides. If in this case, the opposite wheel was already in underslip and gets in addition a torque reduction, the whole axle will be underbraked. This leads to an underbraked vehicle, which is of course during an ABS braking situation something highly dangerous.
- The electric motor increases braking torque to increase the braking slip of one wheel and due to this, it increases the torque also on the opposite wheel (which was already on a high slip situation). This leads to an instable rear axle and thus to an instable vehicle. If this happens on a curve, this is also a very dangerous situation.

Due to these possible situations, the Stability Margin function forbids regenerative braking if the vehicle is close to the limit. Until the beginning of this project, all Hybrid cars where common street vehicles which do not reach the vehicle dynamics limit very often. This is why, even if the SoC of the battery is low, one can supress the energy regeneration during braking in these situations without having a real impact on the Range of the vehicle.

2 The Analysis

2.1 Energy balance in one Lap

The first analysis that was made was an energy balance during one Lap on the track in order to know how much braking energy is available at all (braking work) and how much energy (again braking work) is available at the rear axle. All this calculation was made in Simulation based on real measurements driven on the same track with a similar vehicle (See Fig.5). This had to be done this way because, during this concept phase, there was no representative prototype available. By simulating different measurements of different drivers in the same track (with the same vehicle) the conclusion was, that the braking energy on the rear axle was very dependent on the driving style and this had to be taken into account.

The approximate Braking energy on the rear axle for one Lap varies from about 1.2kWh to 0.9 kWh depending on which driver was behind the steering.

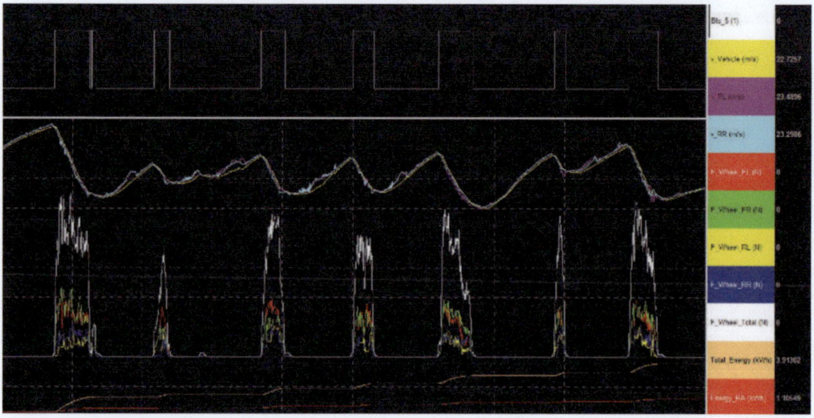

Fig.5 – Basic calculation of total and rear axle braking energy through Simulation

The customer delivered all the boundaries regarding energy delivery and energy regeneration related to electric motor and battery because at this stage of the project, only a theoretical model of the car was available.

Based on the simulated data and the supplied boundaries, the definition of how much energy could be theoretically regenerated on each lap was calculated assuming a battery with 0% SoC. This assumption had to be made just to have the maximum charging current.

2.2 Solution Approaches

Once the energy balance was calculated and the electric boundaries were defined, there were three different concepts that had to be discussed in order to reach the defined amount of energy that the customer needed in order to reach the 10 Laps of full performance:

- •Concept 1 (Race track application): The most simple of the three concepts is basically the tuning of the complete Stability Margin parameters in order to allow regenerative braking in a wider range of situations. In order to optimize this concept, there is also a minor software modification, which ignores the front wheel slip for the building of the stability margin. This would then allow that the front wheels go into ABS without shutting down the regenerative braking.
- •Concept 2 (Regeneration during EBD intervention): this concept adds to the first concept, another software modification that allows regeneration during an EBD (electronic brake distribution) intervention. This allowed torque is limited to a maximum tuneable value and is a select low of the braking torques of both rear wheels. This would allow regeneration even by entering the corner while braking. The problem with this concept is that in most of the simulated Laps, during entering the corner in a very dynamic way, one of the rear wheels goes shortly into ABS and this would still shut down the regenerative braking.
- •Concept 3 (Regeneration during ABS): This last concept adds to the previous ones, the feature that it allows regenerative braking during full ABS control on the rear axle. This is achieved through a major software modification, which makes a select low between both braking torques of the rear axle. In order to allow ABS to control the wheel slip independently with enough precision, this logic also calculates a safety margin between the regulation torque of the ABS on the wheels and the torque done by the electric motor. This torque has to be still controlled by the ESP Hydraulic.

In order to understand the impact of all three concepts on regenerative energy during an aggressive braking into the corner, there was a Simulation of such a manoeuvre with all three concepts. It is possible to see in Figures 6, 7 and 8, the amount of regenerative torque allowed by each of the Concepts in one aggressive manoeuvre.

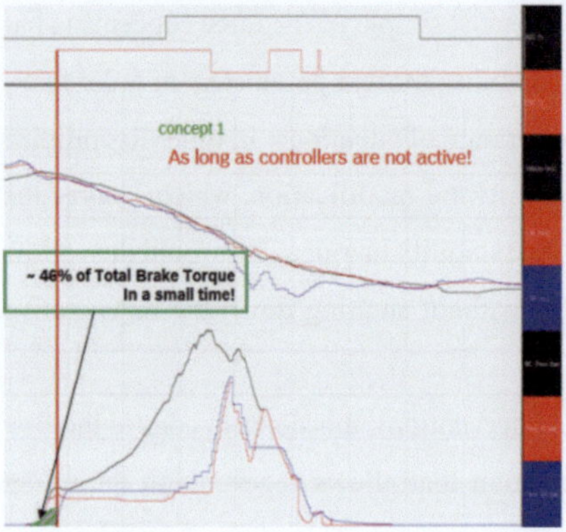

Fig. 6 – Braking Forces on the rear axle and allowed regenerative braking force by Concept 1.

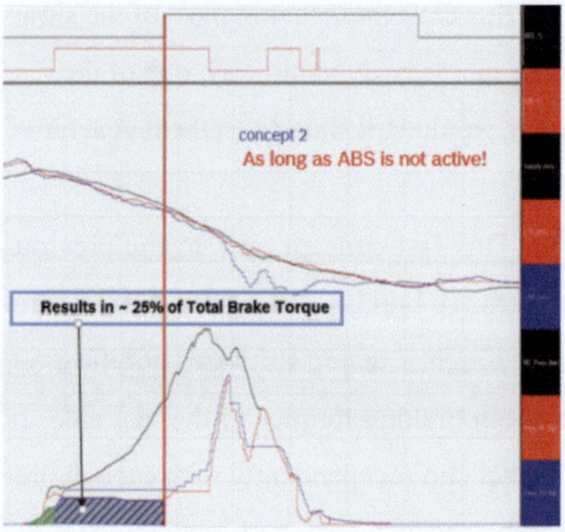

Fig. 7 – Braking Forces on the rear axle and allowed regenerative braking force by Concept 2.

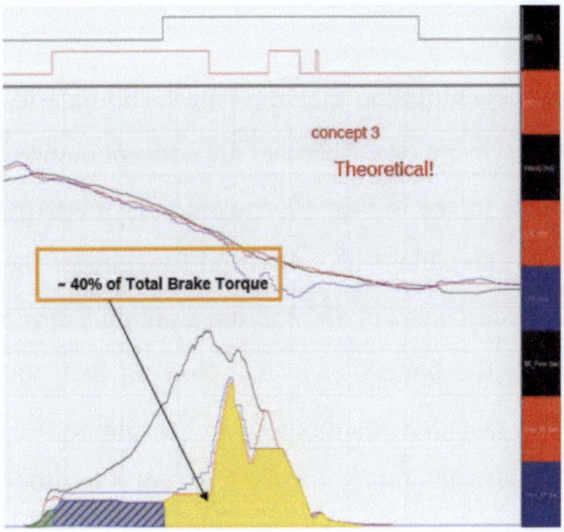

Fig. 8 – Braking Forces on the rear axle and allowed regenerative braking force by Concept 3. Notice that the safety margin torque is still not included; it is only the select low of both wheels (theoretical).

The simplified Graph on Figure number 9 explains the safety margin in terms of brake fluid pressure.

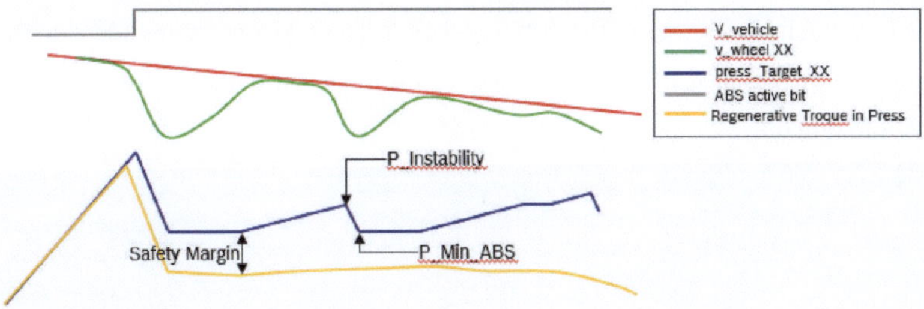

Fig. 9 – Safety Margin to accurately control ABS

3 The Solution

Once the three concepts where analyzed in the simulation including all the boundaries of the electric motor and the different driver styles, it was clear that the concept number three was the path to follow. As it is possible to see in Fig. 10, the amount of energy that the rear axle is braking during a whole Lap cannot be fully used to recharge the battery due to the boundaries given by the electric system 8Maximum Braking Torque of the electric motor, maximum current that the battery can take, etc.). In fact, the amount of energy that these boundaries allow is approximately 30% of the total available energy on the rear axle. This leads to a clear advantage for the Concept 3 in order to be able to reach the targets of energy recovery.

After developing the Software of concept three and integrating it into the ESP® Software, the application proccedure started with real Hybrid prototypes. The advantages of this concept where confirmed very fast, already after a short periode of parameter tuning, the target of the 10 Laps on the defined track was acchieved. Stability was also guaranteed in all situations withuout compromizing the regerative braking. The only siatuation in which the regenerative braking had to be still shut down during ABS intervention was during driving on snow and ice. This was done this way due to the extreme low braking pressures of the ABS regulation. These low pressures added to the stability margin resulted in such small torques for the electric motor, that it was not worth it to turn on.

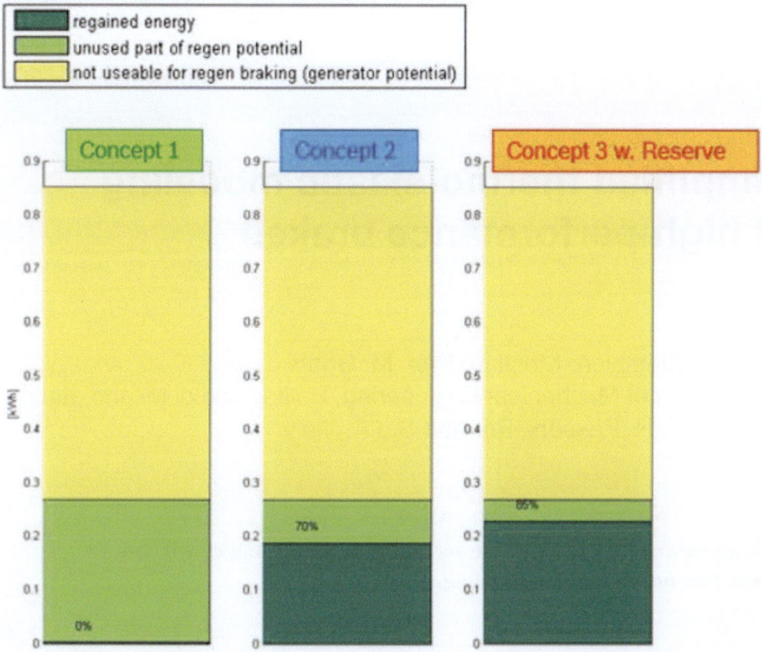

Fig. 10 – Different Concepts of regenerating and the impact on the total ammount of regenerated energy after one Lap.

Simplified thermo-elastic modeling of high-performance brakes

Prof. Dr. Giampiero Mastinu, Prof. M. Gobbi,
Department of Mechanical Engineering, Politecnico di Milano, Italy;
C. Cantoni, R. Passoni, Brembo S.p.A., Italy

**This manuscript is not available according to publishing restriction.
Thank you for your understanding.**

P.E. Pfeffer (Ed.), *8th International Munich Chassis Symposium 2017*,
Proceedings, DOI 10.1007/978-3-658-18459-9_41

Future braking tests for the Periodical Technical Inspection (PTI) – Is the braking effect test still up-to-date?

Dipl.-Ing. (FH) Rafael Frankenstein

P.E. Pfeffer (Ed.), *8th International Munich Chassis Symposium 2017*,
Proceedings, DOI 10.1007/978-3-658-18459-9_42

As a result of the continuous development and integration of new, innovative system components in automotive manufacturing, the requirements for periodic technical vehicle inspection have also changed significantly in recent decades. There have been significant developments in the testing of brake systems with regards to the new requirements that have come about as a result of the wide range of safety-related electronic vehicle systems.

1 The history of the development of brakes and brake testing

At the time of its first public outing in 1886, the Patent Motor Car No. 1 was braked using a band brake made of leather. With a top speed of just 16 km/h (10 mph), the brakes of a vehicle were nowhere near as important then as they are today. In the following years, vehicle manufacturers predominantly focused on improving the efficiency and power of the engines and powertrain systems. The first significant development in the field of brake technology, and therefore in vehicle deceleration, came at the beginning of the 20th century, with the introduction of the disc brake and the drum brake. Up until around 1909, the function, structure and mode of operation of vehicles and their components was dictated by the manufacturers. In 1909, the Motor Vehicle Act (Gesetz über den Verkehr mit Kraftfahrzeugen) came into force. This act introduced regulations relating to the vehicle itself, as well as the handling and driving of motor vehicles. From a historical perspective, this legislation formed the basis for the current German Road Traffic Act (StVO), and therefore, also the German Road Traffic Licensing Act (StVZO). In the following years, a department for testing vehicles and their drivers was created at the "Gesellschaft zur Überwachung und Versicherung von Dampfkesseln" (association for inspecting and insuring steam boilers [a predecessor organisation of TÜV SÜD]), in order to ensure compliance with the legal regulations. The appraisers working in this department were trained by Benz & Co. The following years saw the introduction of new technologies, such as the four-wheel brake with hydraulic actuation, which was the most important development during this period, and was a direct consequence of the increasing engine power in vehicles. Even then, it was realized that the assemblies and components of vehicles were subject to a certain amount of wear during their service life and that the correct functioning of the vehicles would not always be guaranteed. 1938 saw the introduction of the German Road Traffic Act, which was developed from the "Reichs-Straßen-Verordnung" of 1934, which already contained §29 for the inspection of braking systems. The introduction of the new Road Traffic Act meant that the licensing authority was now permitted to summon vehicles for an inspection. Due to the intended use of vehicles at that time, this inspection was mainly focused on assessing the vehicles suitability for use in war. After the end of the Second World War, it was another six years before the periodical technical inspection became mandatory for all road users in 1951. During this time, the effectiveness of the braking system was tested using a method known as the "Sandstraßentest" (dirt road test). From 1965 onwards, it was possible to perform the test using a roller brake tester or a plate brake tester. Alternatively, a deceleration measurement could be performed using a corresponding testing device. The scope of the braking system was extended as a result of the introduction of the Anti-lock Braking System (ABS) in 1978 and the Electronic Stability Program (ESP) in 1995. The following years saw the introduction of a wide range of assistance systems. This resulted in the use of technologies such as Adaptive Cruise Control (ACC) and Automatic Emergency Braking (AEB) in serial production. Since these systems are heavily dependent on the effectiveness of the braking system, testing of these system components has become increasingly important. Since 2015, the PTI Adapter has been used throughout Germany as part of the periodical technical inspection (PTI). By using the PTI Adapter, the braking effect can be verified very effectively and efficiently using reference braking forces. This method enables a direct link between the braking actuation and the actual braking force acting upon the wheel. [1–5]

2 FSD – Central Agency for PTI

Fahrzeugsystemdaten GmbH was founded by the German inspection institutes in 2004 as a non-profit-organisation. The company is involved in the creation of testing procedures, specifications and innovative testing technology for the periodic technical inspection. In this context, the PTI Adapter was developed in cooperation with the inspection institutes, testing device manufacturers and universities.

Picture 1 PTI Adapter

For the inspection of the components and the system, the PTI Adapter communicates with the control units in the vehicle via the OBD interface. The PTI Adapter also features an acceleration and gyroscopic sensor. This creates a wide range of applications for testing procedures.

3 Reference braking force test

Up until April 2015, the braking effect was verified on the brake tester by reaching the locking limit. This procedure involved determining the deceleration using the maximum braking force achieved and the maximum authorised total mass for the vehicle (see formula (1)).

$$z = \frac{F_B}{zGM} \tag{1}$$

However, the vehicles being inspected mostly had a vehicle weight that was close to the unladen weight. This meant that a great deal of additional effort was required to load the vehicle up to the maximum authorised total mass. Other weaknesses of this procedure are the failure to include the friction coefficient between the tester rollers and the tires of the vehicle and the failure to assess the distribution of the braking force between the individual axles.

In accordance with ECE R13 H, a testing procedure with a reference value is preferable as proof of effectiveness. As part of the scope of its duties, the FSD – Central Agency for PTI has developed an algorithm that performs a reference braking force test with the help of the PTI Adapter and the user application. The braking force determined on the brake tester is evaluated using a reference value. The reference value is read out via the OBD interface. The FSD – Central Agency for PTI provides the inspection institutes with the setpoint values required for the reference value.

The diagram in Figure shows the minimum requirements for the reference braking force test. This is calculated from the legal minimum deceleration based on the maximum authorised total mass. For the assessment of the total braking forces of the individual axles, "Total braking force x braking force proportion FA" from the front axle (FA) and "Total braking force x braking force proportion RA" from the rear axle (RA), the minimum deceleration of each axle is determined using the braking force distribution. Here, the actual calculated values for the front axle and the rear axle must be on the corresponding boundary line at a minimum.

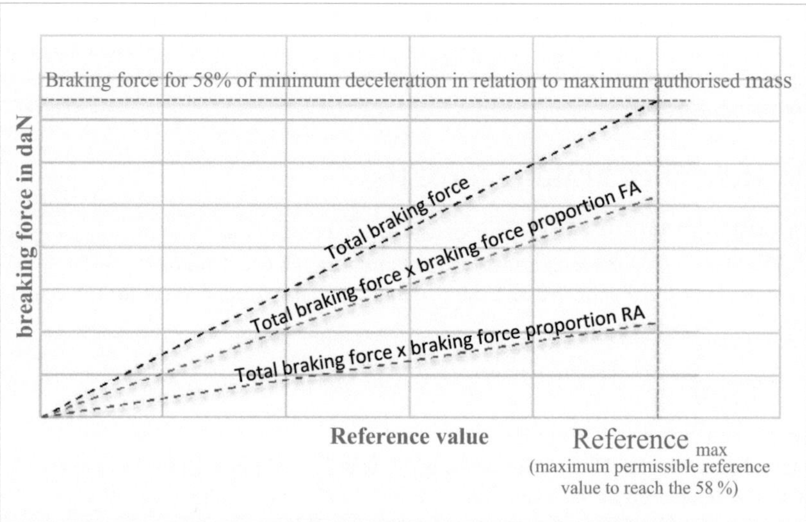

Figure 2 Diagram showing the minimum requirements based on the reference braking force

This enables a direct assessment of the friction between the brake disc and the brake pad. By using the almost linear relationship between the braking actuation and the actual braking force acting upon the wheel as a basis, the officially recognised appraiser or test engineer (ORA/TE) can control this up to any point below the locking limit.

$$z = \frac{F_B}{F_N} \tag{2}$$

In order to pass the test, the actual determined braking force must be higher than the minimum values displayed in the characteristic diagram. If this is the case, then the minimum deceleration is considered to have been verified. In addition to the minimum deceleration, it is possible to determine and assess the distribution of the braking forces between the individual vehicle axles by using the braking actuation as a reference. Therefore, there are three criteria that must be fulfilled to pass a reference braking force test:

- Minimum deceleration with reference value
- Braking force distribution between the vehicle axles
- Difference in braking forces on the left-hand side and right-hand side of an axle

4 Deceleration measurement

Another way in which the braking effect can be assessed is by performing a deceleration measurement using the PTI Adapter. Here, with the help of the internal measuring technology of the PTI Adapter, the acceleration curve during the braking process is drawn, evaluated and transferred to the control unit. For the assessment of the braking effect, the mean fully developed deceleration is calculated in accordance with ISO/TR-13487.

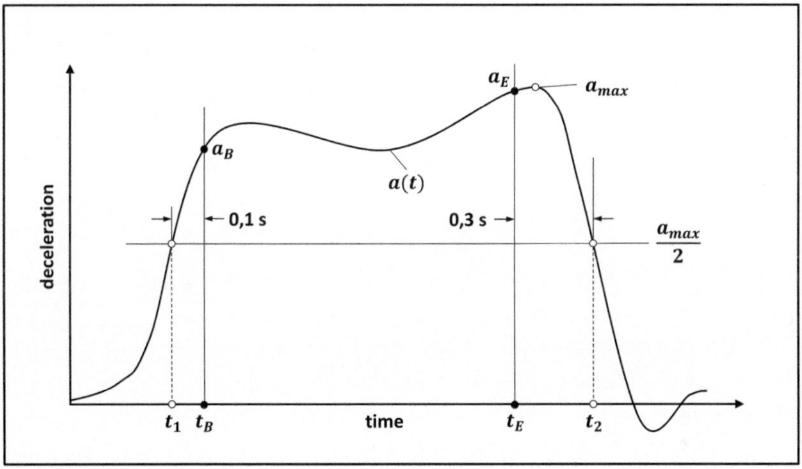

Figure 4 Typical deceleration curve showing the lower and upper limits in accordance with ISO/TR-13487

As with the measurement performed using the brake tester, the deceleration value calculated using this method must relate to the maximum authorised mass of the vehicle. As was the case with the old brake testing procedure, this method has problems with regards to differences between the actual testing mass and the maximum authorised total mass of the vehicle used in the calculation of the minimum deceleration. Furthermore, this method cannot be used to evaluate the distribution of the braking force. Friction between the tyres and the road surface also play a crucial role in the feasibility and attainability of the minimum deceleration. Therefore, the formula for the calculation of minimum deceleration needs to be simplified:

$$z = \frac{a_x}{g} \tag{3}$$

In order to make this method more efficient and effective, investigations are taking place at the FSD – Central Agency for PTI, which are looking at ways to include a reference value in the deceleration measurement method, and thereby, enable the determination of a left/right brake imbalance, as well as the evaluation of braking forces on individual axles during the braking process. Here, the movement of the vehicle body throughout the braking process is determined using the acceleration rate and gyroscopic sensor technology integrated in the PTI Adapter.

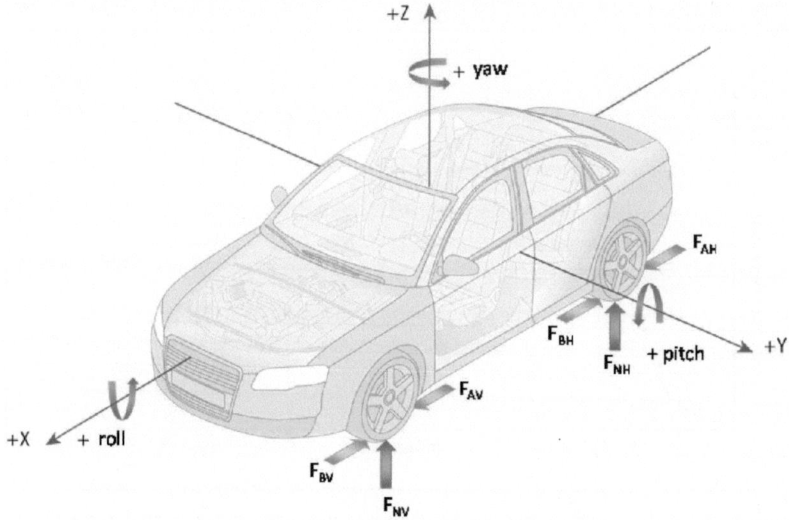

Picture 2 Vehicle coordinate system in accordance with DIN ISO 8855

The yaw rates around the X and Z axles of the vehicle (shown in Picture 2) are used to assess whether there is a left/right brake imbalance. Here, it should be determined whether the vehicle – in relation to the yaw response – moves away from the balance of forces on the individual axles as a result of the limited braking effect of an individual degraded wheel brake. The roll direction – in relation to a degraded braking effect – is determined based on the position of the defect. Therefore, in the event of degradation on the front axle, the vehicle rolls towards the wheel with the impaired braking effect, and in the event of degradation on the rear axle, the vehicle rolls away from the wheel with the impaired braking effect.

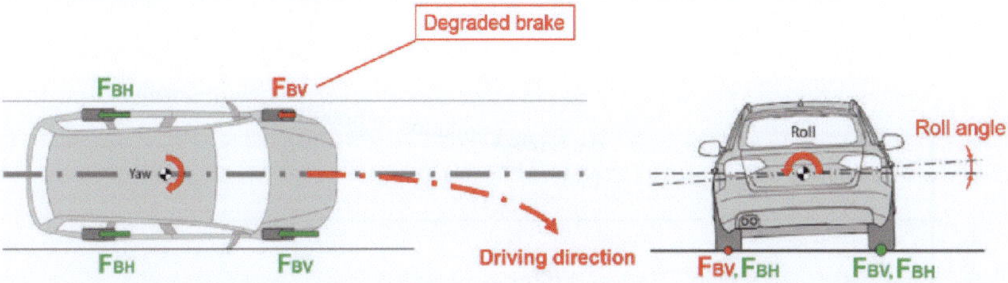

Picture 3 Example of single wheel degradation

Picture 3 illustrates the behaviour of the vehicle body in relation to an front left (FL) single wheel degradation of the braking effect. There are significant combinations of yaw rates around the X and Z axes of the vehicle for degradations of the other three single wheel brakes (see Table 1).

Table 1 Explanation of negative/positive signs for impaired single wheel braking effects

X axis	Z axis	Impaired wheel braking effect
Negative	Negative	front left (FL)
Positive	Negative	rear left (RL)
Negative	Positive	rear right (RR)
Positive	Positive	front right (FR)

In order to assess the distribution of braking force between the individual vehicle axles, the third characteristic yaw rate of the vehicle body movement is analysed. The rotation around the Y axis of the vehicle, also known as the pitch, is mainly dependent upon the axle braking forces during this test procedure. Since each vehicle has a factory-installed braking force distribution, this is used as a specification, in the same way that the specifications are used for the roller brake tester and plate break tester. By comparing this characteristic pitching behaviour with the braking actuation, you can see a linear relationship, which can be used as a reference.

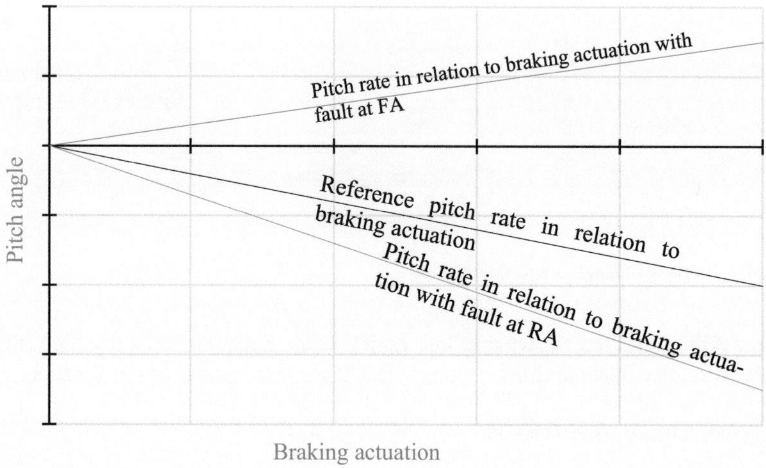

Figure 6 Characteristic pitch rate curves

The lines shown in Figure show the general behaviour of a "Reference pitch rate in relation to braking actuation" that results from the factory-installed braking force distribution. Both the "Pitch rate in relation to braking actuation with fault at FA" and "Pitch rate in relation to braking actuation with fault at RA" lines, show the extreme cases at 0% braking effect for the relevant axle. This shows a more negative pitch rate under the same level of brake actuation at 0% braking effect on the rear axle ("Pitch rate in relation to braking actuation with fault at RA"). This is opposite to the pitch behaviour in the event of a total failure on the FA ("Pitch rate in relation to braking actuation with fault at FA"); here, the failure of the braking effect causes the pitch rate to change from negative to positive, similar to the effect caused by actuating a conventional parking brake, which only acts on the rear axle.

Investigations were also performed for cases involving only slight impairment of the wheel braking effect, as is frequently the case in real life.

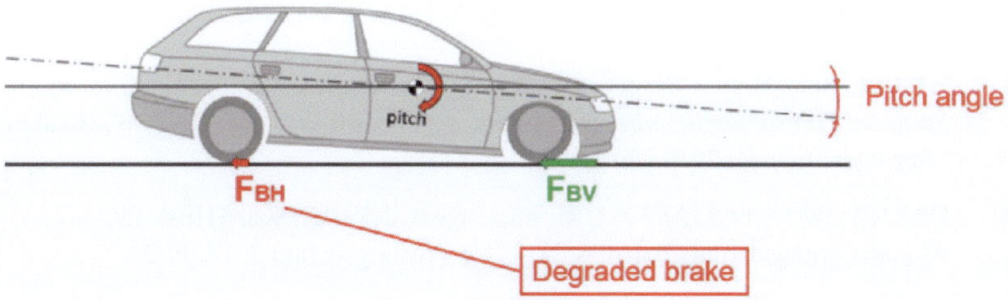

Picture 4 Pitching characteristics with an impaired RA braking effect

The example shown in Picture 4 should clearly demonstrate that a vehicle with an impaired braking effect on the rear axle dips much more than usual, and therefore causes different pitching behaviour compared to braking with a fully functional braking system. The case of a degraded front axle was also looked at. In this case, the dipping of the front axle reduces, creating a deviation in the opposite direction to that shown in the example.

5 Summary

The reference braking force test is a development of the braking effect test that is required due to the constantly increasing demands that are placed on the braking systems of modern vehicles. These demands are a result of the increasing number of assistance systems used today and the development of autonomous driving functions. In addition to developments in vehicle technology, testing technologies based on the real-life requirements and load cases of these systems must also be developed.

The approach towards further developing the deceleration measurement process represents another significant step in this direction. It should be noted that this method has only been performed on a few test vehicles and still needs to be assessed in terms of practical suitability.

List of references

[1] BREUER, Bert (Hrsg.): *Bremsenhandbuch : Grundlagen, Komponenten, Systeme, Fahrdynamik ; mit 53 Tabellen*. 4., überarb. und erw. Aufl. Wiesbaden : Springer Vieweg, 2013 (Praxis)

[2] DEKRA: *90 Years SAFETY.* URL http://www.dekra.de/Share/Historie/. – Aktualisierungsdatum: 2016-06-28 – Überprüfungsdatum 2017-04-21

[3] BRAUN, Heribert ; KRAUTSCHEID, Rainer: *Die Hauptuntersuchung : § 29, die dazugehördenden nationalen und internationalen Vorschriften sowie Richtlinien der StVZO und angrenzenden Vorschriften : ein Leitfaden für amtlich anerkannte Sachverständige / Prüfingenieure, Werkstätten, Fahrzeugverkäufer, Verwaltungsvbehörden, Polizeibeamte, Fahrlehrer, Fuhrparkleiter und alle Interessierten an Techn.-rechtlichen Fragen aus dem Bereich der regelmäßigen technischen Überwachung der Farzeuge und angrenzender Rechtsgebiete*. 22. Auflage, Stand: November 2014. München : Vogel, 2014

[4] *Unternehmensgeschichte | Bosch in Deutschland*. URL http://m.bosch.de/de/de/our_company_1/history_1/history.html. – Aktualisierungsdatum: 2017-04-21 – Überprüfungsdatum 2017-04-21

[5] *Benz Patent-Motorwagen.* URL https://www.mercedes-benz.com/de/mercedes-benz/classic/museum/benz-patent-motorwagen/ – Überprüfungsdatum 2017-04-21

BRAKE WEAR PARTICLE EMISSIONS AND MATERIALS

Development of a commonized methodology for measuring brake wear particles – current status within the PMP IWG

Theodoros Grigoratos, G. Martini,
Joint Research Centre (JRC), European Commission (EC), Italy

This manuscript is not available according to publishing restriction. Thank you for your understanding.

P.E. Pfeffer (Ed.), *8th International Munich Chassis Symposium 2017*,
Proceedings, DOI 10.1007/978-3-658-18459-9_43

Investigation possibilities of brake particle emissions on a brake dynamometer

Dr.-Ing. S. Gramstat, R. Waninger – AUDI AG, Ingolstadt

Dr.rer.nat. D. Lugovyy, M. Schröder – Horiba Europe GmbH

P.E. Pfeffer (Ed.), *8th International Munich Chassis Symposium 2017*,
Proceedings, DOI 10.1007/978-3-658-18459-9_44

1 Introduction

Since particle emissions from the internal combustion engine have been reduced during the last decades impressively, non-exhaust emission sources such as the brake wear particles and tyre wear particles are gaining more interest by environmental organizations and the EU commission. [1]

In general, brake particles are oriented on exhaust emissions in terms of the risk of ultra-fine particles. This means, that diameters smaller than 2,5 µm are respirable and therefore critical. The lower limit of particle sizes is investigated actually by many researchers and might be in the range of 5 to 10 nm. [2]

Unlike the powertrain, a conventional vehicle friction brake is more or less an open system and the sampling of the emitted particles is a very complex process, which is investigated by many researchers. [3]..[5]

Up to now no common or standardized sampling method does exist. Particle collection directly around the brake leeds in most of the cases to significant variations of the measurement results. Hence this possibility is not considered to be productive and therefore it is an objective to develop a robust and reliable measurement and sampling method.

Additionally a representative test cycle has to be established, considering also the influence of the ventilation system in terms of intake air quality and volumetric flow rate. Another objective focuses on the investigation of the influence of sub 23 nm particles. These particles are considered to be relevant since they are respirable and represent an increased potential risk for the human health. Additionally it is known that demanding brake applications shift particle diameters towards smaller values and more particles in the sub 23 nm range are emitted.

The presented results are aimed contribute to a better understanding of brake particle emission measurement and provide an essential base for a standardized test procedure in future.

2 Experimental Setup

A brake dynamometer, used to evaluate the performance of brake systems and test the brake parts under various environments (vehicle weight, velocity, brake force, wet-dry conditions), is the experimental base of the investigations. The Horiba GIANT 5200 has a customized design and operates below atmospheric pressure, in order to perform brake tests under specific ambient conditions. The dyno offers an inertia simulation range from 10 kgm² to 200 kgm² with a max. speed of 2500 rpm at a max. power output of 190 kW.

One of the key points of BPE investigations is the sampling procedure along with the sampling location. Due to the complexity of the particle emission behavior around the caliper, it was decided to estimate the emitted particles in the exhaust duct. A schematic drawing of the experimental setup is presented in Figure 1.

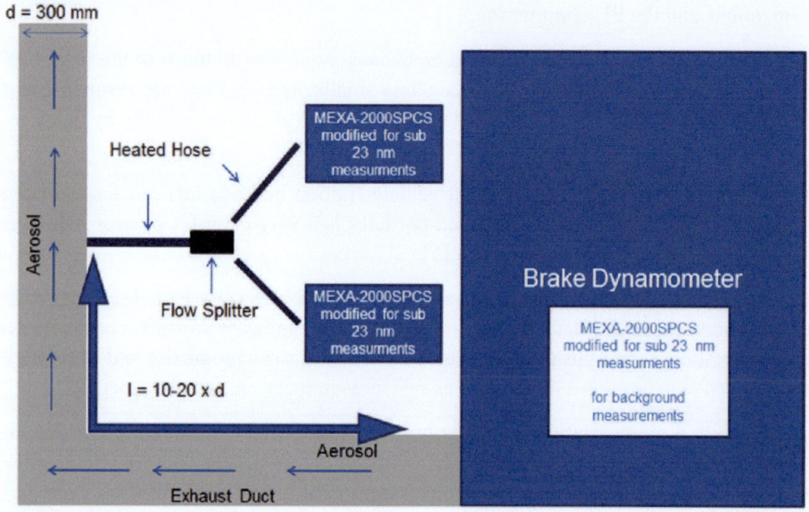

Figure 1. Scheme of the used experimental setup

As background reference, an additional sampling location inside the test chamber was chosen as well. Using this setup it is feasible to investigate, if the emitted brake particles are removed sufficiently from the brake or if a significant amount of particles remain in the chamber.

Additionally, it was decided to optimize the measurements by the consideration of a housing around the brake assembly, as it can be seen in Figure 2. The design of the housing allows to orientate the air flow directly around the brake and avoids the dissipation of emitted particles inside the dynamometer test chamber. It can be seen that the housing was sealed carefully to ensure an optimal result and leads almost all ultrafine particles to the exhaust duct.

Figure 2. Brake assembly without (left) and with housing (right)

For further investigations, in terms of the influence of volatile emissions, a flow splitter was installed after the sampling point in the exhaust duct. Using this flow splitter, it was also possible to investigate the particle size distribution. In this case the MEXA systems were replaced by an EEPS and OPS system provided by TSI GmbH, Germany.

In order to provide and ensure a homogenous mixture of the characterized airflow, the distance L from the chamber outlet to the sampling location was defined in dependency to the dilution tunnel diameter D. According to GRPE-48-11 regulation (with L = 10..20xD) the sampling location is defined as L = 14xD.

Different intake air flow rates were defined – 930 m^3/h, 2700 m^3/h and 3300 m^3/h.

Two particle number counting devices were installed into the existing testing environment. Both units are based on the MEXA 2000 SPCS system by HORIBA, but modified to detect sub-23 nm particles. The measurement range was adjusted to detect particles between 10 nm and 2.5 µm.

Due to the experiences from previous investigations a reduced AK Master profile was selected as a suitable test cycle – the sections 1-3, 5-7, 11 and 14-15 were excluded [5]. Even if this cycle creates higher particle emission rates compared to real driving profiles, it ensures a sufficient amount of particles for those fundamental characterizations. Additionally it can be stated, that stable friction and particle emission conditions are obtained in a reasonable testing time.

In regards of the specimen a 17" two piston frame-design caliper in combination with cast iron rotors and low steel brake pads were used.

3 Preliminary Investigation Results

3.1 Cooling Air Quality

To investigate the *role of the cooling air quality,* measurements with and without an additional pre-filter system were carried out.

During first tests series it was observed, that the particle emissions showed a dependency on time. Since the cooling air was sucked from a workshop next to the dynamometer, particle emissions varied during the day, especially during working hours. In order to stabilize the background particle level, a H13 filter system was added to the ventilation system for intake air, close to the test chamber. A direct comparison of background particle concentrations (measured in the exhaust duct) is shown in Figure 3.

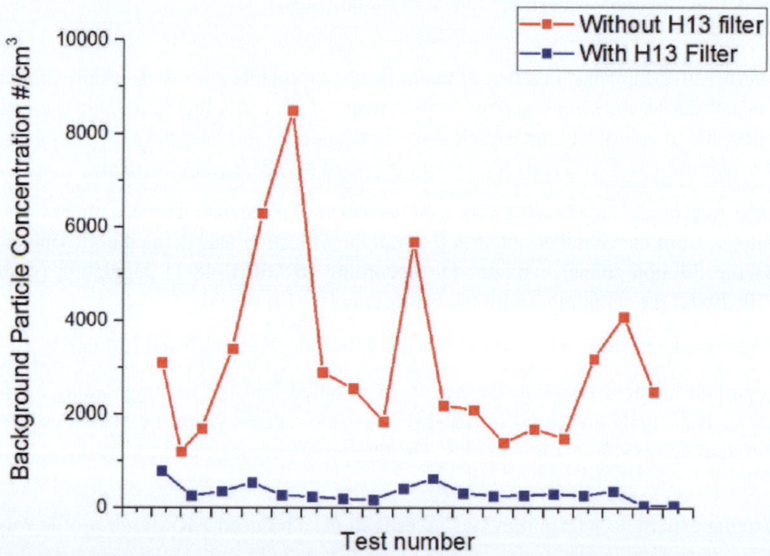

Figure 3. Comparison of background particle concentration with and without H13 filter

It can be observed easily, that the particle concentration in the inlet air without the H13 filter is considerably higher than with the filter. The application of the H13 filter leads also to a stabilization of the background level at very low particle concentrations and does reduce the fluctuation during the test series.

3.2 Stabilization Behavior

After ensuring the appropriate cooling air quality, an analysis of an extensive series of measurements is presented. The total particle concentration of each test is shown in Figure 4. A complete standard AK Master was carried out before the test series to ensure a bedded friction couple. During the test series the volumetric air flow rate was constant.

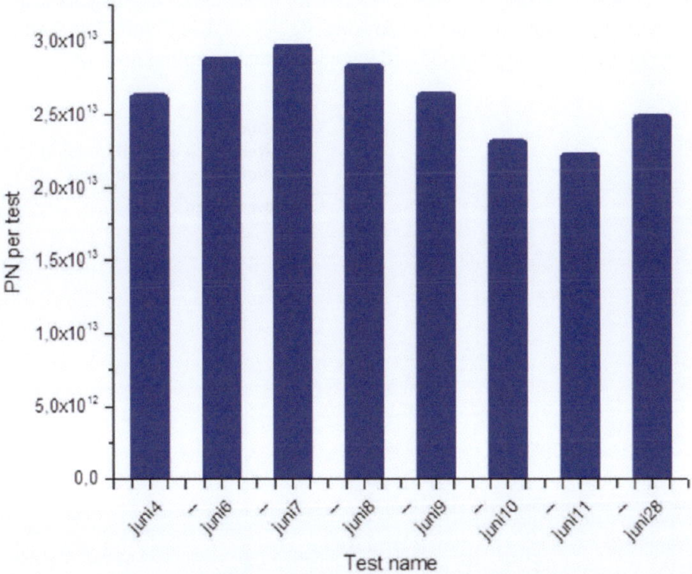

Figure 4. Overview of total PN for a test series with constant air flow rate (2700 m^3/h)

It can be seen that a stabilized behaviour is reached after the pre-conditioning by using a complete AK Master test cycle. This underlines the importance of convenient tribological bedding and confirms, that the selected AK Master is a suitable cycle for bedding purposes.

Additionally, a deviation of PN in the range of 20..30% can be observed. In terms of particle emission testing, this result is satisfactory. Therefore, it can be concluded, that a robust behaviour with good repeatability is achieved.

Finally, no clear trend of decreasing or increasing PN within the test series is obtained. This leads to the conclusion, that there is no indication of altering friction behaviour due to excessive wear of the friction partners. Hence, this method is able to ensure a comparable behaviour and extensive test series can be carried out.

3.3 Integration of an Housing Concept

The proposed housing concept helps to improve particle losses inside the test chamber and reduces the PN-level. The particle concentrations for reference measurements inside the dyno test chamber with and without the housing are depicted in Figure 5 as well as the concentration in the duct (the proposed and intended actual sampling position).

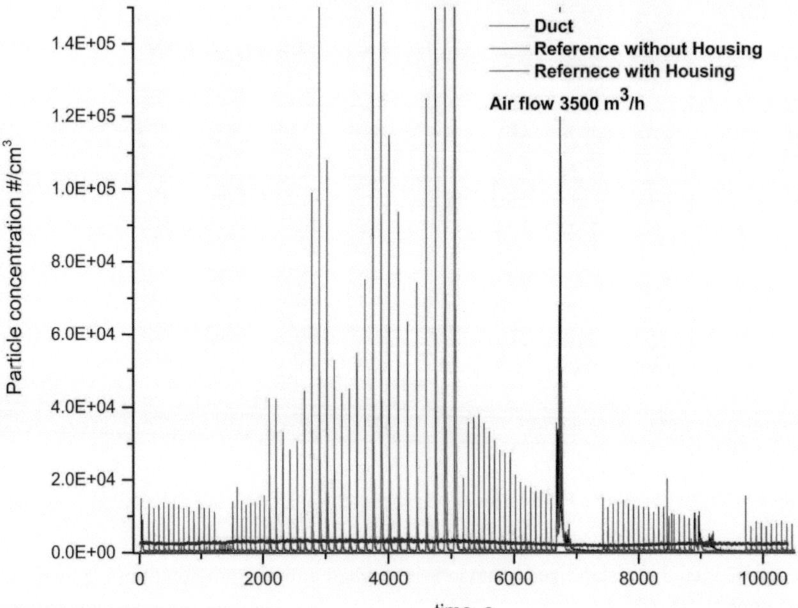

Figure 5. Particle concentration over time for measurements in the duct, with and without housing

It can be stated, that inside the dyno test chamber some particle losses can be seen when no housing is used. Obviously and despite the direct air flow around the brake, a remarkable amount of particles still gets lost inside the chamber. As a consequence, the each single applications can be observed also at the reference sampling point.

Using the housing on the other hand shows a much better picture: the particles losses can be avoided and the measured particle concentration is more or less constant during entire test cycle – a stabilized concentration is obtained.

Beside the positive aspects of the particle loss avoidance inside the test chamber, such an housing approach reveals also some questions. It has to be clarified if any negative influence on the temperature of the brake exist, what could lead to an impact on tribological behaviour.

Therefore a standard thermocouple was integrated in the brake lining matrix, 3 mm below the friction surface. A second thermocouple was placed inside the housing (on the housing wall) with the aim to understand, if thermal radiation of the brake cause a synthetic heating of the housing that in the end affects the brake again. In regards of the role of the applied volume flow the lowest and highest flow rates were adjusted.

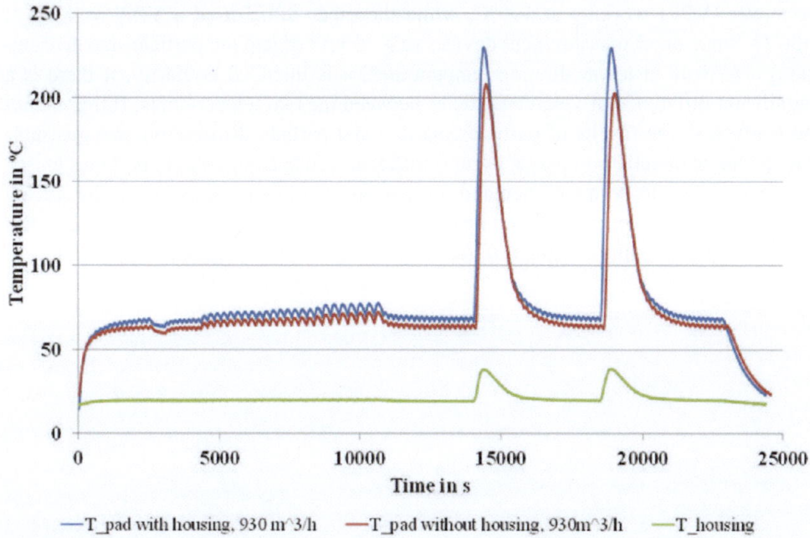

Figure 6. Temperature behaviour of brake lining and inside the housing for an air flow rate of 930 m^3/h

In Figure 6 the temperature behaviour of the brake pad with and without the housing is depicted as well as the temperature inside the housing for an air flow rate of 930 m^3/h. It must be mentioned that during the sections 9 and 12.2. the cooling air flow rate was reduced, what leads to the significant temperature increase in the signals.

The qualitative temperature behaviour of the brake pad with and without the housing is absolutely comparable, most of the sections reveal a difference of only $\Delta\vartheta = 5$ K. And even if the maximum temperature level is higher, the difference is only $\Delta\vartheta = 20$ K and negligible. Also the demanding sections 9 and 12.2 (almost) without cooling do not reveal any significant influence of the housing, what leads to the conclusion, that no negative thermal radiation has to be stated.

3.4 Volatile Emissions

In the exhaust emission community the discussion about the role of volatile emissions has been ongoing for a couple of years. Furthermore, in the field of non-exhaust emissions the relevance of those volatiles is debated as well. Hence, a comparison with two SPCS devices operating in parallel, fed by an air flow splitter close to the exhaust duct sampling point, was carried out. One SPCS was equipped with a Volatile Particle Remover (VPR) working at 190°C, while the other SPCS used a VPR working at 350°C. Since other measurement devices (e.g. to investigate the particle size distribution) also work at lower dilution temperatures, it is intended to clarify, if there is a significant difference of volatile particles between the two temperatures. If this cannot be confirmed, the results of particle counting and particle distribution size measurements can be directly compared, as no significant influence of volatile particles exists. However, it has to be mentioned and underlined, that this study is only valid for the investigated low steel brake lining materials in combination with cast iron rotors. Other friction partners might show different behaviour in terms of volatile emissions.

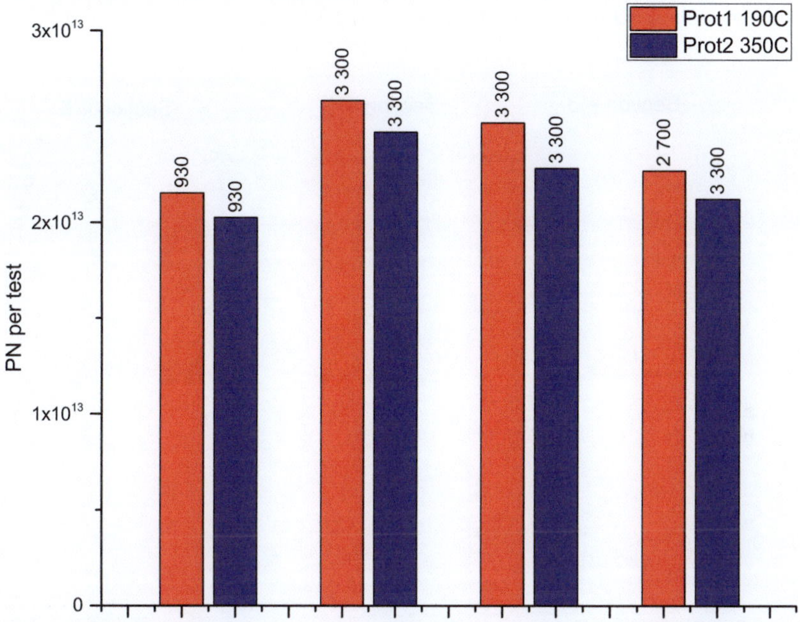

Figure 7. Comparison of PN using different dilution temperatures

Figure 7 shows a comparison of several tests. The numbers 930, 2700 and 3300 represent the actual air flow rate in m^3/h.

It can be observed that the differences between the two dilution temperatures are negligible. Hence, it can be concluded, that applying a VPR temperature of 190°C does not affect the amount of detected particles and, therefore, no falsification of the measurement results exists.

4 Particle Emission Characterizations

4.1 Particle Size Distribution (PSD)

Due to the availability of measurement devices, the characterization of particle size distributions was carried out without the housing concept. Of course, it is intended to complete the PSD analysis in the future also with the housing concept.

In Figure 8 the PSD of the AK Master sections 4.3 to 4.5 is shown for two different air flow rates.

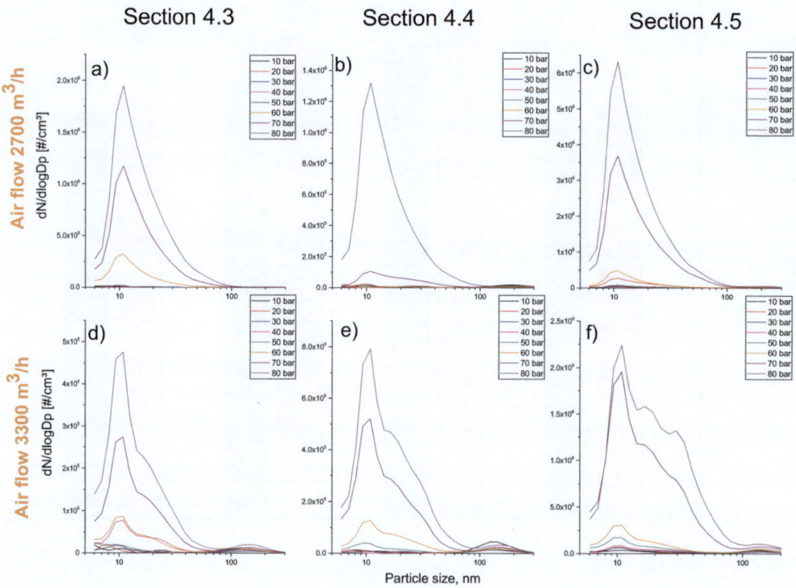

Figure 8. Particle size distributions for AK Master sections 4.3 - 4.5 and an air flow rate of 2700 m^3/h (a)-c)) and 3300 m^3/h (d)-f))

Firstly, it can be stated that a certain pressure influence exists. Especially for pressure levels above 50 to 60 bar, significant amounts of ultra-fine particles with a peak in the size distribution at around 10 nm can be confirmed. It should be mentioned that the effect is directly linked to a temperature level of $T \geq 170°C$.

Another effect reveals the initial brake velocity. An increase of the vehicle speed comes along with an increase of the emitted particles. It must be mentioned that not only the maximum amount of particles is affected but also the shape of the curve which becomes wider for higher velocities.

Finally, also the influence of increasing air flow rate can be seen. Applying a higher air flow rate results in a decrease of the number of emitted ultra-fine particles. The authors expect, that a higher volumetric flow leads to an improved dilution, what has to be confirmed.

However, for both air flow rates a bi-modal particle size distribution is suspected. The first and dominant peak can be found around 10..20 nm, while the second peak seemed to be around 150 and 190 nm.

For a deeper analysis a zoomed PSD diagram is shown in Figure 9.

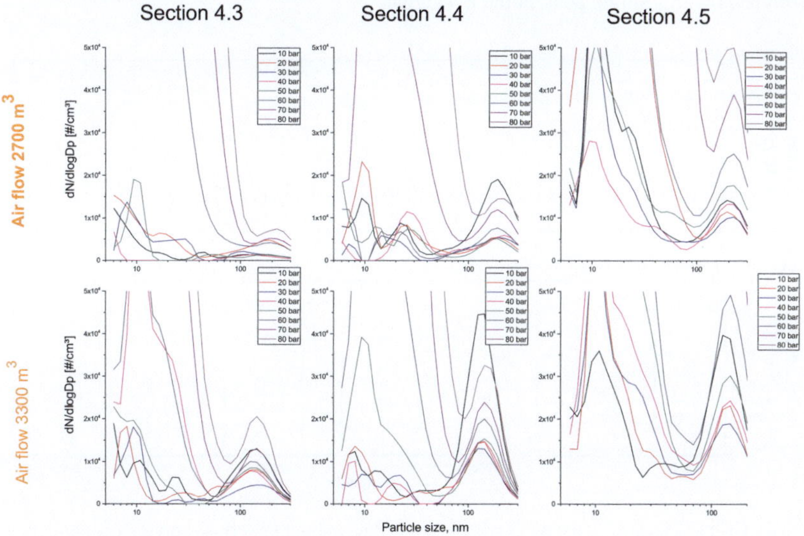

Figure 9. Zoomed particle size distributions for AK Master sections 4.3 - 4.5 and an air flow rate of 2700 m^3/h (a)-c)) and 3300 m^3/h (d)-f))

As it was assumed, a second peak between 150 and 190 nm can be confirmed clearly for both air flow rates. However, unlike to the peaks at 10..20 nm, the concentration of particles between 150 and 190 nm is higher for 3300 m^3/h than for 2700 m^3/h.

It can be summarized, that demanding brake applications lead to a significant increase of ultra-fine particles in the range of 10..20 nm. Moreover, the air flow rate can affect not only the obtained particle concentration but also the relation between the two peaks (at 10..20 nm and 150..190 nm) in the particle size distribution.

4.2 Influence of Load History

A direct comparison of section 8 (characteristic value) reveals the influence of the load history. This is illustrated in Figure 10 presenting the PN measurement results of section 8 – speed / pressure sensitivity sections (4.1 – 4.5). Each single brake event correlates with a single peak in the PN signal.

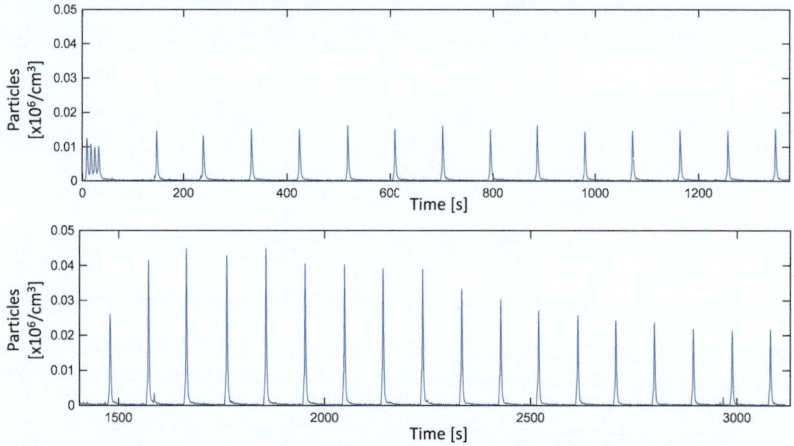

Figure 10. Time history of PN for section 8 before (above) and after (beneath) the speed / pressure sensitivity sections (4.1 – 4.5)

Although the charts of coefficient of friction are comparable (see also Figure 11), the emission of particles is different.

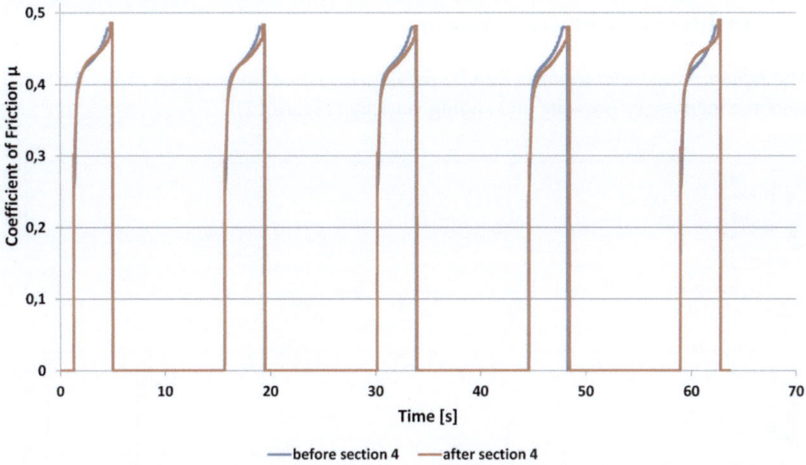

Figure 11. Comparison of µ for section 8 before and after section 4.1-4.5

It is expected that this is directly linked to the load history of sections 4.1 to 4.5. It can be concluded, that the previous load applications play an important role in regard to brake particle emissions. This observation presents interesting new opportunities for tribological characterizations. Since traditional parameters like µ and temperature (not shown here) do not reveal these load history effects, the determination of the PN-time behavior might serve as a tool for finding explanations for other phenomena, such as brake squeal (NVH), as well.

4.3 Influence of Friction Power

The *influence of friction power* can be recognized, since the sections (4.2 – 4.5) consider the increase of pressure and sliding velocity (Figure 12).

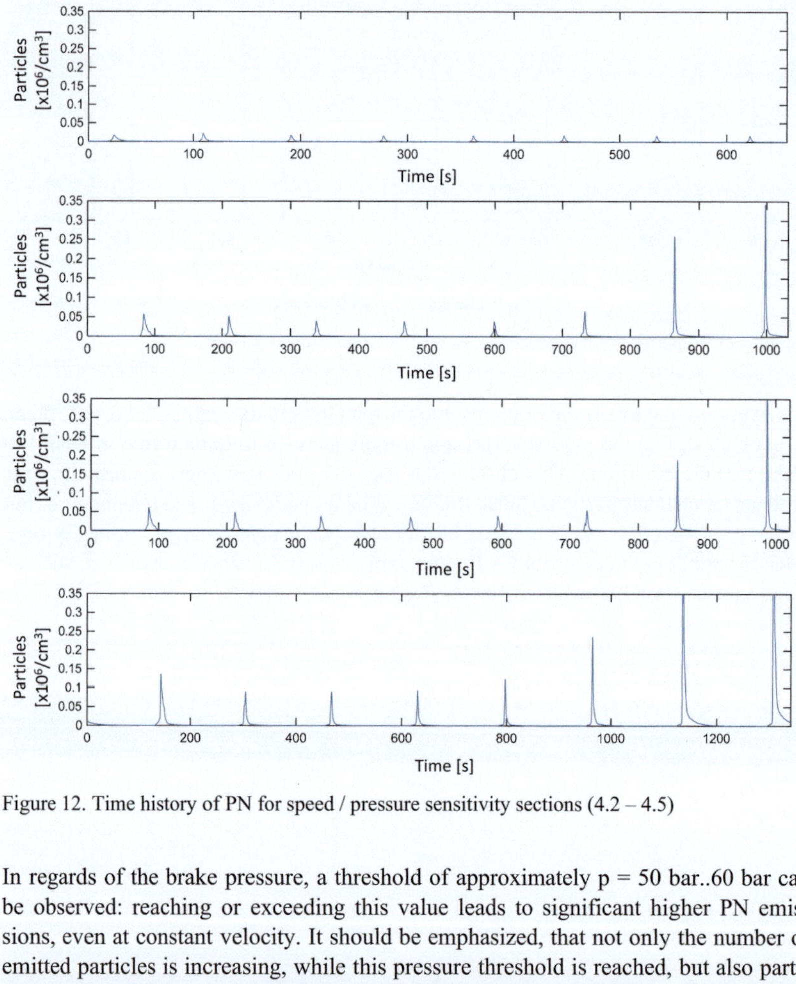

Figure 12. Time history of PN for speed / pressure sensitivity sections (4.2 – 4.5)

In regards of the brake pressure, a threshold of approximately p = 50 bar..60 bar can be observed: reaching or exceeding this value leads to significant higher PN emissions, even at constant velocity. It should be emphasized, that not only the number of emitted particles is increasing, while this pressure threshold is reached, but also particle size distribution is changing, as it was already shown. Significant emissions of particle between 10 and 20 nm can be stated, what also underlines the importance of an adjusted test equipment in comparison to exhaust measurements.

Furthermore, the influence of the velocity can be observed: the PN concentration increases with speed, while the pressure is constant.

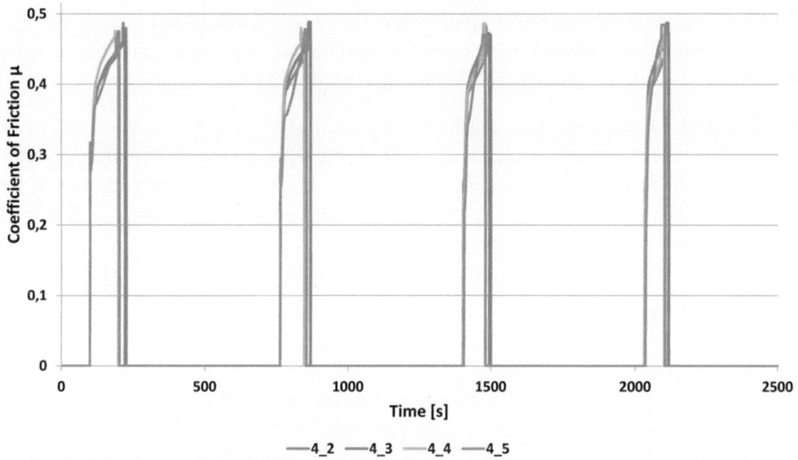

Figure 13. Comparison of µ for sections 4.2-4.5

This effect is very important since the coefficient of friction is similar for brake applications and no correlation to µ can be observed (Figure 13).

In summary, it can be stated that higher values for brake pressure and especially for velocity (and friction power, respectively) lead to higher PN emissions.

5 Summary and Conclusions

In summary it can be stated, that brake particle emissions are getting more and more important as a non-exhaust emission source and BPE are discussed intensively. Since no standards in terms of experimental characterizations exist, the authors present an approach, which is considering a full brake dynamometer with additional particle measurement equipment and an adapted test cycle.

It was shown, that the preconditioning of the cooling air by using a H13-filter is strictly recommended to ensure robust measurements in terms of repeatability. Moreover a housing concept is introduced and in terms of its advantages characterized. Due to the fact, that is does not affect the thermal conditions of the friction couple, it is recommended for future measurement to take this concept into account.

The characterization of the particle size distributions reveals, that ultra-fine particle emissions can be obtained, when certain operating parameters are reached. Considering brake pressures above 50 bar and temperatures above 170°C (rotor temperature!), it leads to significant emissions of particles in the range of 10..20 nm. That underlines the importance of adjusted measurement equipment, since conventional exhaust measurement systems own a lower particle size limit of 23 nm.

Finally it should be mentioned, that brake particle measurements in general and particle number detections in particular seem to be a proper approach to determine preload or load history effects, especially of the brake lining. The characterization of PN confirmed those effects, while the well-known standard parameter, the coefficient of friction, has not indicated any difference.

6 References

[1] Amato, F. et al., “Traffic induced particle resuspension in Paris: Emission factors and source contributions”, Atmospheric Environment 129 (2016) 114-124, Elsevier Ltd., doi: 10.1016/j.atmosenv.2016.01.002.

[2] Olofsson, U., Nosko, O., “On the formation and physical properties of airborne particles from disc brakes”, EB2016-MDS-037, Proceedings EuroBrake 2016, Milan.

[3] Mattia, A., Perricone, G., Olofsson, U., Söderberg, A., Wahlström, J., Ciotti, A., “A Proposed Dyno Bench Test Cycle to Study Particle Emissions from Disc Brakes”, EB2014-SE-001, Proceedings EuroBrake 2014, Lille.

[4] Wirtz, M., Sachse, H., Lienerth, P., Lugovyy, D., Augsburg, K., „Inertia Brake Dynamometer Dust Particle Measurement“, EB2015-EFP-008, EuroBrake 2015, Dresden.

[5] https://www2.unece.org/wiki/download/attachments/25264603/PMP-35-15%20ITT%20Motion%20Presentation%20final.pdf?api=v2, request on 1st April 2016

HORIBA
Automotive Test Systems

Investigation Possibilities of Brake Particle Emissions on a Brake Dynamometer

Dr.-Ing. Sebastian Gramstat, Robert Waninger – AUDI AG
Dr.rer.nat. Dmytro Lugovyy, Matthias Schröder – Horiba Europe GmbH

2 AUDI AG I/EF-51 Dr.-Ing. Sebastian Gramstat 24/05/2017

Agenda

Motivation
Test Procedure
Experimental Setup
Preliminary Investigations
Particle Emission Characterizations
Summary / Outlook

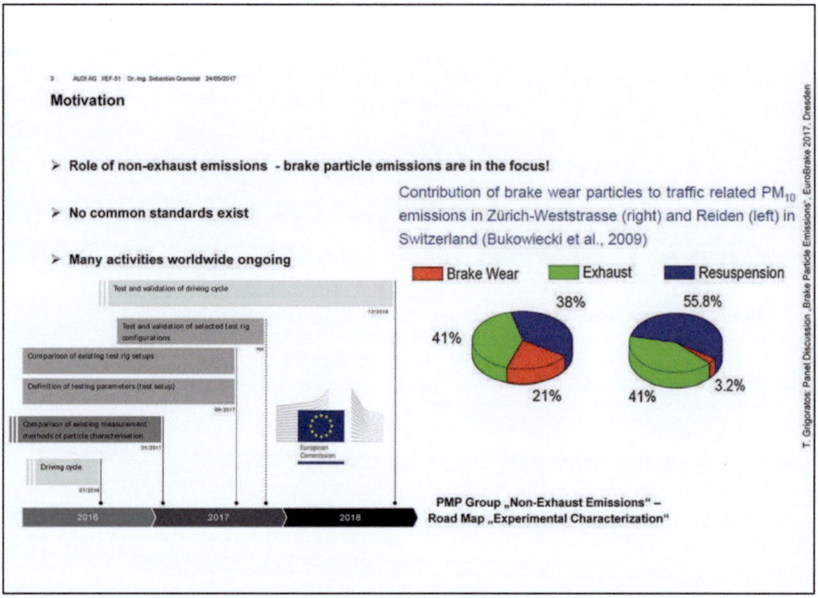
3 AUDI AG I/EF-51 Dr.-Ing. Sebastian Gramstat 24/05/2017
Motivation
➢ Role of non-exhaust emissions - brake particle emissions are in the focus!
➢ No common standards exist
➢ Many activities worldwide ongoing
Test and validation of driving cycle
12/2018
Test and validation of selected test rig configurations
Comparison of existing test rig set ups
Definition of testing parameters (test set up)
08/2017
Comparison of existing measurement methods of particle characterisation
01/2017
Driving cycle
07/2016
2016
2017
2018
European Commission
Contribution of brake wear particles to traffic related PM10 emissions in Zürich-Weststrasse (right) and Reiden (left) in Switzerland (Bukowiecki et al., 2009)
Brake Wear
Exhaust
Resuspension
41%
38%
21%
55.8%
41%
3.2%
PMP Group „Non-Exhaust Emissions" –
Road Map „Experimental Characterization"
T. Grigoratos: Panel Discussion „Brake Particle Emissions", EuroBrake 2017, Dresden

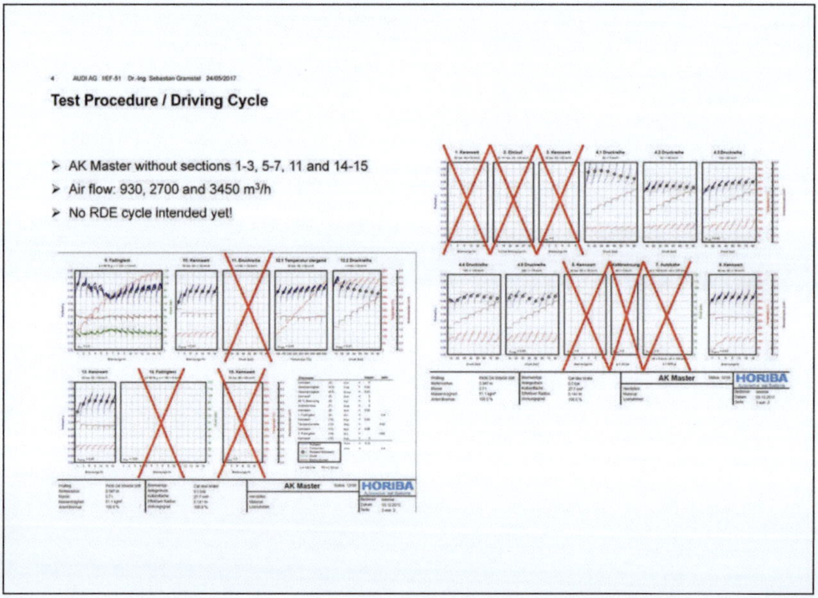
4 AUDI AG I/EF-51 Dr.-Ing. Sebastian Gramstat 24/05/2017
Test Procedure / Driving Cycle
➢ AK Master without sections 1-3, 5-7, 11 and 14-15
➢ Air flow: 930, 2700 and 3450 m³/h
➢ No RDE cycle intended yet!
AK Master
HORIBA
AK Master
HORIBA

5 AUDI AG I/EF-51 Dr.-Ing. Sebastian Gramstat 24/05/2017

Experimental Setup

- ✓ **2 brake systems:**
  - ✓ 17" 2 piston frame-design caliper, low steel brake pads
  - NEW ✓ 19" 6 piston fixed caliper, copper-free low steel brake pads (*inertia variation*)
- ✓ **2 sampling points**
  - ✓ Inside testing chamber
  - ✓ Exhaust duct
- ✓ **Use of SPCS system (2x)**
  - ✓ Sub-23 nm systems
- NEW ✓ **Use of EEPS & OPS (PSD)**
  - ✓ Support by TSI
- NEW ✓ **Housing concept**

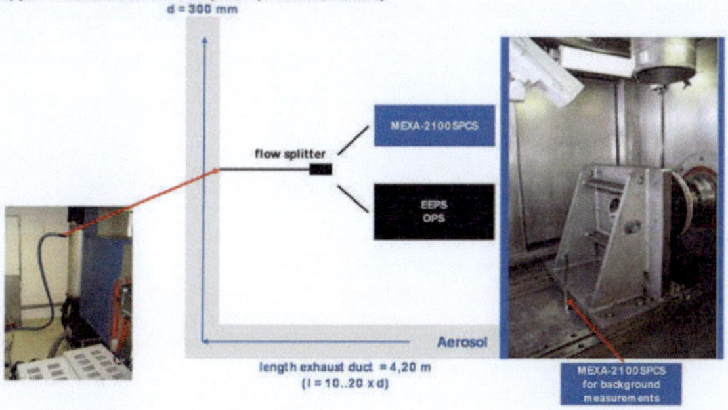

6 AUDI AG I/EF-51 Dr.-Ing. Sebastian Gramstat 24/05/2017

Preliminary Investigations

- ✓ **Cooling air quality**

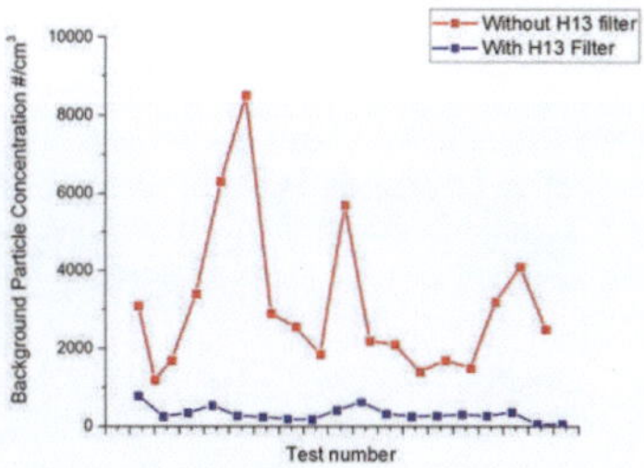

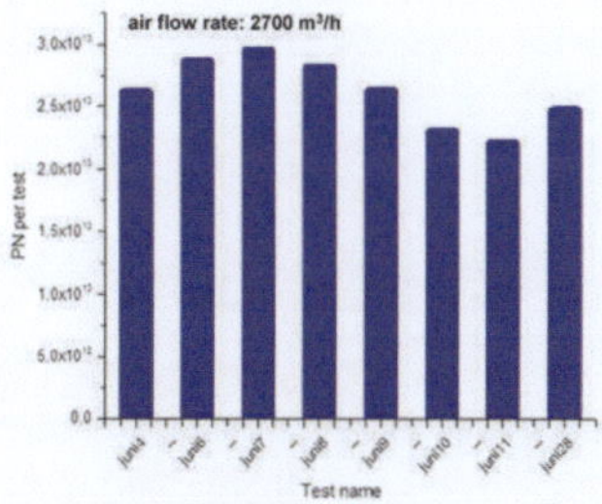

- ✓ **Stabilization behavior**

7 AUDI AG I/EF-51 Dr.-Ing. Sebastian Gramstat 24/05/2017

Preliminary Investigations – Housing Concept

Objective: avoid the dissipation of emitted particles inside the dynamometer test chamber

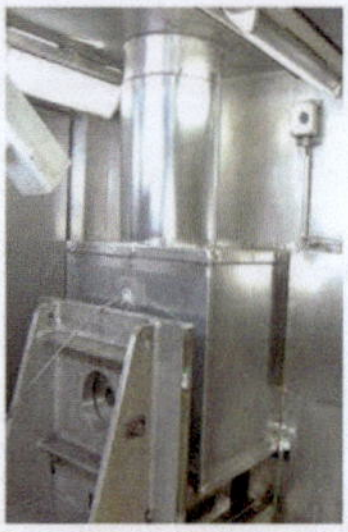

housing is sealed carefully
→ direct air flow leads almost all ultra-fine particles to the exhaust duct

8 AUDI AG I/EF-51 Dr.-Ing. Sebastian Gramstat 24/05/2017

Preliminary Investigations – Housing Concept

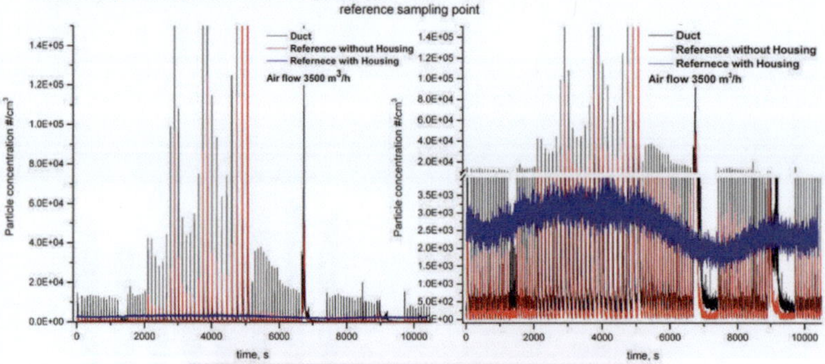

Without housing: particle losses, single brake applications can be observed
With housing: low and stable particle concentration (NO H13-filter!)

Question: thermal condition of the brake?

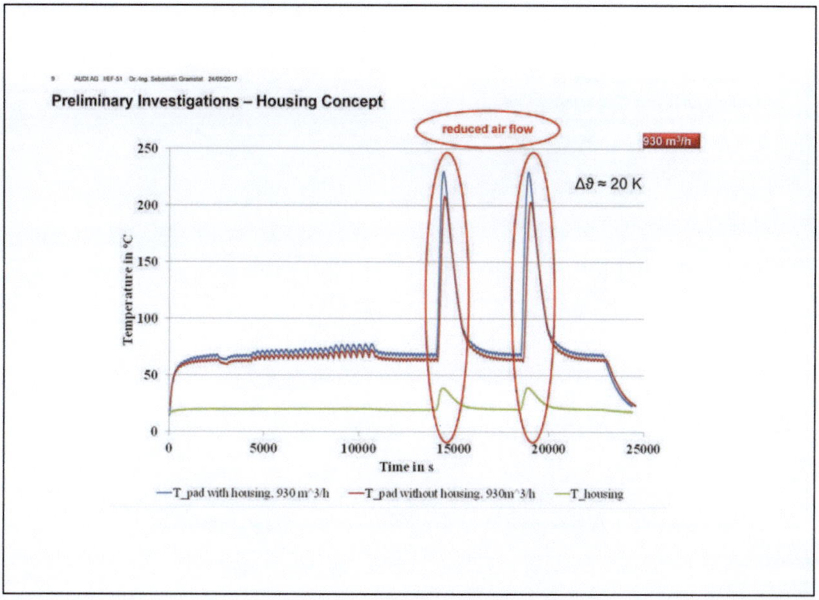

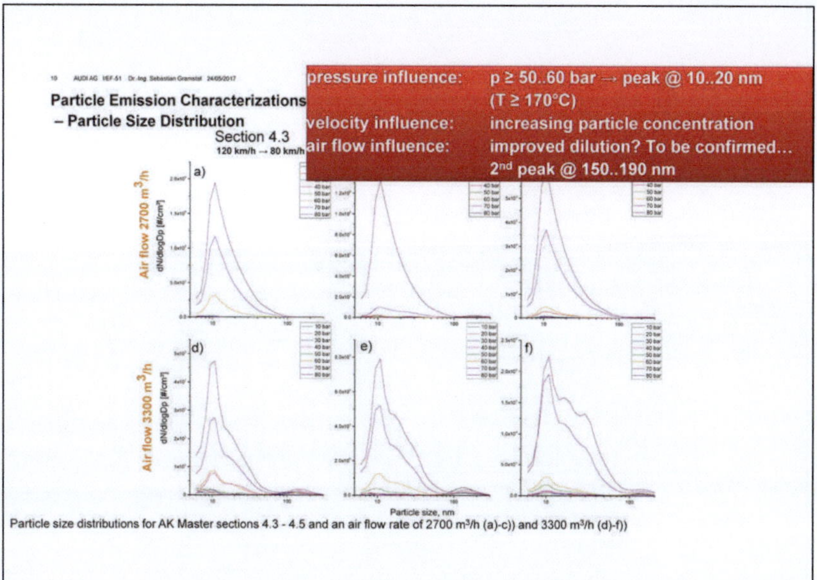

Particle size distributions for AK Master sections 4.3 - 4.5 and an air flow rate of 2700 m³/h (a)-c)) and 3300 m³/h (d)-f))

11 AUDI AG I/EF-51 Dr.-Ing. Sebastian Gramstat 24/05/2017

Particle Emission Characterizations
– Particle Size Distribution

2nd peak @ 150..190 nm for both air flow rates
higher concentration for higher air flow
→ ventilation can affect dilution and PSD

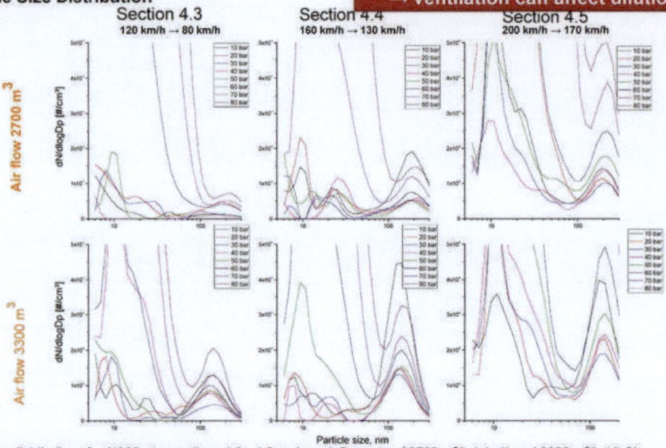

Zoomed particle size distributions for AK Master sections 4.3 - 4.5 and an air flow rate of 2700 m³/h (a)-c)) and 3300 m³/h (d)-f))

12 AUDI AG I/EF-51 Dr.-Ing. Sebastian Gramstat 24/05/2017

Particle Emission Characterizations
– Load History

- ✓ **Test sequence of**
  - ✓ Characteristic value section (moderate)
  - ✓ Pressure-velocity-sensitivity section (demanding)
  - ✓ Characteristic value section (moderate)

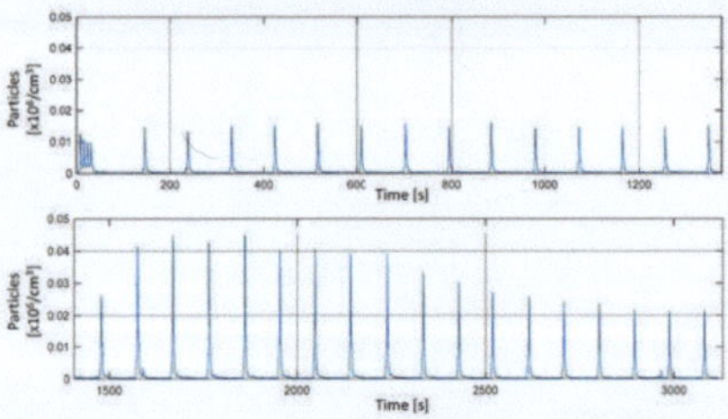

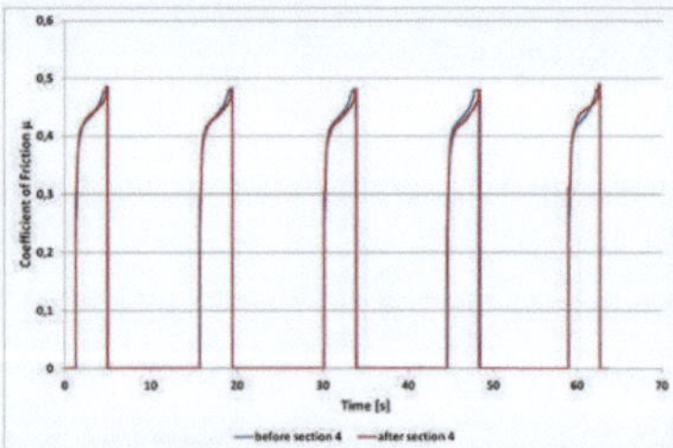

13 AUDI AG I/EF-51 Dr.-Ing. Sebastian Gramstat 24/05/2017

Particle Emission Characterizations – Friction Power

✓ Pressure-velocity-sensitivity section (only 4.2 – 4.5 are shown!)

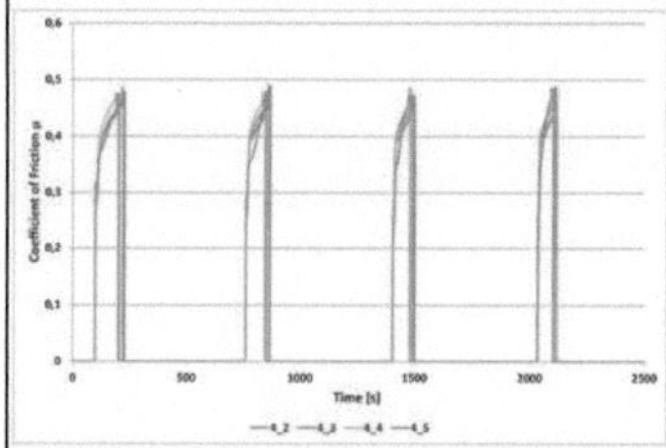

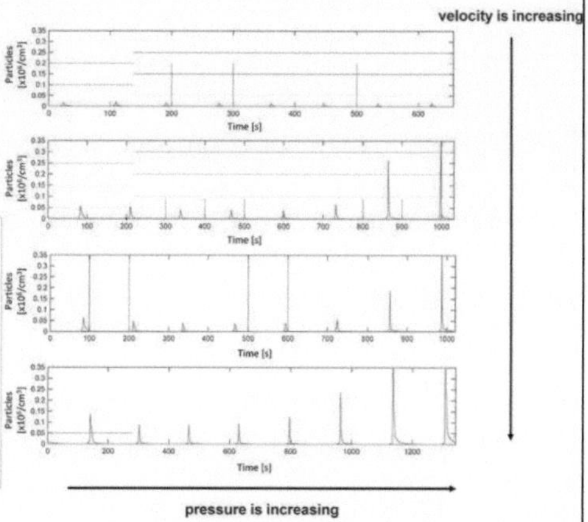

14 AUDI AG I/EF-51 Dr.-Ing. Sebastian Gramstat 24/05/2017

Summary / Outlook

- ✓ Role and importance of BPE
- ✓ Introduction of test setup and cycle
- ✓ Housing concept is recommended for BPE measurements
- ✓ PSD measurements underline the role of ultrafine particles (10..20 nm!)
- ✓ BPE testing as a tool for brake lining characterization (load history!)

- ✓ Refinements of the presented test setup
- ✓ Investigations of additional friction couples
- ✓ Standardization work

15 AUDI AG I/EF-51 Dr.-Ing. Sebastian Gramstat 24/05/2017

Thank you very much for your attention!

Thermal spray ceramic coatings as friction surfaces of brake rotors

Septimiu Popa, Prof. Dr. Dr. R. Gadow, Prof. Dr. A. Killinger,
Institute for Manufacturing Technologies
of Ceramic Components and Composites (IFKB),
University of Stuttgart

This manuscript is not available according to publishing restriction. Thank you for your understanding.

P.E. Pfeffer (Ed.), *8th International Munich Chassis Symposium 2017*,
Proceedings, DOI 10.1007/978-3-658-18459-9_45

BRAKE SYSTEMS OF THE FUTURE

A new wheel-brake concept in the context of mobility trends

Paul Linhoff, Continental Teves AG & Co. oHG

This manuscript is not available according to publishing restriction. Thank you for your understanding.

P.E. Pfeffer (Ed.), *8th International Munich Chassis Symposium 2017*, Proceedings, DOI 10.1007/978-3-658-18459-9_46

Downsizing potential of wheel brakes in electric vehicles

Daniel Wagner, M.Sc.
Institute of Automotive Technology, Technical University of Munich

Dr.-Ing. Jens Hoffmann
Continental Teves

Prof. Dr.-Ing. Markus Lienkamp
Institute of Automotive Technology, Technical University of Munich

P.E. Pfeffer (Ed.), *8th International Munich Chassis Symposium 2017*,
Proceedings, DOI 10.1007/978-3-658-18459-9_47

Abstract

In most cases of daily driven trips electric vehicles allow to cover decelerations up to 0.3 g only through regenerative braking. The usage of the conventional friction brake is massively reduced.

Nowadays classic performance criteria are still the foundation of dimensioning friction brakes. These performance criteria include driving down mountain passes and consumer tests like sequenced stoppings. Since with a fully charged or not conditioned battery, regenerative braking is not possible. The recuperative braking system officially is not part of the service brake and is switched off during the test maneuvers.

A vehicle with an electric powertrain that offers a competitive driving range must be equipped with a heavy battery and therefore has a higher overall mass compared to a vehicle with an internal combustion engine with comparable performance. With higher overall weight, the conventional friction brake has to be dimensioned larger to fulfill the same decelerating requirements. The conclusion is that with today's way of dimensioning electric vehicle with less use of brakes must have larger brakes.

The paper analyzes the potential of downsizing wheel brakes in electric vehicles. It will regard methods, which could avoid the overheating of smaller dimensioned wheel brakes while driving down mountain passes. The number of the sequenced stoppings of the consumer tests will be reduced.

Self-developed thermal models of disc brakes with different materials are added to the vehicle models. Furthermore, models of constant energy sinks are implemented to the electric powertrain.

The objective is to analyze the effect of regenerative braking and reduced requirements for the design of wheel brakes.

1 Motivation

The development and construction of wheel brakes for motor vehicles aims at meeting established performance requirements for the complete vehicle. The performance tests have been established throughout the years or have been introduced by the press for comparison. The High Speed Fading test, the AutoMotorSport test, and the Grossglockner downhill drive are a few examples to be mentioned here.

The tests do not differentiate between different vehicle concepts and dimensions. Differences in terms of the amount of stress applied to the wheel brakes arise from vehicle-inherent characteristics, such as the total weight, top speed and acceleration performance. Due to this equalization, all electric vehicles are assessed with the same criteria. A huge discrepancy in the daily load spectrum of the wheel brake and the extreme braking action of the performance tests can already be observed in conventional vehicles. There is even a greater difference with electric vehicles, leading to additional difficulty. Electric vehicles have the possibility of regenerative braking and thus an additional option to decelerate the vehicle. The electric machine is operated as a generator and therefore produces electric energy from mechanical rotary energy. Besides the powertrain, the degree of recuperation in current electric vehicles depends on the operating strategy and the philosophy of the manufacturer. Common practice is recuperating in propulsion operation, where the recuperation takes place with partly graded intensity as soon as the driver lifts his or her foot from the accelerator pedal. Another way is to recuperate as soon as the driver applies the brake pedal. If that is not sufficient, the conventional braking system is engaged. For this process, the brake system has to be able to blend, as the driver should not notice the switchover.

The possibility to decelerate with regenerative braking is currently not seen as part of the brake system, as necessary backup solutions have not been established for safety reasons [1]. In addition, the potential of regenerative braking is, along with the mechanics, electric motor, power electronics and onboard power system, also restricted by the capacity of the battery. The battery can be restricted through the state of charge or its temperature. Due to the weight of the battery, electric vehicles are usually heavier than vehicles with internal combustion engines and comparable performance [3, p. 8]. Vehicles built with increased constructional efforts and financial expense regarding the lightweight construction in order to compensate for the disadvantage of the heavy battery can be an exception. As the weight of the vehicle is a significant factor for the necessary brake power, electric vehicles consequently would have to be equipped with a bigger brake system similar to comparable vehicles with internal combustion engines.

Regardless of how the recuperation is implemented in the vehicle, decelerations up to 3 m/s^2 are usually possible. This deceleration performance covers the so-called "comfort zone" into which 85 % of all brake procedures fall and within which the ordinary driver brakes [2, p. 172].

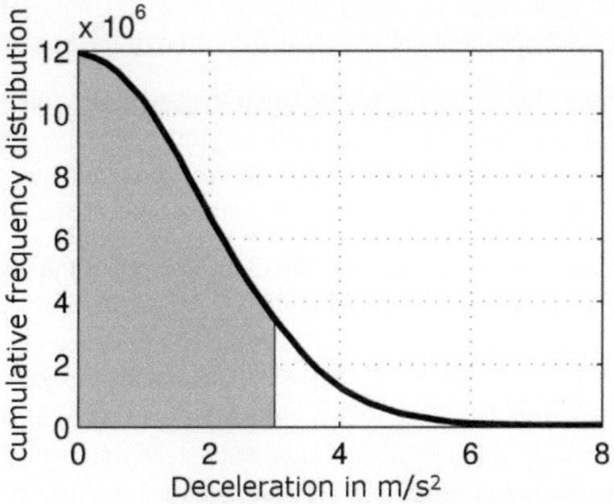

Figure 1-1 Cumulative frequency distribution of Deceleration [2, p. 172]

If the electric machine takes over most of the braking work in daily use, the usage of the conventional friction brake system will be reduced in the same way [4]. The brakes, even bigger because of the vehicle weight, are not used sufficiently. The condition of the friction brakes gets effected [5]. Consequently, additional corrosion on the brake discs and vitrification of the brake pads can be observed. Furthermore, the resistance of the operation mechanism can increase. As a result, diminished brake performance and lifetime as well as enhanced residual torques are possible. Moreover, in daily life, the oversized brakes are associated with higher costs and a higher tire sprung mass. This paper analyzes the requirements on wheel brakes for electric vehicles and examines possible alternatives.

2 Simulation environment

For the necessary study, two electric vehicles in the compact class were designed in the IPG CarMaker. The CarMaker driver's model is used. The performance tests have been reconstructed in the simulation environment. The multiple stop maneuvers have been reconstructed on a plane road. For the downhill scenario, the course of the real Grossglockner alpine road were taken from GPS data.

The vehicles have been built up with the data shown in Table 2-1 and complemented with a analytical brake system. Additionally, a simple antilock braking function (ABS) and an electronic brakeforce distribution function (EBD) have been implemented. These can be activated optionally or a fixed distribution of the brake force can be used.

For the model of the electric machine, usually a map with the full load torque output and the mechanical and electric losses is used. For this paper this engine was replaced by an idealized one with a constant power and a limited torque output. The advantage is that the analyses do not get distorted by the characteristic of the engine.

The battery is not modeled. The reason is the assumption that the battery might be not available to absorb the electric energy generated by regenerative braking. If the vehicle should decelerate by regenerative braking, there have to be other ways with guaranteed electric performance for absorbing the energy. To consider this possibility, the model of the powertrain was supplemented with a virtual energy sink in the DC sector. For the first analyses, the energy sink can be activated with steady performance. The aim is to simulate a defined amount of regenerative braking taking into account the model of the engine.

The focus lies on the vehicles in the strong-selling compact class. Both a front wheel and rear wheel driven version were implemented to analyze the recuperative influence of both axles.

	Version FWD	Version RWD
driven axle	front	rear
weight (empty)	1200 kg	1200 kg
weight distribution empty	47.5 % front / 52.5 % rear	52.5 % front / 47.5 % rear
overall weight	1700 kg	1700 kg
weight distribution fully loaded	45.3 % front / 54.7 % rear	48.8 % front / 51.2 % rear
engine performance	100 kW	100 kW
max. torque	250 Nm	250 Nm
transmission ratio	11	11
tire dimensions/rolling radius	205/55 R16 / 315 mm	205/55 R16 / 315 mm
top speed	150 km/h / 14 000 rpm	150 km/h / 14 000 rpm
wheel base	2420 mm	2420 mm
brake front	ventilated disc, 280x22x8 mm	ventilated disc, 280x22x8 mm
brake rear	disc, 280x9 mm	disc, 280x11 mm

Table 2-1 Simulation vehicle data

Subsequent to the simulation of the whole vehicle, the braking torque, the rotational wheel speed, and the speed of the vehicle is used as incoming data for a one-dimensional thermal simulation of the wheel brakes. The model induces the brake power from both sides over the complete surface of the friction ring. Convection as well as radiation are considered. The heat transfer is dissolved through a finite differential method [6, p. 218]. The model calculates the temperature of the individual layers of the brake disc with variable discretization steps. In the first version, the thermal simulation draws on non-varying parameters of the material. The heat transfer coefficient of the convection is speed-dependent. Only the friction rings of the brake discs are considered. Simulation models for the entire disc brake, for drum brakes and a universal model for other concepts are also implemented, but not validated yet.

The simulation framework of the different models offers the possibility to analyze the effects of parameter variation on the temperature of the wheel brakes in a fast and effective way. Quick adaptations of vehicle models and test requirements can be considered.

3 Current performance tests

For the dimensioning of the wheel brakes, the High Speed Fading test, the AutoMotorSport test, and the Grossglockner downhill drive are used. They are described in the following section. The brakes should be dimensioned in a way which ensures that the permissible range of temperature for the respective brake disc material is not exceeded in the different tests and therefore no decrease of the brake performance occurs. The friction coefficient of a conventional grey cast brake disc decreases rapidly above 700 °C [7, p. 44]. A steady braking torque can only be reached through a proportional rising brake pressure. As the braking pressure is limited and other problems could occur due to the temperature, an exceedance of this temperature level needs to be avoided.

The drop in the friction coefficient is not depicted in the model due to the strong dependency of the friction combination of the brake disc and the brake pad. The exceedance of the fading limit will be evaluated as a failure.

3.1 High Speed Fading test (HFT)

The High Speed Fading test (HFT) serves to assess the deceleration and the tendency for fading of the vehicle at the higher speed range. The test consists of 15 subsequent braking actions in which the vehicle is decelerated from the initial speed, which is 90 % of the top speed of the vehicle with a deceleration of 6 m/s^2 to 90 km/h. Following this deceleration, the vehicle is accelerated with the maximal performance at its disposal. The HFT is conducted with half of the vehicle load capacity.

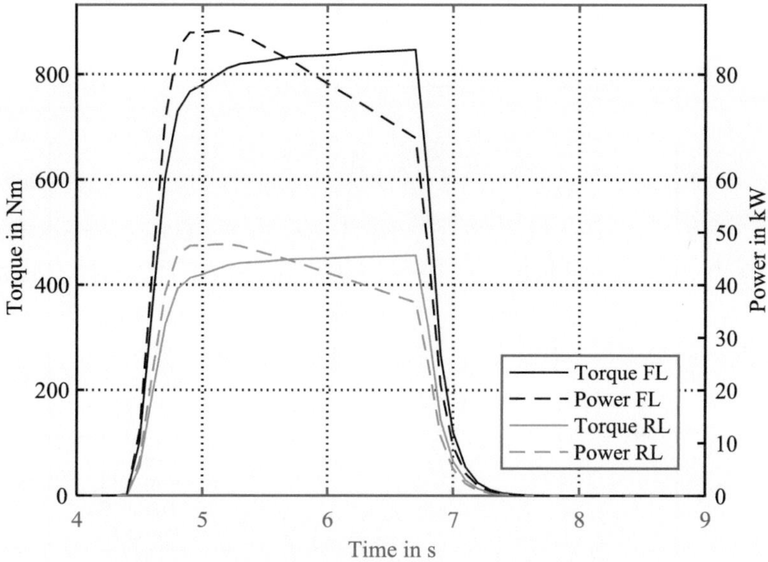

Figure 3-1 Brake torque at one cycle of HFT (FWD)

In Figure 3-1 the brake torque and performance at the front left (FL) and the rear left (RL) wheel brake of the vehicle with front wheel drive is depicted. The HFT is simulated with a fixed distribution of the brakeforce. The front brakes have to brake with approx. 850 Nm, the back ones with approx. 450 Nm. The needed break performance goes up to 90 kW in the front and 50 kW at every rear brake.

3.2 AutoMotorSport test (AMST)

The AutoMotorSport test (AMST) comprises a sequence of ten full brake applications until standstill, between which full acceleration proceeds again. There are two alternatives with regard to the starting speed and the vehicle load capacity. With the old testing procedure, the vehicle was decelerated from 100 km/h and the vehicle was fully loaded. In the new AMST, the vehicle is decelerated from 130 km/h, but is only loaded with half of the maximal load capacity. Vehicles that have a maximum speed below 180 km/h are still assessed with the old testing procedure and so are the simulated vehicles [8].

The ABS function adjusts during the braking process due to the full braking.

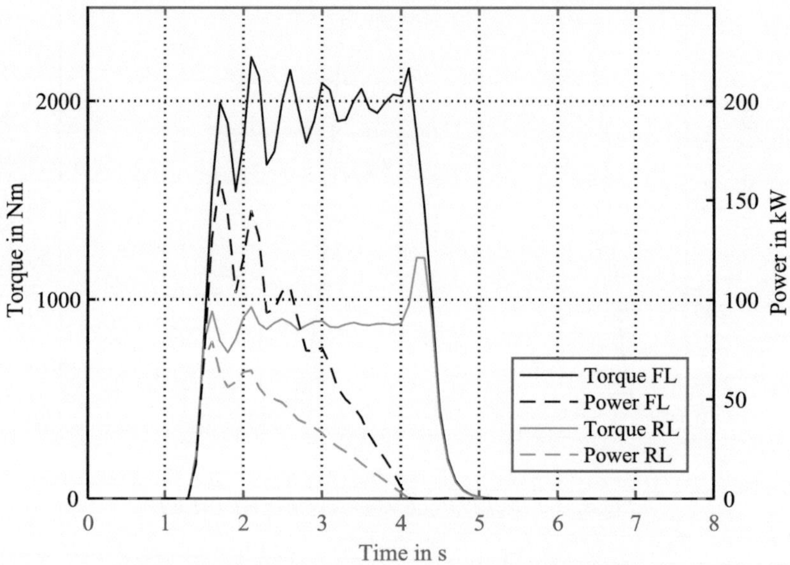

Figure 3-2 Brake torque at one cycle at AMST (FWD)

In the AMST, depicted in Figure 3-2, the front brakes have to brake with approx. 2200 Nm in the peak while the back brakes each brake with a maximum of 940 Nm. The needed peak break performance goes up to 160 kW in the front and 80 kW at every rear brake. There are big jumps of the brake performance due to the simple ABS controller. Because of the high deceleration and the implemented electronic distribution of the braking force, the distribution is approximately ideal.

3.3 Grossglockner downhill scenario (GG)

While the brake is stressed with a higher performance at a relatively short interval in the HFT and AMST, it is the long continuous input of a moderate performance which strains the wheel brakes during the Grossglockner downhill drive. This downhill drive is a route that is 15 km in length with an average incline of 9.5 %. The difference in altitude is 1429 m. The test maneuver is conducted in two speed variants, a so-called

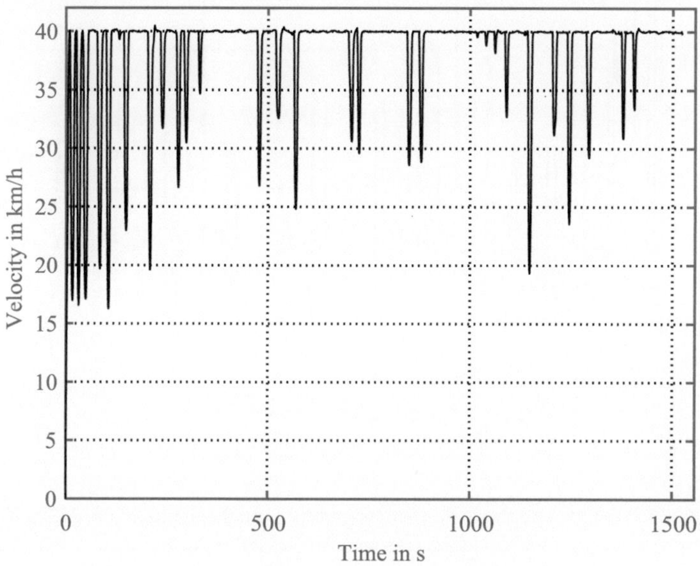

Figure 3-3 Velocity at GG

"*Schleichabfahrt*" at 20 km/h and a "*Touristenabfahrt*" at 40 km/h. As the *Touristen-abfahrt* has higher thermal load in the temperature of the brakes in this wheel brake model, the *Schleichabfahrt* will not be considered in this paper.

In Figure 3-3 the velocity of the vehicle at the GG is depicted. The driver's model has been preset with a target speed of 40 km/h in the simulative implementation. The maximum tolerated lateral acceleration is 4 m/s², according to this guideline, the driver's model decelerates in the curves [9]. The velocity drops down due to this limitation of the lateral acceleration. The Grossglockner downhill drive is driven with maximal load capacity.

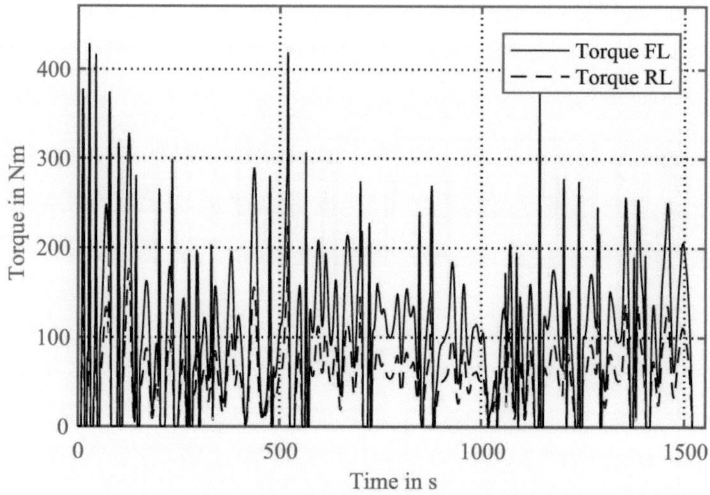

Figure 3-4 Brake torque at GG (FWD)

During the GG, the load on the brakes varies greatly. The mean brake torque is 107 Nm at each front wheel and 58 Nm at the rear wheels. Nevertheless, there are also peaks up to 430 Nm at each front wheel and 230 Nm at the rear wheels. In an idealized simulation with a constant slope of 9.5 %, the need brake torque at each front wheel is 118 Nm and 64 Nm at each rear wheel.

4 Analysis and aim

The three presented tests were analyzed with different vehicle parameter settings in order to draw comparisons in terms of the effects of the brake performance and the necessary dimensioning.

AMST	Overall vehicle weight, performance, number of subsequent brakings
HFT	Overall vehicle weight, performance, top speed, number of subsequent brakings
GG	Overall vehicle weight

Table 4-1 Analysed aspects

The effect of the real continuous driving performance of the latest electric vehicle has been observed by using the speed progression of two HFT driving tests as a driving target for the driving model in the simulation.

To consider the potential for deceleration of the electric vehicle in Section 5, the effects of virtual energy sinks in the powertrain were taken into account. The energy sinks can incorporate constant performance in the model and were analyzed in 10 kW steps.

The aim of the examination is to analyze the potential of downsizing the necessary brake performance in electric vehicles. The diminution of the requirements would enable a smaller dimensioning of the previous brake concepts and therefore move the constructional aim closer to the performances that are necessary in everyday life. The expected result would be to reduce the consequences, which emerge from low frequented actuations.

4.1 Overall weight

It is obvious that the overall weight has a direct influence on the necessary brake performance. In the GG, this influence can be observed directly in Figure 4-1. The overall weight was varied from 40 % above to 40 % below.

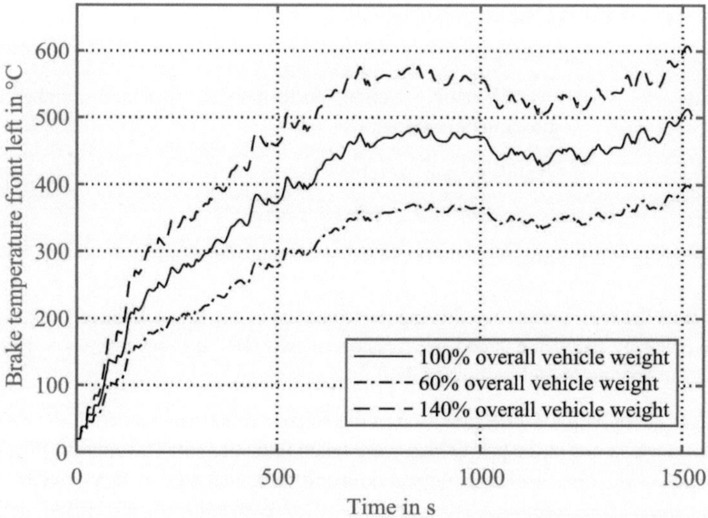

Figure 4-1 Brake temperature at GG with different weight (FWD)

The bigger or smaller mass has to be decelerated, and what follows is a steady increase or decrease of the temperature of the brake discs. The original temperature at the end of the test ist 513 °C. The variant with only 60 % of the original weight ends with 399 °C (-22 %) and the vehicle with 140 % overall weight ends with 607 °C (+18 %).

In contrast to the GG, the overall vehicle weight in the HFT and the AMST shows both, negative and positive effects. More brake performance is needed for the deceleration of the vehicle. Due to the changed weight, the period of the acceleration phases between the braking changes, and the wheel brake has more time to cool down. Both influences are depicted in Figure 4-2.

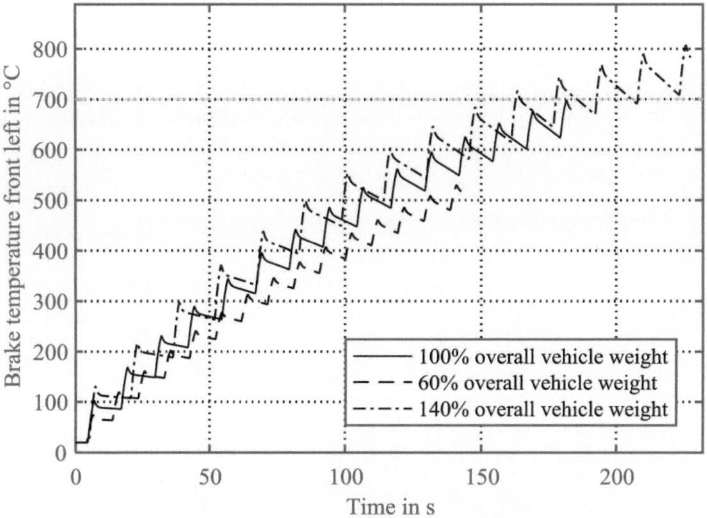

Figure 4-2 Brake temperature at HFT with different weight

The changed period of the whole trial is illustrated in Table 4-2.

	HFT cycle time	**peak temper-ature**	**AMST cycle time**	**peak temper-ature**
100% weight	186 s	698 °C	110 s	684 °C
60% weight	144 s (- 22 %)	530 °C (- 24 %)	97 s (- 12 %)	514 °C (- 25 %)
140% weight	231 s (+ 24 %)	807 °C (+ 16 %)	131 s (+ 19 %)	792 °C (+ 16 %)

Table 4-2 Simulation time and temperature for different weight

The consequence of a longer acceleration time is more clearly apparent during the HFT, since the vehicle needs to be accelerated within a higher speed range in which, due to the influence of the aerodynamic drag, the required performance increases to a greater extent.

The analysis shows a higher level of temperature that, however, turns out to be less drastic than the differences regarding the heating of the single braking processes in Figure 4-2.

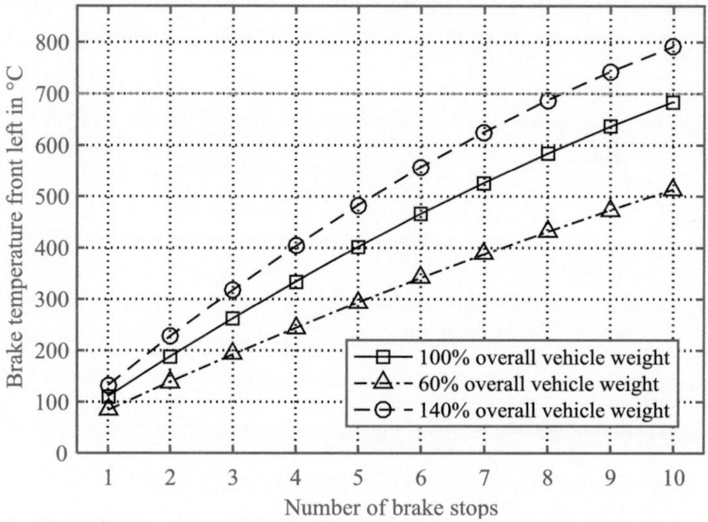

Figure 4-3 Brake temperature at AMST with different weight (FWD)

A similar picture can be seen with the AMST in Figure 4-3.

It can be stated that the overall weight of the vehicle influences the wheel brake temperature proportionately, despite the opposed influence at the multiple stop tests.

4.2 Engine performance

The driving power of a vehicle influences the ability to accelerate. This influence has to be compared with the influence of the overall weight at constant braking performance demand.

	HFT cycle time	peak temperature	AMST cycle time	peak temperature
100% power	186 s	698 °C	110 s	684 °C
60% power	320 s (+ 72 %)	547 °C (- 22 %)	160 s (+ 46 %)	626 °C (- 8 %)
140% power	149 s (- 20 %)	752 °C (+ 8 %)	102 s (- 7 %)	704 °C (+ 3 %)

Table 4-3 Simulation time and temperature for different power

As the amount of acceleration in GG is marginal, no effects can be observed through the altered vehicle performance. Effects on HFT and AMST are shown in Table 4-3 and in Figure 4-4.

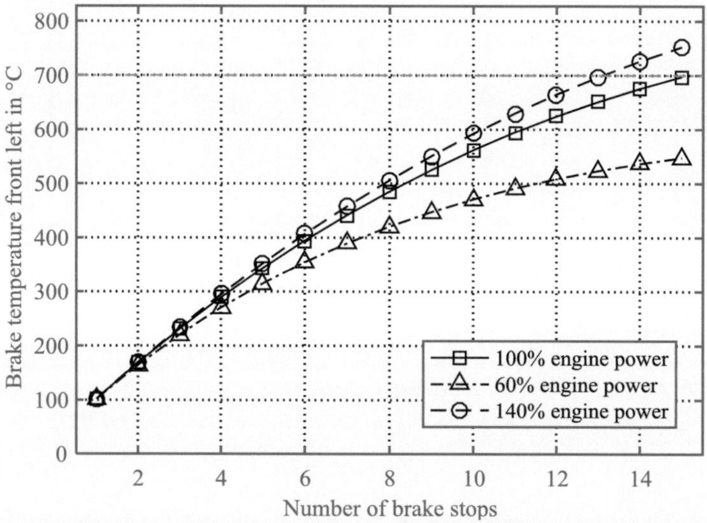

Figure 4-4 Brake temperature at HFT with different power (FWD)

In HFT and AMST, the differences in the time of the maneuver are directly reflected in the behavior of the temperature. The longer acceleration times allow more time for the wheel brake to cool down. As an irregularity, the AMST variant with 140 % power has to be identified. This can be ascribed to the fact that the tire model is overstrained with the traction torque and thus the driver's model downsizes the driving power.

For the multiple stop tests, the driving power is included proportionately in the expected wheel brake temperature.

4.3 Top speed

In HFT, the initial speed of the succeeding decelerations are directly connected to the top speed of the vehicle. It can be expected that this has a considerable influence on the brake temperature.

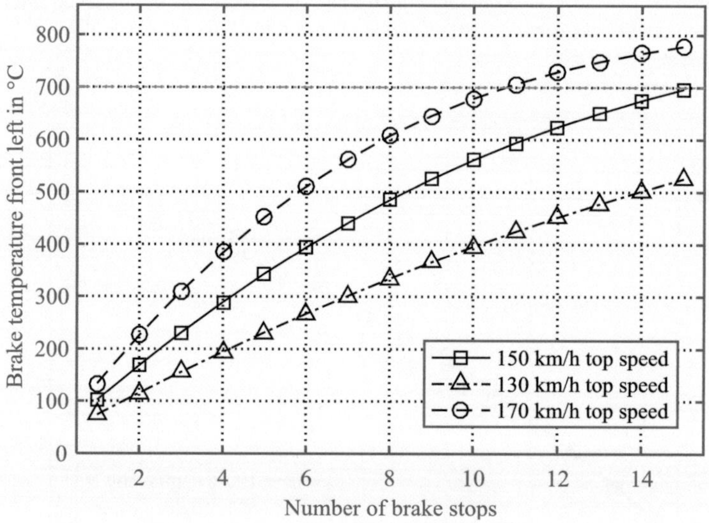

Figure 4-5 Brake temperature at HFT with different top speed (FWD)

The times of the individual maneuvers are depicted in Table 4-4. In this trial, the transmission ratio of all three vehicles was set to 9.7 to reach the higher top speed of 170 km/h.

	HFT cycle time	peak temperature
150 km/h top speed	179 s	700 °C
130 km/h top speed	118 s (- 34 %)	525 °C (- 25 %)
170 km/h top speed	266 s (+ 49 %)	782 °C (+ 12 %)

Table 4-4 Simulation time and temperature for different top speed

The simulation affirms the expected strong influence of the top speed on the brake temperature in the HFT.

4.4 Energy sinks/recuperation

Since a battery model is not integrated, but the important influence of regenerative braking on the friction brakes should be considered, a virtual energy sink was implemented in the DC high voltage sector and varied in 10 kW steps.

The reason for this consideration is the assumption that the battery has to be seen as non-receptive in the worst case and alternative options need to be reconsidered in order to take in the electric energy. In terms of sinks, consumers such as the heater, the climate compressor and the radiator fan come into consideration. A detailed description of possible implementations are not included in this paper, however the topic is revisited in Section 6.

While considering the energy sinks and the possibility of recuperation, the focus lies on the GG and the HFT. The AMST is not considered, as the ABS controller is activated in the maneuver and this activation cannot be realized without adaptation regarding the chosen typology of the powertrain with central engine. Solutions for this issue would be one engine for every wheel or a torque vectoring solution for the single engine. Additionally, the implementation of using the regenerative braking to bring a basic level of braking up to both drive wheels and to implement the overlapping control of the ABS function with the friction brake should be considered. There are many approaches for this topic in [10]. The braking torque at the rear axle has already been adapted to the wheel with the lower level of friction at the select-low regulation in some current braking systems. However, this leads to losses with regard to the braking distance, since the potential cannot be fully tapped into [7, p. 142].

Constant energy sinks are best suited to cover a low continuous power as needed in downhill driving. Figure 4-6 depicts the GG downhill drive with activated energy sinks.

In the charts shown in this section, it needs to be factored in that, in consideration of the rear axle, the vehicle model was used with a rear wheel drive, since otherwise it could not be recuperated at the axle. The temperature level of the rear axle in the RWD vehicle is higher than the one in the FWD vehicle, since this vehicle model has more weight on the rear axle, there is also a larger portion of the whole brake power on the rear axle.

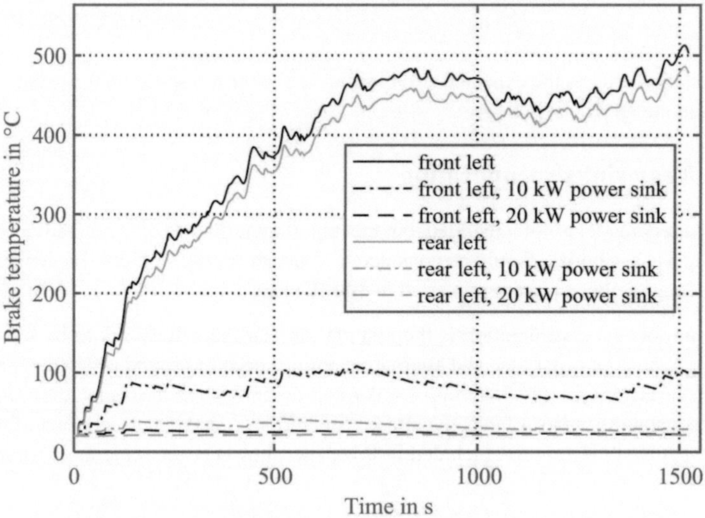

Figure 4-6 Brake temperature at GG with different power sinks

It becomes apparent that already a small recuperative performance is sufficient to considerably reduce the temperature level of the respective affected axle. From 20 kW on, only a minor rise in temperature can be observed on the front axle; on the rear axle, this is already easily reached starting from 10 kW.

Figure 4-7 and Figure 4-8 illustrate the HFT with energy sinks.

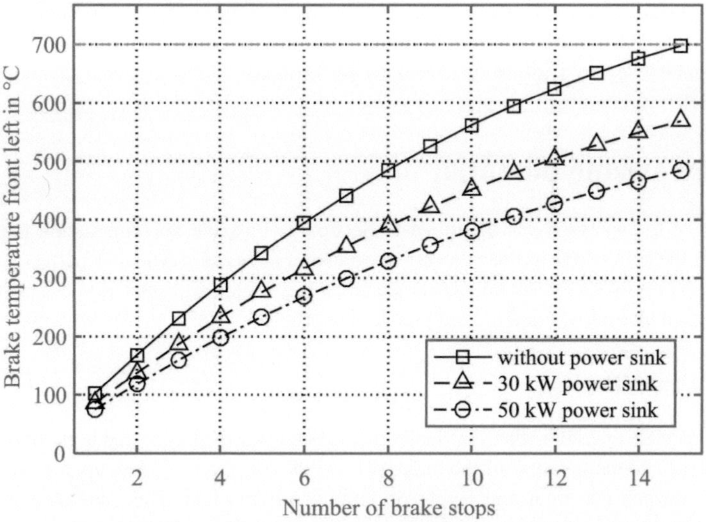

Figure 4-7 Brake temperature at HFT with different power sinks (FWD)

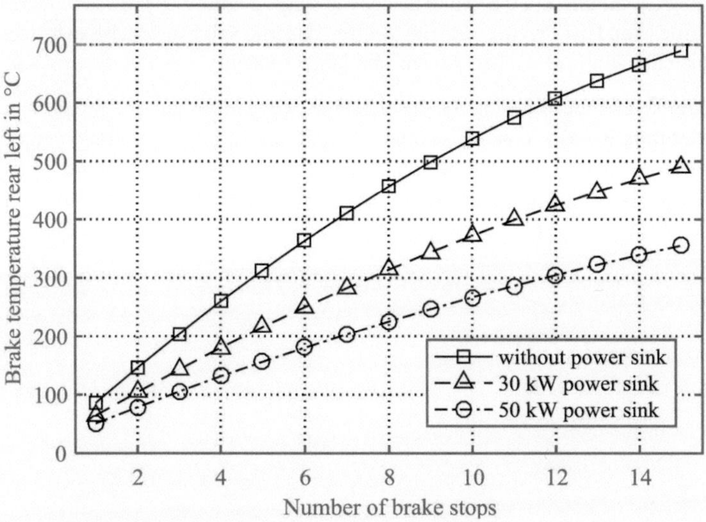

Figure 4-8 Brake temperature at HFT with different power sinks (RWD)

As was expected, the brake temperature reduces depending on the performance of the energy sinks. The relative reduction is larger on the rear axle because the absolute brake performance is lower. The absolute breaking performance of every wheel is depicted in Figure 3-1.

5 Downsizing potential

The aim of the analysis was described in Section 1. Possible implementation is presented in the following. In this respect, the question arises as to whether performance tests in the previous form, especially for electric vehicles, are still appropriate. The possibility of an alternative requirement profile is examined.

5.1 Initial situation

With the AMST and HFT, the test ideally includes an identical, repeated input of braking energy. The temperature of the brake discs rises with every follow-up stop until a balance between the input and output performance is reached. This saturation is far above the acceptable limit temperature.

It can be observed that in driving tests, real electric vehicles cannot perform repeated accelerations as required in the whole maneuver range. Figure 5-1 shows the recreated speed profile of an HFT driving test with a current electric vehicle, which can be closely approximated with the rear wheel driven simulation model.

The progression of speed during the decelerations was known, but the progression of the speed during the accelerations was not. As the times of the acceleration could be concluded from the data, the progression of acceleration could be approximated mathematically. The resulting speed profile has been used as guideline for the driver's profile.

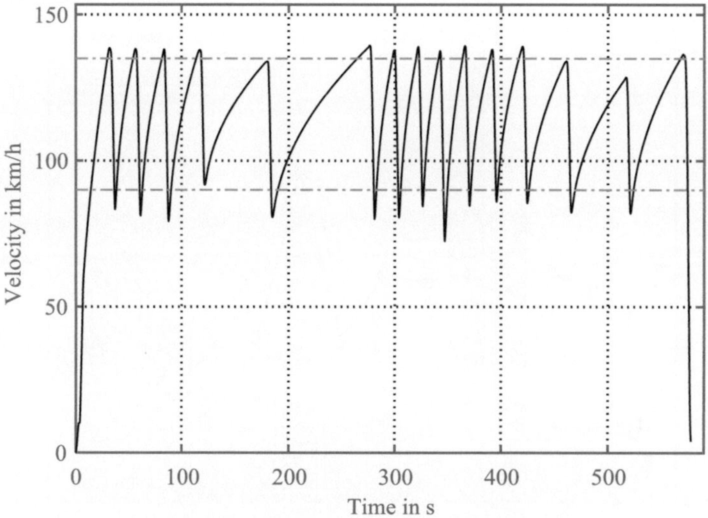

Figure 5-1 Velocity at HFT driving test

It can be seen the available engine performance of the vehicle cannot provide a continuous acceleration for the multiple stop tests. The vehicle accelerates irregularly, although the test driver applies the accelerator pedal completely during each repetition. The assumption for the reason for the shown behavior of the vehicle is that this is due to the derating of the electric powertrain. The required times for the individual accelerations are depicted in Figure 5-2. The evaluation starts with the acceleration after the first deceleration.

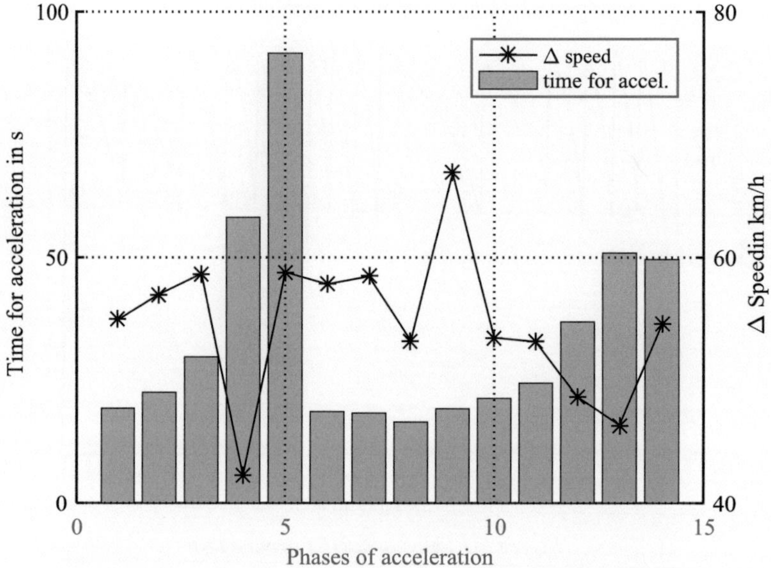

Figure 5-2 Time for acceleration and speed delta at HFT driving test

Along with the time, the value of the speed, which was surpassed at the respective phases of acceleration, are listed. As it is a driving test, there are fluctuations in the initial and final speed.

To deduce it to the existing performance, the simulation model was adapted to the tested vehicles and the engine performance was varied. Since the electric motor and losses in the powertrain are not depicted explicitly and validation has not taken place, absolute values have to be refrained from. The relative fluctuation of the available engine performance is, however, remarkable.

	Acceleration # 5	Acceleration # 8
time for accel.	92 s	17 s
min. speed	81 km/h	85 km/h
max. speed	140 km/h	138 km/h
speed delta	59 km/h	52 km/h
estimated engine power (torque: max. 250 Nm)	26.5 kW	64 kW

Table 5-1 Time for acceleration at # 5 and # 8

Figure 5-2 and Table 5-1 show that the available performance of the observed vehicle fluctuates with 60 % downwards. The consequence is the progression of the temperature after the respective stops, which is depicted in Figure 5-3.

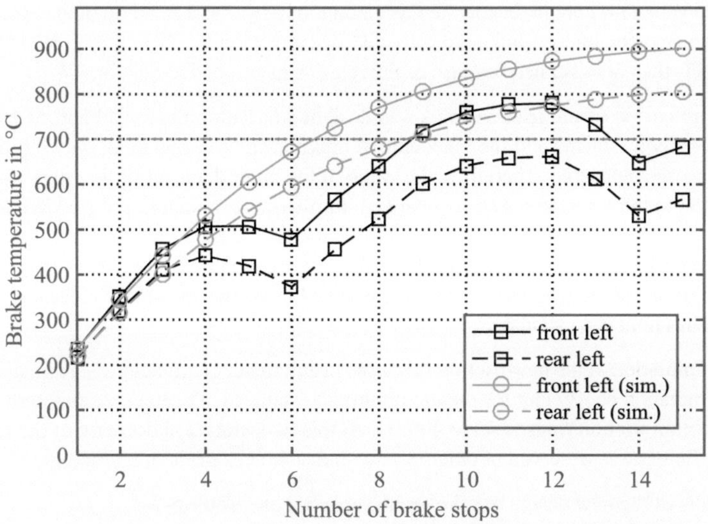

Figure 5-3 Brake temperature at HFT driving test and simulation

After the fourth and eleventh deceleration, the temperature stagnates or drops again. From the ninth deceleration on, the front axle brakes moves in a temperature range that is too high. In addition to the temperature resulting from the speed profile of the driving test, the temperature of a simulation with the same vehicle model and an engine with a constant power of 64 kW is depicted. With this performance at full disposal, the wheel brakes of the front axle would exceed the temperature limit at the seventh deceleration.

The decrease of the powertrain performance can also be observed in driving tests with other electric vehicle. The characteristic of the behavior can be different.

5.2 Process and results

A new requirement specification was generated that is better aligned with the properties of electric vehicles. The number of stops the car has to pass without reaching the temperature limit in the AMST was reduced from 10 to 5 decelerations. The number of the multiple stops in the HFT were reduced from 15 to 7. The GG downhill scenario should be left unchanged as part of the new specification.

A 10 kW energy sink was implemented in the vehicle models analogous to Section 4.4. Thus, the vehicle is able to recuperate this amount of power as a function of the electric machine map. The energy sinks are activated when driving the GG and the HFT. For the AMST, they are disabled because of the reasons mentioned in Section 4.4.

A new dimension of the friction rings was designed based on the model and with the new lower requirements and the regenerative braking. In the simulation, only the friction rings are considered. Therefore, the value of the inner diameter of the rings as the connection to the wheel hub was not changed. Furthermore, aluminum was added as an alternative disc material. The use of an aluminum alloy for the friction ring could minimize the corrosion problems arising from the low frequented actuations in everyday use. The proportion of the friction rings is oriented to known disc brakes, because the stiffness has not been calculated.

For the simulation of the new friction rings, the heat transfer coefficient was maintained and the radiation coefficient for the aluminum was reduced. The limit temperature for the aluminum friction rings is set to 400 °C. At this temperature, a decrease of the friction coefficient was observed in other tests with disc brakes made of aluminum.

The results of the new design are shown in Table 5-2 and Table 5-3.

	FWD front	FWD front new	FWD rear	FWD rear new
material	grey cast	aluminum	grey cast	grey cast
outer diameter	280 mm	290 mm	280 mm	260 mm
inner diameter	177 mm	177 mm	177 mm	177 mm
thickness	22 mm	18 mm	9 mm	7 mm
cooling channel	8 mm	-	-	-
weight	4.8 kg	2.3 kg	2.4 kg	1.5 kg
reduction	→	-53%	→	-40%

Table 5-2 New brake dimension FWD vehicle

	RWD front	RWD front new	RWD rear	RWD rear new
material	grey cast	grey cast	grey cast	aluminum
outer diameter	280 mm	280 mm	280 mm	270 mm
inner diameter	177 mm	177 mm	177 mm	177 mm
thickness	22 mm	8 mm	11 mm	14 mm
cooling channel	8 mm	-	-	-
weight	4.8 kg	2.1 kg	3.0 kg	1.4 kg
reduction	→	-56 %	→	-53 %

Table 5-3 New brake dimension RWD vehicle

The simulation results shown in Figure 5-4 to Figure 5-7 confirm that the new friction rings are sufficient for the new requirement specification. The rings do not exceed the limit temperature. Attention should be paid to the difference in the temperature limit.

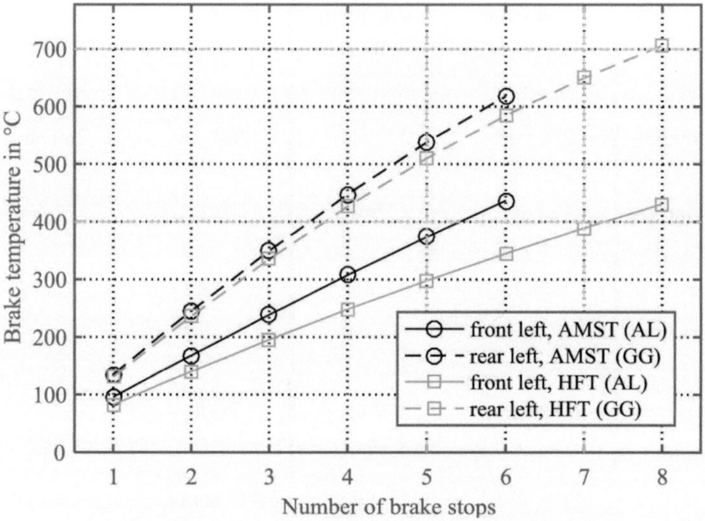

Figure 5-4 Brake temperature for the FWD vehicle

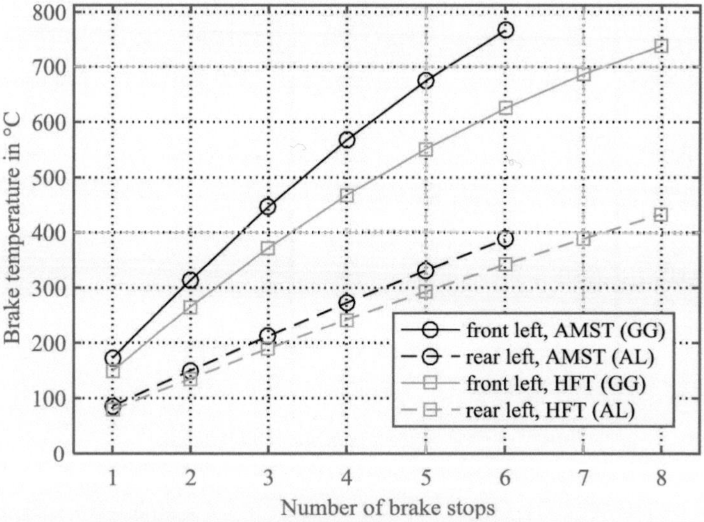

Figure 5-5 Brake temperature for the RWD vehicle

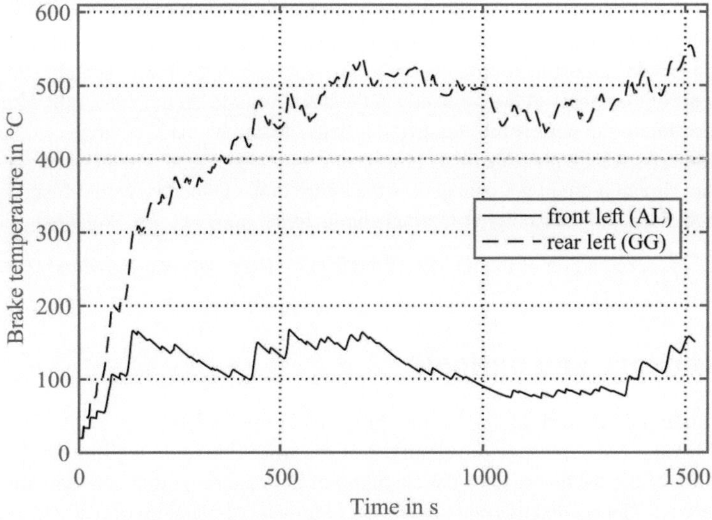

Figure 5-6 Brake temperature at GG with FWD vehicle

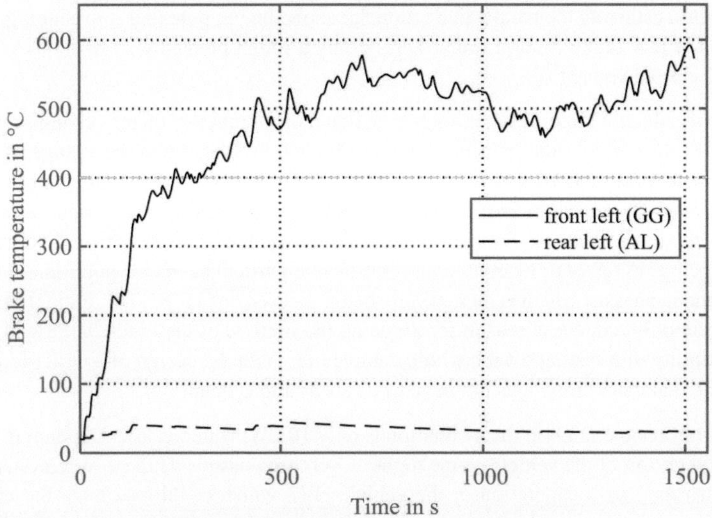

Figure 5-7 Brake temperature at GG with RWD vehicle

At the HFT and the HFT, the wheel brakes with the new friction rings can perform 5 and 7 stops without reaching the temperature limit. At the GG, the 10 kW energy sink is enough to take almost the entire energy at the rear axle in the RWD vehicle and the biggest part of the energy at the front axle in the FWD vehicle. At the other axle without the connection to the powertrain, the friction rings have to be made of grey cast, since the aluminum discs are not able to absorb the energy without reaching the temperature limit. This situation could get changed, if the entire brake geometry would be considered, since aluminum has a higher thermal conductivity value [11, pp. 769-774] [13].

Nevertheless, a reduction of the weight of the friction rings between 40 % and 56 % is possible.

6 Summary and outlook

The simulation framework offers the possibility to consider a defined amount of regenerative braking in the advance development of the wheel brake. In addition to the dimensioning of the friction disc or the changing of the material, other concepts should be considered. The needed thermal models are in process of planning or ready to work, but not validated yet. The aim is to find the best brake concept for the given parameters of a new car.

As an initial estimate, the results in Section 5.2 show the big potential for reducing the mass of the disc brake friction rings. This is only a rough prediction, which has to be proven by prototypical tests.

Instead of reducing the dimension of conventional disc brakes or merely changing the material of the brakes, other concepts could be an option. Because of the change in the requirement specifications and the use of the electric motor as a generator, concepts, which previously caused thermal problems, could now be used. Some closed designs, such as the drum brake, or completely other concepts should be an option. The huge advantage lies in fewer or even no emissions of fine particulate matter pollution in the environment, an issue which is increasingly being focused on [12, p. 117]. If the friction wear is the only remaining reason for reducing the lifetime of the brake, there will be the possibility of it lasting for the whole vehicle life. With the design of a new friction brake, the maintainability does not have to be taken into account.

The level of recuperation in the evaluation is only 10 kW, which is low. Depending on the configuration of the vehicle, some higher levels are possible. If there were systems for electro thermal recuperation in the vehicle [14], which could maximize the efficiency of the car, they could also be used as an energy sink in the case of a full or not conditioned battery.

The other mentioned option of bigger energy sinks would be interesting if the electric vehicle had all-wheel drive with two or more machines. The vehicle could then use the regenerative braking at both axles. This could be the next big step, especially for the Grossglockner downhill drive.

All of these considerations for downsizing the conventional friction brake affect the safety issues of the vehicle. The safety requirements and the reliability of the single components of the brake system are higher than for the components of other systems [1]. Extending these requirements from the brake system to large parts of the powertrain or even to auxiliaries is not a good option. The question is whether there is the necessity to do this. The friction brakes are designed to accommodate several emergency stops without the support of regenerative braking. If there is a failure in one of the systems, which are necessary for the regenerative braking, the system can limit the motor power and indicate the problem to the driver.

Contributions and acknowledgement

This research is funded by Continental Teves. The author would like to thank them for their support.

Markus Lienkamp made an essential contribution to the conception of the research project. He revised the paper critically for important intellectual content. Mr. Lienkamp gave final approval of the version to be published and agrees to all aspects of the work. As a guarantor, he accepts responsibility for the overall integrity of the paper.

References

[1] C. Directive, *71/320/EEC on the approximation of the laws of the Member States relating to the braking devices of certain categories of motor vehicles and their trailers*: Brussel.

[2] B. Heißing, *Fahrwerkhandbuch: Grundlagen, Fahrdynamik, Komponenten, Systeme, Mechatronik, Perspektiven,* 3rd ed. Wiesbaden: Vieweg+Teubner Verlag / Springer Fachmedien Wiesbaden GmbH Wiesbaden, 2011.

[3] Johannes Dietz, Prof. Dr. Eckard Helmers, "Ökobilanzierung von Elektrofahrzeugen," 2015.

[4] Y. Gao, L. Chen, and M. Ehsani, "Investigation of the Effectiveness of Regenerative Braking for EV and HEV," in : SAE International400 Commonwealth Drive, Warrendale, PA, United States, 1999.

[5] D. B. Antanaitis, "Effect of Regenerative Braking on Foundation Brake Performance," *SAE Int. J. Passeng. Cars – Mech. Syst.*, vol. 3, no. 2, pp. 14–30, 2010.

[6] H. D. Baehr and K. Stephan, *Wärme- und Stoffübertragung,* 9th ed. Berlin, Heidelberg: Springer Vieweg, 2016.

[7] B. Breuer and K. H. Bill, *Bremsenhandbuch*. Wiesbaden: Vieweg+Teubner Verlag, 2012.

[8] auto motor und sport, *So testet auto motor und sport.* [Online] Available: http://www.auto-motor-und-sport.de/bremsentest-so-testet-auto-motor-und-sport-11-798987.html. Accessed on: May 01 2017.

[9] M. Wegschweider and G. Prokop, "Modellbasierte Komfortbewertung von Fahrerassistenzsystemen," in vol. 1900, *VDI-Berichte*, VDI Verlag GmbH, Ed., 2005, pp. 17–36.

[10] M. Rosenberger, *Regelung radnaher elektrischer Einzelradantriebe während der ABS-Bremsung*. Zugl.: München: Techn. Univ., Diss., 2013, 1st ed. Göttingen: Cuvillier, 2014.

[11] F. Ostermann, *Anwendungstechnologie Aluminium*. Berlin/Heidelberg: Springer Berlin Heidelberg, 2014.

[12] A. Hainsch, "Ursachenanalyse der PM10-Immission in urbanen Gebieten am Beispiel der Stadt Berlin," 2004.

[13] *DIN EN 1561/Gußeisen mit Lamellengraphit*. Berlin: Beuth, 1997.

[14] E. Sawazki, M. Brüll, S. Deinhard, and C. Baumgärtner, "Potenziale zur Reichweitensteigerung von Elektrofahrzeugen im Winter," *ATZ Automobiltech Z*, vol. 117, no. 1, pp. 36–41, 2015.

Integrated Power Brake – modular set extension for highly automated driving

Dipl.-Ing. Urs Bauer
Dipl.-Ing. Mirko Brand
Dr. rer.nat. Thorsten Maucher

Robert Bosch GmbH

P.E. Pfeffer (Ed.), *8th International Munich Chassis Symposium 2017*, Proceedings, DOI 10.1007/978-3-658-18459-9_48

1 Introduction

With the introduction of vehicle stability systems like ESP®, and passive safety systems like seat belts and airbags, driving safety has been improved and the number of road traffic fatalities and severe injuries have been significantly reduced in the last decades. Although these safety systems are state of the art for new passenger cars, in Germany the number of fatalities remains on a similar level in the latest years. At the same time, the need of the people for individual mobility is still increasing and this is resulting in higher traffic density, traffic complexity, and increasing stress for the drivers. Statistical data shows that the driver remains as an essential source for accidents.

To further increase the driving safety and to relieve the driver of the complex task of vehicle guidance, today nearly all OEMs and suppliers are dealing with ideas for self-driving vehicles or are engaged with highly automated driving (HAD) functionalities.

These trends towards new technologies for highly automated driving, the same as electrification for CO_2 reduction and energy efficiency, are resulting in new challenges and requirements for braking and steering architectures.

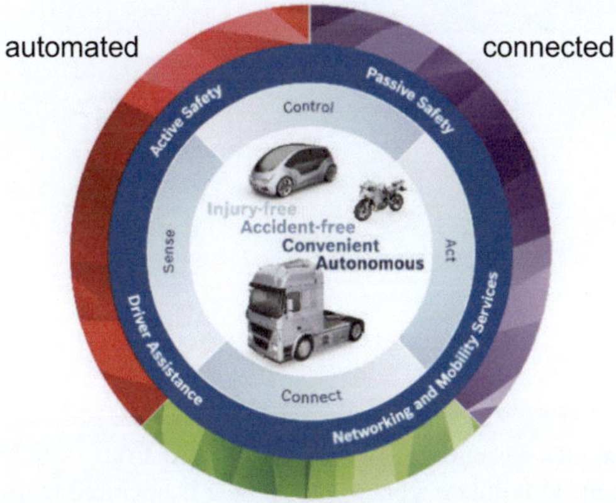

It is expected that these highly automated driving functions will be introduced into the market in the next years. Vehicles shall be able to provide capabilities to drive without helping hand or guidance by the human driver in at least several use cases.

The variety of these use cases where the vehicle shall operate on its own is currently very high and often driven by new business models that come up with the newly available technology.

Beyond the area of assisted functions where the driver has at any time the full responsibility for vehicle operation and guidance, some specific use cases are in focus to be supported by automated driving functions.

The first series introductions for automated driving are currently planned for highway use cases like driving in traffic jam situations or even operating a vehicle on a highway like a human driver does. The advantage for starting with these use cases is primarily the limited vehicle dynamics during these driving situations, and unexpected events are rather rare or excluded by definition of the environment. For example, no pedestrians are crossing the streets without regard to the traffic, no cyclists may change lanes, and no kids are playing near the roadway which might lead to the need for emergency operations. In traffic jam situations, the vehicle and environmental dynamics are reduced to a minimum which is ideal to start with such applications.

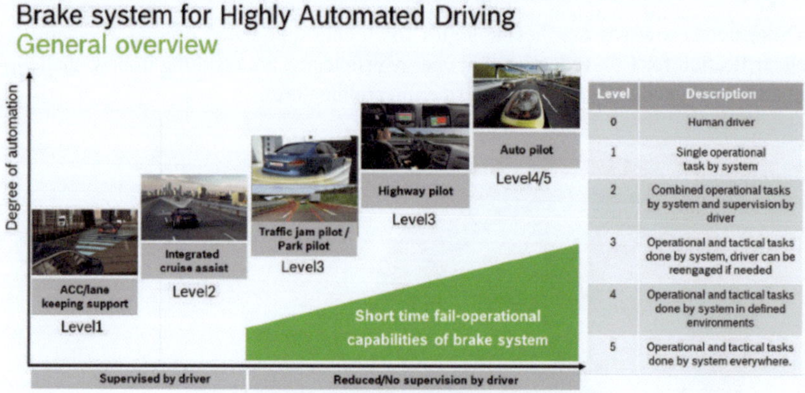

Figure: Different levels of automated driving

Currently applications are in focus that allow the driver to be released from vehicle guidance for a certain time as long as no failure occurs or the vehicle is operated within the defined boundaries (e.g. on released highways or in traffic jam situations). However, the driver must still be there to take over control in case of a failure of the assistance system, but not immediately when the failure event happens. A similar use case with limited vehicle dynamics and unexpected situations might be the automated parking in defined situations. This might result in applications where a vehicle is capable to operate a parking maneuver without a driver at the driver's seat but in the sphere of action to monitor the vehicles environment or to be able to react accordingly in case of any failure.

The next evolutional steps, which are already on track, are applications that allow a car to be operated in urban environments without any human driver in the driver's seat. The more difficult situation dynamic, caused by traffic, or unexpected events, might be covered with higher requirements to the system architecture (e.g. increased number of sensors).

2 State of the art

2.1 Brake system architectures

Through electrification of the drive train and the need for optimized energy-efficiency, new braking systems have been introduced into the market. In addition to conventional vacuum-based servo-brake systems there are new energy-efficient brake systems available like electro-mechanical brake booster (e.g. iBooster), or power-brake systems (e.g. Integrated Power Brake – IPB) which integrate the functions of brake boost and wheel individual modulation into one unit.

For integrated brake systems it is mandatory to consider both legal requirements (e.g. ECE R13-H) and functional safety aspects (e.g. ISO26262). In case of a failure in the brake system, it is assumed that the driver always can control the vehicle by means of the mechanical backup and adapt his driving behavior after a correspondent warning. From functional safety analysis, for example, there is a requirement on the achievable deceleration during transition to mechanical backup: The driver must be able to achieve a minimum vehicle deceleration of $4{,}8 m/s^2$ with a pedal force of 500N (ASIL-A) in a failed brake-boost situation. This higher deceleration requirement in mechanical back-up for integrated brake systems, compared to servo-brake systems, results from failure probabilities for a failed brake-boost situation due to missing redundancy.

Integrated brake systems decouple pedal feel aspects from design of mechanical backup performance. Therefore, it is possible to reduce the diameter of the tandem master cylinder (TMC) without negative side-effects on pedal feel and reach higher hydraulic pressures in the TMC in hydraulic backup with the same input force. Since the pedal framework is rather fixed in today's vehicles, neither the pedal ratio nor the maximum pedal travel can be adapted. The possible TMC stroke stays therefore almost similar and the reduction of TMC diameter results in a reduction of available brake fluid volume in the TMC needed especially for braking in mechanical backup mode.

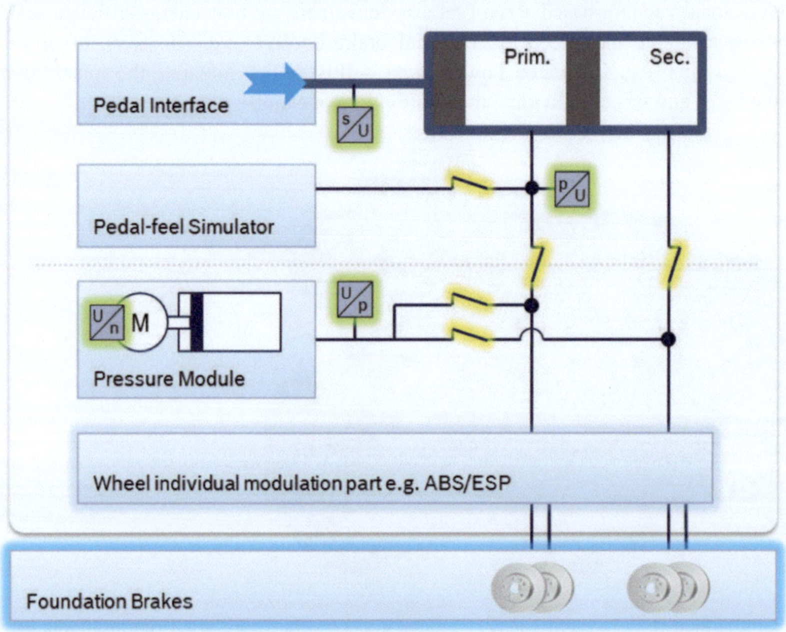

Figure: Schematics of an integrated power brake (IPB)

During brake system sizing of mechanical backup, according to ECE R13-H, it is necessary to consider besides the pedal force for integrated brake systems also the robustness in brake-fluid volume. This is different compared with servo-brake systems where usually for mechanical backup the force criteria is considered only because the conventional

TMC sizing is oriented on volume consumption requirements for brake fading situations and provides therefore enough brake fluid volume for the hydraulic backup.

Before in a potential mechanical backup situation of IPB due to limited brake-fluid the legal minimum deceleration (>2,44m/s²) would not be reached anymore, the driver must be informed (e.g. by warning lamp or text message) that the brake system must be serviced. The driver must be aware that in case of a failure in brake boost then there might be a risk for a reduced brake performance.

For an integrated brake system and volume robustness in brake-fluid, it is important to consider variations of the foundation brake (e.g. taper wear, air, knock-back) over lifetime and at the same time that the driver is not warned too early (prevention of false positive) respectively that the availability of the vehicle is not needlessly reduced.

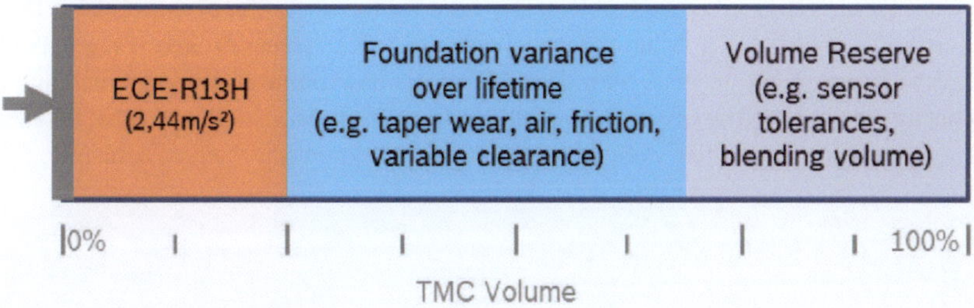

Besides the volume robustness on foundation changes over lifetime it must be furthermore considered in brake system sizing that the sensors used for this evaluation have measuring inaccuracies, which might lead to too early driver warning (false positives).

These considerations result in a conflict of goals for the warning limit:

- No reduced brake performance without driver warning in advance results in the need for reliable and early warning of the driver.
- No false positive warnings respectively no reduced vehicle availability due to warning lamps results in the need of very robust and low residual probability for warning.

Therefore, a certain volume reserve exceeding just the volume to ensure legal minimum deceleration in mechanical backup (failed brake boost) must be considered in brake system sizing. With this reserve in brake fluid volume a good compromise in robustness, safety and availability is possible.

In a vehicle's perspective IPB is only one component of the overall braking function chain. All other components of the braking function chain like pedal ratio, pedal travel or foundation brakes are outside of the system boundaries of the IPB, but also influence the braking performance in mechanical backup. Therefore, the OEMs have the responsibility to ensure that the total functional chain for braking is fulfilling all legal, safety and functional requirements.

2.2 From driver assistance towards highly automated driving

Today we already have very powerful brake systems. These systems are able to decelerate vehicles by command of an external function that assists the driver in several use cases like driving on the highway and keeping the distance to the vehicle driving ahead, changing lanes or also during parking maneuvers. These demands are supported with high dynamics and a very good NVH.

However, although the functionality has reached a very high level of comfort and reliability, the driver is still responsible at any time for the operation of the vehicle. This is especially needed if a failure happens with the brake system or also if the overall assisted function has an issue with detection of the environment, driving situation, or trajectory planning. The driver is responsible to control the vehicle and must be able to reach a safe vehicle state even with a failed brake system (mechanical backup).

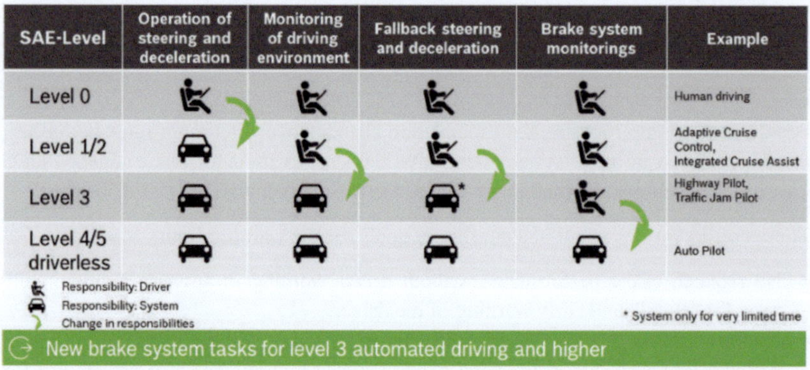

On our way from assisted functions to automated, highly automated or even fully automated systems, the driver is taken step by step out of the responsibility.

The first step from assisted functions (<=Level 2) to automated functions (Level 3) is, that the driver must still be in the driver's seat, but can hand over the complete operation of driving to the automated system for defined use cases. In case of a failure or if the area of the defined use cases is left, the driver must take over the control. In very rare cases, when the driver cannot take over control the automated system must provide capabilities to bring the vehicle to a safe state.

For applications supporting Level 4/5 of automation the driver is no longer mandatory to be in the driver's seat and therefore the vehicle has to maintain operation without support of the driver and bring the vehicle to a safe state.

3 Requirements for highly automated driving

Both, driver assistance functions and highly automated functions result in changed and new requirements for brake systems and brake system architectures. Oriented on aircrafts there are higher needs on the availability of the basic function "braking" because the driver might not take over the task to control the vehicle again after a failure has occurred.

Current brake systems are designed as "fail safe" devices according to the available legal requirements (e.g. ECE R13-H). This means, that failures in the brake system are tolerable as the driver has at any time the responsibility and the capability to take over the control by means of the brake pedal. That is why today a brake pedal with mechanical push through is a mandatory design element in power brake systems. Due to the push through the driver can handle all electrical failures without losing the capability for modulation of the deceleration.

Moreover, today's systems might even make use of the driver's availability, as the driver might be "part" of the functional designs or might take over some functional aspects of functional concepts even without detailed knowledge of this (e.g. driver has to press the brake pedal in order to be able to release the gear latch). Also for mechanical and safety related designs the driver is taken into consideration as it can be assumed that during each braking maneuver the driver might be able to support by applying the brake pedal.

In fact, the driver is the highly available backup for today's brake system and his limitations in performance are accepted. To use the driver as backup and design the brake system as "fail safe" provides a very high availability and accepts on the other hand that single failures might lead to a complete loss of wheel individual modulation or brake boost capability. This approach to use the driver as backup allows at the same time a certain reduction of monitoring demands as some monitoring duties are transferred to the driver (e.g. indication of specific noise or feeling during brake pedal apply).

3.1 Brake system requirements derived from state of the art

In general, it is possible to derive the requirements for the backup mode from the capabilities of a driver operating a brake system. These capabilities, but also restrictions, are known, accepted, and define the state of the art.

The capabilities of a brake system to be able to operate in backup mode without any interaction by a driver must therefore be able to provide a similar performance as a human driver in similar failure situations.

If in a state of the art braking system the ESP unit fails, the driver is fully in charge to operate the vehicle until a safe state is reached. In this situation it might happen that no support by the system is possible so the main modulation capabilities provided by a braking system must be covered by the human driver by brake pedal operation.

Therefore, today's state of the art braking system have a backup that is the human driver.

Backup functionality for HAD

▶ Backup requirements for HAD brake system derived from state of the art

Required functionality	Vac. Booster + ESP Use case: driver braking		Vac. Booster + ESP Use case: ACC, TJA, LKA, ...		HAD ready brake Use case: HAD braking	
	if VB fails	if ESP fails	if VB fails	if ESP fails	if secondary fails	if primary fails
Brake boost/deceleration	Available (HBC by ESP)	Available	Available	Task to driver (w/ boost support)	Required	Task to brake system
Vehicle stability (EBD)	Available (EBD by ESP)	Task to driver	Available	Task to driver	Required	Task to brake system
Vehicle steerability (ABS)	Available (ABS by ESP)	Task to driver	Available	Task to driver	Required	Task to brake system
Wheel individual (VDC)	Available (VDC by ESP)	Not possible	Available	Not possible	Required	Not required

state of the art — state of the art

Requirements transfer: **New system** has to cover **same traffic situations** in backup

➔ Required functionality for HAD brake system can be transferred from state-of-the-art

Nevertheless, it is accepted that in such failure situations the wheel individual modulation capabilities are limited (e.g. a 1-channel modulation by driver to maintain the vehicles stability or provide steer-ability).

This means that for similar failures in automated driving situations comparable strategies or even performance should be sufficient as it is in line with the state of the art.

Correspondent to this state of the art driver it can be derived, that capabilities for the braking system are adequate deceleration, keeping vehicles longitudinally stable (lateral stability control is also not available by human driver in similar failure situation) and also to provide adequate steer-ability on a similar level of a human driver.

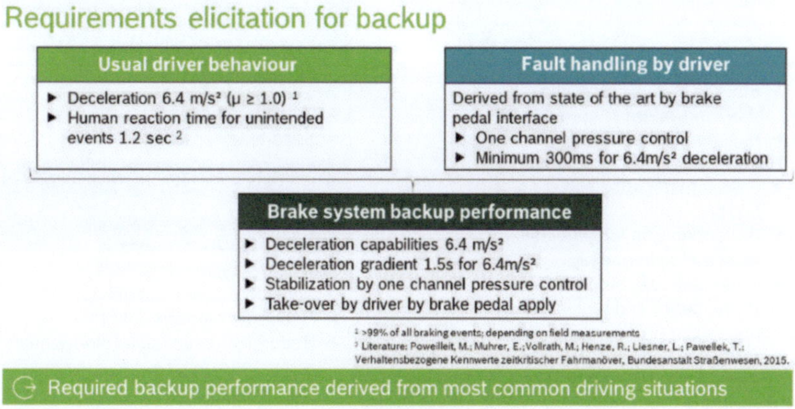

From field data it is known that with a minimum deceleration capability of 6,4m/s² (service brake performance) about 99% of all braking event are covered. So field data are an additional valuable source to derive requirements for backup performance. In most of the situations it is additionally recommended to reduce the speed of a vehicle.

3.2 Brake system requirements from highly automatic driving

Due to the fact that the demands for automated driving rose very fast during the last two years, the development of products according to respective needs and requirements have been intensified.

That is why for automated driving the current approach is to reuse the known and existing fail safe devices and bring some enhancements in to fulfill the requirements for the visions and concepts of the OEM.

However, this means that it is not sufficient to simply double the brake systems actuators and bring some pressure build up or longitudinal stability control capabilities in. In addition, these systems must be enabled to support use cases where no driver is available to take over the responsibility from the previously known and reliable driver (e.g. in terms of monitoring, brake boost or even as part of an overall functional concept).

In fact the biggest challenge along the alley of automated driving is the development of requirements for failure treatment and to extend the fail safe designed devices in order to fulfil the requirements for automated driving.

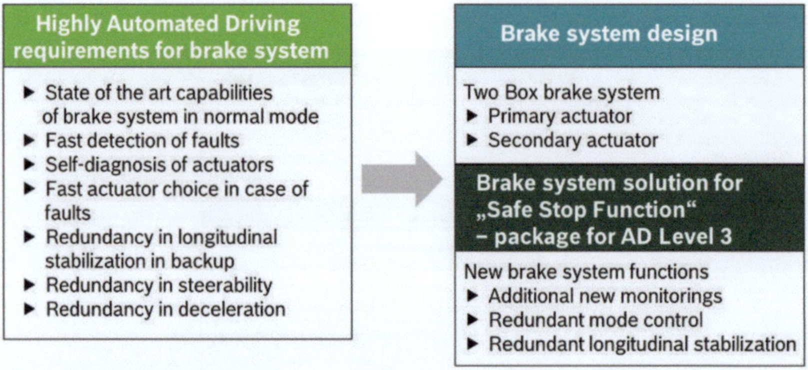

This leads to a change from fail-safe brake to fail-operational brake system architectures.

For automated driving (especially for automated driving >=level 4) the duties covered by the driver must be shared within the system. Therefore, it must be worked out during design of the vehicle system, where duties originally covered by the driver can be allocated.

3.3 Requirements for stabilization in HAD backup

As already discussed state of the art braking systems rely in mechanical backup on the human driver's capabilities. Therefore, in backup mode for automated driving lateral stability control is not required – as it is not available by average human drivers in similar failure situations.

A second argumentation line results from Germany's statistics on road traffic accidents. With assumptions for the distance to the next safe place to stop (e.g. guidelines for road build), and a very worst case assumption for not available lateral stability control function it can be shown, that the statistical benefit to avoid severe injuries with a lateral stability control intervention during backup operation in highly automated mode would be very low.

ESC Effect in Field for 100% market penetration (based on data from Germany 2014)		
Potentially avoided accidents with injured people projected to 100% ESC penetration	~ 37650	Bosch CR/AEV1 Accident Research*
Mileage driven by 44.4 million cars	~ 627 billion km	Federal statistic office
Distance to safe state in automated driving	< 5 km	Worst case assumption
Assumption on failed lateral stabilization functionality	< 2.5 %	Worst case assumption
Share of mileage driven in HAD backup for 100% penetration and usage of highly automated driving	< 0,0009%	Calculated on above worst case assumption

Yearly 37650 serious accidents potentially avoided by ESC

Full System Mode ESC (99,9991% ~ 36649,66)

< 0.0009% (0.33) serious accidents per year in HAD backup not avoided (under w.c. conditions)

* Source: "Evaluation of effectiveness of ESC" (2013) and 2015 updated values for study

All the available statistical data is based on vehicles that are operated by human drivers. Because these drivers are also a source for accidents, it can be concluded that this evaluated forecast is rather conservative. If considered further that automated driving functions will have a lower probability for faulty behaviour than humans, traffic rules and physical limits are usually correctly interpreted, the statistical benefit seen from today's data is higher than it is expected for highly automated driving.

3.4 Brake system requirements to E/E-architecture

The E/E-architecture must also prevent common-cause failures, which might impact the primary brake system as well as the secondary brake system. A solution for this might be the separation of functionality and availability for e.g. the wheel speed sensors.

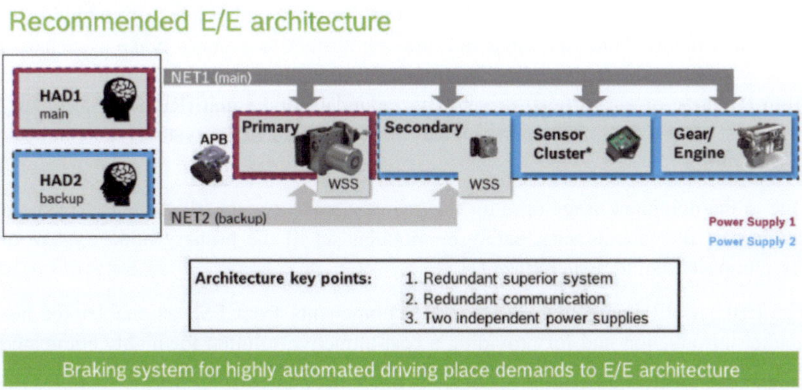

Redundancy is a possible solution but to just simply double or even triple every component (like e.g. in planes or nuclear power plants) would result in a very high level of complexity and high cost for a single vehicle.

Therefore, a tradeoff between availability of the braking functionality (even in failure situations), safety aspects, and cost is needed.

With two independent actuators redundancy in braking capability for Level 3 and Level 4/5 applications is possible. Nevertheless the availability of braking function is limited also by other vehicle components like sensors, communication network or power supply. Therefore, the braking system in highly automated vehicles requires elementary infrastructure in order to operate in a safe and reliable way. Two independent power supplies should be available and redundant communication networks between the braking system actuators and the redundant HAD guiding unit. Redundancy in essential sensors is also mandatory to fulfill the requirements for braking systems for automated driving vehicles.

By having this elementary redundancy the braking system for automated driving can be designed to operate as one virtual unit with a very lean interface to the superior HAD guiding unit handling the needed redundancy for braking on its own.

3.5 Requirements related to integrated brake system

Independent of the current evolution towards electrical and highly automated driving it is assumed that the brake system will remain on a hydraulic technology basis in the next years.

To ensure that safety relevant brake systems are accepted in the market the residual failure rates for the capability to brake must be very low (independent from highly automated driving). Thus, the integrated braking system as primary brake system will usually build up the requested brake pressure independent if the brake request comes from a human or an artificial driver. The redundant brake unit (RBU) as secondary unit must ensure the braking functionality if the primary brake systems fails, but this is very rare.

Due to the dominant usage time for the primary brake system (IPB) the RBU should not impact the functionality, safety or modular set of the primary brake system or these impacts should be minimized.

The RBU itself should reuse established components from ESP as enabler for fast market introduction and for providing a cost-attractive solution for highly automated vehicles.

4 Concepts for IPB with redundant brake unit

4.1 Possible Hydraulic architectures

For highly automated driving functions, integrated brake systems like IPB have to be extended by a redundant brake unit (RBU) in order to ensure active deceleration, longitudinal stabilization and steer ability of the vehicle in all single failure conditions.

The following figure shows the basic hydraulic architecture of IPB and several different possibilities of how to connect a RBU with IPB

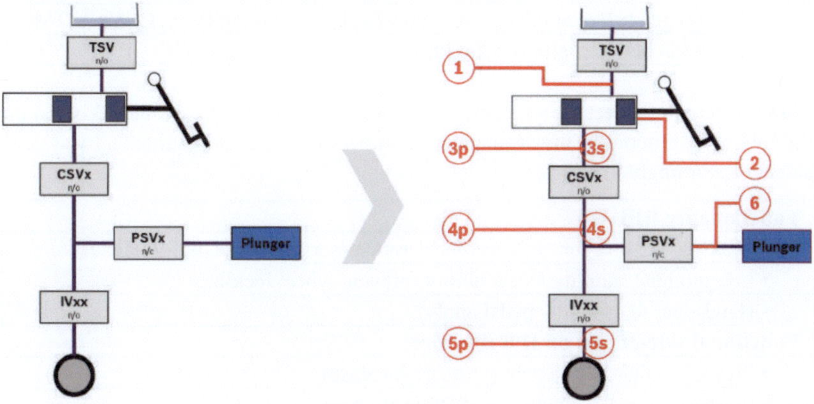

To distinguish the brake fluid supply-path of a potential RBU the concepts are labeled with "p" if the supply-path is connected directly to the brake fluid reservoir. The "s" is used to indicate an inline supply-path through the TMC. The circuit separation valves (CSVx) are normally open and can connect or disconnect the TMC from the wheel circuits. Together with the plunger separation valves (PSVx) which are normally closed it can be chosen if either by-wire mode or mechanical backup mode is active. The inlet valves (IVxx) and not shown outlet valves are conventional ABS valves to modulate the wheel pressures. With an additional test separation valve (TSV) it is possible to close the connection between TMC and the brake fluid reservoir.

A description of different possibilities to connect the RBU and IPB is shown in the following table.

	Connection of RBU to IPB	**Volume supply path**
1	Feed in above TMC (TSV must be closed)	Additional path to reservoir
2	Feed in TMC, move TMC1 piston	Usage of TMC backup path
3p	Feed in between TMC and CSVx	Additional path to reservoir
3s	Arranged between TMC and CSVx	Through IPB TMC
4p	Feed in between CSVx, PSVx and IVxx	Additional path to reservoir
4s	Arranged between CSVx, PSVx and IVxx	Through CSVx and TMC
5p	Feed in between inlet valves and wheels	Additional path to reservoir
5s	Arranged between inlet valves and wheels	Through IVxx, CSVx & TMC
6	Feed in between plunger and PSVx	

These different connection arrangements of RBU with IPB were evaluated based on the following criteria. In the evaluation the importance of each requirement was considered accordingly.

Functionality RBU
• Braking pressure for >6,4m/s² deceleration (service brake performance)
• Longitudinal stability / steer ability (prevent wheel locking)
• Hand-over to driver by pedal apply
Functional side-effects on IPB (by-wire)
• Pressure build-up dynamic (pressure increase)
• Pressure decrease dynamic & ABS application
Safety impact on IPB (mech. backup mode)
• Additional leakage paths in mechanical backup mode
Complexity
• Electrical and communication network
• Design RBU and impact on IPB modular set
• System architecture mixed platforms w/ APB
• Hydraulic complexity

Because the RBU is active for hydraulic pressure build-up only when the primary brake unit (IPB) has failed it is obvious that the activation time of a RBU over vehicle lifetime is very low. Usually the IPB unit performs brake requests and therefore this leads to the conclusion that the side-effects of a RBU on the IPB functionality and safety should be minimized. From a functional point of view, this is relevant especially for pressure build-up and pressure decrease dynamics

Low complexity in electrical and communication network has also to be considered with high relevance, especially for mixed platforms (with/without HAD) to avoid high variance in E/E architecture. Furthermore the consistency of architectures with automated parking brake (APB) for mixed platforms has to be evaluated.

Beside the design and packaging impact on IPB modular set, also the functionality of the RBU itself is relevant. Service brake performance, longitudinal stabilization and steer ability like a human driver can do must be available. It is further necessary to detect at any time that a driver intends to take-over the control by applying the brake pedal.

4.2 Hydraulic RBU Design

Based on QFD analysis the concept 3s is a suitable solution for the extension of IPB modular set.

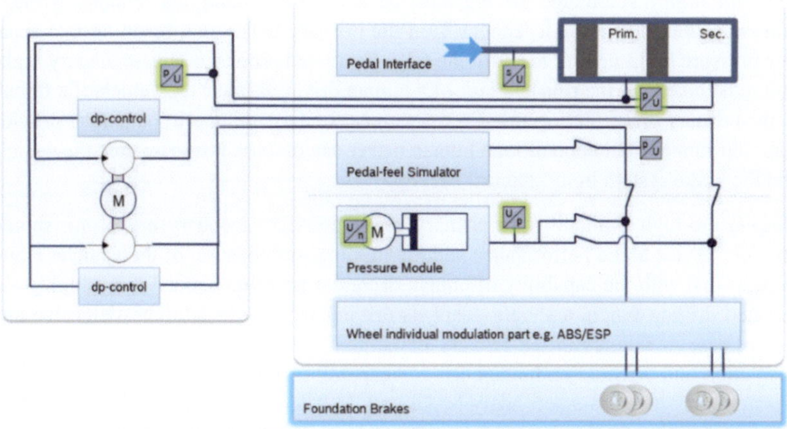

The main advantages are:

- IPB core function is not affected, minimum impact on IPB software and hardware
- Minimized side-effects on IPB from safety perspective
- Capability to apply hydraulic pressure at four wheels for service brake performance
- Longitudinal vehicle stability, wheel locking prevention and steer-ability
- Detection of driver take-over by pedal apply at any time
- No function sharing between IPB and RBU necessary (low networking complexity)
- Prevention of different architectures for automated parking brake (APB) in mixed platforms (with and without HAD)
- Scalable design for HAD and non-HAD vehicles with minimized effort

For the compatibility within the modular set of IPB certain premises have to be considered. An IPB with additional hydraulic ports for the connection of a RBU should have no differences in components, the same ECU/HU-interface to use the same ECU, same positions of the wheel ports and the same outer shape for packaging reasons.

It is possible to determine design solutions to fulfill these design requirements and to extend the modular set of IPB with minimum variance.

The design of RBU can reuse established components from ESP like valves, pumps and electronics as well as the experience from high volume and high quality manufacturing and assembly processes.

5 Summary and outlook

The brake system of passenger cars is safety relevant and therefore brake unit extensions for highly automated driving must be safe, controllable and reliable. A brake unit extension must be independent from the primary brake unit, resulting in a separate pressure build-up unit to fulfill the fail-operational property. This secondary brake unit must take over the role and task of a human driver to brake the vehicle if a failure in the primary brake unit occurs. Therefore, a secondary unit has to be able to provide a similar range of functionalities a human driver can do if he has to operate the vehicle and the brake system in mechanical backup.

Despite the high availability of primary brake units, a secondary brake unit should provide service brake performance and longitudinal stabilization of the vehicle. However, in line with the capability of human drivers it is recommended that a highly automated driving function should adapt its driving strategy to be more defensive and safe in case of failures in the primary brake unit. By this intelligent driving strategy adaptation and oriented on human driver's capabilities in these kind of very rare failure situations it seems a secondary brake unit does not need to provide a total redundancy in functionality of the primary brake and stabilization unit.

The extension of an integrated power brake system (IPB) results in the concept of the redundant brake unit (RBU). To connect the IPB and the RBU additional hydraulic brake pipes are needed.

Generally, the RBU is active only if the IPB unit fails. This is very rare and therefore the RBU should not affect the IPB in functionality or safety. This might consequently result in some conflict of goals and compromises in RBU functionalities have to be considered. The RBU approach outlined in this report shows minimum side-effects on the IPB's functionality and safety. The additional ports to connect the RBU with the IPB are further considered in the design and the variance of the modular set.

A further integration of the RBU into the IPB is often discussed. Unfortunately the independence by redundancy cannot be realized for all components of an IPB (e.g. difficult for gears or seals) so that a single failure might affect both paths of redundancy when integrated in one box and result finally in a total loss of braking capability for highly automated vehicles. From today's point of view the experience from field data on the reliability of these components, which cannot be realized in a redundant way, is not deep enough. Before doing this potential next step of integration it is therefore necessary to determine high-grade quality data on the reliability of these potential common cause failure components based on field data, to ensure conformity with ASIL-D requirement from highly automated driving function.

TIRE.WHEEL.TECH SECTION

DEVELOPMENTAL AIMS AND OBJECTIFICATION

Passenger car vs. truck tires – different development priorities

Stephan Brückner, Hankook Tire Europe GmbH

This manuscript is not available according to publishing restriction. Thank you for your understanding.

P.E. Pfeffer (Ed.), *8th International Munich Chassis Symposium 2017*, Proceedings, DOI 10.1007/978-3-658-18459-9_49

Prediction of driver's handling assessment using a general regression neural network

Saskia Monsma
HAN University of Applied Sciences, Arnhem, The Netherlands

P.E. Pfeffer (Ed.), *8th International Munich Chassis Symposium 2017*,
Proceedings, DOI 10.1007/978-3-658-18459-9_50

Abstract

This paper describes a method for predicting the driver's subjective assessment of tire handling, based on vehicle dynamics measurements taken during driving a double lane change. Often used methods for regression have limitations that prevent their proper implementation for tire handling data. A General Regression Neural Network (GRNN) is developed in this research to model the driver's subjective assessment, as this modelling technique does not have these limitations and additionally, offers a clear interpretation of the method and results. The results show that the GRNN can predict the subjective handling aspects scores well, even for handling aspects not directly related to driving the double lane change. Analysis of the important measures used by the GRNN provides additional information on the vehicle dynamics behavior relevant for the driver. For general handling behavior, regarding all aspects, this showed that metrics based on lateral velocity, steering wheel moment and vehicle slip angle were most relevant.

1 Introduction

What defines good vehicle handling is the topic of many discussions, not only in technical and scientific context, but also in normal day life. Although it is a broad concept, there is common agreement on the fact that it has to do with desired 'cornering feel' of the vehicle and that it contributes to the vehicle's active safety [1]. Next to being functional, handling is also a significant part of the ‘fun factor’ of driving a vehicle [2]. Tires have a significant influence on vehicle handling, often referred to as tire handling. To develop tires that contribute to good vehicle handling is therefore an important objective for tire manufacturers. Handling aspects like stability, response times or sensitivity can be measured objectively to see if they match desired values, but in the end, the driver decides how the vehicle handling ‘feels’. This makes handling subjective: similar vehicle behavior can be perceived differently by drivers, depending on their experience, preferences and expectations. In early research, it was even stated [3] that 'the driver plays by far the most important role in vehicle handling'. Therefore, tire manufacturers use driver assessments as the important final handling test (Figure 1). The tire test drivers are in general highly trained professionals able to perform tests consistently and having good skills for perceiving even small handling differences. Unfortunately, these subjective assessments are expensive, time consuming and can only be performed when a tire is manufactured.

Figure 1 Tire cover-up for tire handling subjective assessments.

Although subjective assessment provides the tire manufacturer with the driver's opinion, they gain limited insight in what specific vehicle behavior influences this opinion. An obvious way to find out seems to be 'ask the driver', but this is not so straightforward.

While test drivers can add additional comments to their assessments, it is difficult to put 'tire feeling' into words. More insight could help the tire manufacturer in understanding or predicting the driver's subjective assessment and including this knowledge earlier in the tire development process.

This paper describes a method derived in previous research [4], for predicting the driver's subjective assessment of tire handling, based on vehicle dynamics measurements. Often used methods for correlation or regression of objective and subjective data have limitations that prevent their proper implementation for tire handling data, as explained in the next section. A General Regression Neural Network (GRNN) is used in this research to model the driver's subjective assessment, as this modelling technique does not have these limitations and additionally, offers a clear interpretation of the method and results. The results are presented and discussed in section 3 and conclusions are given in the last section.

2 Method

2.1 Subjective Tire Handling Experiment

For this research, two professional tire test drivers evaluated six different tires. The tires were selected based on their handling performance derived from data, test results and experience. Figure 2 shows the relative expected handling performance for all six tires.

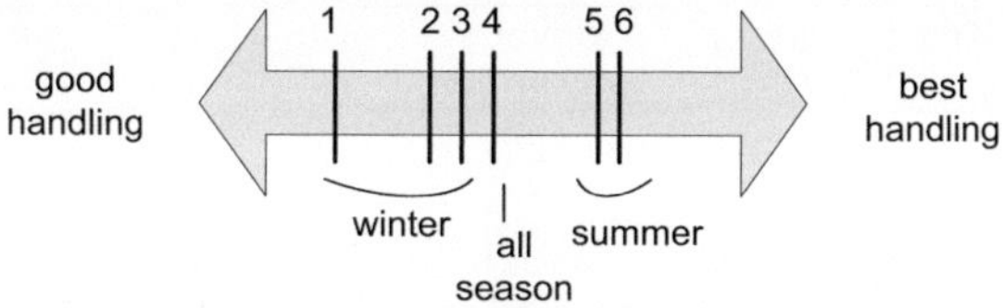

Figure 2 Relative expected handling performance for all six tires.

The tire testing was held on a proving ground with an instrumented vehicle, measuring the relevant vehicle dynamics like (angular) velocities and accelerations in x,y,z vehicle directions, steering wheel angle and moment. In the first part of the test, the driver could steer freely with varying speed. In the second part, a double lane-change maneuver was set up with cones through which the drivers drove several times during one test. The offset between the lanes was set to 5.5 m, instead of the 3.5 m prescribed in the ISO double lane change [5] to make the maneuver more challenging for the drivers without having to drive with very high speed.

The drivers were free to adapt their speed but were instructed to drive in third gear with maximum velocity in the double lane change without hitting any cones.

Immediately after each test, the driver filled in the assessment form for that tire by scoring several handling aspects from 1 to 10 (Table 1).

The subjective assessments were performed with so-called blind testing, i.e., the driver had no information of which tire was tested. Similar tires were assessed in groups, called batches. This resulted in three batches that included the following tires:

- Batch A: tires 1 and 2
- Batch B: tires 2, 3 and 4
- Batch C: tires 5 and 6

Tire 2 was included in both batch A and batch B, to get a good comparison. Both drivers performed a total of seven assessments (6 tires, tire 2 twice), identified with batch letter and tire number from A1 to C6. During each assessment 3 to 5 double lane changes were driven.

2.2 Research Approach

Seen from a system modeling view, the assessment of the driver can be defined as the process of interest, with as the input the vehicle handling behavior and as output the driver's subjective opinion of the vehicle handling, as visualized at the top part of Figure 3. Regression models are used for predicting output data based on input data. Consequently, the input and output must be converted to suitable data.

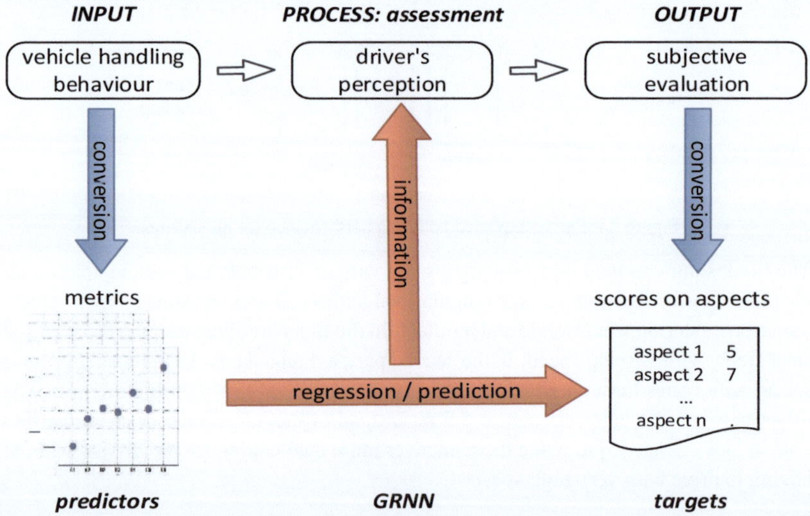

Figure 3 Overview of the research approach.

For the input data, metrics are derived from the vehicle dynamics measurements (left-hand side of Figure 3). Ideally, these metrics should include characteristic data that are distinctive for the different tires. As example, Figure 4 shows the two time plot of the steering angle signal of the double lane change driven by the same driver for tire 1 and tire 6. From this plot, the maximum peak value of the steering wheel angle is clearly different for the two tires and could therefore be a suitable metric. The metrics that are used for prediction are referred to as 'predictors'.

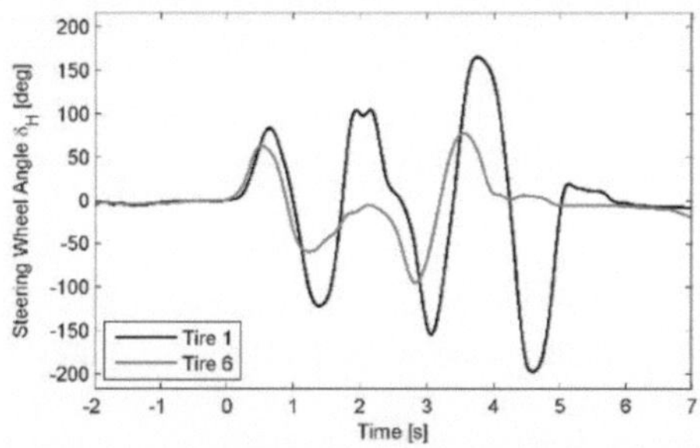

Figure 4 Two steering wheel angle signals belonging to double lane changes driven by the same driver with different tires

For this research, only measurements of the double lane changes are used. Having well-defined boundaries and driven at comparable speeds, tire distinctive metrics could be derived much better than for the free driving part, where the drivers were allowed more freedom to use the throttle and choose their path. Certain handling aspects regarding stability while cornering with no throttle change and with throttle change, were specifically determined while free driving, as stated by the test drivers. Therefore, the assumption is that these aspects will not be predicted as well as the aspects related to driving the double lane change. (Remarkably, the results invalidated this assumption; the GRNN was also able to predict these aspects well from only double lane change metrics.) In total 29 suitable metrics were derived from the measurements. These values for all tire assessments of both drivers were taken and in addition, the mean value of the drivers for every assessment was taken. To eliminate the negative effects of scales and units, for each metric the values are standardized to zero mean and unity standard deviation. This gives an input metric dataset of three dimensions: 3 x 7 x 29, for every driver (driver 1, driver 2 and driver mean), for every tire assessment (A1 to C6), 29 metric values could serve as predictors in a regression.

The conversion to output data is done by the drivers through scoring the handling aspects (right-hand side of Figure 3), which are given in Table 1.

Table 1 Handling aspects for the subjective assessment of tires.

1	Steering precision while cornering
2	Stability while cornering (no throttle change)
3	Stability while cornering (throttle change)
4	Yaw overshoot
5	Predictability
6	Yaw delay
7	Steering angle
8	Grip
9	Controllability
10	Overall judgment

These scores are referred to as 'targets'. This gives a similar output score dataset of three dimensions: 3 x 7 x 10, with 10 aspect scores that serve as targets in a regression. The scores are treated as absolute values, although this is not entirely justified. The then applicable tire testing allowed scores to be compared for one driver within one batch and not between drivers and batches, making the ratings absolute within one batch of one driver, but relative between batches and drivers. For a regression model, it is important to have a large data set available to obtain a reliable regression model. If the ratings were treated as relative ratings, so a regression could only be performed within one batch, this would reduce the data set too much, having only two or three tire assessments per driver per batch. On the other hand, the scores were also not entirely relative: the summer tires were scored higher than the winter tires, but not as much as could be expected if all tires in all batches would be rated in absolute scores. Driver dependence of the ratings is largely removed by standardizing the scores for each aspect for every driver to zero mean and unity standard deviation.

2.3 General Regression Neural Network

Applying a regression model requires in general as first step estimating a model structure and as second step fitting the model parameters from the data. This leads to two problems for this tire data set. For estimating a model structure, mostly knowledge about the process is used, like the order and linearity. In this case, such knowledge is not available, which makes it difficult to choose a proper model structure. Another problem is the limited data set available. We can have up to 29 predictors and 10 targets, resulting in a high dimensional space, but one tire assessment, being a combination of values for the predictors and targets, can be seen as one data point in such a high dimensional space. For a regression model, there are seven data points available with 29 metrics that can be used as predictors for each driver. Seven data points is a too small dataset for proper regression in combination with so many possible predictors and leads to an underdetermined model, unable to fit the model parameters uniquely. Eliminating or combining (many) predictors to solve this, would increase the risk of losing information and making the relationship less clear for interpretation.

A regression method that does not have the previously mentioned restrictions is a GRNN, which is a specific artificial neural network. An introduction and detailed motivation for choosing this specific method is given in [4]. In short, it does not require choosing a regression model. Instead, it empirically determines the appropriate relationship from the data as a probability density function. Additionally, it does not need a large data set to perform well. Although more data points will improve its performance, even with a limited data set a GRNN is able to converge to the underlying (linear or nonlinear) regression hypersurface, with a smooth transition between data points [6]. In addition, the architecture of the GRNN is relatively simple, having only one hidden layer in a feedforward network structure.

In the construction phase, the model structure and model parameters of the GRNN are fully determined by the data points, which are given as example to the GRNN. In addition, no iterative learning phase is present. This makes the construction of a GRNN simple and fast. After construction, the GRNN can predict the outputs, based on new input data, unknown to the GRNN. For this prediction phase, only one parameter must be estimated.

Figure 5 shows the GRNN that is developed for the tire handling assessment. For every driver this GRNN is constructed as follows. The hidden layers consists of so-called 'neurons', each representing a tire assessment (or data point), like A1, A2, …C6 which is used as example for constructing the network.

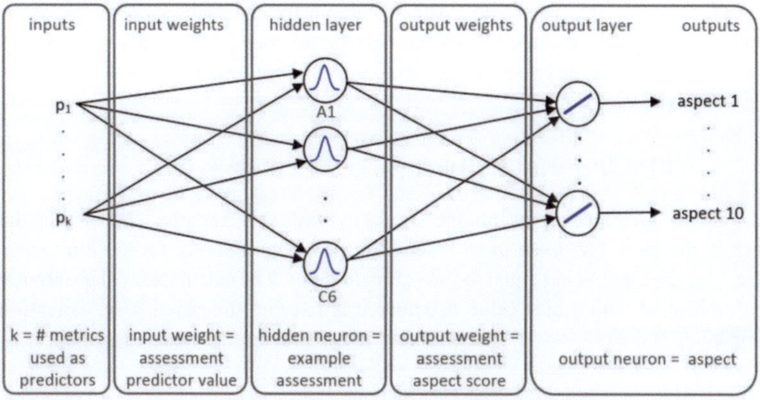

Figure 5 GRNN for handling assessment.

All the inputs are connected to all the neurons in the hidden layer. These connections have so-called input weights. Each input weight is equal to the metric value of the data point that the hidden neuron represents. All the hidden neurons are connected to all the neurons in the output layer. Each output neuron represents one output aspect of a data point used for constructing the network, making the number of output neurons equal to the number of aspects. These connections have output weights. Each output weight is equal to the score value for that output of the data point that the hidden neuron represents.

The hidden layer neurons have a radial basis function; the output layer neurons have a linear function. Figure 6 shows a radial basis neuron in the hidden layer of the GRNN. This hidden neuron represents one of the examples for training/constructing the GRNN.

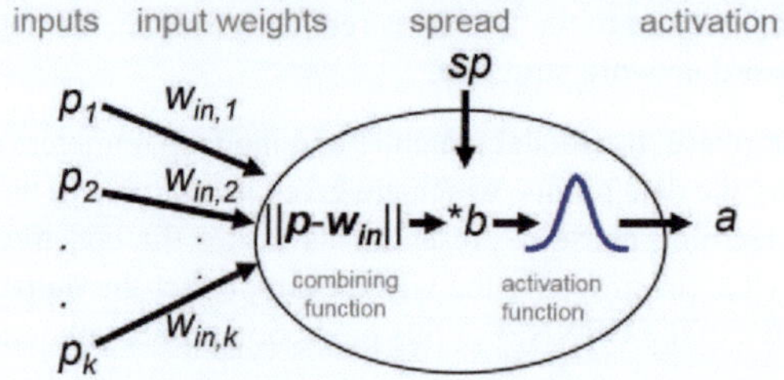

Figure 6 A radial basis neuron in the hidden layer.

This radial basis neuron has k inputs or predictors, the input vector $\boldsymbol{p}$. Each connection between one predictor and the neuron has an input weight $w_{in,j}$, $j = 1, \ldots, k$, giving k values in the input weight vector $\boldsymbol{w_{in}}$. This weight vector equals the input vector of the example represented by the neuron. The combining function for this neuron calculates the distance r between the two k-dimensional points represented by the weight vector $\boldsymbol{w_{in}}$ and the input vector $\boldsymbol{p}$

$$r(\boldsymbol{p}) = \|\boldsymbol{p} - \boldsymbol{w_{in}}\| = \sqrt{(p_1 - w_{\text{in},1})^2 + .. + (p_k - w_{in,k})^2} \tag{1}$$

The distance r therefore represents the similarity with the example. The smaller the distance r, the more the given input resembles the example input, for the limit case, where $r = 0$ the input being equal to the example input. The calculated scalar output r is then multiplied with a bias value b before it is fed into the radial basis activation function φ

$$\varphi(r) = exp(-r^2) \tag{2}$$

giving the output or activation a of the neuron

$$a = \varphi(r \cdot b) \tag{3}$$

with bias value b

$$b = \frac{\sqrt{ln(2)}}{sp} \tag{4}$$

with sp the spread value, which is a parameter defined by the user. This spread value determines the distance at which the output of the radial basis activation function equals 0.5. It determines the sensitivity of the radial basis function. This can be visualized as making the radial basis function narrower or wider, as shown in Figure 7.

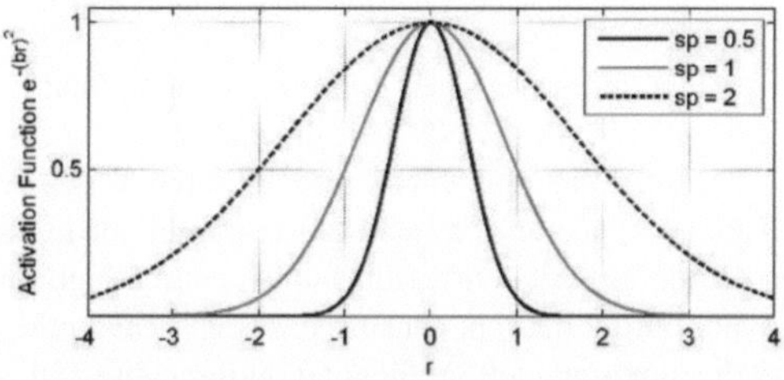

Figure 7 Three activation radial basis functions for different values of the spread sp.

In this figure, three radial basis functions are plotted, each with a different spread value sp. For the same distance r, the activation radial basis function with the largest spread will produce the largest output value. A small value of spread results in an activation function that only produces an output significantly larger than 0 when $|r|$ is small, indicating that the input is quite similar to the example input. For $|r| = sp$ the activation function outputs 0.5.

Figure 8 shows a linear neuron from the output layer of the GRNN.

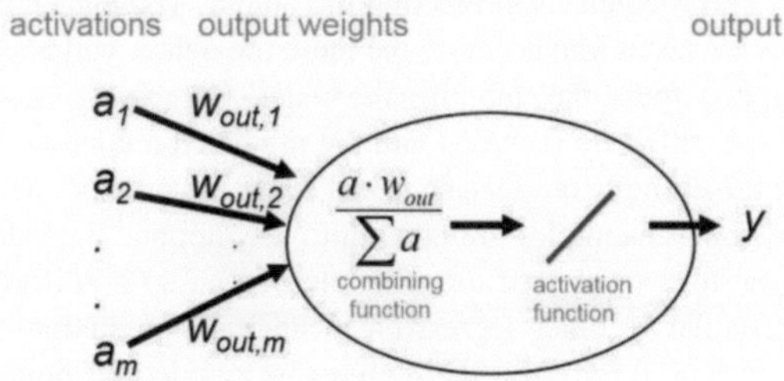

Figure 8 A linear neuron in the output layer.

This linear neuron has m inputs, which are the outputs or activations from the m neurons in the hidden layer, the activation vector $\boldsymbol{a}$. Each connection between an activation and the neuron has a weight $w_{out,j}\ j = 1, \dots, m$, giving m values in the output weight vector $\boldsymbol{w_{out}}$. An output weight value is equal to the output value of the example, which the hidden neuron where the activation comes from, represents. The combining function for this neuron calculates the dot product of the weight vector $\boldsymbol{w_{out}}$ and the activation vector $\boldsymbol{a}$ and normalizes this result with the sum of activations. The activation function for this neuron is the identity function, so the output y is calculated by

$$y(\boldsymbol{a}) = \frac{\boldsymbol{a} \cdot \boldsymbol{w_{out}}}{\sum \boldsymbol{a}} = \frac{a_1 w_{\text{out},1} + \cdots + a_m w_{out,m}}{a_1 + \cdots + a_m} \tag{5}$$

Summarizing, in the prediction phase, when presented with an input vector consisting of new unknown metric values, the GRNN will give an output vector with predicted scores for all aspects. The more the input vector resembles the weight vector of a hidden neuron (representing the input vector of a 'learned' example), the higher the activation output of this hidden neuron. Subsequently, this hidden neuron high activation value is one input value for every output neuron, causing to weigh heavily the example output value, represented by the output weight, in the total output value of this output neuron. For an input vector, this is done in parallel for all hidden neurons or examples, giving all their activation values to all output neurons. The actual output vector being a weighted average of all activation values multiplied with the output weights or examples output weight vectors. In general, the more an input vector resembles an example input, the more the output vector will resemble the example output. This does not imply that if a GRNN is presented with the exact same input vector of one of the examples used for constructing the GRNN, the GRNN will output the exact same output vector of this example. This depends on the spread value sp, which determines how many similar examples are used.

The only GRNN parameter spread sp thus determines how much 'similarity' of the examples is taken into account for predicting the output. The higher the spread value, the more examples are taken into account, the more the output will be a weighted average of more example outputs, the smoother the underlying regression surface. For high spread values, all examples are included and the predicted output value will approach the mean value of the example outputs, so it will act as the mean model. For very small spread values, no similar examples are taken into account and the value of the output vector will be equal to the value of the example output, so it will act as the nearest neighbor model. Similar to other regression techniques, prediction outside the data range of the examples used for constructing must be avoided. Although it will not behave badly outside this range, as the output will be between the mean and nearest neighbor values, with such extreme input, it cannot rely on learned similar examples.

Previous research [4] showed that using all metrics as predictors did not result in the best performance. Therefore, the number of predictors for the GRNN will be added as an additional model parameter, which will be analyzed together with the other parameter spread in section 3. The selection of which metrics should be used as predictors for an output aspect (given a certain number of predictors) is based on the absolute strength of the linear relationship between a metric and that output aspect given by Pearson's correlation coefficient r [7], calculated for a group of N values (x, y)

$$r = \frac{cov(x,y)}{s_x s_y} \tag{6}$$

with s the (sample) standard deviation defined as

$$s = \sqrt{\frac{\sum_{i=1}^{N}(x_i - \bar{x})^2}{N-1}} \tag{7}$$

and with the covariance between x and y:

$$cov(x,y) = \frac{\sum_{i=1}^{N}(x_i - \bar{x})(y_i - \overline{y})}{N-1} \tag{8}$$

with $\bar{x}$ and $\bar{y}$ the mean values. For $r = \pm 1$ the relationship is perfectly linear, for $r = 0$ there is no linear relationship. As an example, two scatter plots of standardized metric – aspect 10 values for all assessments for driver mean are shown in Figure 9. The plot on the left shows metric v_y-V, which has the highest value for Pearson's correlation coefficient r. The plot on the right shows metric a_y-V, which has the lowest r with the aspect 10 scores. For each output aspect, the metrics are added as predictors from highest to lowest absolute r-values. So, for a GRNN with five predictors, only the five highest correlating metrics for a certain output aspect are used as predictors for that output.

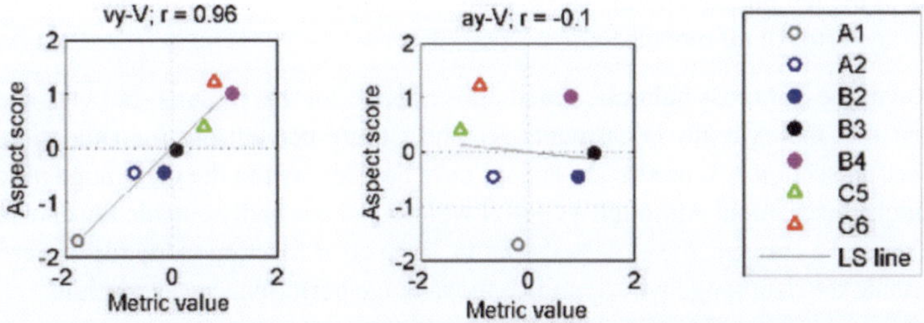

Figure 9Two example scatter plots of standardized metric – aspect 10 values for all assessments for driver mean with the highest (left) and lowest (right) absolute r-value.

During usage, new examples can easily be added to a GRNN, including this information in future predictions, but also using the existing examples information. For this GRNN, the examples are new tire assessments, which can be extra assessments for the same tire or new, unknown to the GRNN, tires. With more examples, the GRNN will learn more about the underlying relationship between predictors and handling aspects, the better the GRNN prediction will become. Likewise, metrics can be added. If new measurements become available from which metrics can be derived, these can be added to the GRNN as inputs. If this metric will be used as predictor, the hidden neurons of existing examples have no input weights for this predictor, because they have no values for this new predictor. By setting these input weights to not-a-number value, this predictor will be ignored by this hidden neuron. The activation output of this hidden neuron will then be calculated as before, based on the other known predictors. The same method can be used when metrics are removed, but for the existing examples still should be available. This can be because they are not relevant or are no longer derived as predictors for new assessments. This also accounts for aspects. New assessments with more or less predictors and/or more or less aspects than the existing examples can be incorporated in the GRNN, adding their information, while the existing assessments information is also kept.

3 Results and Discussion

A practical way of testing a model based on such a limited data set, is cross validation, with as special case the 'leave-one-out' method [7,8]. This method implies that one data point is left out of the data set to construct the model. After the model is constructed based on the remaining data points (examples), the model is used to predict the left out data point outputs, based on the left out data point inputs. This procedure is repeated until all data points are once left out for validation. The performance of each model is evaluated by applying an error function between the predicted outputs and the actual outputs of the left out data point. The overall model performance is evaluated by taking the average error of all models.

This leave-one-out cross validation method is applied for this research but with the adjustment that the extreme assessments A1 and C6 are not left out for validation. As explained in section 2.3, prediction should only be done within the data range used for the model construction. Although a GRNN will not behave badly outside the data range of the learned examples, the expected relative large error function value for this prediction outside the data range will dominate the average performance calculation.

As performance measure of the GRNN, the Root mean Squared Error (RMSE) between the actual aspect scores given by the driver and the GRNN predicted aspect scores is taken. The lower the RMSE value, the higher the prediction performance. The results of the GRNN prediction will also be analyzed in comparison with the mean model results, because this is often used as baseline model to compare other models with. Also here, the mean model predicts the left-out target or aspect score, by taking the mean of all the remaining example assessments target values.

For prediction, the GRNN parameters, number of predictors used and spread, must be assigned a value. These values will be estimated by the values that result from the leave-one-out cross validation, that result in the lowest average error of the left out test assessments. The contour plot of the average RMSE of test assessments A2 to C5 are presented for driver 1 (Due to limitations of pages in this paper, only the results for driver 1 are shown.)

The figure shows that the lowest average RMSE related to best GRNN prediction performance is realized in the area with a low number of predictors used and a low spread. This best prediction area when looking at all drivers results is positioned between 1 to 8 predictors used and .3 to 3.0 as spread value with the restriction that increasing number of predictors is combined with an increasing spread.

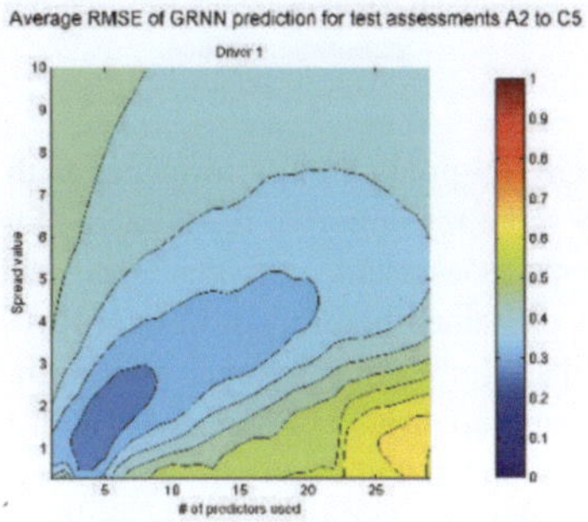

Figure 10 Result of the leave-one-out cross validation: contour plot of the average RMSE of GRNN prediction for test assessments A2 to C5 for all drivers.

The parameter values corresponding to the lowest RMSE for a driver are the suggested values to use when there are no targets available. To analyze the GRNN prediction performance when using these suggested parameters, these are applied for the GRNN for predicting the intermediate test assessments for all drivers. The GRNN now uses the same suggested parameter values for all test assessments for a driver. The results for driver 1 are shown in Figure 11.

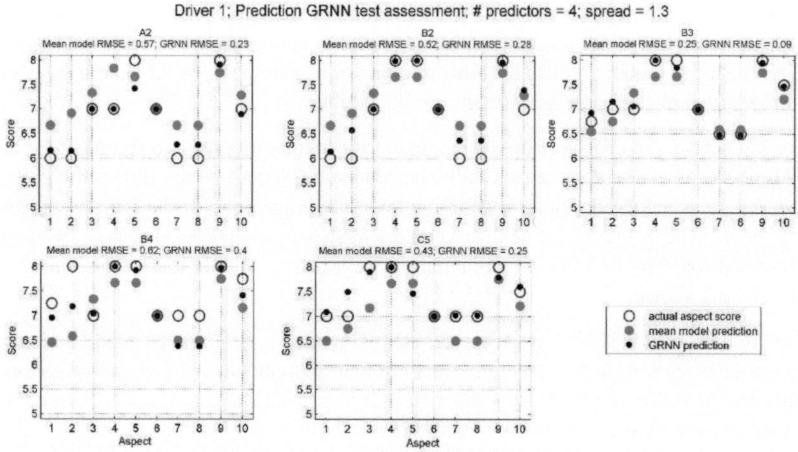

Figure 11 Driver 1 plots showing the actual, mean model and GRNN prediction aspect score for the intermediate test input assessment when using the leave-on-out cross validation suggested parameter values.

These plots show good GRNN prediction performance with all RMSE values around or lower than the drivers score resolution of 0.25, except for test assessment B4. This is due to being the 'extreme assessment' for batch B and as batch C is rated relatively low, also for batch C, as explained before. All GRNN RMSE values are much lower than the mean model RMSE. Interesting to see, is that even handling aspects 2 and 3, stability while cornering with no throttle change and with throttle change, are also predicted well. The test drivers stated that these aspects were specifically determined during free driving and not during the double lane change. Apparently, the GRNN is able to predict handling behavior outside the double lane change based on double lane change metrics.

Analyzing the used predictors provides information on what vehicle dynamic behavior is relevant for the driver. For general handling behavior, regarding all aspects, this showed that metrics based on lateral velocity, steering wheel moment and vehicle slip angle were most relevant.

4 Conclusions

The GRNN showed good prediction of the driver's subjective assessment scores of tire handling aspects covering driving a handling track, based on vehicle dynamic measurements taken only during the double lane changes. The GRNN can even perform well on predicting handling aspects not related to driving the double lane change. Analysis of the important measures used by the GRNN can provide information on the vehicle dynamics behavior relevant for the driver. For general handling behavior, regarding all aspects, this showed that metrics based on lateral velocity, steering wheel moment and vehicle slip angle were most relevant.

Additional advantages of using a GRNN compared to other regression methods are: it works well with a limited dataset, does not need a-prior knowledge about the data or underlying model, has good prediction between data points. In addition, constructing the GRNN is easy and adding new examples, with possibly different in- and outputs can be done during usage, incorporating new knowledge when it becomes available.

Acknowledgement

This research has been supported by Apollo Tyres Global R&D, Enschede, Aalto university, Finland and HAN university of Applied Sciences.

References

1. Heissing B, Ersoy M. Chassis Handbook: Fundamentals, Driving Dynamics, Components, Mechatronics, Perspectives. Vieweg+Teubner; 2011.

2. Ooki M, Sakai H, Muragishi Y, Fukui K, Ono E. Vehicle Transient Response Based on Human Sensitivity. AVEC. 2008.

3. Bergman W. Measurement and subjective evaluation of vehicle handling. SAE Pap. 1973;730492.

4. Monsma S. Feel the Tire - Tire Influence on Driver's Handling Assessment. Aalto University; 2015.

5. ISO. ISO 3888-1:1999(E) passenger cars - Test track for a severe lane-change manoeuvre- Part 1: Double lane change. 1999.

6. Specht DF. A general regression neural network. Neural Networks, IEEE Trans. 1991;2:568–76.

7. Field A. Discovering statistics using SPSS. Sage Publications Limited; 2009.

8. Hyötyniemi H. Multivariate regression: Techniques and tools. Helsinki University of Technologie; 2003.

Application of objective vehicle dynamics methods in the tire development process

Author & Speaker: Dipl.-Ing. Christian Cramer
(Continental Reifen Deutschland GmbH, Hannover,
Test Method Development & Vehicle Dynamics)

Co-author: Prof. Dr. Burkhard Wies
(Continental Reifen Deutschland GmbH, Hannover,
Vicepresident Tire Line Development Worldwide)

P.E. Pfeffer (Ed.), *8th International Munich Chassis Symposium 2017*,
Proceedings, DOI 10.1007/978-3-658-18459-9_51

1 Abstract

Vehicle dynamics requirements are of major interest in the tire development process. An efficient and target-oriented development of vehicle dynamics relevant tire characteristics can be supported by using objective methods of vehicle dynamics evaluation. This article explains the application of objective vehicle dynamics evaluation methods during the tire development process of a premium tire in HP/UHP segment at Continental.

At first customer demands are transferred into vehicle dynamics requirements. After this the vehicle dynamics simulation is run in combination with a handling regression model, from which a tire characteristic target setting is derived. The target tire characteristics can be approached by several means during the tire design phase. FEM-aided selections of promising tire designs are then realized as prototype tires. Finally, the evaluation of vehicle dynamics properties takes place based on subjective assessment and objective measurements in combination with a regression model. This is used to evaluate whether the vehicle dynamics requirements are fulfilled and to refine the process chain as well as the regression model.

2 Introduction

Improving todays tire performance is coupled with a lot of target conflicts in the tire development. For instance: to reduce the vehicles CO2-emissions, improvements in rolling resistance are of major interests while not decreasing wet braking performance. Electric vehicles are going to set a focus on comfort criteria like mechanical and acoustical noise behavior of tires.

Although it's the aim of autonomous driving to reduce accidents, it can only be realized by increasing handling specific tire development targets. Forces and moments are transmitted by the tires' footprints and are the basis for a stable and appropriated vehicle reaction. The number of different vehicles on the market is increasing and the complexity of tire design as well. This leads to an efficient and target oriented tire development process. Objective vehicle dynamics methods are able to support the tire developer to realize promising handling related tire designs in less development time. The following sections describe the application of objective vehicle dynamics methods in the tire development process at Continental.

3 The Tire Development Process

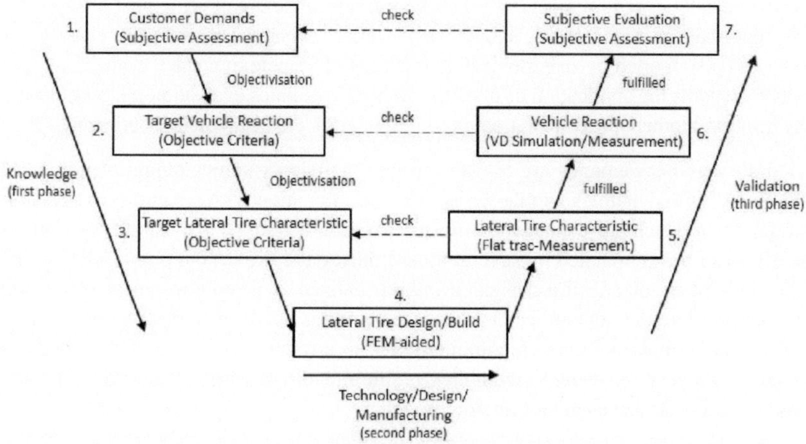

Figure 1: V-shape organigram of the handling specific tire development process

The handling specific tire development process can be visualized by a v-shape organigram (cf. Figure 1). At first a knowledge based phase takes place. Within this the first step (1) is to clarify customer demands for the new tire. For the development of a new replacement tire, these demands are internally set after benchmarking the actual tire generation in comparison to its competitors on the market and with respect to the latest technology steps. In contrast to that, for the development of an oe-tire, the customer has set targets for the handling properties of the new vehicle. He analyzes the handling properties of his current vehicle in comparison to its main competitors' vehicles on the market.

Almost every handling dependent target can be reached by optimizing vehicle subsystems, e.g. steering-system, suspension, mass distribution as well as the vehicle's moments of inertia. Yet these vehicle subsystems in general are also linked with non-handling development criteria (e.g. comfort, cost efficiency) in a diametrical way. A suspension system, which is optimized for less dynamical wheel-load changes, leads to compromises in comfort – which presents a target conflict in development. This can be solved by an active suspension system, which correlates with more complexity and less cost efficiency. Alternatively, handling demands can be reached by a vehicle-specific tire characteristic. From the vehicle manufacturer's point of view, virtual degrees of freedom in the vehicle development process are generated.

To use these degrees of freedom efficiently, the weighted customer demands are transformed to changes in objective vehicle dynamics criteria (2) (e.g. increasing gain of yawrate/steering-wheel angle). To find a target lateral tire characteristic, vehicle-specific boundary conditions have to be taken into account. Vehicle-specific boundary conditions are for example weight and moments of inertia, positon of the center of gravity as well as kinematics and elasto-kinematic properties. Tire-specific boundary conditions are tire size (round or mixed fittment) and expected inflation pressure on front and rear axle.

Vehicle clusters concerning these boundary conditions are made and all vehicles are classified into these vehicle clusters. Taking these vehicle- and tire-specific boundary conditions and the target vehicle reaction into account, a target lateral tire characteristic is set (3). In general, within each vehicle cluster there exist several local optima in lateral tire characteristics, but in only one global optimum.

Depending on the vehicle-cluster, the target lateral tire characteristic to achieve the same changes in the vehicle reaction is completely different. Although the target lateral tire characteristic differs within one vehicle cluster – caused by individual customer demands, changes in lateral tire characteristic show similar effects in the lateral vehicle reaction.

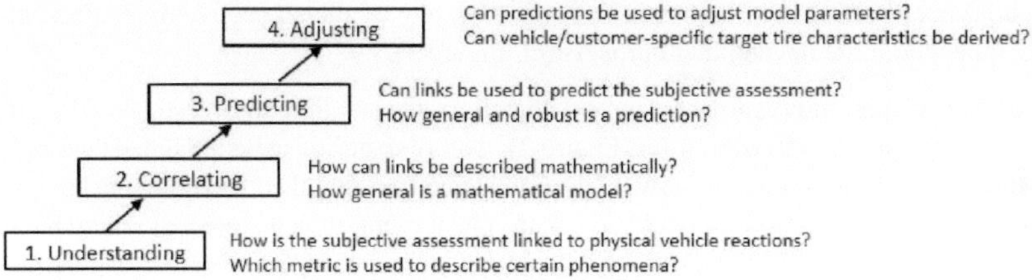

Figure 2: Hybrid objectivization method with four major steps from understanding the perceived vehicle behaviour to application of this knowledge in the tire development process

Finally an FEM-aided lateral tire design takes place (4). The main purpose is to simulate which changes in construction, contour, pattern and compound fulfil the target lateral tire characteristic.

The second phase in the tire development process is the tire manufacturing (4). The third and last phase is the validation phase. At first flat trac measurements of the new tire are done to measure the lateral tire characteristic (5). If the actual lateral tire characteristic is different from the target tire characteristic, the tire design is adapted to match the target characteristic. The tire matching the target lateral tire characteristic is tested objectivly on the vehicle (6). In an early stage this may be done on the predecessor vehicle, afterwards on a prototype vehicle. The purpose is to check whether the desired changes in objective criteria of the vehicle response are realized. If the changes are not realized the target lateral tire characteristic will be adapted, which also results in changes in the tire design.

The tire which meets the expected demands in the vehicle reaction is evaluated by a subjective evaluator (7). The main focus is to clarify whether the handling behaviour demanded by the customer has been reached. Otherwise - in accordance with the customer - the demands will be adapted, which has effects on the target vehicle reaction as well as the target lateral tire characteristic and the tire design.

4 Hybrid Objectivization Method

The objectivization of subjective evaluation is of major interest, because it is the least understood part in the vehicle dynamics development scheme (cf. Figure 1). From tire design to tire characteristic and resulting vehicle reaction, changes can be described and predicted by physical models. On the one hand, the subjective assessment of vehicle behaviour seems to be "weak" because it cannot be quantified that easily. On the other hand, it is a "strong" method, because it has been established in the automobile and tire industry for decades and has proven valid to discuss and rate vehicle behaviour. Nevertheless, uncertanties in the understanding of the subjective assessment affect the whole tire development process (cf. Figure 1).

The aim of this study is to use a hybrid objectivization method to investigate four different steps of objectivization (cf. Figure 2). The first step is called "Understanding". Main research questions are how subjective evaluation is linked to vehicle behaviour and which metric is used to describe certain phenomena in vehicle behaviour. This is realized by online measurement of vehicle movement during subjective evaluation. The second step is named "Correlating". The main purpose of this step is to find mathematical models which describe links between subjective evaluation and objective criteria in standardized objective tests. Statistical methods are used to correlate characteristic values with subjective assessment. Manoeuvres used in subjective evaluation and synthetic objective tests (e.g. transfer function or ramp-steer manoeuvres) differ. Compared to the first step the extracted unterstanding is indirect.

1. Stufe	2. Stufe				3. Stufe	
Bewertung	Bewertung	Mangel	Wahrnehmbar durch	Note		
Eigenschaft im Industriestandard	optimal	nicht wahrnehmbar	ausgebildete Beobachter	10		
	sehr gut	kaum wahrnehmbar	ausgebildete Beobachter	9		
	gut	äußerst gering	ausgebildete Beobachter kritische Kunden	8	8,5	8,25 7,75
	noch gut	sehr gering	kritische Kunden	7	7,5	7,25 6,75
	befriedigend	gering	kritische Kunden	6	6,5	6,25 5,75
	genügend	gut	kritische Kunden und Normalkunden	5	5,5	
Eigenschaft unter Industriestandard	mangelhaft	unangenehm, Reklamation, Verbesserung erforderlich	Normalkunden	4		
	schlecht	nicht akzeptabel Bauteil fehlerhaft	alle Kunden	3		
	sehr schlecht	nicht akzeptabel, Bauteil bedingt funktionsfähig	alle Kunden	2		
	völlig ungenügend	nicht akzeptabel Bauteil ohne Funktion	alle Kunden	1		

Figure 3: Standardized scale of subjective assessment in lateral dynamics as used in the automobile industry, based on [5]

Nevertheless statistical links form the basis for the third "Prediction"-step of objectivization. A robust and generalized mathematical model allows a prediction of the subjective assessment by the input of objective criteria from vehicle dynamics measurements. The fourth and last step of objectivization is called "Adjusting". It links the "Prediction"-model with a monte-carlo simulation to optimize the predicted rating of subjective assessment by adjusting the tire-dependent parameters in the vehicle dynamics model. The output of step four are several promising target lateral tire characteristics, which provide the input for the tire design.

5 Subjective and Objective Vehicle Dynamics Assessment

At Continental subjective and objective vehicle dynamics evaluation are traditionally separated. In this research the main purpose is to accept subjective evaluation as an established test method. In contrast to other research, the evaluation sheet is not modified. This enables the evaluator to rate the different sets of tires as he is used to. Moreover, the findings of this research can easily be applied to standard test procedures of Continental. The marks of the subjective assessment of dry handling evaluation is based on the standardized scale in the automobile industry, which is described in [5] and illustrated in Figure 5. The rating is done with marks between 1 and 10. The marks are coupled to an absolute level of acceptance in a certain evaluation criteria. A mark of 10 means that even experts cannot recognize a deficit, whereas the mark 1 means that the deficit is recognized even by a normal driver and is perceived as insufficient. Half and quarter marks are introduced to rank different sets of tires relatively. In addition to this, at Continental a standardized dictionary is established to describe deficits in each category as they are perceived during subjective evaluation (e.g. “flat” for too less vehicle reaction).

On the objective side, different vehicles are equipped with measurement equipment (cf. Figure 4). Laser height sensors (A) are used to measure the distance and orientation of the vehicle’s body with respect to track surface. REDI-sensors (B) fixed at the innerliner of the tire measure temperature and inflation pressure during the test run. Infrared temperature sensors (C) monitor the current tread temperature of the tires. To measure the steering wheel angle and steering wheel torque investigations have shown that external measurement equipment should be preferred to internal steering wheel sensors. [4] has shown that subjective evaluation of steering feel depends on the haptic properties of the steering wheel. For this reason a sensor which can be applied to the vehicle’s original steering wheel is chosen (D). As rubber and tire properties depend on boundary conditions (e.g. temperature), the track surface temperature (E) is measured with an infrared sensor as well. The ambient temperature is recorded to receive information about the general climatic boundary conditions on the testday. The vehicle’s position on the track as well as the velocity are measured with a RTK-GPS-system (F). Accelerations in the vehicle’s longitudinal, lateral and vertical direction as well as the rotational velocities around these axes are measured with a strapdown platform (G). An optical correvit sensor is used to measure longitudinal and lateral velocity. Knowing the correvit sensor’s position, its measured velocities and the rotational velocities of the vehicle body, the slipangle at the center of gravity can be calculated. Filtering, delay correction and coordinate transformations are done in the data preprocessing.

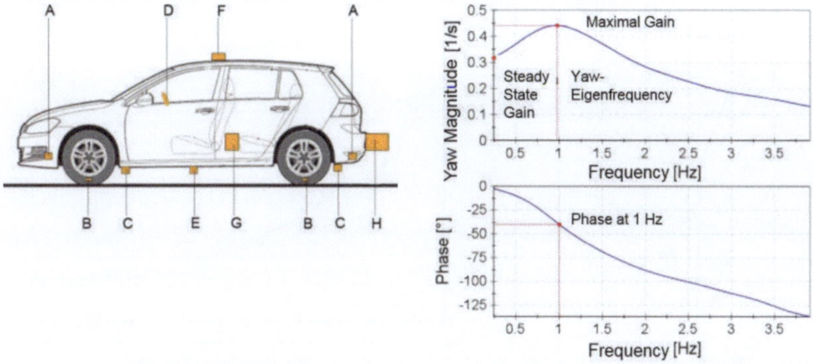

Figure 4: An example of the used measurement equipment (left) and derived characteristical values from objective manoeuvers such as magnitude and phase from transfer functions (right)

After this a manoeuver specific evaluation method takes place to extract tire depending information from the measured vehicle behavior (cf. Figure 4). For instance the transfer function of yawrate/steering wheel angle is discussed. The test is driven with a constant velocity and a sinusoidal steering wheel input with constant amplitude (adjusted to a steady state lateral acceleration of 3 m/s^2) varying the frequency randomly distributed between 0.2 and 4 Hz. Typical characteristic values derived from this manoeuver are steady state gain, yawratio – which is defined as the quotient of maximum gain to steady state gain -, damped yaw eigenfrequency or phase of lateral acceleration at 1 Hz. Furthermore, additional transfer functions and characteristic values as well as characteristic values from other linear and nonlinear tests are built.

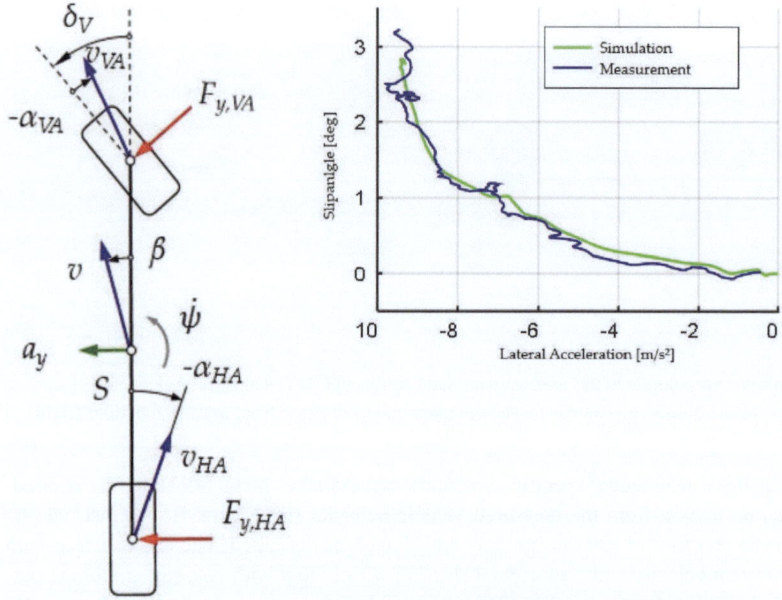

Figure 5: Parameterization of a nonlinear single track model (left) by vehicle dynamics measurements. For instance this can be realized by using ramp steer maneuvers (right).

6 Parameterization of Vehicle Dynamics Simulation Models

In addition to the derived characteristic values from objective driving manoeuvers in section 5 parameterization of generic vehicle dynamics simulation models is done. This initiates two different possibilities. On the one hand characteristic values from a multitude of different driving manoeuvers (including typical manoeuvers of subjective assessment) are built. On the other hand the influence of different virtual sideforce characteristics can be compared.

The differential equations of the nonlinear single track model are defined by:

$$\mathrm{J}_{ZZ}\ddot{\psi} = F_{YV} \cdot l_V \cdot \cos(\delta_V) - F_{YH} \cdot l_H \qquad (1)$$

$$\mathrm{m} \cdot \mathrm{a}_{\mathrm{Y}} = F_{YV} \cdot \cos(\delta_V) + F_{YH} \qquad (2)$$

A magic formula sideforce model is adapted to extend the validity of the simulation model from linear to nonlinear sideforces and cover transient lateral tire behaviour:

$$F_{Yi} = f(\alpha_i) - \frac{\sigma_i}{v} \cdot \frac{dF_{Yi}}{dt} \tag{3}$$

$$f(\alpha_i) = D \cdot \sin\left(C \cdot \arctan\left(B \cdot (1-E) \cdot \alpha_i + E \cdot \arctan(B \cdot \alpha_i)\right)\right) \tag{4}$$

On even roads and quasi-stationary driving manoeuvers this model has proved to match the measured vehicle behavior sufficiently (cf. Figure 8).

Dynamic wheel load changes as a result of rolling, pitching or road disturbances as well as elasto-kinematic effects are neglected or indirectly covered by the identified side force characteristic.

As this model is used to evaluate relative differences in the lateral tire characteristic these compromises can be accepted for a reduced model complexity and less effort in parameterization. For detailed studies including the vehicle body movements (e.g. driver's perception) more complex simulation models are used.

7 Regression Analysis

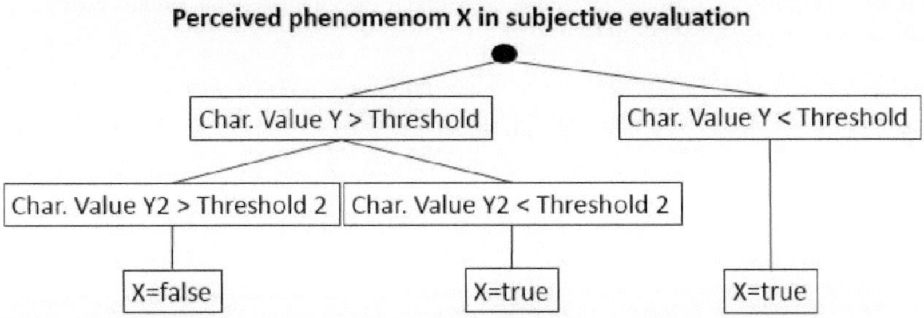

Figure 6: Nonlinear data mining methods (like the decision tree above) are trained by objective criteria from vehicle dynamics measurements and the subjective evaluation to identify links between both of them.

After generating characteristic values from vehicle dynamics measurements and simulation, a regression analysis takes place. The purpose is to find mathematical dependencies between characteristic values of vehicle movement and subjective assessment. As presented in section 5 the subjective evaluation of tire handling behavior can be subdivided into marks in different evaluation categories and a standardized deficit description. In this approach both sides are analyzed. Most published approaches as [3], [4] and [6] focus on numerical models – mainly on multiple linear regression. This implies the evaluator's perception – more precisely the given marks in different categories – being linearly correlated to physical effects. In contrast to that [13] uses a quadratic regression model and [1] a neural network. Both show that nonlinear models match the marks of subjective evaluation better than linear models. Most studies focus on only one vehicle, one subjective evaluator and a small database. For these cases, multiple linear regression shows promising results. Yet it is uncertain, how general the extracted information is. The conclusion of nearly all studies is that studies with larger databases and different subjective evaluators show less promising correlation results. One focus of this approach is to analyze the information included in the perceived deficit. Modern nonlinear datamining algorithms allow an analysis of dependencies between measured characteristic values and perceived deficits in vehicle behavior, neglecting the mentioned linear model behavior. Let P(X) be the probability that the perceived deficit X in subjective assessment occurs over the attributes i. The entropy H(X) – which can be seen as an information value – over the attributes i is defined as:

$$H(X) = -\sum_{i=1}^{n} P(x_i)\log P(x_i) \tag{5}$$

The first step is the calculation of the entropy of the deficit X over the attributes i. In the next step the conditional entropies of X given the characteristic values Y of vehicle measurements is calculated:

$$H(X \mid Y) = \sum_{i,j} P(x_i, y_j)\log \frac{P(y_j)}{P(x_i, y_j)} \tag{6}$$

From this the information gains of all characteristic values Y over all instances for the deficit X are computed:

$$IG(X,Y) = H(X) - H(X \mid Y) \tag{7}$$

On the measure of information gain this data mining algorithm splits up the instances. It is trained by using a 10-fold cross validation method to obtain generalized dependencies and prevent overfitting (cf. Figure 9). The result is a decision tree for each perceived deficit in vehicle reaction (cp. Figure 10). From these decision trees links between characterstic values from measured vehicle behavior and the subjective assessment can easily be derived.

Although this technique neglects the numerical scale of subjective evaluation, it shows several advantages over established regression methods used for objectivization of subjective evaluation in vehicle dynamics: it does not require linear behavior between marks in subjective evaluation and differences in physical effects of vehicle behavior. Furthermore, the measured vehicle behavior is linked to the evaluator's perception and not to his evaluation/rating expressed in marks; thus showing promising results in different vehicle clusters, different subjective evaluators and a large database.

8 Application of the Objectivization Method

In this part the presented objectivization method is applied in the tire development process. On the basis of the predecessor tire a benchmark was made to identify important development criteria. The boundary conditions for the tire development is a new 18 inch high performance/ultra high performance tire in the replacement market. A representative car of the executive car segment is chosen. The development targets of a new tire can be subdivided into different categories.

The focus is set on lateral handling development targets, neglecting other categories such as noise/comfort, rolling resistance or wet braking.

Table 1: Handling targets for the development of the new tire are defined by relative marks in certain criteria compared to the existing predecessor tire.

Lateral Handling Development Targets (criteria – perceived deficit)	Relative Marks compared to the predecessor tire
On Center Response - Steering Precision	>=+0.50
Steering Wheel Torque - Centerpoint-Feeling	>=+0.50
Linearity of Sideforce build up – Over All	=
Rate of Sideforce build up – Over All	=
Weaving (Slalom) – Over All	=
Lane Change – Over All	>=+0.25
Behaviour at Griplimit – Over All	=

As a result of this, lateral handling development targets for the new tire are set (cf. Table 1). The criteria mentioned in this table are based on the Continental intern subjective evaluation sheet. The first two criteria describe the on-center behaviour of the tire in terms of vehicle reaction and steering torque. The targets for the new tire are a significant improvement in the on-center steering precision (>=0.5 marks). Vehicle response is supposed to feel more direct with less time lags compared to the steering wheel input. Concerning the perceived steering torque the centerpoint is to be exactly defined and more pronounced compared with the predecessor tire (>=0.5). The next two lateral handling criteria focus on the vehicle response at larger steering wheel angle input within the area of linear vehicle response ($a_y<4\ m/s^2$). The transient behaviour – linearity of sideforce build up - as well as steady state behaviour – rate of sideforce build up – is to stay on the same level compared to the predecessor tire.

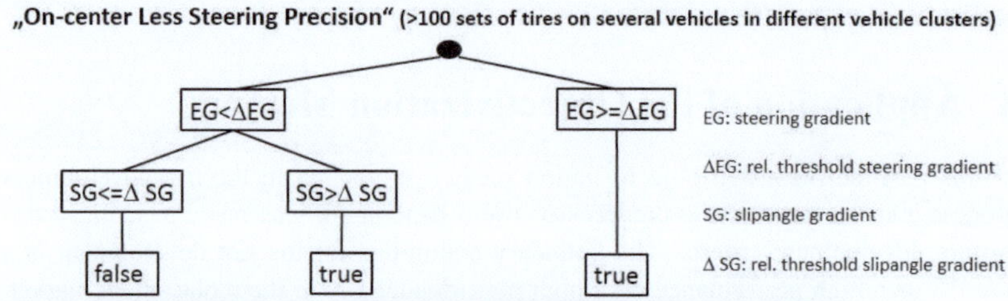

On-center Steering Precision	Subj. „less steering precision"	Subj. „target steering precision"
Predicted „less steering precision"	>0.75	<0.25
Predicted „target steering precision"	<0.25	>0.75

Figure 7: From the decision tree of the criteria on-center steering precision a general link to the steering gradient (EG) and the slip angle gradient (SG) can be derived. The confusion matrix underlines a significant and general dependency on a large data base.

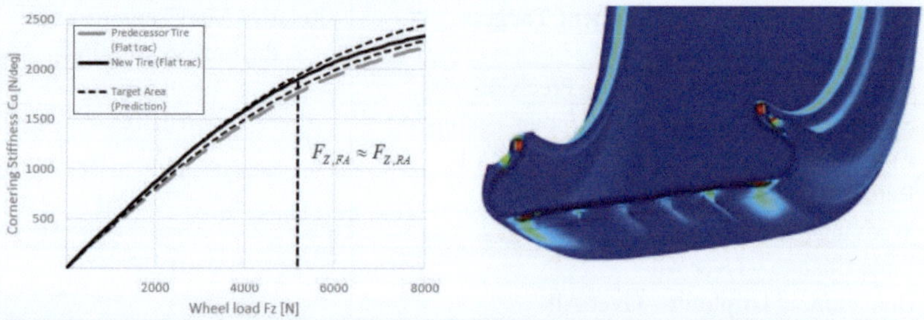

Figure 8: Targets for the cornering stiffness versus wheel load are derived by coupling the decision tree with the vehicle dynamics simulation (left). An FEM-aided tire design takes place to fulfil the target tire characteristic (right).

In the nonlinear lateral dynamics an appropriate fast vehicle reaction and a high rear axle stability are of major interest. The predecessor tire shows an appropriated vehicle reaction in the nonlinear slalom and the single lane change manoeuver. The new tire development targets keep these safety related properties and slightly increase the agility during the lange change manoeuvre (>=0.25 marks). Finally the behaviour at griplimit of the predecessor tire is perceived as expected. The vehicle behaves slightly understeeringly and is easy to handle.

These development targets based on the subjective assessment have to be translated into characteristic values of vehicle reaction (cf. Figure 1). In this study more than a hundred sets of tires on different vehicles (e.g. sports cars, fwd cars,...) and with different subjective evaluators are observed, and - based on these data sets - different regression models have been analyzed.

Based on the introduced decision tree algorhythm Figure 9 shows the result of the perceived deficit “less steering precision” for all instances. According to this result “less steering precision” occurs when the steering gradient (EG) exceeds a certain threshold (ΔEG). This can be interpreted as an amplitude caused steering imprecision. A higher steering gradient means that a the driver’s steering effort to reach a certain amount of vehicle reaction is increased. The other way around this link does not seem to be that clear. Staying below this threshold in steering gradient (ΔEG) does not clearly result in a “higher steering precision” in subjective assessment. If the slipangle gradient is too high (SG>ΔSG) the vehicle response is perceived as imprecise as well. This can have two reasons – the deviation between the vehicles orientation and the direction of velocity and increased time lags between steering input and vehicle reaction. As the steering wheel effort of the predecessor tire has not been evaluated as too high (cf. Table 1 “Rate of Sideforce”), the development target for the new tire is set to the same amount of steering gradient as its predecessor.

The second development target for an increase in “steering precision” is to reduce the slipangle gradient by at least ΔSG. As an accompanying effect of decreasing the slip angle gradient an improved rear axle stability in lane change is predicted.

The next step using the methodology of objectivization described above is to define the vehicle cluster. The used vehicle can be seen as a representative car for a typical vehicle cluster in the executive car segment. The vehicle-specific boundary conditions for this cluster are defined by the vehicle weight (2,2 t), the wheel load distribution (approx. 50% front axle, 50 % rear axle) and the wheelbase. Moreover, tire-specific boundary conditions in this case are set by the tire size (18 inch), a round fittment and an equal inflation pressure of 2.4 bar on front and rear axle tires. A combination of vehicle dynamics simulation and the regression model predicts the promising lateral tire characteristics by using a monte carlo approach (cf. Figure 8). The solid and the dashed curve represent the cornering stiffness over wheel load curve measured on the flat trac. The dotted curves predict an area of promising cornering stiffness curves. At the vehicle’s static front and rear axle wheel load an increasing amount of cornering stiffness is needed to fulfil the targets in vehicle reaction. The curvature of the cornering stiffness curve is not changed significantly to avoid changes in other categories of subjective evaluation (e.g. linearity of sideforce build up).

The next step is the FEM-aided tire design phase (cf. Figure 8). The main aspect is how to modify the tire design to increase the overall performance of the tire and adjust specific tire characteristics (in this case cornering stiffness versus wheel load curve) to the target characteristic. For this, tire modifications in the contour, construction, pattern and compound are realized. For instance the contour is adapted to realize an optimized footprint area with equally distributed tensions. Furthermore, the optimized tread pattern design delivers an increased block stiffness. During the tire design phase FEM is able to predict whether changes in tire design result in the target lateral tire characteristic. After the tire design is set, the tire is built and the lateral tire characteristic is measured on the flat trac (cf. Figure 8 solid lines). The measured cornering stiffness vs. wheel load curve validates changes in the tire design delivering the proposed result. After that a vehicle dynamics simulation using a complex multibody vehicle model takes place to predict the influences of the measure lateral tire characteristic on the vehicle reaction. An objective measurement of the vehicle reaction is done (cf. Figure 9). The adjusted cornering stiffness characteristic leads to almost the same steady-state yaw gain (cp. Figure 13) as well as the same steady state lateral acceleration gain (A). The steady state steering wheel effort remains the same compared to its predecessor. Moreover, a significant change is obtained in the damped yaw eigenfrequency (B) which is increased from approx. 1.3 to 1.6 Hz. Up to 1 Hz a reduction in the time delay between steering wheel input and the first (yawrate) and second vehicle reaction (lateral acceleration, cf. Figure 14 (C)) is measured. This indicates a benefit in the responsiveness of the vehicle (cf. [11]).

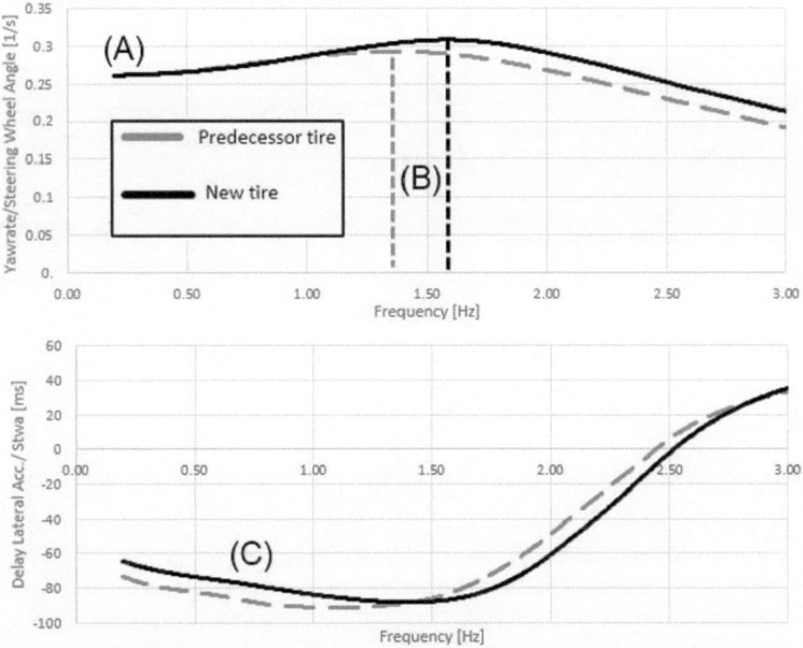

Figure 9: The magnitude of yawrate/steering wheel angle shows a significant higher yaw eigenfrequency; moreover a reduction of delay between the lateral acceleration and steering wheel angle is realized by the new tire.

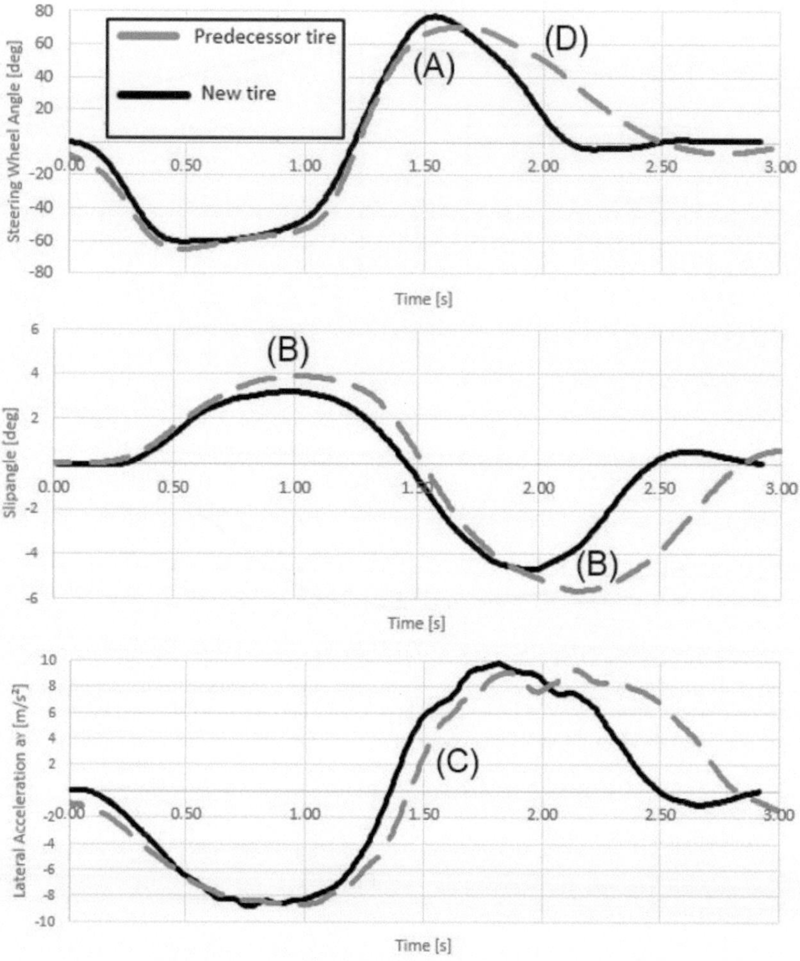

Figure 10: Comparing single lane change manoeuvers from the subjective evaluation show a reduction of maximal slip angle values and a faster lateral acceleration response which lead to improvements in subjective assessment.

Finally, the difference between both sets of tires in nonlinear lateral dynamics is investigated. Therefore subjective assessment is accompanied by the same measurement equipment used for the objective tests (cf. Figure 10). The manoeuver is a free driven single lane change manoeuver at a velocity of 120 kph. On both sets of tires, a steering wheel angle and steering wheel velocity based comparison of all single lane changes in the first 1.5 seconds is computed choosing the best matching manoeuvers of the subjective evaluation (A). The last 1.5 seconds of the manoeuvers are neglected because the steering wheel input of the driver turns out to be already influenced by the vehicle's reaction during this phase. Comparing the vehicle reaction on both sets of tires the maximum slip angle is significantly reduced (B). As a result of less needed vehicle movement to build up the stabilizing lateral forces on the rear axle the vehicle feels more responsive entering the second lane. This subjective impression is sustained by a measured reduction in time delay to build up the lateral acceleration (C). During the second change of the vehicle's direction the same physical effect takes place. The predecessor tire needs higher slip angle values and therefore also more time to build up the stabilizing rear axle side forces. Interestingly, the driver's reaction to this delay and higher amount of slip angle can be seen at the steering wheel angle input (D). The driver holds the steering wheel angle approximately 0.4 s longer to force the vehicle's turning into the second lane.

Table 2: The subjective assessment of the new tire shows improvements in the criteria on center steering precision and lane change as well. The results exceed the handling development targets (cf. Table 1).

Subjective Assessment (criteria – perceived deficit)	Relative Marks compared to the predecessor tire
On Center Response - Steering Precision	+0.75
Steering Wheel Torque - Centerpoint-Feeling	+0.75
Lane Change – Over All	+0.50

As a result of this the whole manoeuver time for the single lane change in subjective evaluation is reduced by approximately 0.5 s with the new tire compared to its predecessor.

The subjective evaluation result of both tires show the proposed result. The steering precision and the centerpoint feeling of steering wheel torque is perceived as improved by +0.75 marks and the mark in lane change by +0.5 marks (cf. Table 2). On the one hand this result even exceeds the targeted results in vehicle reaction (cf. Table 1). On the other hand it is used to enlarge the database and refine the objectivization method in general.

9 Summary and Outlook

In this paper the application of objective vehicle dynamics methods in the tire development process at Continental is presented. After this a hybrid objectivization method is introduced to describe links between handling related subjective assessment and objective vehicle dynamics measurements. The aim of this method is to differentiate between researches concerning perceived phenomena in vehicle behavior and statistical models to correlate the subjective assessment to characteristic values of objective tests. The first point is realized by online measurement during subjective evaluation to increase the understanding of drivers' perception. For the second point different linear and nonlinear handling regression models are analyzed. Modern data mining techniques have shown promising links on a large database in different vehicle clusters.

An application of this method is realized during the tire development process of a HP/UHP tire for an executive car. The decision tree method is used to show links between subjective assessment and objective vehicle dynamics measurements in the criteria on center steering precision. Relative thresholds in the characteristic values steering gradient and slip angle gradient are determinated for this criteria.

A nonlinear single track model is parameterized by measurement data and coupled with the regression model. Using a monte carlo based optimization method a target area for promising lateral tire characteristics is predicted. The flat trac measurement validates that the target is reached after tire design and tire building. Finally objective vehicle measurements and subjective evaluation show that the handling specific tire development targets are exceeded.

Next steps in this research are increasing the data base to cover currently excluded vehicle clusters as well as increasing the statistical significance.

10 References

[1] Ash, H.A.S.: Correlation of Subjective and Objective Handling of Vehicle Behaviour, University of Leeds, Leeds, 2002

[2] Chen, D.C.: Subjective and Objective Handling Behaviour, University of Leeds, Leeds, 1997.

[3] Gutjahr, D.: Objektive Bewertung querdynamischer Reifeneigenschaften im Gesamtfahrzeugversuch, Karlsruher Institut für Technologie, Institut für Fahrzeugsystemtechnik, Karlsruhe 2014.

[4] Harrer, M.: Characterisation of Steering Feel, University of Bath, Department of Mechanical Engineering, Bath, 2007.

[5] Heißing, B., Ersoy, M.: Fahrwerkhandbuch, 4. Auflage, Springer Vieweg Verlag, 2013

[6] Huneke, M.: Fahrverhaltensbewertung mit anwendungsspezifischen Fahrdynamikmodellen, Technische Universität Braunschweig, Institut für Fahrzeugtechnik, Braunschweig 2010.

[7] King, R.; Crolla, D.A.; Ash, H.A.S.: Identification of Subjective-Objective Vehicle Handling Links Using Neural Networks for the Foresight Vehicle, SAE paper 2002-01-1126, Detroit, 2002.

[8] Kobetz, C.: Modellbasierte Fahrdynamikanalyse durch ein an Fahrmanövern parameteridentifiziertes querdynamisches Simulationsmodell, Technische Universität Wien, Institut für Mechanik, Wien, 2003.

[9] Kollreider, D.: Identifikation der Reifeneigenschaften als Grundlage zur Fahrdynamikbewertung, TU Graz, Fakultät für Maschinenbau und Wirtschaftsingenieurwesen, Graz, 2009.

[10] Koegeler, H-M.; Schick, B.; Ecker, J.: Continiuous development of steering feel using objective and model-based methods, 7th International Munich Chassis Symposium 2016, München, 2016.

[11] Mimuro, T.; Ohsaki, M.; Yasunaga, H.; Satoh, K.: Four Parameter Evaluation Method of Lateral Transient Response, SAE Technical Paper 901734, 1990.

[12] Mitschke, M.; Wallentowitz, H.: Dynamik der Kraftfahrzeuge., 5. Auflage, Springer Verlag, 2014.

[13] Schimmel, C.: Entwicklung eines fahrerbarsierten Werkzeugs zur Objektivierung subjektiver Fahreindrücke, Technische Universität München, Institut für Maschinen- und Fahrzeugwesen, München, 2010.

[14] Witten, I; Eibe, F; Hall, M: Data Mining – Practical Machine Learning Tools and Techniques, 3rd Edition, Morgan Kaufmann Publishing Company, Cambridge, 2011.

TIRE TESTING AND SIMULATION

Advanced method for virtual tire and braking distance simulation

Carlo Lugaro,
Senior Application Engineer (TASS International, The Netherlands),

Dr.-Ing. Florian Niedermeier,
Development Engineer (BMW Group, Germany),

Bernd Wassertheurer, CAE Engineer (BMW Group, Germany),

Sonny Huisman, (TASS International, The Netherlands),

Jürgen Schüling (TASS International, Germany)

P.E. Pfeffer (Ed.), *8th International Munich Chassis Symposium 2017*,
Proceedings, DOI 10.1007/978-3-658-18459-9_52

Abstract

Being the force transmitter between road surface and vehicle chassis, tyres play a key role in the vehicle dynamics, influencing important aspects as the directional stability, braking distance and comfort of passenger cars. This fact, in combination with the Automotive industry's everlasting strive for reduced development times, explains why tyre modelling for vehicle simulations is becoming of increased importance.

Although Anti-Lock Braking Systems (ABS) are on the market since 1978 and even compulsory on all new passenger cars within the European Union since 2004 [1], simulation of an ABS braking manoeuvre with an accurate prediction of braking distance is still far from common in the Automotive industry. To a certain extent, tyre models are a bottleneck in this process, as during highly dynamic braking multiple levels of modelling complexity are addressed. Further, for ABS system validation Hardware in the Loop (HiL) tests are being performed which set the requirement of being capable of running in real-time on industry standard platforms. Therefore, the level of detail and complexity of the tyre model that can be used is limited to keep the computational intensity within acceptable bounds. In this specific study, a methodology for representing tyre forces and torques in ABS simulations is developed by making use of the currently existing tyre modelling theory.

First of all the paper addresses the different tyre modelling aspects involved in representing the tyre forces and torques during an ABS braking manoeuvre. Besides the important slip characteristics of the tyre, a specific in-depth focus is given to the highly non-linear transient response of tyres. A high level overview of how the transient tyre behaviour is modelled by the TASS International MF-Tyre/MF-Swift tyre model [13] is given. Furthermore, the influence of both the tyre testing procedure and the test rig (indoor vs. outdoor) is discussed. The investigations towards the appropriate tyre modelling theory and parameterization manners lead into a methodology that permits to perform accurate ABS braking simulations. The obtained ABS braking simulation is compared with experimental results acquired with an instrumented test vehicle to determine and discuss the accuracy. In conclusion, the study addresses future points of improvement for accuracy and usability of tyre models for simulating ABS braking manoeuvres. This article is an extension and continuation of the results presented in [11].

Introduction

The performance of an ABS control system depends on all the involved components. Namely the tyre, wheel, suspension, vehicle chassis, ABS sensors, ABS control algorithms and ABS actuators characteristics play a role in this process. As a result, ABS control algorithms must be tuned accordingly to the specifications of the other components interconnected in the loop. Common practice in the automotive industry consists in tuning the ABS parameters based on vehicle tests; the effect of a parameter change is directly evaluated in the field in a test vehicle. Considering the large amount of operating conditions and interactions within further vehicle control systems, the testing effort becomes remarkable. In this context, numerical simulations can help to significantly reduce the expensive and time consuming testing effort.

The tyre is one of the ABS control loop components that highly determines the performance. Considering also the severe operating conditions under which an ABS braking manoeuvre is conducted (e.g. large levels of slip, dependency on different road surfaces), accurate prediction of the forces and moments generated by the tyre is challenging. An often used approach is to model the longitudinal behaviour of tyres by measuring the steady state slip characteristics on artificial sandpaper road surfaces and modelling these characteristics with the Magic Formula [2]. Further, the transient tyre response is normally represented as a linear system. Generally, ABS braking simulation results obtained by this method show deficiencies such as wheels locking up rapidly and large overshooting of the longitudinal slip after releasing the brake pressure. Therefore the BMW Group and TASS International have jointly investigated what the most important tyre modelling and parameter identification strategies are that have to be taken into account to produce accurate and reliable simulations of an ABS full braking manoeuvre. This is accomplished with the following steps:

- evaluation of the typical operating conditions that characterize an ABS braking manoeuvre;
- evaluation of the dominant tyre modelling aspects under the given operating conditions;
- analysis of different tyre modelling and parameter identification strategies to determine what suits best in the given context;
- conduction of vehicle tests and simulations to analyse the accuracy of the proposed method.

After this introduction, the operating conditions are examined and an analysis is subsequently conducted to determine the dominant modelling aspects. A review of the theory that focuses on the determined aspects is given. A discussion about the different modelling and parameter identification strategies follows. Then results are presented. Finally conclusions are drawn and recommendations are given.

Operating conditions during an ABS braking manoeuvre

An ABS braking controller is typically applied in vehicles either as a stand-alone system or as a component of a stability control system (e.g. DSC). Usually an optimum target slip (reference slip) is calculated during all driving conditions; the ABS system controls the braking moment in order to minimize the difference between the actual estimated wheel slip and the reference slip. In general the reference slip is a trade-off between braking deceleration and vehicle stability and depends, for each wheel, on the actual driving conditions (e.g. vehicle yaw rate, lateral acceleration, friction between tyre and road). Considering driving conditions on different road types and in different climates in addition, it results that in general the tyre operating conditions during ABS braking can cover a vast range of configurations.

To narrow the scope of this study, a typical manoeuvre has been selected for the investigations: straight line braking from 100km/h to standstill on dry asphalt in normal European climate conditions. This permits to restrict side slip and camber angles to small values and also limit the range of tyre and road temperature. This manoeuvre is also commonly used as a standard to evaluate the performance (e.g. braking distance, control quality, etc.) of a vehicle under ABS braking conditions. Table 1 reports typical values that can occur during such a manoeuvre.

Table 1 Typical operating conditions during straight line ABS braking for a compact-medium size passenger vehicle in normal European climate conditions

Parameter	Range	Units	Description
V_x	[0, 100]	[km/h]	Longitudinal speed
κ	[-1, 0]	[-]	Longitudinal slip
α	[-1, 1]	[deg]	Side slip angle
γ	[-2, 0]	[deg]	Camber angle
F_z	[0, 8000]	[N]	Vertical load
T_t	[40, 100]	[degC]	Tyre temperature
T_r	[10, 30]	[degC]	Road temperature

The longitudinal slip covers the range of braking conditions up to full lock ($\kappa = -1$), the large range of the vertical force is due to large longitudinal accelerations and the tyre temperature increases during braking. It also has to be noted that the braking manoeuvre can occur on various road types (e.g. types of asphalt) and that each of them performs differently depending on the conditions (e.g. wear, dirt, humidity, etc.).

Dominant modelling aspects for the tyre, wheel, vehicle and ABS controller in the loop

To study the interactions that occur within the wheel and the ABS controller during ABS braking, the control loop for one wheel is analysed as a stand-alone system. In Figure 1, C refers to the ABS feedback controller, W to the wheel (tyre and all rotating parts), E to the sensors and algorithms that measure the wheel slip. Block W is further decomposed in the mass and inertial effects (M) interconnected with the tyre response (T). This simple model neglects interactions between vehicle and wheels: the longitudinal speed of the wheel and its vertical load, both function of the vehicle velocity and acceleration, are constant in this case.

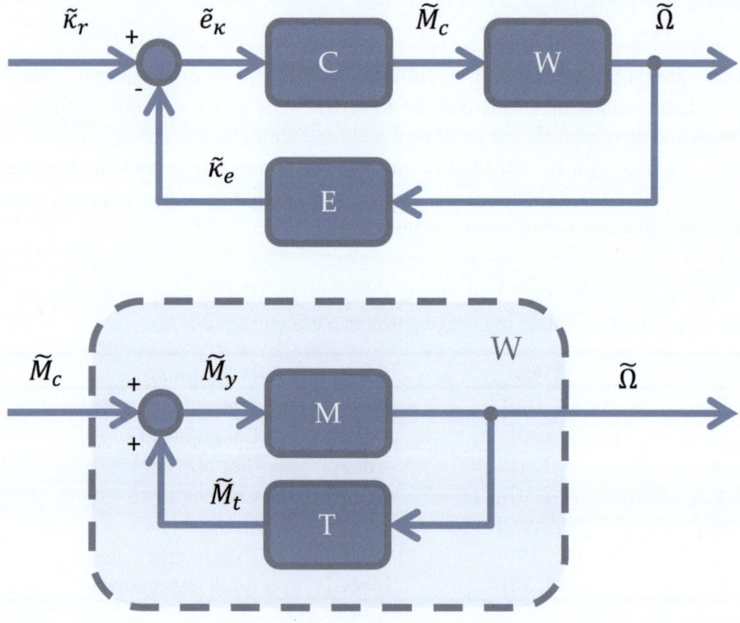

Figure 1: Top level view of the ABS control loop of one wheel (according to [11])

In Figure 1, κ_r is the reference slip, κ_e the estimated actual slip, e_κ the control error, M_c the control moment and Ω the wheel angular velocity. The ~ operator indicates the value of x relative to the operating point x_0:

$$\tilde{x} = x - x_0 \tag{1}$$

Mass and inertia effects determine the wheel angular velocity when a moment M_y is applied around its spin axis:

$$\tilde{\Omega} = \overbrace{\frac{1}{s\, I_y}}^{M} \tilde{M}_y \tag{2}$$

where I_y represents the inertia moment of the mechanical parts rotating with the wheel and s the Laplace complex variable.

The tyre can be represented as a steady-state slip characteristic interconnected with a transient response. Its response can be split in a pure gain when the input signals are constant (a) and a gain and phase change for frequencies larger than 0 (b).

$$T = \overbrace{K_{st}}^{(a)} \cdot \overbrace{H(s)}^{(b)} \tag{3}$$

Figure 2 depicts an example of a tyre longitudinal transient response. At low frequencies it holds:

$$\begin{cases} |H(0)| = 1 \\ atan(H(0)) = 0 \end{cases} \tag{4}$$

In this case T is equal to the steady-state gain K_{st} and zero phase delay. As the frequency of the input signal increases, $H(s)$ introduces a phase delay and a gain change.

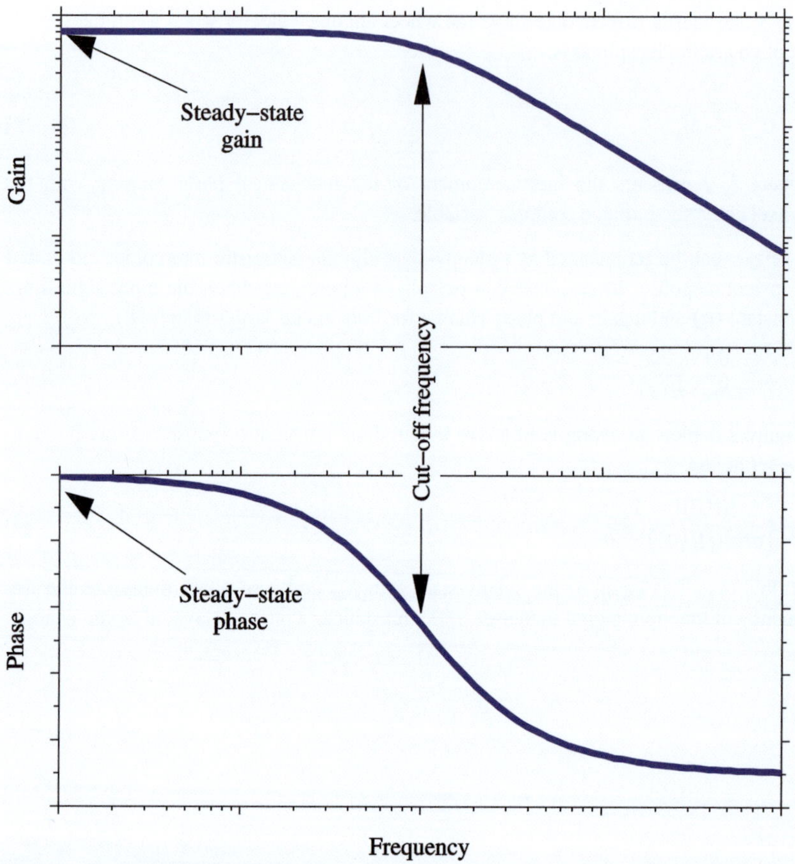

Figure 2: A typical tyre longitudinal transient response

From Figure 1 it results:

$$\widetilde{\Omega} = \frac{\overbrace{M}^{W}}{1-M\,T}\,\widetilde{M}_c \tag{5}$$

Substitution of T and M in (5) yields:

$$\widetilde{\Omega} = \frac{1}{s\,I_y - K_{st}\,H(s)}\,\widetilde{M}_c \tag{6}$$

The steady-state gain can be decomposed in three terms:

$$K_{st} = \frac{\partial M_t}{\partial \Omega}\Big|_{\Omega_0} = \overbrace{\frac{\partial M_t}{\partial F_x}\Big|_{F_{x0}}}^{1} \overbrace{\frac{\partial F_x}{\partial \kappa}\Big|_{\kappa_0}}^{2} \overbrace{\frac{\partial \kappa}{\partial \Omega}\Big|_{\Omega_0}}^{3} = \overbrace{-R_l}^{1}\ \overbrace{C'_{F\kappa}}^{2}\ \overbrace{\frac{R_e}{V_x}}^{3} = -\frac{C'_{F\kappa} R_e R_l}{V_x} \tag{7}$$

where R_l is the loaded radius, R_e the effective rolling radius and $C'_{F\kappa}$ the longitudinal slip stiffness at the operating point (κ_0). The parameters V_x, R_l, R_e and $C'_{F\kappa}$ influence the response of W. The dynamic response is also influenced by I_y and $H(s)$.

I_y, V_x, R_l and R_e can be measured with a good accuracy. The accuracy of $C'_{F\kappa}$ relies on the identification of the steady-state slip characteristic, the accuracy of $H(s)$ relies on the modelling of the transient response. Both play a dominant role in the overall loop performance.

Accurate identification of the steady-state slip response

The brush model, originally developed by Fromm and Julien (see [2] and [3] for references), can be employed to explain the most important phenomena that occur in the contact patch of a rolling tyre when a longitudinal slip is applied. For this situation Figure 3 depicts the tread longitudinal deflections in the contact patch. When a tread element gets in contact with the road at the leading edge, it is in an undeflected state. As it further travels along the contact patch, its deflection increases; the tread deflection (ρ_{px}) multiplied by its longitudinal stiffness (c_{px}) generates the longitudinal tension q_x (adhesion zone). As soon as the longitudinal tension exceeds the maximum transmissible tension μq_z, the tread starts sliding (sliding zone). It results that, at low levels of slip, the force is mostly determined by the tyre structural characteristics (e.g. stiffness of the treads) and, at large levels of slip, the force is mostly determined by tyre-road contact friction properties.

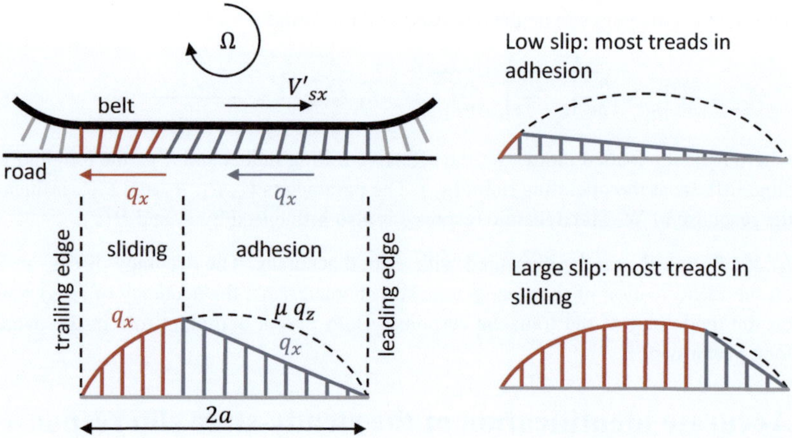

Figure 3: The brush model, case of a rolling tyre subjected to longitudinal slip (according to [2])

Figure 4 depicts an example of a typical steady-state slip characteristic in case of a braking force ($\kappa < 0$). According to the brush model, three areas can be distinguished:

1. linear characteristic around the origin: in this region the tyre force increases proportionally with the slip; the amount of increase is controlled by the longitudinal slip stiffness ($C_{F\kappa}$) which is mostly determined by the tyre structural properties (e.g. contact patch length, treads stiffness);
2. peak friction: in this region, the interaction between the tyre structural properties and the friction at the interface tyre-road plays a dominant role;
3. sliding area: in this region, the friction at the interface tyre-road plays a dominant role.

An ABS braking system operates in a region between areas 1 and 2. It is therefore very important that both the tyre structural and frictional characteristics are accurately modelled.

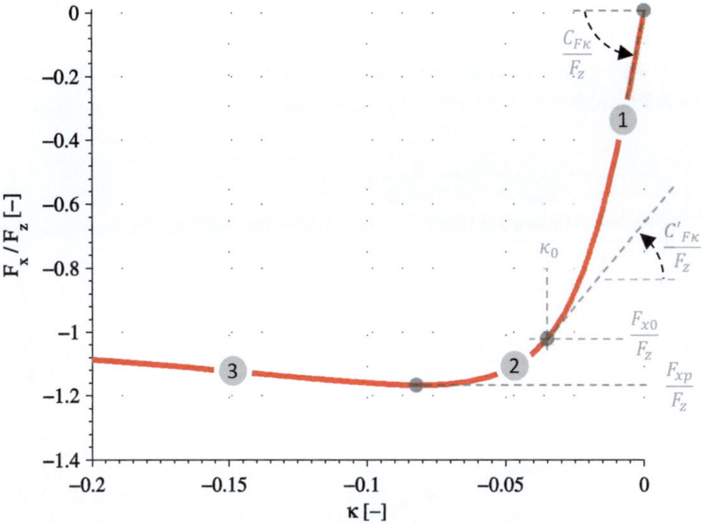

Figure 4: A typical tyre steady-state slip response (according to [11])

The figure also shows the longitudinal slip stiffness at the operating point (κ_0):

$$C'_{F\kappa} = \left.\frac{\partial F_x}{\partial \kappa}\right|_{\kappa_0} \tag{8}$$

Obviously, when $\kappa_0 = 0$ it results $C'_{F\kappa} = C_{F\kappa}$.

Several options are available to measure the steady-state slip characteristic of a tyre. They can be divided in two main categories: indoor and outdoor test facilities. Indoor test facilities offer the advantage of highly reproducible results but at the cost of testing conditions that are not fully representative of real driving situations. In particular it is very challenging to accurately reproduce real road characteristics [4]. Flat belt systems are the most common indoor test facilities and make the tyre roll on a steel belt overlaid with sandpaper. Also, drum rigs exist where a real asphalt road can be mounted on, but the road is not flat in this case. Alternatively, outdoor test facilities commonly consist of a vehicle or a truck trailer where an extra test wheel can be mounted. By means of load cells and encoders, the tyre characteristic is measured. These facilities offer results that are more subjected to environmental conditioning (e.g. ambient temperature, humidity) than an indoor test facility, but allow the tyre to be measured while rolling on a real road (e.g. wet and dry asphalt, concrete, snow and ice). It results that it is important to carefully assess which type of test facility should be employed in a certain situation.

The steady-state slip characteristic of one tyre has been measured by the BMW Group with both an MTS FlatTrac test rig [5] and by TASS International with the TASS tyre test trailer [6]. This allows a direct comparison of the two test methods. The results are depicted in Figure 5.

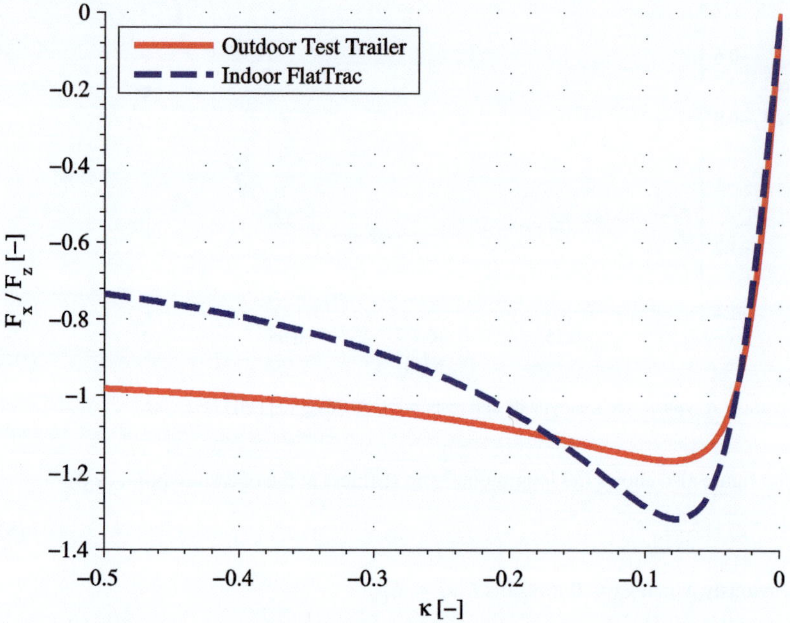

Figure 5: Comparison of the steady-state slip characteristic for a 17" tyre, measured with the MTS FlatTrac and the TASS International tyre test trailer (according to [11])

It is observed that, for a low level of slip, where the response is dominated by the tyre structural characteristics, both methods offer comparable results. However, at a large level of slip, where the friction becomes important, the results are less comparable.

The friction generated at the interface between road and rubber compound is determined by several phenomena, the two most important being hysteretic friction and adhesion [7]. Hysteretic friction is generated by the rubber compound being viscous and excited at a given frequency; adhesion is generated by attractive forces at molecular level. Hysteretic friction plays a dominant role in general, especially when the contact interface is not completely dry or clean.

Hysteretic friction is determined by the road excitation amplitude and frequency (i.e. road roughness power spectrum density and longitudinal velocity of the wheel) and the rubber compound characteristics (which also depend on the temperature) [7], [8], [9]. TASS International has measured the roughness of a typical asphalt road and a sandpaper sample with a 3D laser scanner with a resolution of 50 μm. Figure 6 shows the results; in general it is observed that the asphalt road shows a larger roughness than the sandpaper.

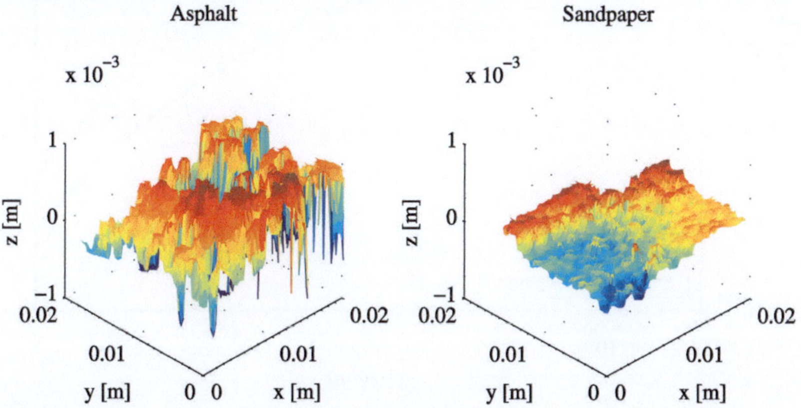

Figure 6: 3D scan of a typical asphalt road (left) and sandpaper (right) (according to [11])

Based on the 3D scan measured data, the roughness power spectrum density for both surfaces has been calculated and depicted in Figure 7.

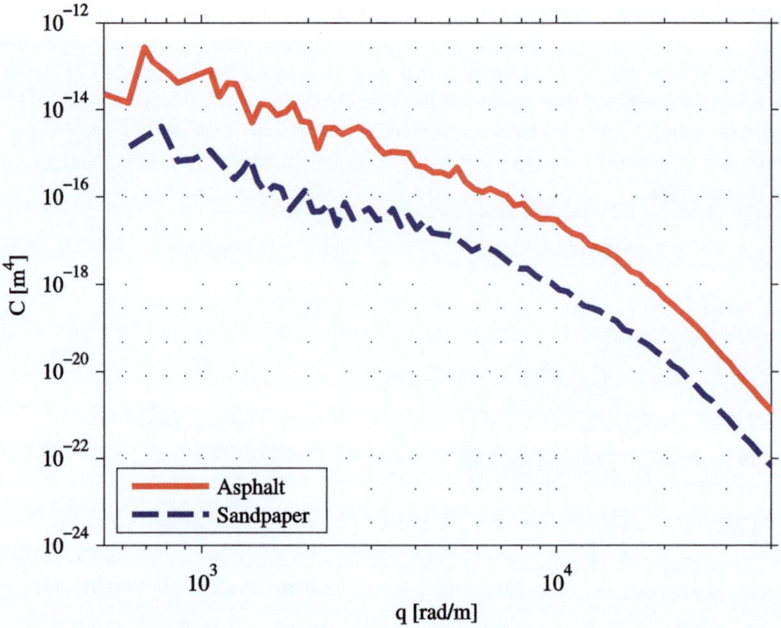

Figure 7: Power spectrum densities of the surfaces (asphalt and sandpaper) measured with the 3D scanner (according to [11])

It is observed that the roughness of the asphalt is larger at all spatial frequencies. This explains, to a large extent, the different steady-state slip characteristics measured with the MTS FlatTrac and the TASS International tyre test trailer depicted in Figure 5.

Modelling the tyre transient response

A common approach to take into account the transient response in the tyre model is to apply a low pass filter to the input slip quantities (linear transient model [2]).

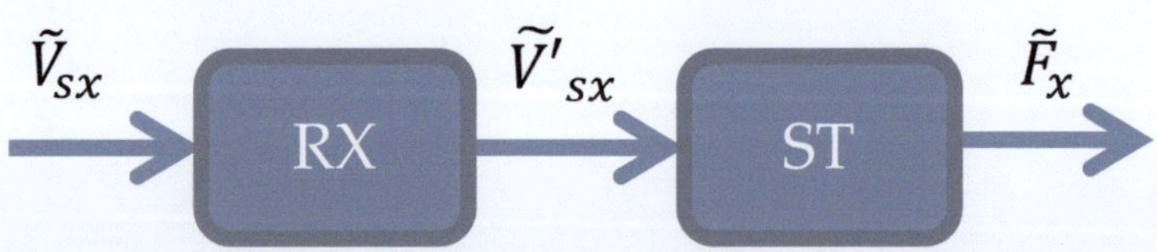

Figure 8: Block scheme of the tyre linear transient model (according to [11])

In Figure 8, ST refers to the steady-state sliding velocity characteristic ($F_x(V'_{sx})$) and RX represents the low pass filter that models the transient response. V'_{sx} is the sliding velocity that actually generates a force. The perturbation of the longitudinal force around an operating point reads:

$$\tilde{F}_x = \left.\frac{\partial F_x}{\partial \kappa}\right|_{\kappa_0} \cdot \tilde{\kappa}' = -\left.\frac{\partial F_x}{\partial \kappa}\right|_{\kappa_0} \cdot \frac{\tilde{V'}_{sx}}{|V_x|} = \overbrace{-\frac{C'_{F\kappa}}{|V_x|}}^{ST} \tilde{V}'_{sx} \tag{9}$$

The low pass filter can be implemented with a first order transfer function:

$$\tilde{V}'_{sx} = \overbrace{\frac{1}{\frac{s\sigma_x}{|V_x|}+1}}^{RX} \tilde{V}_{sx} \tag{10}$$

where $\sigma_x = {}^{C_{F\kappa}}/_{C_x}$ is the relaxation length and C_x is the tyre longitudinal stiffness. This simple approach offers accurate results in most situations, however with some limitations.

The transfer function of the complete linear transient model is obtained by combining equation (9) and (10):

$$\tilde{F}_x = -\frac{C'_{F\kappa}}{\sigma_x} \frac{1}{s+\frac{|V_x|}{\sigma_x}} \tilde{V}_{sx} \tag{11}$$

which at a standstill ($|V_x| = 0$) becomes:

$$\tilde{F}_x = -\frac{C'_{F\kappa}}{\sigma_x}\frac{1}{s}\tilde{V}_{sx} = \frac{C'_{F\kappa}}{\sigma_x} R_e \frac{\tilde{\Omega}}{s} = \frac{C'_{F\kappa}}{\sigma_x} R_e \tilde{\theta} \tag{12}$$

where R_e is the effective rolling radius and θ is the rotation angle of the wheel around the spin axis. Equation (12) shows that, at a standstill, the linear transient model produces the equivalent of a pure spring system, making the whole wheel system undamped. To overcome this issue, an artificial damping is commonly introduced in the tyre model when standstill is approached.

Experimental evidence shows that the tyre transient response depends on the actual operating point (i.e. longitudinal slip level): the larger the slip the faster the response [2]. This effect is commonly reproduced by artificially changing σ_x to match the same response that is measured at different levels of slip.

To overcome these issues a different approach can be applied. The tyre transient response can also be predicted by introducing a physical model of the carcass stiffness and damping that governs the contact patch dynamics (enhanced non-linear transient model [2]).

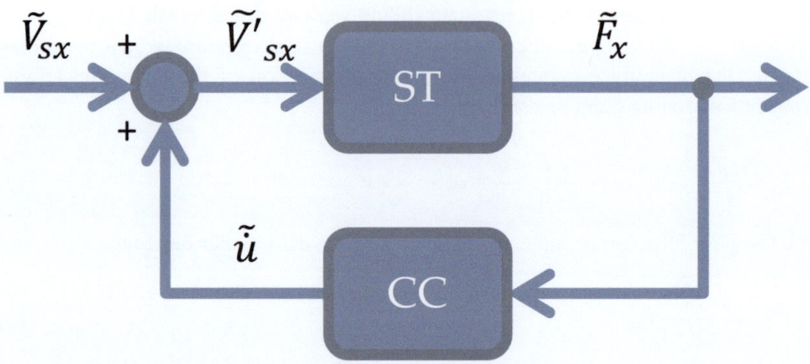

Figure 9: Block scheme of a tyre non-linear transient model

In Figure 9, CC represents the contact patch dynamics. u is the carcass deflection generated by the force F_x, $V_{sx} + \dot{u}$ produces the slip velocity of the contact patch V'_{sx}. The equilibrium of the longitudinal forces acting on the contact patch reads:

$$\tilde{\dot{u}}\, k_c + \tilde{u}\, c_c = \tilde{F}_x \tag{13}$$

where c_c is the carcass stiffness and k_c the damping. The mass of the contact patch is neglected. Equation (13) permits to determine the transfer function that models the contact patch dynamics:

$$\tilde{\dot{u}} = \overbrace{\frac{s}{s\, k_c + c_c}}^{CC} \tilde{F}_x \tag{14}$$

From Figure 9 it results:

$$\tilde{F}_x = \frac{ST}{1 - ST \cdot CC} \tilde{V}_{sx} \tag{15}$$

Substitution of ST and CC in (15) produces the transfer function of the tyre model including carcass stiffness and damping:

$$\tilde{F}_x = \frac{\frac{C'_{F\kappa}}{|V_x|}\left(\frac{s\, k_c}{c_c}+1\right)}{\frac{s\left(k_c+\frac{C'_{F\kappa}}{|V_x|}\right)}{c_c}+1} \tilde{V}_{sx} \tag{16}$$

Equation (16) shows that in this case the frequency response is a function of $C'_{F\kappa}$ and $|V_x|$; either the lower $C'_{F\kappa}$ or the larger $|V_x|$, the faster the overall response. This is in accordance with experimental evidence. Furthermore, equation (16) explicitly introduces the carcass stiffness and damping in the model; both of them correspond to physical quantities of the tyre that can be experimentally measured.

Because the contact patch has a finite length, its response to a change of slip also shows a transient behaviour. This effect can be included in the model by empirically reproducing it as a low pass filter.

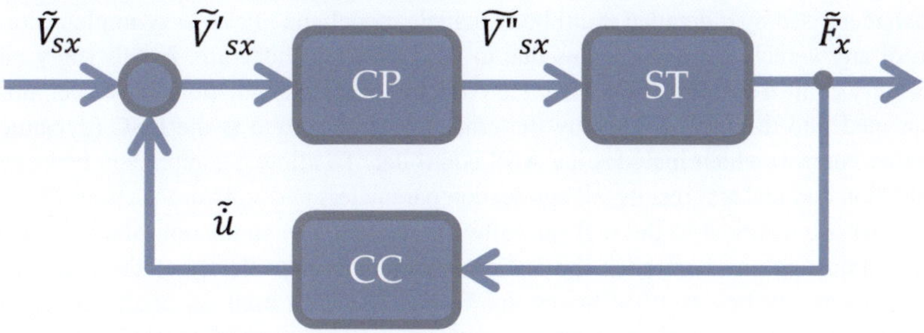

Figure 10: Block scheme of a tyre non-linear transient model, with the contact patch transient included (according to [11])

In Figure 10, CP represents the contact patch transient. This can be modelled with a first order transfer function:

$$\tilde{V}''_{sx} = \overbrace{\frac{1}{\frac{s\,\sigma_c}{|V_x|}+1}}^{CP}\tilde{V}'_{sx} \tag{17}$$

where σ_c is the contact patch relaxation length, its value is approximately equal to the half of the contact patch length [10]. This approach produces a correct model response also in the special case of standstill.

Model quality and comparison to vehicle tests

The criterion to evaluate the quality of the presented simulation method is the braking distance from 100km/h to standstill as this is one of the standard test procedures for ABS braking performance. Based on the positive results gained in [11], a more extensive test setup was derived: in total eleven different ABS controller settings (application variants) are generated and tested in vehicle braking manoeuvres. These ABS settings differ from each other in terms of e.g. the pressure built-up characteristics. All other vehicle parameters (tyres, vehicle mass, etc.) are kept constant during those tests. To identify any effects on the braking distance, the test vehicle was equipped with a conventional braking distance measurement systems. It has to be noted that the vehicle measurements were performed on a different proving ground than the measurements for the identification of the steady-state tyre characteristic, however the friction characteristics seem to be similar, see also [11].

As described in [11], the vehicle simulation was performed in the virtual simulation environment called ISAR (Integrierte Simulationsumgebung Fahrdynamik Regelsysteme) [12]. This simulation platform is built up in MATLAB/Simulink and has been developed by the BMW Group. ISAR is based on a five-body two track model which is parameterised by a detailed multi body vehicle model and allows easy implementation of any vehicle control systems due to its modular architecture. In this study all necessary control systems for performing ABS full braking simulations have been implemented into the model. Thereby the crucial control system is the DSC (dynamic stability control) which includes the ABS controller. To allow a comparison between simulation and real test results, all application parameters and software versions of the controller are identical to the real car software version. The simulation platform also allows adjusting any controller application parameters, equivalently to the real controller. As it will be described below this feature has been used for predicting ABS braking distances for various controller settings. For all simulations MF-Tyre/MF-Swift 6.2 [13] has been used, with the non-linear transient model and a steady-state characteristic measured on a real outdoor asphalt road. High speed in-plane cleat measurements have been used to identify the longitudinal carcass stiffness and damping [14]. These excitations are closer to the ones during ABS full braking manoeuvres, compared to an identification at standstill or low speed.

At first, the ABS base variant is used as an example to qualitatively compare the simulation with the measurement result, see Figure 11.

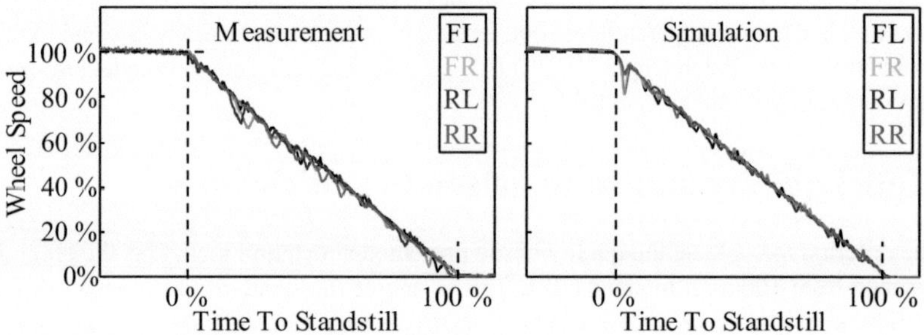

Figure 11: Comparison of an ABS-braking simulation and its corresponding measurement on a normal asphalt road

The slip characteristics of simulation and measurement are similar and already indicate a good model quality. By having a closer look on the simulation readings of Figure 11 it can be noted that the wheel speeds of the left and right tyres are slightly different, which is an improvement to the simulation results shown in [11] which were run on an ideal flat road. This refinement was achieved by the integration of an OpenCRG road [15] file (created using a measured asphalt road) and using the 3D road contact model that offers MF-Tyre/MF-Swift [13].

As a next step, the resulting braking distances from vehicle measurements are compared to those achieved from simulation, the results are shown in Figure 12.

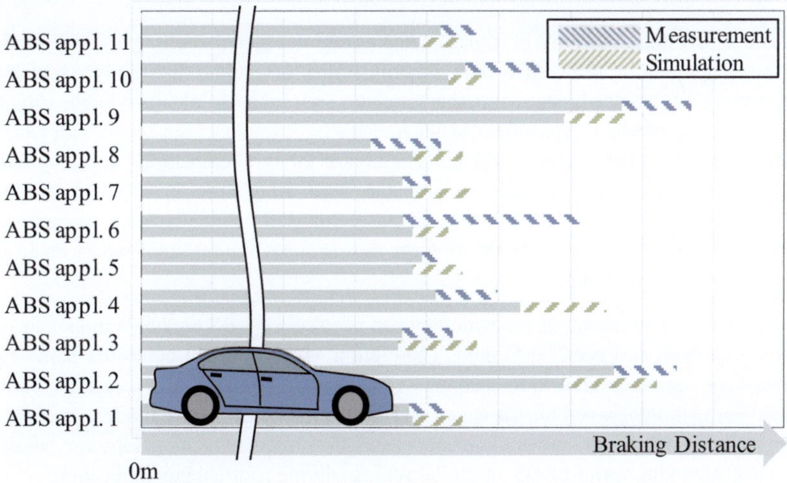

Figure 12: ABS-braking distances of eleven different ABS controller settings, comparison between measurement and simulation

Special focus should be put on the variation shown in the simulation results, as this is not obvious at a first glance. Thereby slightly different braking manoeuvres are simulated, all of them having different starting times of the virtual braking manoeuvre; the rest of the manoeuvre parameters are exactly the same (e.g. pressure built-up gradient). The ABS controller has a fixed clock rate, by slightly changing the start time of the braking manoeuvre within this delay loop, this results in slightly different braking results. This is even intensified by the slightly different braking positions on the virtual OpenCRG track, which is, as already mentioned earlier, derived from a real asphalt road and therefore not homogenously distributed. By evaluating the results the simulated braking distance variation is actually comparable to real braking distance measurements.

All in all, the results are satisfying as for all variants (except ABS appl. 4) overlapping values of the braking distances between measurement/simulation can be found. One can see that two out of the eleven ABS-variants (ABS appl. 2 and appl. 9) have remarkable longer braking distances than the other ones. Both times the simulation model predicts the effect of the ABS controller settings on the braking distance. The larger variation in the measurement of ABS appl. 6 can be explained by possible external effects.

Summary and outlook

Various investigations have been performed to improve tyre modelling and tyre model parameterisation to allow ABS braking simulations. It resulted that the tyre transient response and the steady-state slip characteristic are crucial to obtain reliable simulations. The commercial tyre model MF-Tyre/MF-Swift 6.2 has been tuned with particular attention to these aspects. The contribution of the tyre temperature and longitudinal speed and their interactions with the road roughness is a very complex phenomenon that could not be addressed yet in this study. Since both the temperature and longitudinal speed variations are large during ABS braking, this aspect could be investigated in the future.

To prove that this advanced methodology can be used for ABS braking simulations, an extensive study has been performed. Therefore a set of different controller parameter setups has been measured, simulated and subsequently compared. The results indicate a good correlation between measurement and simulation by using tyre models created by the methodology shown above. As a conclusion, the presented methodology can be used for benchmarking various ABS controller settings during a virtual pre-application.

Besides the possibility to assess the efficiency of different ABS controller settings (relative prediction), also the absolute accuracy is remarkable, bearing in mind the always present effect of asphalt and environment influences. In the future, further investigations will be taken regarding the effects on different vehicles.

Nomenclature

a	Half of the contact patch length
c_c	Carcass longitudinal stiffness
c_{px}	Longitudinal stiffness of the treads per unit length
e_κ	Slip control error
k_c	Carcass longitudinal damping
q_x	Longitudinal force of the treads per unit length
q_z	Vertical force on the treads per unit length
u	Carcass longitudinal deflection
$C_{F\kappa}$	Longitudinal slip stiffness
$C'_{F\kappa}$	Longitudinal slip stiffness at an operating point
C_x	Longitudinal tyre stiffness
F_x	Longitudinal tyre force
F_{xp}	Longitudinal tyre force at peak friction
F_z	Vertical tyre force
I_y	Inertia moment the parts rotating with the wheel around the spin axis
M_c	Control moment
M_t	Tyre moment
M_y	Wheel moment
R_e	Effective tyre rolling radius
R_l	Loaded tyre radius
T_r	Road temperature
T_t	Tyre temperature
V_{sx}	Sliding velocity of the wheel
V'_{sx}	Sliding velocity of the contact patch
V''_{sx}	Sliding velocity of the contact patch with relaxation length
V_x	Longitudinal velocity component of wheel centre
α	Side slip angle of the wheel
γ	Camber angle of the wheel
θ	Torsional carcass deflection
κ	Longitudinal slip of the wheel
κ_e	Estimated longitudinal slip of the wheel
κ_r	Reference longitudinal slip of the wheel
μ	Friction coefficient
ρ_{px}	Longitudinal deflection of the treads
σ_c	Relaxation length of the contact patch
σ_x	Relaxation length of the tyre
Ω	Angular velocity of the wheel around the spin axis

References

1. Robert Bosch GmbH: From innovation to standard equipment. 30 years of safe braking with Bosch ABS. Online: http://www.bosch.co.jp/en/press/group-0807-05.asp, access: 08 May 2017

2. Pacejka, H. B.: Tire and Vehicle Dynamics. Oxford: Butterworth-Heinemann, 2012

3. Hadekel, R.: The mechanical characteristics of pneumatic tires. In: S&T Memo 10/52, British Ministry of Supply, TPA 3/TIB, 1952

4. Bhoopalam, A. and Kefauver, K.: Using Surface Texture Parameters to Relate Flat Belt Laboratory Traction Data to the Road. In: SAE Technical Paper 2015-01-1513 (2015), doi:10.4271/2015-01-1513

5. MTS: Flat-Trac Tire Testing Systems. Online: https://www.mts.com/en/products/producttype/test-systems/simulation-systems/tire/flat-trac/index.htm, access: 15 February 2017

6. TASS International: Dedicated on-road test laboratory. Online: https://www.tassinternational.com/tyre-testing, access: 15 February 2017

7. Persson, B. N. J.: Theory of Rubber Friction and Contact Mechanics. In: Journal of Chemical Physics (2001), p. 3840

8. Klueppel, M.; Heinrich, G.: Rubber Friction on Self-Affine Road Tracks. In: Rubber Chemistry and Technology 73 (2000), No. 4, pp. 578-606

9. Grosch, K. A.: The Relation between the Friction and Viscoelastic Properties of Rubber. In: Rubber Chemistry and Technology 37 (1964), No. 2, pp. 386-403

10. Zegelaar P.W.A.: The Dynamic Response of Tyres to Brake Torque Variations and Road Unevennesses, Delft University of Technology, 1998.

11. Lugaro, C., Niedermeier, F., Wassertheurer, B., Schüling, J.: Method for Virtual Tyre and Braking Distance Simulation. In: ATZ worldwide (2017), No. 6 (submitted and approved)

12. Kvasnicka, P.; Prokop; G.; Rettinger, A.; Stahl, H.: Durchgängige Simulationsumgebung zur Entwicklung und Absicherung von fahrdynamischen Regelsystemen. VDI-Berichte (2006), No. 1967, pp. 387-403

13. TASS International: MF-Tyre/MF-Swift (Version 6.2). Online: https://www.tassinternational.com/delft-tyre, access: 01 March 2017

14. SAE International: Dynamic Cleat Test with Perpendicular and Inclined Cleats (2006), SAE J2730

15. OpenCRG® – managing the road surface ahead. Online: http://www.opencrg.org, access: 18 April 2017

Decoupling the physical effects of transient tire behavior based on optical measurements of carcass deformation

Dipl.-Ing. Pavel Sarkisov, Technische Universität Dresden

Prof. Dr.-Ing. Günther Prokop, Technische Universität Dresden

Dr.-Ing. Sergey Popov, Bauman Moscow State Technical University

Dipl.-Ing. Jan Kubenz, Technische Universität Dresden

P.E. Pfeffer (Ed.), *8th International Munich Chassis Symposium 2017*,
Proceedings, DOI 10.1007/978-3-658-18459-9_53

Abstract

In order to know tire transfer function precisely enough for automated driving in dynamic situations and even to be able to use tire as a sensor instead of black box, it is necessary to understand several phenomena of the tire transient behaviour. In this research tire is considered as a physical system – a composition of particular physical properties. With help of eight in-rim cameras carcass lateral stiffness and damping as well as its bending behaviour were analysed and described in a physical simulation model. The main result is understanding of several physical processes of a rolling tire and their influence on its transient handling behaviour, which is necessary nowadays for further development of driver assistance systems and autonomous driving.

1. Motivation

Development of autonomous driving requires precise and efficient control of vehicle motion based on a number of real-time measurements. Even for advanced driver assistance systems it is already a challenge. Such systems like ESP, Emergency Steering Assist and Active Rollover Prevention consider tire as a black box. Trying to use as much grip potential as possible, these systems load tire up to 100 % slip in both longitudinal and lateral directions, and they load it dynamically up to 30 Hz (Fig. 1).

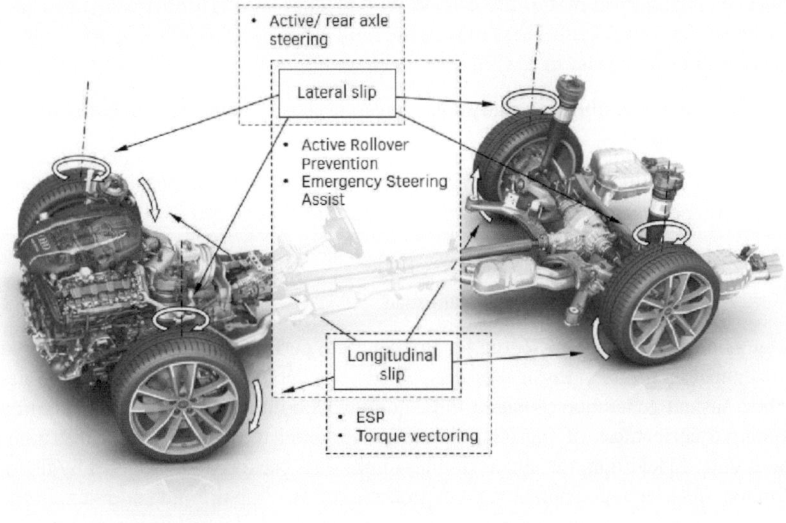

Figure 1. Impact of some assistance systems on wheels (source of background picture: [2]).

Due to this complex motion mode the traditional empirical description of a tire as a black box [1] becomes less and less suitable for this task. In order to develop more safe and efficient control algorithms for assistance systems and even be able to use tire as a sensor, it is required to understand the physical background of tire transient behaviour. In the tire mechanics there was performed a lot of research work, but most of published projects were targeted to other applications.

2. State of the art

Description of tire handling performance with link to tire physics is traditionally divided into group of simple physical simulation models (Treadsim [1], Brush-and-Ring-Tire [3], TameTire [4-5]) and group of detailed Finite-Element-Models [6]. The models of the first group are targeted to real-time simulation, this is why they have simplified structure and empirical consideration of transient effects. Consequently they are not suitable for the new challenge. The FE-models provide precise simulation results, but their creation, parameterization and calculation are too time- and effort-consuming for the task of rolling on a flat rigid surface. Hence, there is a necessity of a physical simulation model, which considers more relevant physical effects, than existing simple models do.

The known models generally divide whole tire body into two elements – tread (isotropic rubber) and cylindrical part of tire carcass with breaker layers (further called just carcass, anisotropic structure). These models are built up on basis of most important physical effects of carcass and tread (Fig. 2):

- Carcass lateral flexibility, usually considered via elastic connection between tire carcass and rigid rim;
- Carcass bending, usually taken into account via flexible one-dimensional elements like string or beam, or just approximated with a given assumed function;
- Tread shear, usually simulated via brush model [1];
- Friction properties between tread and road.

As far as these effects are essential for tire handling performance, they are considered in this paper as primary physical effects. Their consideration makes it possible to describe transient generation of lateral force quite well, but they do not provide required qualitative correctness of transient generation aligning torque, especially in a non-steady state. In addition, the understanding of physical background, provided by them, is limited. In order to get more insight in processes of a rolling tire and understand force and torque generation wider, it is necessary to go one level of detail further and to take into account more physical effects and properties. In [7] there was investigated influence of contact patch shape, which changes with rolling conditions, on tire handling behaviour. In [8] there was analysed bending behaviour of tire carcass in sense of orientation of its lateral cross-sections. For the given tire there was shown that carcass bending features high shear angle, hence this bending behaviour does not comply with Euler-Bernoulli Hypothesis and can be described with Timoshenko theory [9]. These physical effects are considered in this paper as secondary ones.

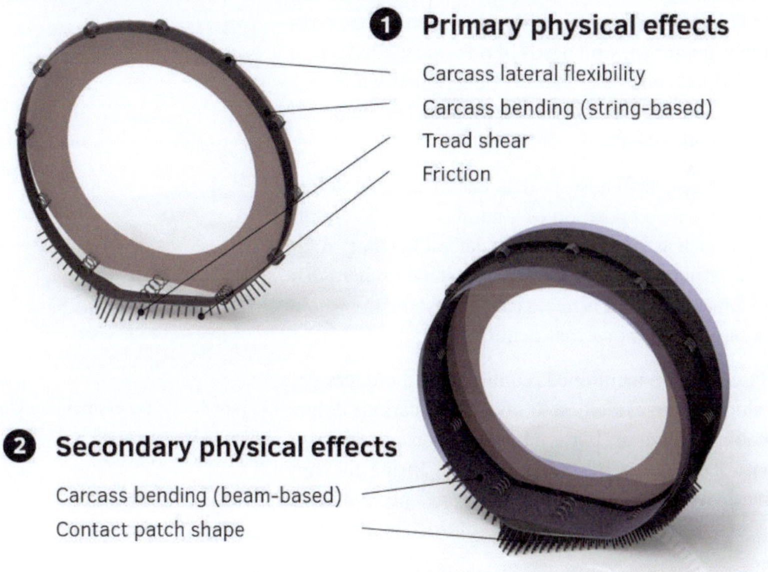

Figure 2. Primary and secondary physical properties of a tire in terms of handling behaviour.

A significant importance for these carcass bending behaviour represents bending line of tire carcass. This deflected form of tire carcass was partly investigated in [10-11], but it was not analysed detailed enough yet in terms of transient tire behaviour. Hence it defines a mission statement of this research.

3. Mission statement

The global goal is to analyse physical effects of a rolling tire experimentally, to describe them in a physical model and to estimate the role of each of them in tire force and torque generation. The focus of this paper is lateral deflection and bending of the tire carcass in a non-steady state, which precisely describes lateral force and aligning torque.

4. Experimental methodology

The first challenge is that the lateral tire deformation is a sum of its carcass deflection and tread shear. The tread in this sense is circa one order of magnitude stiffer, as the carcass (Fig. 3). The whole deformation of a tire can be calculated based on rolling kinematics, but it remains unknown, which part of it is carcass deflection and which part is tread shear. Consequently it is necessary to measure at least one of these two deformational components experimentally.

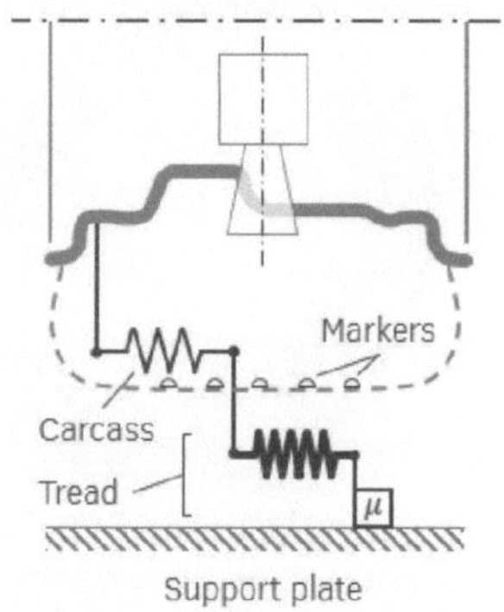

Figure 3. The necessity to look inside the tire to decouple carcass and tread deformation.

The second difficulty is that lateral carcass deformation is a composition of a number of different effects, such as lateral deflection, lateral bending, rotation around radial axis and some other. To understand them it is necessary to observe the whole circumference of carcass centreline.

Based on two mentioned requirements there was developed a novel method to investigate carcass deformation behaviour (Fig. 3). It is based on a special rim with the eight windows (Fig. 4, a). Inside the rim there were mounted cameras, one per window (Fig. 4, b). On the inner surface of tire carcass there were glued markers (Fig. 4, c).

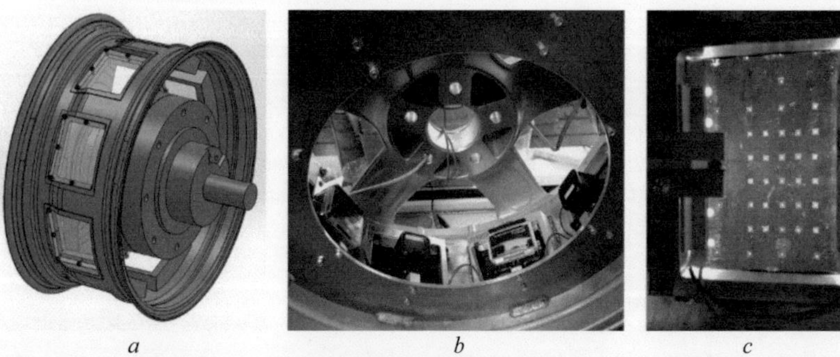

Figure 4. The concept of optical measurement of carcass deformation.

The developed system has made it possible to perform measurements targeted to explanation of different physical effects, which will be described in separated subchapters.

5. Measurement results

As far as lateral deformation of tire plays significant role in lateral force and aligning torque generation, it is important to understand the connection between deformation and force, which is stiffness and damping. For that purpose there were performed measurements of motionless tire excited in lateral direction, torsional direction, and measurements of rolling tire excited with slip angle.

5.1. Carcass lateral properties

Analysis is based on measurements of tire lateral stiffness: the tire stands still on the flat plate, the plate is moving in lateral direction with different speed (Fig. 5). The eight cameras have made it possible to follow the deflection of the whole circumference of carcass.

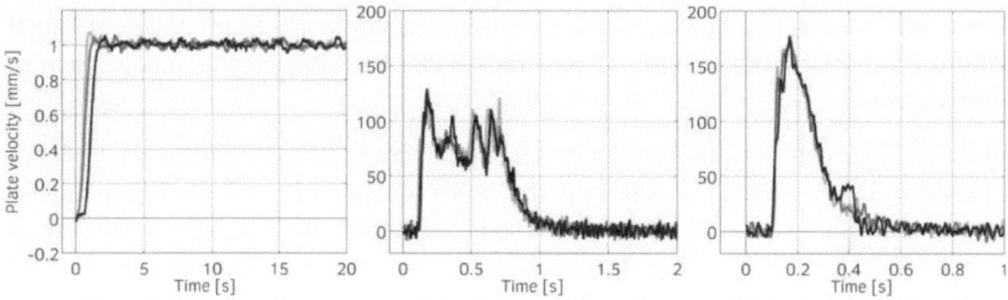

Set velocity: 1 mm/s *Set velocity: 100 mm/s* *Set velocity: 200 mm/s*

Figure 5. Actual plate velocity for three values of set plate velocity for tire lateral excitation (three repetitions per diagram).

The analysis of linear area of force-displacement curve has the advantage that in this area there is practically no slide in the contact patch, hence the carcass stiffness and damping can be investigated directly. For three mentioned measurements of lateral stiffness there was analysed the carcass deflection: To visualize the results, for each of eight cameras there was selected one central marker. These eight central markers are distributed along the carcass circumference with the step of 45°. The displacement of these markers depending upon the lateral force is depicted on the Fig. 6. On the background of this figure there are shown grey curves of displacement of the plate depending upon the lateral force. Even though the plate displacement causes the force, the diagram is inverted for the sake of clarity: For the same value of lateral force there are shown displacement of the plate, which causes this force, and displacement of carcass, which has taken place at the same moment of time.

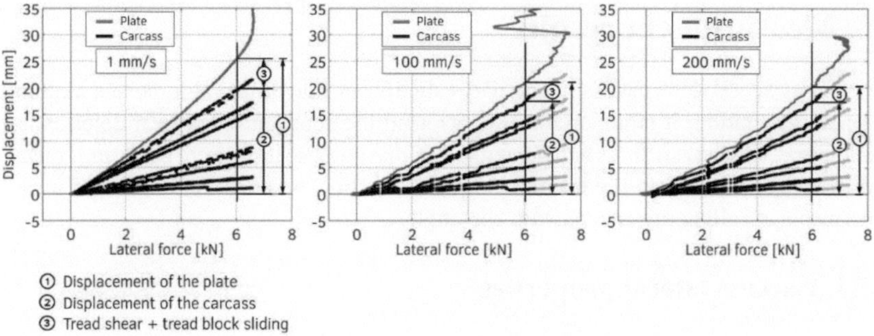

Figure 6. Displacement of the plate and tire carcass in eight points depending upon the lateral force.

The diagram of static deformation shows linear connection between carcass displacement and lateral force. Consequently, for given tire and in a working range of lateral deformation the lateral stiffness of tire carcass can be considered as independent of displacement.

The difference between plate displacement and carcass deflection (dimension 3, Fig. 6) in the non-linear range (15 – 30 mm) stands for the sum of tread shear and sliding, including lateral overrolling of the tire. The higher plate speed is, the smaller the difference is. Physical explanation of the effect is frequency-dependent friction: For the given tire and the given road surface slide friction coefficient accounts for approximately 0.93 by 1 mm/s sliding speed, and for 1.08 by 100 mm/s. Hence, higher friction force is achieved with higher speed. A consequence of higher force is higher deflection of tire carcass. This logical chain confirms that the analysis of carcass deflection is qualitatively correct. Fig. 6 depicts that the linearity of carcass displacement curve varies with the deformation speed. To analyse this difference there were taken measurements of carcass displacement in one central marker – the one which is in the middle of contact patch (this is a last point to start sliding). These three curves were compared in a traditional force-displacement diagram (Fig. 7), not an inverted one (not like Fig. 6).

Thanks to the trend line it is visible that static deformation causes linear force response to the carcass deflection. For high-speed deformation the same carcass deformation (let say, 10 mm) causes perceptible higher force (3.15 kN to 3.70 kN). The slope stays generally the same, so stiffness properties are the same. Therefore for the given tire and wheel load range it is fair to conclude: Tire carcass exhibits slight frequency-dependent properties in lateral direction.

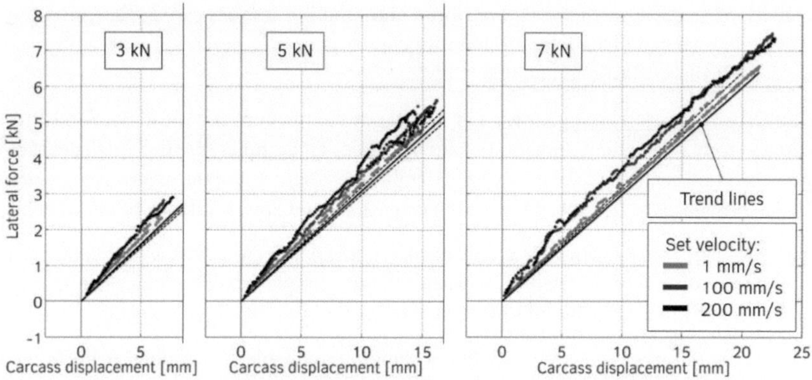

Figure 7. An overview of flexible and damping properties of tire in lateral direction.

Numerical determination of damping coefficient is limited because of two factors: Firstly, the deformation speed was not constant (Fig. 5). Secondly, the damping coefficient may also depend upon speed. For the investigated case the damping share of lateral force achieves up to 20 %.

Analysis of measurement results for three different wheel loads has proven the responsibility of described effects (Fig. 8). It has additionally shown that carcass lateral stiffness can be considered as independent upon the wheel load in the investigated range.

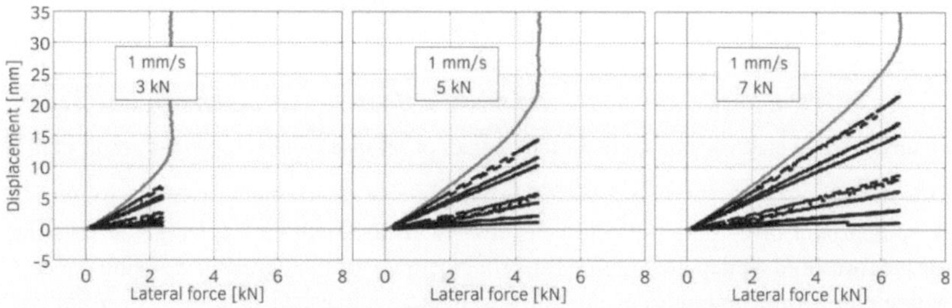

Figure 8. Comparison of displacement of plate and carcass for three different wheel load values.

5.2. Carcass bending properties

Speaking about carcass bending it is important to define the terms. In sense of handling, namely of generation of tire lateral force and aligning torque, carcass lateral deformation is essential. This lateral flexibility can be generally described with a physical model "Beam on elastic foundation".

The term "carcass bending" means in this paper the phenomenon of deflection of the carcass body in its lateral direction under the forces and torques, applied in planes, tangent to cylindrical body of the carcass: lateral force, aligning torque, sidewall elasticity. In another words, it is the bending of the carcass around any radial axis of the wheel.

The term "torsional stiffness of a tire" means in this paper the stiffness of standing still tire in yaw direction: A relation between aligning torque (steering resistance torque) and angle of tire turning around its vertical axis (direction of slip angle) [1].

The measurement of torsional stiffness of standing tire depicts low lateral deflection of tire carcass – below 2 mm for the given conditions. Consequently significant part of resisting torque is caused by carcass bending stiffness: lateral flexibility does not influence it much. Hence, by measuring the torsional stiffness with different yaw turning speed values it is possible to estimate damping effects of carcass bending (Fig. 9).

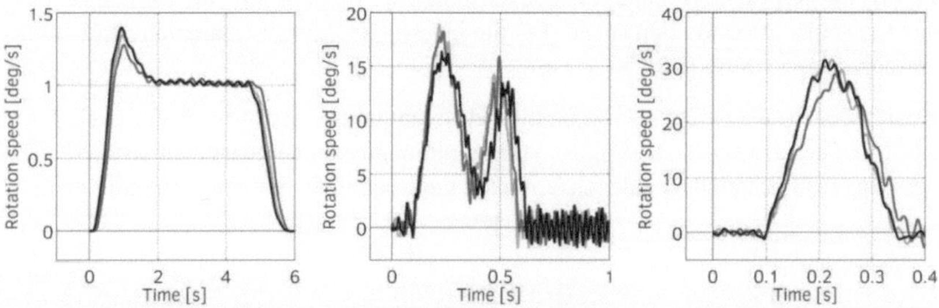

Set speed: 1 deg/s *Set speed: 10 deg/s* *Set speed: 25 deg/s*

Figure 9. Actual plate rotation speed for three values of set speed (three repetitions per diagram).

A comparison of carcass deflection, excited with different speed, by given aligning torque threshold, is depicted on the Fig. 10. Accuracy of the optical measuring concept with eight in-rim cameras does not allow to make a numerical statement regarding the difference in the lateral deflection curves because of low deflection values (ca. 2 mm). Still it allows to make a qualitative conclusion that there is no significant difference between three deflected forms of the carcass which correspond to the same aligning torque. Hence, for the given conditions the bending properties of tire carcass do not exhibit perceptible damping.

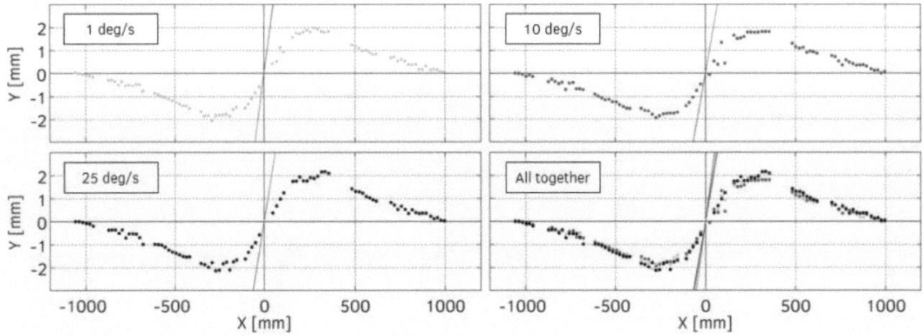

Figure 10. Bending lines of the tire carcass corresponding to aligning torque of 0.4 kNm depending upon the yaw rotation speed.

5.3. Carcass behaviour during cornering

Cornering combines lateral and torsional excitation of tire. It is also influenced by damping properties because of tire rolling. Application of eight cameras makes it possible to capture carcass bending line not only during cornering with fixed slip angle, but also during transient processes. Fig. 11 shows the bending line based on eight groups of markers. With help of these data, deflection of tire carcass in each moment of transient process was integrated and multiplied by constant stiffness coefficient (Fig. 12).

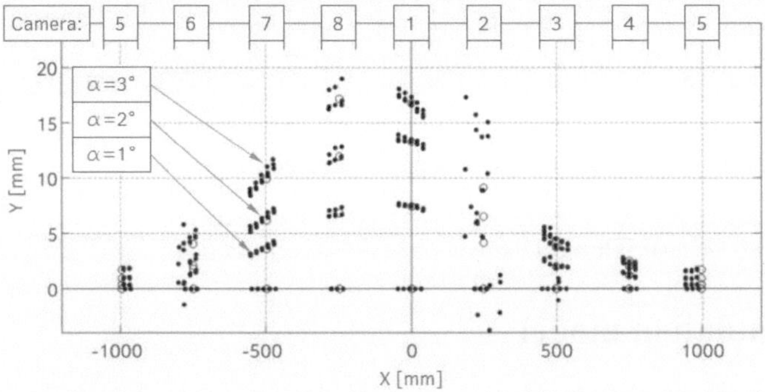

Figure 11. Carcass bending line captured by eight cameras during cornering.

Correlation between measured lateral force and the calculated one (based on integrated carcass deflection and constant stiffness coefficient) is high enough to conclude that coefficient of carcass lateral stiffness can be considered as displacement-independent, also in transient condition.

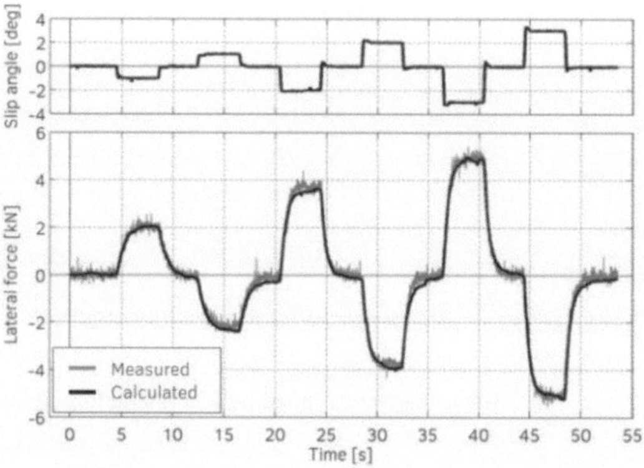

Figure 12. Comparison of measured and calculated based on bending deflection lateral force.

To summarize, optical measurements have provided following important insights for given tire and mentioned conditions:

1. Carcass features damping properties in the range 1-200 mm/s, their force component may achieve up to 20 %, but in the major part of deflection range it is much lower (Fig. 7);
2. Carcass lateral stiffness can be considered as independent upon wheel load (Fig. 8) and displacement (Fig. 12), also in a non-steady state;
3. Bending properties of tire carcass do not exhibit perceptible damping (Fig. 10);
4. Bending lines of carcass in cases of cornering, lateral and torsional excitation are captured, which is useful for model parameterisation.

These insights provide general understanding of the processes, which is necessary to develop an adequate physical model.

6. Simulation model

Based on gained knowledge, there was developed following physical simulation model (Fig. 13): The carcass and breaker are described as a flexible belt, connected to the rim through an endless number of elastic springs. The tread is considered as an array of brush elements, which are constructed on the flexible belt. The model was parameterized with help of measurements of lateral, torsional and cornering stiffness. Next the obtained parameters were corrected considering captured carcass bending lines.

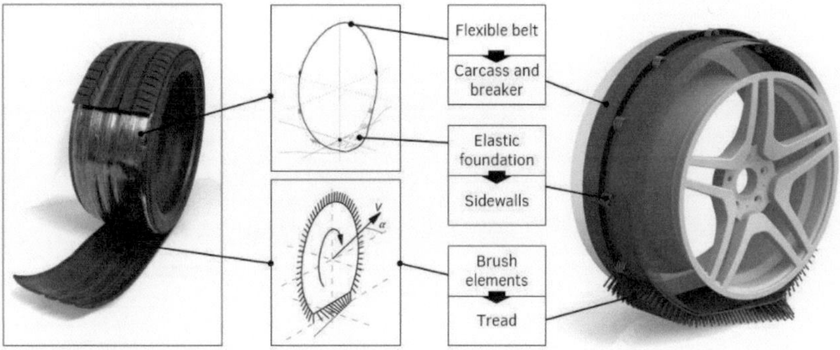

Figure 13. Structural scheme of the physical tire model.

As a validation there were conducted measurements of slip angle step manoeuvre: tire rolls with 3 km/h, slip angle changes stepwise from 0° to 3° and then after 4 s back to 0°. The figure 14 depicts the responses of lateral force and aligning torque to this excitation – measured as well as simulated results. Simulation of tire response to transient wheel load excitation as well as to overlapping wheel load and slip angle excitations also provides qualitatively correct results, although the numerical precision is not high enough (error ca. 20 %).

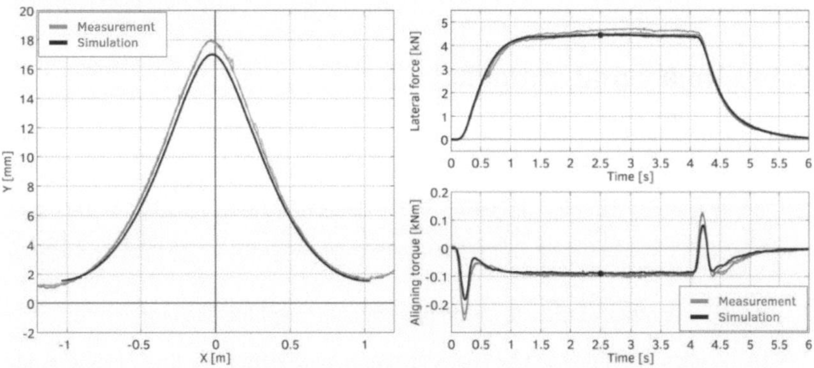

Figure 14. Comparison of measured and simulated responses of lateral force and aligning torque to slip angle step, as well as carcass bending lines.

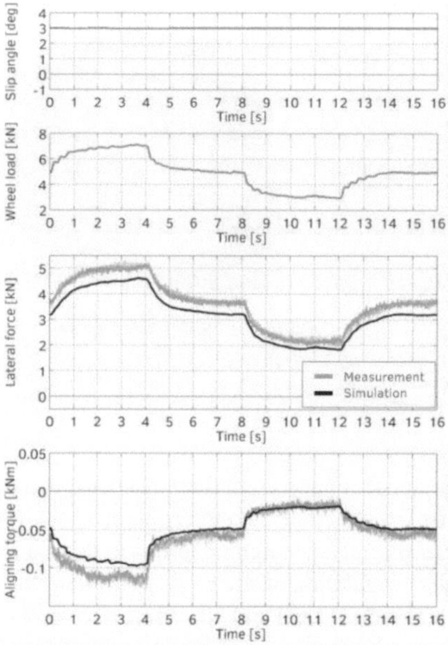

Figure 15. Lateral force and aligning torque responses to wheel load excitation of a cornering tire.

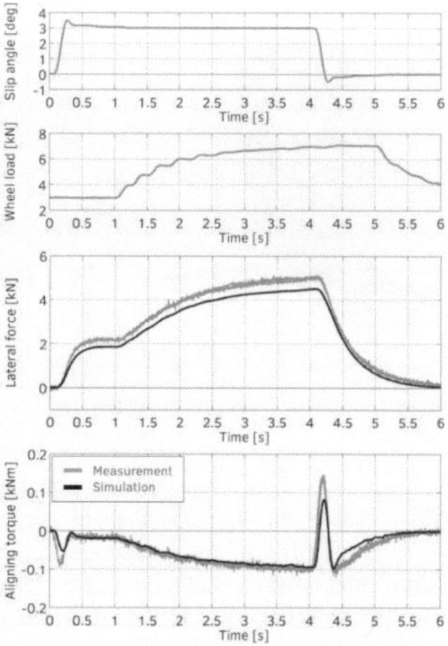

Figure 16. Lateral force and aligning torque responses to overlapping wheel load excitation and slip angle step.

7. Summary and outlook

With help of optical measurements there were gained following insights for the given tire:

1. Carcass features damping properties in the deflection speed range of 1-200 mm/s, their component of lateral force may achieve up to 20 %, but in the major part of deflection range it is much lower;

2. Bending properties of tire carcass do not exhibit perceptible damping;

3. Carcass lateral stiffness can be considered as independent upon the wheel load and displacement, also in a non-steady state.

By means of modelling further knowledge was generated:

1. A physical model, which is based on flexible carcass with wheel-load-independent and displacement-independent lateral stiffness and bending stiffness, but without damping properties, provides qualitatively correct simulation of lateral force and

aligning torque in wide range of modes including transient excitation via slip angle and wheel load. Quantitative accuracy is not high enough though (error ca. 20 %).

2. The model provides a connection between force responses and strain state of the tire body: there were validated not only output curves (lateral force and aligning torque), but also deflections of the physical system of a rolling tire (carcass bending line). Hence, this tool improves an understanding of physical processes underlying tire transient handling behaviour.

Described analysis shows suitability and limits of selected structure of a simple physical model for tire handling simulation. In sense of experimental work, detailed investigation of carcass damping properties can help to improve simulation quality. Considering model itself, reasonable subject of further research are introduction of lateral damping properties to tire carcass, carcass bending description with Timoshenko theory, consideration of variable normal pressure distribution in a contact patch due to carcass deflection and taking into account frequency-dependent friction.

The insights regarding the investigated effects and developed model contribute to understanding of the physical background of tire force and torque generation, which is strongly necessary at the present time for further development of advanced driver assistance systems on the way to autonomous driving and higher road traffic safety.

References

1. Hans B. Pacejka. Tyre and Vehicle Dynamics. Oxford: Elsevier BH, 2002.
2. AUDI AG, Presskit 08/15.
3. Gipser, M., Hofer, R. & Lugner, P. Dynamical Tyre Forces Response to Road Unevennesses. Vehicle System Dynamics, 1997, Band Supplement 27, pp. 94 – 108.
4. Pierre Février, Gérard Fandard. A new thermal and mechanical tire model for handling simulation. VDI-Berichte, 2007, Nr. 2014.
5. Pierre Février et al. Method for simulating the thermomechanical behavior of a tire rolling on the ground. US patent 2010/0010795 A1 issued 2010-01-14.
6. Michael Kaliske. Numerical Modeling in Tire Mechanics. Bamberg, 9. LS-DYNA Forum, 2010.
7. Sarkisov P., Kubenz J., Prokop G., Popov S. Understanding of transient tire behavior by measuring carcass deformation and contact patch shape. In: Bargende M., Reuss HC., Wiedemann J. (eds) 17. Internationales Stuttgarter Symposium. Proceedings. Springer Vieweg, Wiesbaden, 2017.

8. Sarkisov P., Kubenz J., Prokop G., Popov S. Optical measurement of tire deformation focused on transient handling properties. Tire Technology Expo 2017, 14-16.02.2017, Hanover.

9. Feodesjev V. Strength of materials. Bauman Moscow State Technical University, 1999.

10. Tuononen A.J. Optical position detection to measure tyre carcass deflections, Vehicle System Dynamics: 46:6.

11. Erdogan G. Novel wireless tire deformation sensors for estimation of tire slip angle and tire-road friction coefficient. University of Minnesota, Twin Cities Department of Mechanical Engineering.

Acknowledgements

Authors would like to thank Mr. Steffen Drossel, Mr. Stefan Eckert, Mr. Dirk Schlimper and Mr. Erik Deunert for their contribution to project development. Completion grant of the Graduate Academy at the TU Dresden is thankfully acknowledged.

Influence of testing surface on tire lateral force characteristics

Einfluss der Prüfoberfläche auf die Reifenseitenkraft-Eigenschaften

Christian, LUDWIG, Hyundai Motor Europe Technical Center
Chang Su, KIM, Hyundai Motor Company

P.E. Pfeffer (Ed.), *8th International Munich Chassis Symposium 2017*,
Proceedings, DOI 10.1007/978-3-658-18459-9_54

Abstract

An accurate representation of tire lateral forces is vital for vehicle dynamics simulation. At early stage of a vehicle project, the data used to build up a vehicle's simulation model are mostly sourced from component test rigs. Widely used tire test rigs, such as a Flat Track, inner drum or outer drum, are usually equipped with sand paper surfaces to simulate as close as possible the tire/road interface forces.

Extending the approach of [1] this study presents the analysis and discusses the results of real road tests in comparison with indoor flat track tests. For deeper insight, a detailed friction analysis was used for both surfaces based on the mobile friction test rig "LiRep" from FKA Aachen.

Introduction

Motivation

The ultimate target of all simulation based vehicle dynamic analysis is to get as close as possible to the real road behaviour of a car. With regard to the tire this means to measure the tire in conditions, that match as close as possible the ones encountered on real road.

Thanks to the introduction of Flat Track® machines, at the end of the 70's by MTS, it is today possible to reproduce the tire's vertical load distribution similar to the one encountered on real road on an indoor test rig.

The tire is rolling on a pretensioned steel belt on the Flat Track® which is supported by a water bearing. The water bearing is located just underneath the contact patch of the tire, in order to minimize the steel belt's deflection. (refer to picture on the left). A sandpaper surface is usually applied to the steel belt in order to match the friction behavior of tire on real road surfaces.

MTS Flat Track CT plus

Methodology

The studied tire was measured on sandpaper on a Flat Track and on the proving ground tracks of ATP Papenburg using FKA's tire test trailer (FaRep). These measurements were then used to identify Magic Formula parameters for both testing surfaces. The Magic Formula representation of each measurement enables easy comparison of the force and moment curves of the respective tire/surface pairings.

In order to analyze in more detail the frictional behavior of the tire on each surface, water cut samples of the tread rubber have been tested on both surfaces using the mobile linear friction test rig of FKA (LiRep, ref. Chapter 3).

1 Trailer Tests

FKA tire test trailer

The FKA tire test trailer (FaRep) was used to measure the tire's lateral force characteristics on real track (ATP Papenburg, high speed oval).

FKA tire test trailer FaRep

The test rig is equipped with a dynamometric wheel spindle, measuring force and moments. The tire slip angle is measured via an optical DATRON sensor mounted in front of the tire. Camber angle respective to the track's surface is calculated from sensor information giving the test rig incidence, the overall roll angle of the trailer and two height measurements on both sides of the tire.

Being built in a semi-trailer the FaRep shows little reactions on disturbing force inputs (wheel load) due to its dimensions, weight and inertias. FKA further implemented a predictive dynamic wheel load controller in order to limit wheel load variations during the test. Crosswinds can affect the measurements, as the trailer disposes of a huge lateral surface, but for the tests crosswinds were absent.

FaRep Specifications

Wheel diameter	560 – 1240 mm
Max. Wheel load	60 kN
Max. Test speed	100 kph
Camber Angle	±10°
Tire slip angle	±45°, 2°/s
Max. Braking Torque	25kN
▪ camber axis on road surface ▪ dynamic wheel load control for higher measurement accuracy	

Test track: ATP Papenburg high speed oval (OKR)

The two straights of the oval track with a length of 4km each, give enough space even for longer measurement protocols and limit the trailer's own movements. This is vital for obtaining usable tire data suitable for model parameter identification, as already small movements around the yaw and roll axis of the trailer will impact the targeted values in camber and slip angle of the tire test rig.

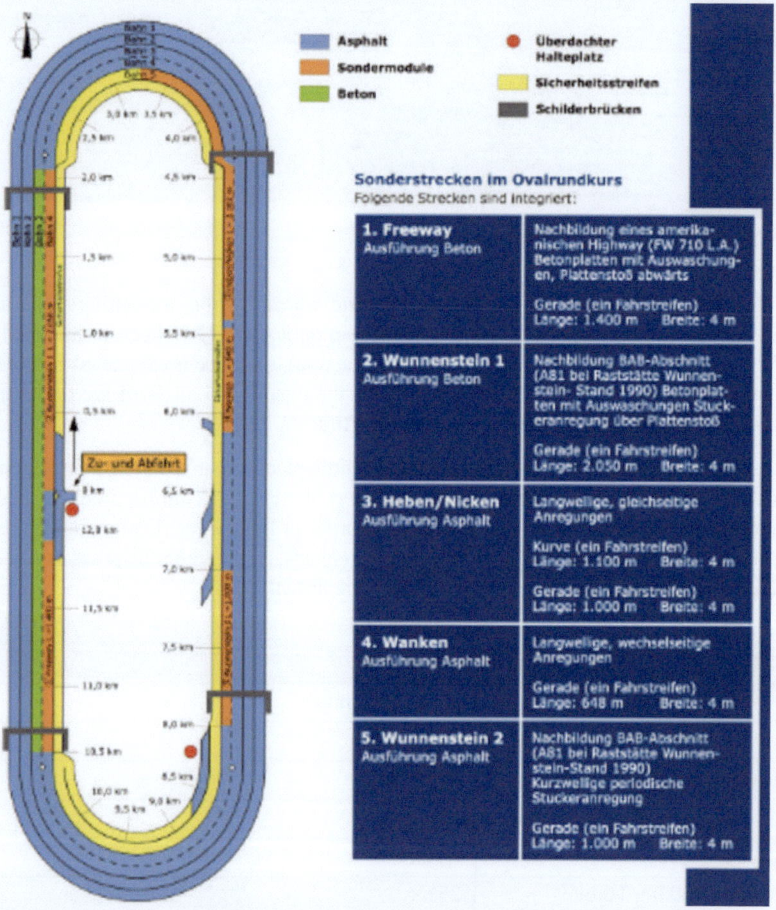

High speed oval track ATP Papenburg (OKR)

Test protocol

The test protocol consists of a simple slip angle sweep +/- 12deg. This sweep was repeated for five different wheel loads.

2 Flat Track Tests

MTS Flat Track

The indoor tire tests were executed on our in-house MTS Flat Track CT, located in our R&D center in South Korea. The rig measures the tire force and moments via a dynamometric spindle, close to the wheel center. Traveling speed can be controlled via the belt speed controller, a huge hydraulic cylinder applies the wheel load, further actuators allow the application of camber and slip angle relative to the belt's surface.

The belt's surface was applied with sandpaper.

MTS Flat Track CT

Standard Flat-Trac CT Specifications:	
Speed	± 250 kph
Vertical Force	25 kN
Lateral Force	± 15 kN
Spindle Torque	± 2800 Nm
Slip Angle	± 30 deg
Inclination Angle	-12 to +45 deg
Tire OD	≤ 910 mm
Tire Width	≤ 400 mm

Flat Track CT Specifications

Test protocol

The test protocol consists of a simple slip angle sweep +/- 15deg. This sweep was repeated for three different wheel loads.

3 Friction Tests

The objective of the friction measurements is to gain a better understanding of the available friction potential on a given surface, referred as grip of the tire.

Grip of the tire

As given in [2] the rubber-road friction on dry surfaces is a combination of indentation and molecular adhesion. Both effects are intimately linked to the visco-elastic characteristics of the tread rubber material. The modulus and the hysteresis of a visco-elastic material vary in function of frequency and temperature.

The solicitation frequency of the tread rubber stress is dependent of the vertical load, which presses the tread rubber blocks of the contact patch into the road profile, and the local sliding speed of each tread rubber block.

Therefore we can define the grip as a function of normal pressure, sliding speed and temperature. The following measurements acquired the rubber sample friction in function of these inputs, whereas the temperature was acquired but could obviously not be controlled (on the proving ground surface). As for the sandpaper surface the temperature were checked in order to match the acquisitions of the proving ground.

Friction test rig LiRep

The mobile linear friction test rig LiRep has been designed and created at FKA [3]. A 60x60mm tread rubber sample is mounted on a 3D force sensor equipped sledge. This sledge can be vertically loaded with different weights from 3-60kg, resulting in different normal pressures ~0.3 – 3.5Bar. While the sledge is dragged with constant velocity above the testing surface, forces and sledge position are recorded.

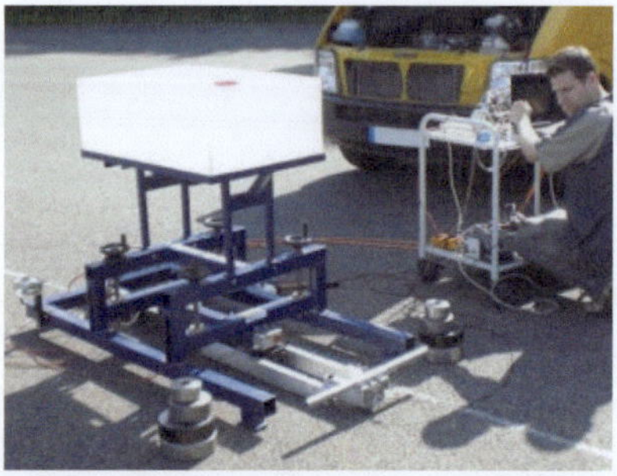

Linear friction tester LiRep

Tread rubber sample

Design of experiment

In order to meet as close as possible the tire's functioning conditions on real track, the friction measurements were done on the same lane as the tire tests [ref. pictures below]

Measurement location of the friction tester on the ATP Papenburg oval track

The acquisition had to take place one day later than the tire lateral force measurements, as for safety reasons it was not possible to access the track at the same moment. As the meteorological conditions stayed stable over the two days, in terms of temperature, humidity, wind and insolation, we expect comparable grip conditions.

The tread rubber sample was tested for 4 different vertical loads and respective 4 sliding speeds, whereas each test case was repeated at least three times for proof of repeatability [ref. to table below].

Test conditions

Normal pressure	[0.9 1.3 1.8 2.2] Bar
Sliding Speed	[0.005 0.01 0.1 1] m/s

4 Analysis

The Analysis of the results was broken down in two parts. After analyzing the results of the Tire measurements on trailer and flat track, we analyzed the respective friction tester data.

Trailer vs Flat Track

After fitting the MF parameters from the trailer and the Flat Track data, we can compare the results of the two tire models under the same input conditions (wheel load, slip angle and camber).

We analyzed the following characteristics of the lateral force curve:

- Lateral Force Maximum, given by the peak factor Dy of the MF
- Decay gradient of lateral force after its maximum value

Lateral Force Characteristics

Max Fy (Dy – peak factor)	Fy decay gradient
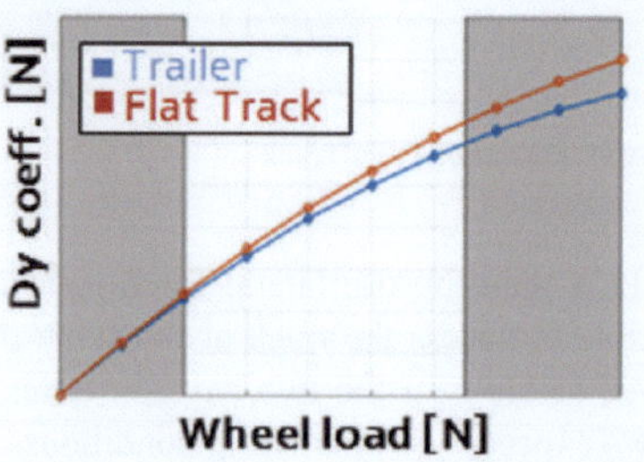	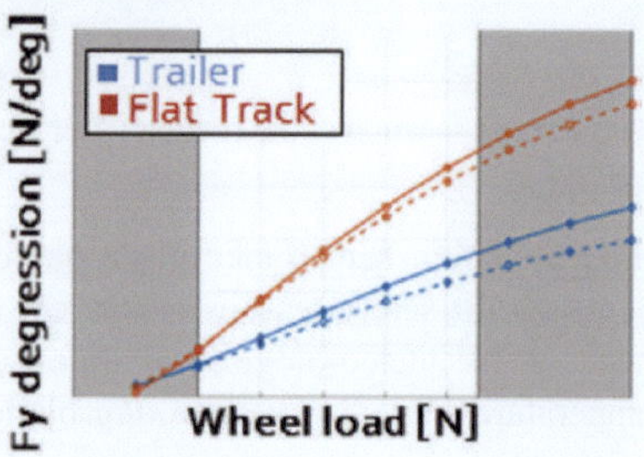

White area – Measured points, Grey area – extrapolated points by MF

We observe a higher maximum lateral force level (grip) for the Flat Track, paired with a stronger decline of the lateral force after reaching the maximum [Fy decay gradient (dFy/dSA)].

Friction measurements

As the tread rubber sample is dragged across the testing surface, the rubber is first sheared under the friction forces until they reach their maximum (μadhesion). Then the rubber samples starts to 'slide'*.

*For many test conditions the sample shows stick-slip behavior, therefore one could not speak of pure sliding.

Calculating the friction coefficient with (1):

$$\mu = \frac{F_X}{F_Z} \tag{1}$$

We extract from each raw measurement the following parameters:

- Maximum μ for that measurement condition
- Average sliding μ
- "Apparent shear stiffness"

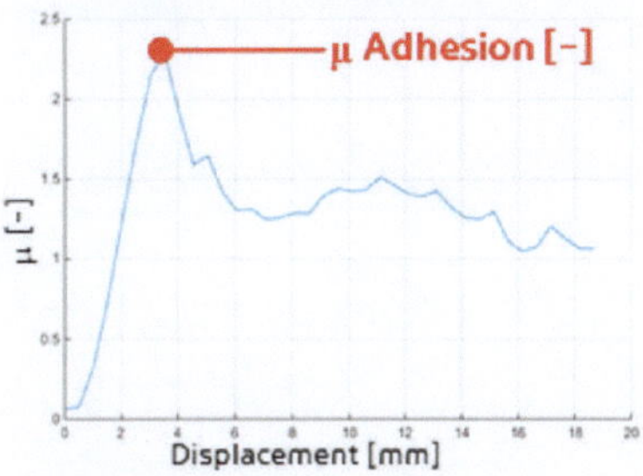

Maximum μ

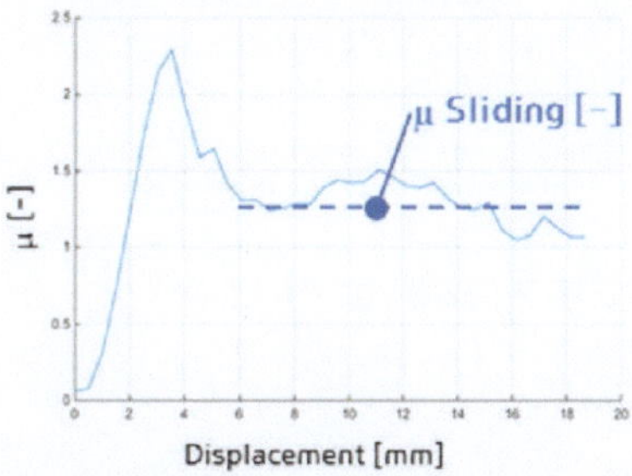

Average sliding μ

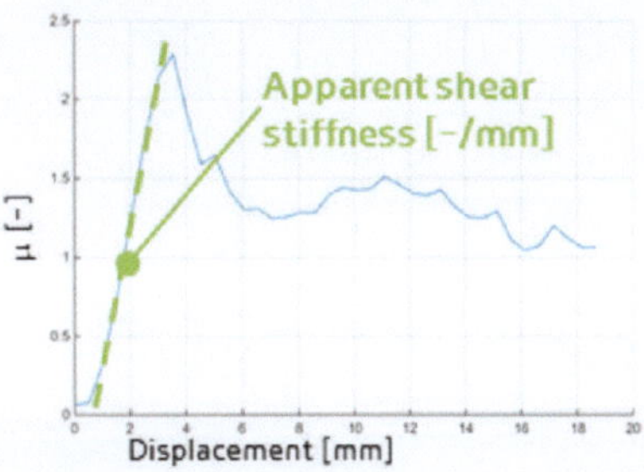

Apparent shear stiffness

5 Results

Based on the parameterized analysis, which we described in chapter 4 we extracted the influence of each input parameter (normal load, sliding speed) on the respective key characteristic.

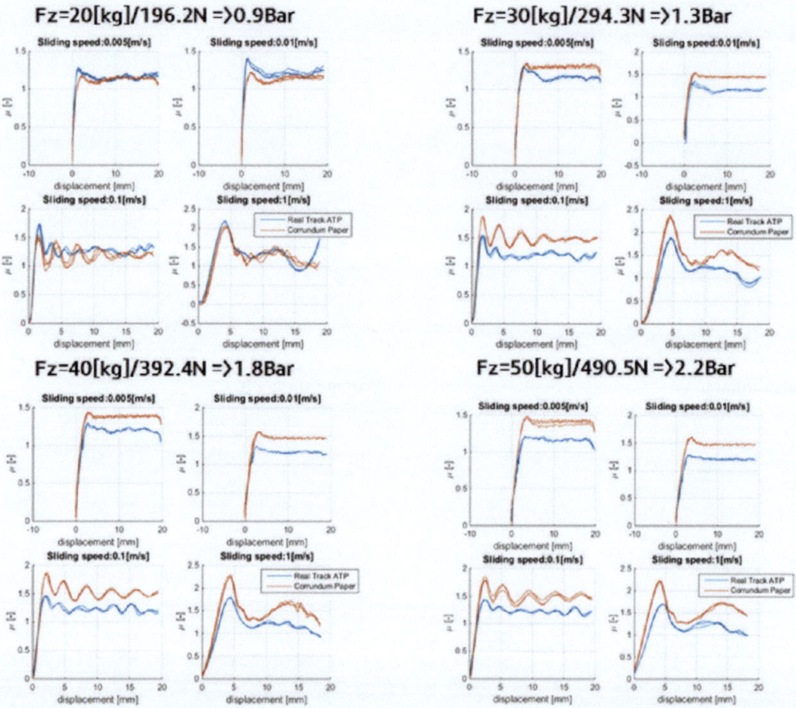

Overview of the pre-processed raw data

5.1 Maximum friction / μ Adhesion

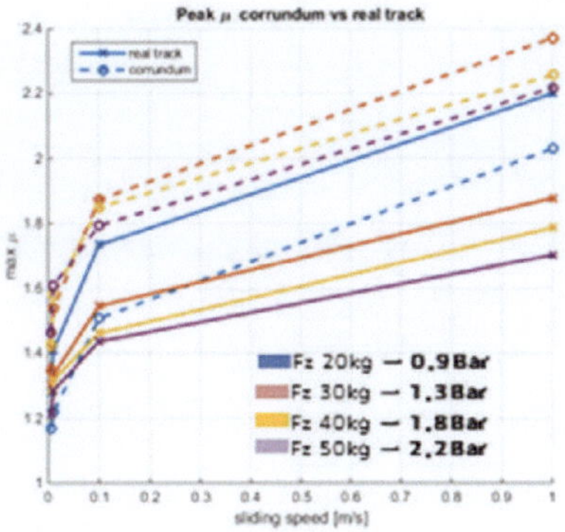

Comparison of Maximum friction on sandpaper vs real track

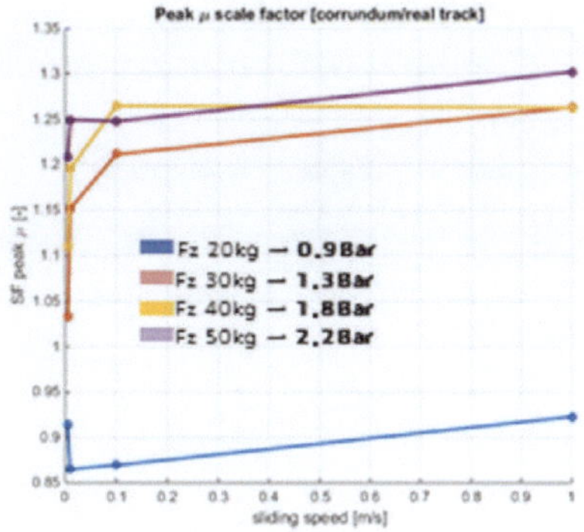

Ratio of $\mu_{Sandpaper}/\mu_{RealTrack}$

Comparing the results of both surfaces, wc observe that the sandpaper measurements over evaluate the maximum friction by ~20%.

This general tendency is broken for the lowest normal load of 20kg/0.9Bar for which the 'real track' measurement shows in average 10% higher grip over the entire speed range. The exact reason for this behavior could not be identified up to now. But it cannot be considered as an outlier as this phenomenon is observed across the entire speed range and all repetitions. Therefore we assume that this relative low pressure favors the friction pairing of this compound with the track surface.

Normal load influence

Further we detect a clear dependency of the friction on normal load, with the maximum friction peaking at 1.3Bar/30kg on both surfaces.

Sliding speed influence

The speed influence shows a similar tendency on both surfaces, maximum grip increases with the sliding speed. But we observe an increasing sur evaluation of the grip level by the sand paper with increasing sliding speed.

5.2 Sliding friction / μ Sliding

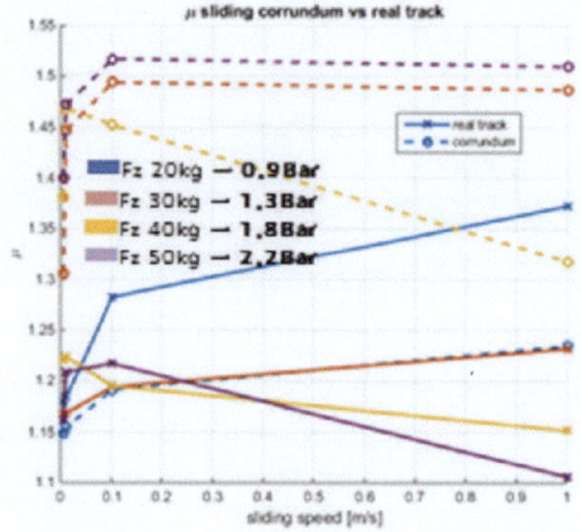

Comparison of Sliding friction on sandpaper vs real track

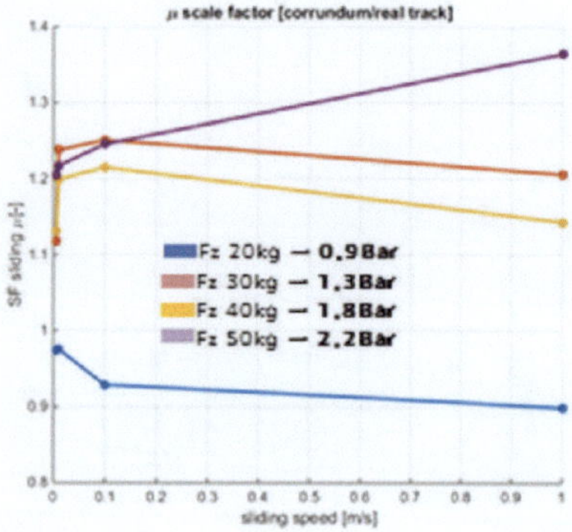

Ratio of $\mu_{\text{Sandpaper}} / \mu_{\text{Real track}}$

Comparing the results of both surfaces, we observe that the sandpaper measurements over evaluate the sliding friction by ~25%.

As seen before this general tendency is broken for the lowest normal load of 20kg/0.9Bar.

Normal load influence

The sliding friction on sandpaper increases strongly with normal load, except for the normal load case of 30kg/1.8Bar which shows decreases in sliding friction for speeds greater than 0.1m/s.

The real track in contrast shows much less sensitivity to the normal pressure, but for the highest speed (1m/s) we observe a linear decline of sliding friction over normal load.

Sliding speed influence

For the sandpaper surface the sliding friction saturates at 0.1m/s for all loads except the load case of 30kg/1.8Bar. Whereas on the real track the speed dependency seems to change with the load level, e.g. the lowest case shows increasing sliding friction values with increasing sliding speeds. Up to the highest load case where the sliding friction declines with increasing speed from 0.01m/s onwards.

6 Conclusion

The analysis of the friction condition in the contact patch of a tire remains indistinct, as complex tire mechanics and rubber-surface interaction govern the overall resulting forces. The studied friction behavior of a rubber sample represents only a very simplified and idealized case of rubber-surface friction. Local pressure and thermal effects could not be reproduced by this study as they are happening in the contact patch of the tire. Therefore we can draw only some general conclusions from this study that shall be confirmed and deepened by future studies.

By the mean of the linear friction measurements, we could measure an over evaluation of the grip level (maximum friction & sliding friction) on the sandpaper surface by roughly 20% compared with the real track results.

This explains the higher lateral force potential of the tire that we measured on the flat track. Whereas the stronger decline in lateral force after reaching its maximum, seems not to be concluding with the friction measurements (especially with the sliding friction results). As the influence of the tire thermal state plays a strong influence on this part of the tire's lateral force, we have planned further studies taking into account the thermal influence.

Looking closer on the results of the sliding friction and especially the measured differences of the sliding velocity dependency on normal load for both surface, one might question the choice of sandpaper surfaces for tire testing.

Reference

[1] Hüsemann, Thomas Christian. Adaption von Labor-Reifenkennfeldern an reale Fahrbahnoberflächen. Forschungs-Ges. Kraftfahrwesen, 2011.

[2] Michelin. The tyre: Grip, 2001.

[3] Hüsemann et al., Mobile Prüfstandstechnik zur Messung von Reifen-Fahrbahn Reibwerten, ATZ 06/2011

WHEEL TECHNOLOGIES AND TRENDS

Possible contributions of the wheel regarding CO_2 emissions or how to increase the range of electrically powered vehicles

Dr. Detlef Kube, H. J. Klamann, M. Timm,
UNIWHEELS Automotive (Germany) GmbH

**This manuscript is not available according to publishing restriction.
Thank you for your understanding.**

P.E. Pfeffer (Ed.), *8th International Munich Chassis Symposium 2017*,
Proceedings, DOI 10.1007/978-3-658-18459-9_55

Influence of rotating wheels on the vehicle dynamics of passenger cars

Maximilian Georg Reisner,

Robert Clauß,

Prof. Dr-Ing. Günther Prokop

P.E. Pfeffer (Ed.), *8th International Munich Chassis Symposium 2017*,
Proceedings, DOI 10.1007/978-3-658-18459-9_56

1 Introduction

The requirements profile of products of the individual mobility is nowadays always subjected to change. The desire for greater wheel dimensions causes increased chassis strains. Also in the course of electric mobility new driveline and chassis concepts will be necessary. In this way a wheel hub drive for example can involve different boundary conditions. Both of these trends are inevitable linked to an increase of the wheel's mass moment of inertia. The question is, how this difference in characteristics affects the vehicle handling?

In a research the effects of rotating wheels during different manoeuvres are studied within the scope of experimental road tests and simulations. The parameters of the axles are influenced through the gyroscopic moments caused by the levelled camber velocity as well as the toe velocity both in stationary and transient conditions.

First of all the measured values on full vehicle level and subsystem level are recorded within a road test. After the evaluation of the experimental data a multibody simulation model is created and validated through test. This model allows to derive relevant axle properties. Subsequently an abstract model approach is developed to analyse the effects more accurate. The integration of these effects in a characteristic based full vehicle model enables to examine the differences in properties on full vehicle level. In this way particular effects can be explained and a prediction of the influence of the vehicle handling is possible.

There have been reported only a few researches regarding the influence of gyroscopic moments on the vehicle dynamics. Bischof et al. [2] investigate the dynamic response of a passenger car to a flywheel-based kinetic energy recovery system. Cossalter [3] analyses the influence of gyroscopic moments on the vehicle handling of motorcycles. Zeng et al. [11] study the effects of wheel set gyroscopic action on linear and nonlinear hunting stability of high-speed trains.

2 Influence of the gyroscopic moments

Gyroscopic moments occur through the rotation of an arbitrary body around a rotation axis in combination with another rotation about an axis, which is not collinear to the first axis. The torque appears on an axis, which is to both perpendicular. The amount of the gyroscopic torque is proportional to the angular velocities of both perpendicular rotational axes as well as the mass moment of inertia of the primary rotation axis ([9], pp. 355 ff.).

For the transfer of this behaviour on a vehicle with spinning wheels, which rotate with an angular frequency ω_y, exist two cases. An positive angular velocity along the longitudinal axis of a rotation wheel Ω_x evokes a negative torque around the vertical axis (see equation 1). A spinning wheel with an angular velocity around the positive vertical axis Ω_z, generates a positive torque around the longitudinal axis of the wheel according to [1] (see equation 2). The orientation of the considered wheel fixed axis system fits the conventions of [4].

$$M_z = -J_{yy}\omega_y\Omega_x \tag{1}$$

$$M_x = J_{yy}\omega_y\Omega_z \tag{2}$$

All bodies, which are rigidly fixed to the wheel, count to the rotating masses. Figure 1 shows the percentage distribution of the mass moment of inertia around the rotational axis of a passenger car wheel, which belongs to a rear axle, with the tire size 255/55 R19. Therefore beside the tire and the rim, also the wheel hub and the break disk belong to the spinning masses of the wheel. Is the wheel driven by a powertrain, the axle journal is calculated to the rotating masses.

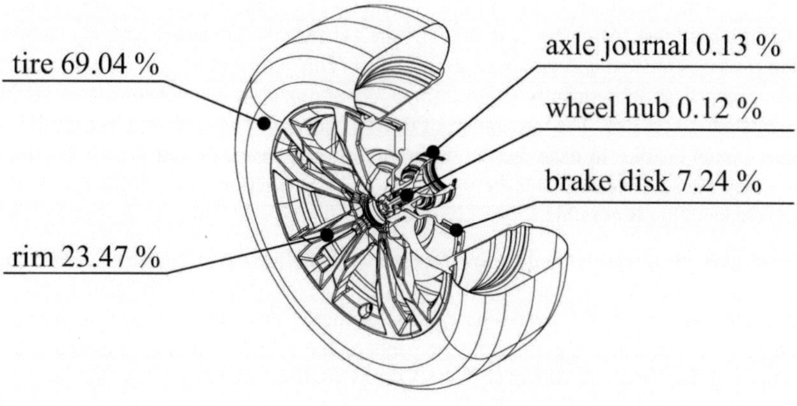

Figure 1: Percentage distribution of a rear wheel's mass moment of inertia

The impact of the gyroscopic moments on a vehicle, generated through rotating wheels, could be divided in two effects: For one thing there is an influence of the torque on the vehicle body, for another thing the gyroscopic moments cause deformations in the wheel suspension, which effects the wheel position quantities.

2.1 Simplified consideration

In the first instance any steering systems are neglected. The body of a passenger car is able to rotate about three axes: Yaw, pitch as well as roll. The wheels have beside their rotational degree, enabled through the wheel bearing, exactly one degree of freedom, assuming that the suspension is kinematically determinated. This degree of freedom consists primary of the wheel travel and very small longitudinal and lateral displacements. The rotational movement of the wheel carrier respectively to the vehicle body are negligible small. With this consideration it is simplified assumed, that the global body angular velocity is similar to the angular velocities of the wheel carriers. The wheel bearing is the connector of the wheel and its carrier. Consequently the body pitch movement is disconnected to the wheel. A yaw rate of the body causes a similar yaw rate of the spinning wheels. Thus a forward driven vehicle is strained with a torque along the vehicle longitudinal axis through the gyroscopic moments. This means, that during a steady-state circular vehicle test with a constant yaw rate $\dot{\psi}$, the body is permanently effected by a roll torque M_x (see left side of figure 2). Contrary to the yaw rate a roll rate $\dot{\Phi}$ occurs only within a transient vehicle manoeuvre. Most of the suspension systems have a reduction of camber loss during a body roll motion to meet a better position of the wheel relatively to the road. This is relatively small compared to the roll angular, so that a part of the body roll rate is enforced on the wheels. The camber rates in combination with the rotation of the wheels around their axis evokes a torque on each wheel around the vertical axis M_z (see right side of figure 2).

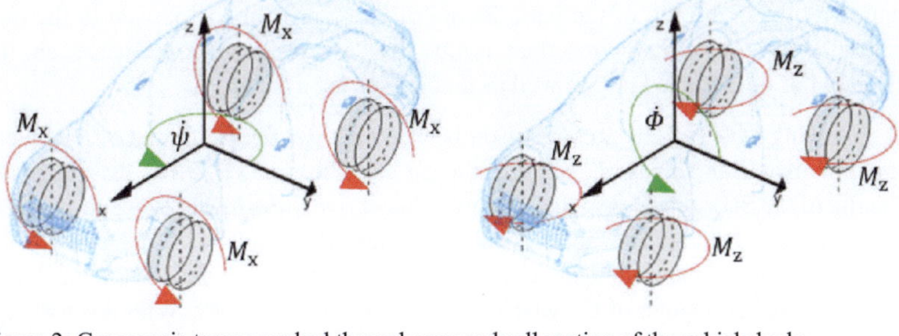

Figure 2: Gyroscopic torque evoked through yaw and roll motion of the vehicle body

2.2 Flexible deformation through gyroscopic moments

The attachment of the wheels on the vehicle body of passenger cars is commonly realized through an single-wheel suspension. This connection causes compliance between the wheels and the vehicle body. In this way torque application on a wheel effects the wheel position quantities, which directly influence the vehicle dynamics. For an example a lateral force, applied in the pneumatic trail, causes a toe torque. Through specific design of the suspension system a steering reaction with the help of the lateral force is enabled ([5], p. 134).

Regarding gyroscopic moments the rotational stiffnesses between the rotating wheel and the vehicle body are of interest. These are the toe torque stiffness as well as the camber toe stiffness, which are crucial for the elastokinematics. Through the undesirable flexible camber loss during an attached lateral force, the suspensions are designed with the aim to reach a as high as possible camber torque stiffness. In contrast, the toe torques are relatively small, because they are primary induced through the shear stress distribution within the tire contact surface and the resulting small lever arms for the lateral and longitudinal tire forces. In fact this entity can be employed to enable a side force steering. Consequently the objective for the characteristics of the toe torque stiffness is not still to reach a as high as possible stiffness, compared to the camber torque stiffness.

2.3 Comparison gyroscopic effect – motorcycle vs. passenger car

The influence of the gyroscopic moments on the vehicle dynamics are especially for one-track vehicles of great significance. Therefore the gyroscopic moments are an issue for the vehicle dynamics of motorcycles, which is shown in [3] and [9]. In many approaches concerning this matter the motorcycles are modelled as a kinematic system without the flexibilities of components, bushings and wheel bearings. This is accounted by the high ratio of the mass inertia of the wheels in comparison to the motorcycle body. Thus the gyroscopic torque has a considerable influence on the movement of the vehicle body as well as on the steering system.

The following table reflects a comparison between the gyroscopic torque of a conventional motorcycle and a passenger car of the car classification SUV. The mass moment of inertia of the SUV wheel is in this example almost five times higher as the motorcycle wheel. Also the dynamic tire radius is valued higher, which involves a reduction of the angular velocities of the passenger car wheels. In this simple example it is assumed, that there is no camber angular of the wheel relative to the vehicle body. With this assumption the vehicle roll angle is equal to the camber angle. However the motorcycle needs for cornering a higher roll angle, which is regarded with the multiplication of the cosine of the ideal roll angle, because the horizontal yaw velocity is considered.

At first a steady-state circular test of each vehicle and the resulting gyroscopic moments around the vehicle longitudinal axis, evoked through the yaw rate, are compared to each other. The comparison shows, that the gyroscopic moment of the vehicle is about factor ten higher. The relativization to the whole primary vehicle mass moment of inertia, referred to the centre of mass, indicates, that the relative gyroscopic moment of the motorcycle is many times higher.

The consideration of a typical body roll motion points out, that a calculated gyroscopic moment of almost 300 Nm, evoked through the roll rate of the four wheels of the passenger car, applies on the vehicle body. The relativization of this yaw torque compared with the mass characteristics around the vehicle vertical axis clarifies, that the amount, compared with the motorcycle, is very small. Usually the camber rates of the wheels exist only during a change of the roll angle, so that the torque applies only for a short period of time. The comparison of similar transient effects of passenger cars to a yaw torque, evoked through a transient gusty cross-wind, is similar to this magnitude in the case of the considered vehicle velocity ([10], p. 53). In fact this effect of the gyroscopic moment is, compared to such aerodynamic effects, linear to the vehicle velocity and not quadratic.

Table 1: Comparison – motorcycle vs. car: gyroscopic torque through yaw and roll

	motorcycle [3]	passenger car
mass moment of inertia (wheel) J_{yy} [kgm²]	0.6	2.8
dynamic tire radius [m]	0.320	0.376
vehicle velocity [m/s]	40	
angular velocity of the wheels [rad/s]	125	106.4
Consideration of a steady-state circular test (left turn)		
cornering radius [m]	200	
yaw rate [rad/s]	0.2	
roll angle [°]	-39.2	3
gyro. torque around longitudinal axis [Nm]	23	238
primary vehicle mass moment of inertia around longitudinal axis [kgm²]	20	950
relative gyroscopic torque [rad/s²]	1.15	0.25
Consideration of a typical body roll motion		
roll rate [rad/s]	0.5	0.25
roll angle [°]	0	
gyro. torque around vehicle vertical axis [Nm]	-75	-298
primary vehicle mass moment of inertia around vehicle vertical axis [kgm²]	50	4250
relative gyroscopic torque [rad/s²]	-1.5	-0.07

3 Experimental road test

In order to analyse the influence of rotating wheels on the vehicle dynamics of passenger cars, an experimental road test is performed. The relevant quantities to capture on suspension level include the loads, which appear in the wheel hub, as well as the positions of the wheels, which are a result of the loads. Furthermore the body motions are of great interest. To capture the loads, a six-degree of freedom load cell is applied. This measured loads are among others necessary for the validation of the multibody simulation model (MBS). With a touchless optical system the wheel positions are captured. A gyro-stabilized inertial measurement system captures the body inertia quantities. Also a steering robot ensures reproducible input variables.

An AUDI Q7 (AU536) with passive suspension serves as test object. The rim consists of a special 19-inch lightweight construction adapter rim, which is suitable for load cells. The mass moment of inertia is almost equal to the original 19-inch forged aluminium rim. Figure 3 shows the test object with mounted measurement systems.

Figure 3: Test object AUDI Q7 (AU536) with mounted measurement equipment

To meet the concerning points for the manoeuvre selection, the causal roots of the gyroscopic moments are focused. For one thing the road tests should be performed by different vehicle velocities, for another thing different yaw- and roll rates should be part of the manoeuvre catalogue. Table 2 shows the selected driving manoeuvres. Four of these are step steer manoeuvres with different vehicle velocities and lateral accelerations referring to [6]. Also double square steps, trapezoidal steps and one fast double lane change during a high vehicle velocity are performed. The aim of the double square step manoeuvres is to reach a as high as possible roll rate.

Table 2: Vehicle driving manoeuvres

	vehicle velocity [m/s]	lateral acceleration [m/s²]
Single steps		
step steer „lws 100 4“	100	4
step steer „lws 100 6“	100	6
step steer „lws 130 4“	130	4
step steer „lws 130 6“	130	6
Multi steps		
square step „sq step”	125	7
asy. square step „asy sq step”	170	5
trapezoid step „trapez“	140	8
asy. trapezoid step „asy trapez“	160	4
Lane change		
lane change “learn”	190	6

The derivatives of the wheel angular positions, the body angular velocities and the rotational speed of the wheels are important for the evaluation of the results. With this measured data and the mass moment of inertia of the wheels, the gyroscopic torque can be calculated. The illustration of figure 4 shows the measured data: rotational speed of the wheels, toe rate, camber rate as well as the calculated gyroscopic torque around the wheel's longitudinal and vertical axis of each wheel about the driving manoeuvres. The abbreviations Lf, Rf, Lr and Rr stand for „left front wheel“, „right front wheel“, „left rear wheel“ as well as „right rear wheel“. The toe and camber rates include the absolute angular velocity relative to the initial axis system. The respective bars show the range between the minimal and maximal measured value. The George cross symbol (+) shows the initial condition during steady-state straight-line driving. The St. Andrew's cross symbol (X) shows the steady-state condition during the manoeuvre after steering input.

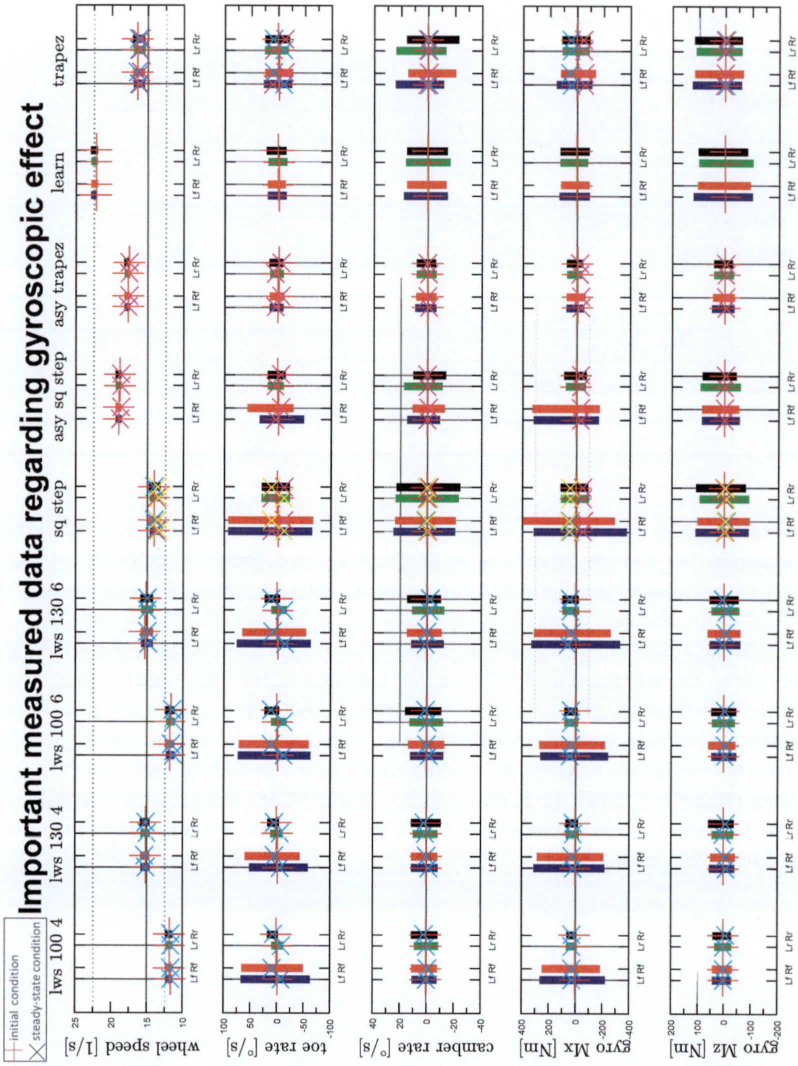

Figure 4: Important measured data regarding gyroscopic effect

The upper plot represents the rotational speed of the wheels of the different manoeuvres. The plot in the second line visualizes the angular velocities of the wheels around their vertical axis. These are calculated by the yaw rate and the derivative of the toe angle. It could be recognized, that the rates of the steered front suspension are higher than the rates of the rear suspension. Especially the manoeuvres with a jump function as input exhibit a very high angular velocity up to 100 °/s. This effects the calculated gyroscopic camber torque. These conditions are regarded as not relevant, because the human driver usually uses smaller steering rates. The trapezoidal manoeuvres as well as the by the driver trained steering input are usual for the human steering behaviour. During these manoeuvres the differences between the gyroscopic camber torque from rear to front axle are significant smaller. The camber rates represent the angular velocities towards the road plane. They include the kinematic and compliance camber rates superimposed by the body roll angular velocity of the vehicle. Following the laws of the gyroscopic moments, this angular velocities evoke a torque around the vertical axis of the wheels. The manoeuvre labelled as "learn", which represents a fast lane change during high vehicle velocity, exhibits, that a toe torque peak value about 100 Nm applies for a short period of time.

3.1 Influence of the toe rate on the wheel position

The Overview of figure 4 shows, that very high toe rates appear during the synthetic fast steered manoeuvres on the front axle. As a result of this, a high gyroscopic torque around the longitudinal axis applies during the change of steering angular. The visualization of the captured elastokinematic camber angle of both front wheels about the time in figure 5 shows, that the influence of the gyroscopic torque is clearly visible in the transient behaviour of these quantities. Furthermore the applied steering angular of the manoeuvre "sq step" is visualized through a black solid line with its y-axis on the right side. The camber torque applies temporary on the vehicle body. An increased roll acceleration is the consequence of this.

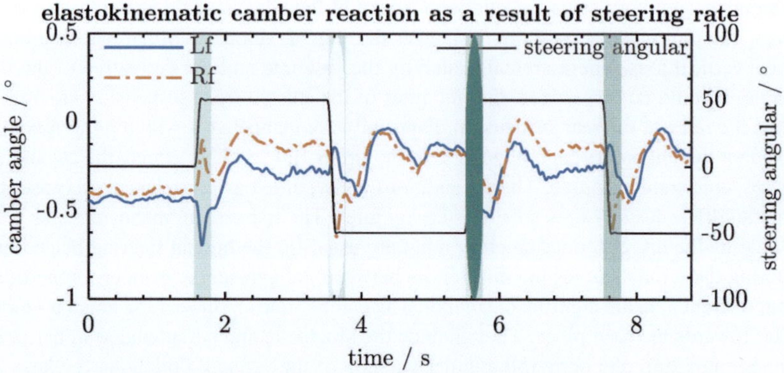

Figure 5: Elastokinematic camber reaction as a result of steering rate

3.2 Influence of the camber rate on the wheel position

Next the chronological sequence of the measured data is analysed. Figure 6 illustrates the camber rate, the toe angle and the torque around the vertical axis of the rear wheels. At the time of the super elevated camber rate, there is a toe angle peak on each wheel. The first square of the double square step is equal to a simple steer step. The rear wheel outside of the corner should steer into direction toe in to achieve a tendency understeering vehicle behaviour. However the gyroscopic moments ensure that the outer wheel steers in direction toe out and the inner wheel steers in direction toe in.

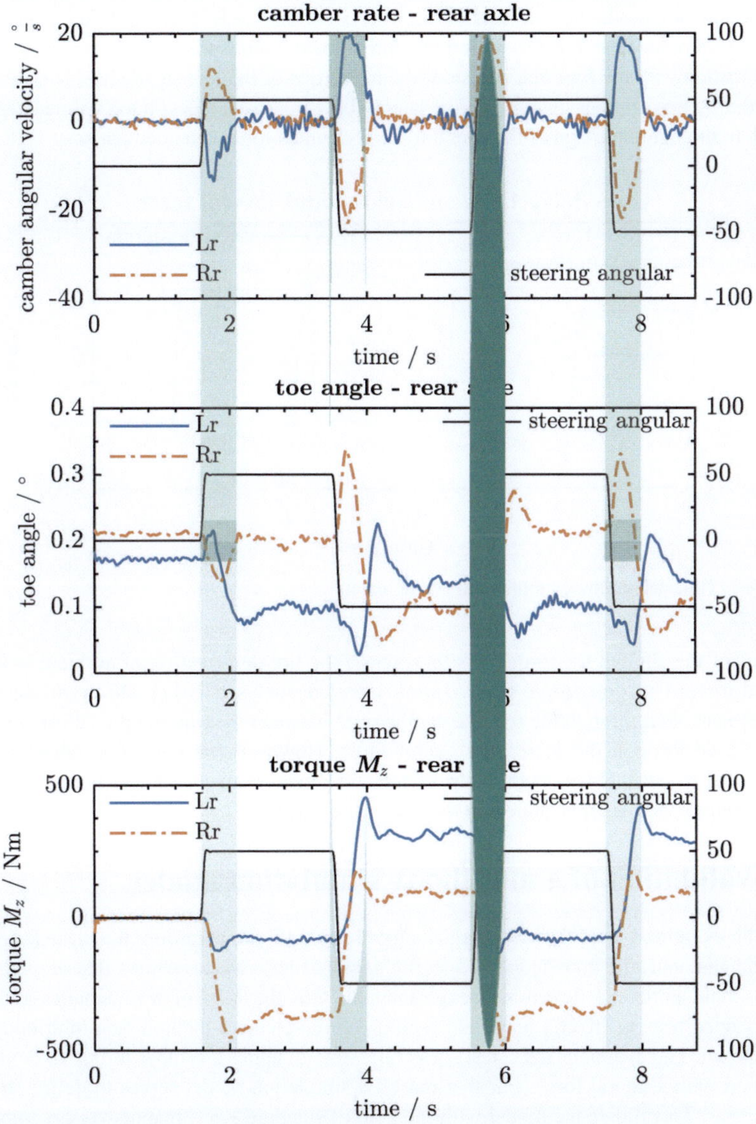

Figure 6: Chronological sequence: camber rate, toe angle and vertical wheel torque

The torque around the vertical axis of the wheels during steady-state condition is based on the pneumatic trail. For this reason the chronological sequence of the lateral force in combination with the captured course of the toe torque of the rear right wheel is shown in figure 7. It is obvious, that before the lateral force increase process a toe torque is applied. In the figure the sign of the lateral force is changed for a better comparison.

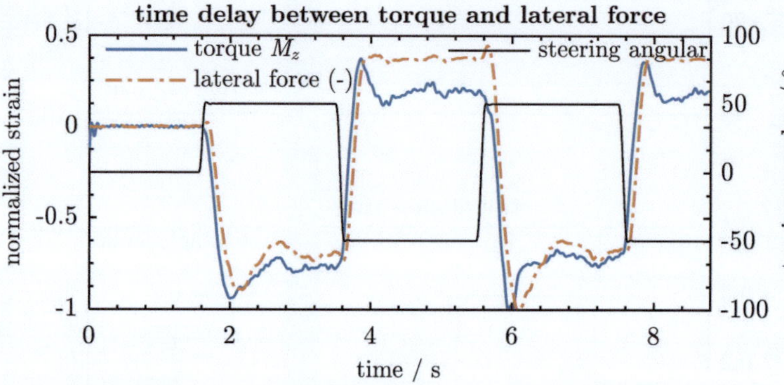

Figure 7: Time delay between torque and lateral force

A part of the applied toe torque has its roots in the tire deformation of an increasing lateral force. This gyroscopic effect is more accurate analysed by [7]. This gyroscopic torque is evoked through the relative camber rate between the rim and the deformable tire caused through the increasing lateral force. However this kind of camber rate could not be captured with the help of the used measurement system, because the measurement points are attached on the rim of the wheel.

4 Validation of a multibody simulation model

For the examination of the effects of the gyroscopic torque regarding the elastokinematic behaviour, the knowledge of specific axle elasticities is necessary. These quantities could either be determined experimental with the help of a kinematics-and-compliance test rig (K&C) or virtual within a finite-element-method or a multibody simulation. The available K&C test rig is only able to apply a camber torque in combination with a lateral force. For that reason a simulation model serves to define the properties. To validate the model with the help of standard K&C manoeuvres, a comparison between the test rig and simulation results is generated. It is noticed within the validation process, that for a good description of the axle properties a very detailed mechanisms-based flexible MBS model is necessary. Figure 8 shows the results of different detailed MBS models (2-5) compared with the K&C measurement (1) for a manoeuvre with a toe torque on the rear right wheel. A kinematic model does not describe any compliance (2). In the case of including bushings and wheel hub stiffness (3) the half of the measured stiffness could be reached. The further substitution of the rigid rods with pure push/pull rods including discrete stiffness (4) improves the model a little bit. However this model does still not fit the measured stiffness. At least the model within the modal reduced finite-element structures of the whole suspension parts (5) enables a realistic description of this axle property. This model is shown in the upper left part of figure 8.

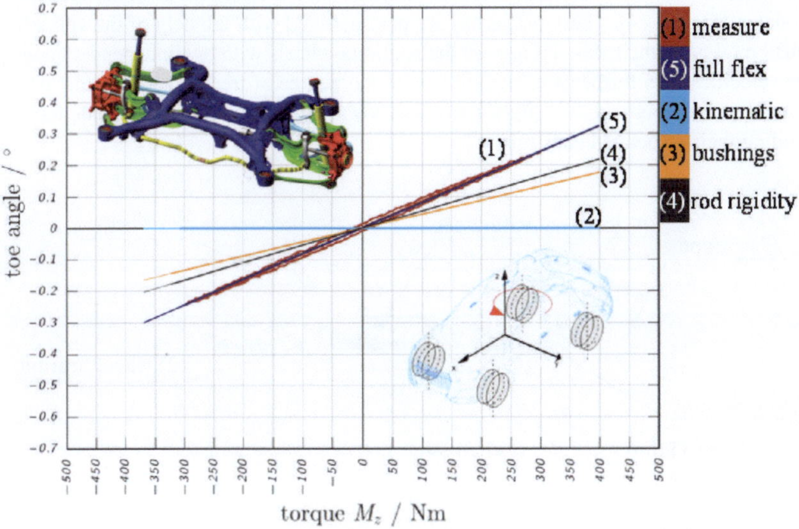

Figure 8: MBS model quality of different detail levels [8]

The determined rotational rear axle properties amount a toe torque stiffness of 1200 Nm/° and 2200 Nm/° for the camber torque stiffness. The camber torque stiffness of the front axle amounts a value of 5500 Nm/°. This one is therefore many times higher as the stiffness of the rear axle. The front axle with a toe torque stiffness of 630 Nm/° has more compliance about the vertical axis compared with the rear axle. These characteristics are regarded as linear for the relevant vehicle dynamics range.

5 Abstract virtual gyroscopic test rig

To examine the influence of the gyroscopic torque on the elastokinematic wheel position, a virtual gyroscopic test rig is created in a MBS environment (see figure 9). The model consists of a chain structure of four rigid bodies, which are coupled with joints and constraints. The wheel (1) is modelled as a rigid body, which is only parametrized by the mass moment of inertia. The wheel is connected with the help of a revolute joint (2), which represents the degree of freedom of the wheel bearing. The spinning rotation is imposed on the wheel by a rheonomous constraint. A ball joint (3) connects the wheel bearing with another joint. Furthermore they are coupled with discrete rotational stiffnesses (4). These properties are parametrized with the help of the validated MBS models of the axles. To impose the spinning wheels with two independent angular velocities, two further revolute joints with the degree of freedom around the longitudinal (5) and vertical axis (6) of the wheel towards the ground (7) are modelled including their associated constraints.

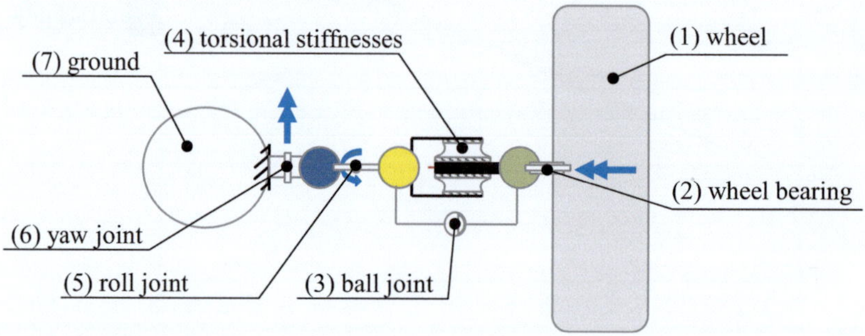

Figure 9: Abstract virtual gyroscopic test rig

For the input signals the during the road test captured angular velocities are attached on the model. Subsequently the rotation angles and the torque, which are evoked by the gyroscopic effect, are evaluated. The influence on steady-state and transient wheel positions can be verified with the help of this model. The comparison of the simulated gyroscopic torque to the calculated gyroscopic torque of the road test implies an adequate agreement. Figure 10 shows the change of the rotational wheel position of the rear axle during the manoeuvre steer step “lws 130 6”, which results only because of the axle compliance. The wheel position entities of the left vehicle side are sign inverted visualized. The camber angle reaches a constant value after the super elevation, because the vehicle rotates around its vertical axis with a constant yaw rate after steering input. The change in toe angle is only a result of the temporary camber rate of the wheels.

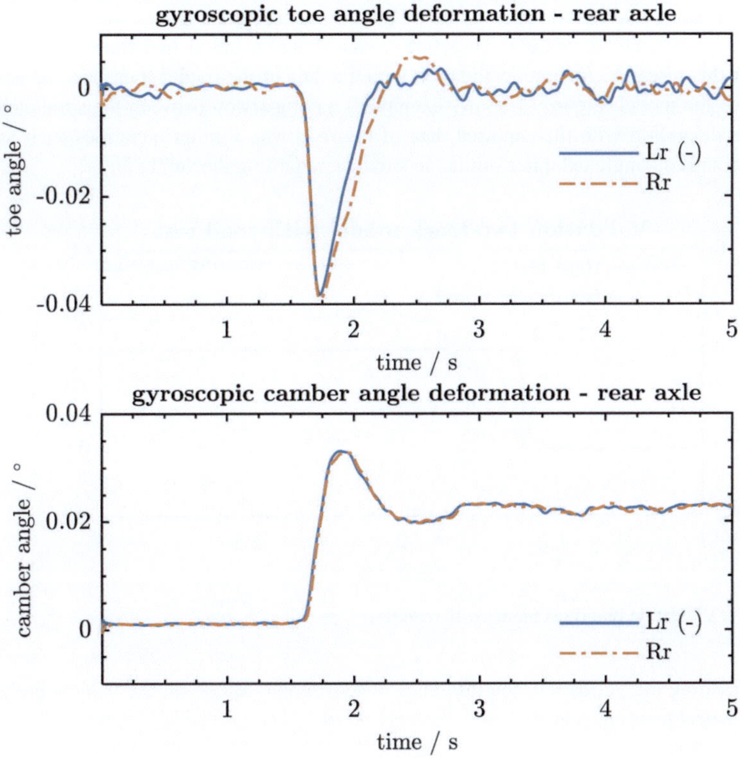

Figure 10: Results of the abstract gyroscopic test rig

The model provides the opportunity to determine the influence of the gyroscopic torque on the elastokinematics. However the influence on the vehicle dynamics could not be directly determined with the help of this approach.

6 Influence on the vehicle dynamics

To examine the influence of rotating wheels on the vehicle dynamics of passenger cars furthermore, a model is needed, which involves the kinematic and elastokinematic suspension characteristics as well as is able to switch separately the gyroscopic effect on vehicle body and wheel position. For one thing a mechanism orientated flexible multibody model could be a conceivable solution. However for this kind of model a very high effort for modelling is needed to serve a behaviour with sufficient precision.

Moreover using a MBS-model, it is not possible to examine the gyroscopic effects of toe and camber rate separately. Therefore a characteristic based two track model is used for this research. At first a validation with the data of the road test ensures the accuracy of the model. Figure 11 shows exemplary a comparison between the simulated lateral acceleration with the captured data of the road test. Further quantities such as yaw rate and roll angle exhibit a similar accordance regarding the validation.

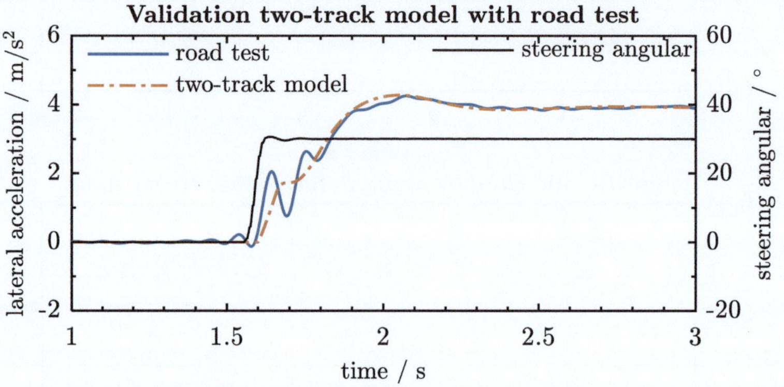

Figure 11: Validation two-track model with road test

After finishing the validation, the influence of gyroscopic torque on the vehicle body and the wheel position is analysed.

First of all the gyroscopic torque is activated and deactivated within the simulation. Hereinafter the variant with all activated impacts is labels as "GyroOn". The simulated condition without gyroscopic moments serves as reference for all further researches. The consideration of the influence of the gyroscopic torque on the vehicle body takes place separately regarding roll and yaw torque as well as both together („RollGyro", „YawGyro", „RollYawGyro"). The influence of the gyroscopic torque on the elastokinematics is also firstly separated in a consideration of toe and camber torque and afterwards both together („ToeGyro" ", „CamberGyro"„ „ToeCamberGyro"). Furthermore the elasokinematic effects between front axle (FA) and rear axle (RA) are varied („ToeFAGyro", „ToeRAGyro", „CamberFAGyro", „CamberRAGyro").

A steering ramp and two sinusoidal sweeps at an vehicle velocity of 100 and 150 km/h serve as simulation manoeuvres. The vehicle dynamics quantities regarding [6] as well as further additional quantities are evaluated for the effect analysis. The discussion of the choice of suitable vehicle dynamic quantities is not part of this paper. For this reason there is no further consideration of this topic. Below the influence of gyroscopic effect on selected objective quantities are exemplary presented. Figure 12 shows the change of yaw eigenfrequency regarding the gyroscopic effect. It becomes apparent, that the whole part of the gyroscopic torque levels the yaw eigenfrequency slightly upwards. The frequency offset is primary effected through the influence of the gyroscopic torque on the vehicle body, whereby the gyroscopic yaw torque is the principal cause. The elastokinematic changes of wheel position on the yaw eigenfrequency, caused through the gyroscopic torque, are very small. The toe torque takes hereby a bigger influence. If the toe torque is varied between front and rear axle, there will be a bigger influence on the yaw eigenfrequency. However both together activated, almost cancel each other.

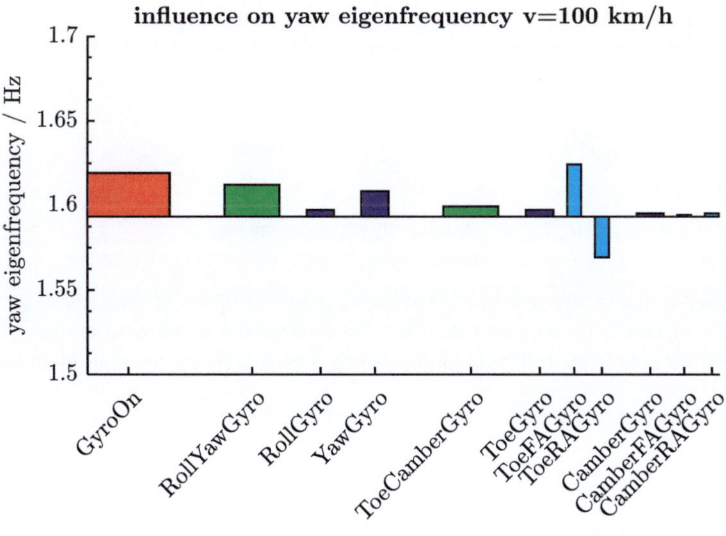

Figure 12: Influence on yaw eigenfrequency

The dynamic gain of the yaw rate relatively to the steering angular during a sine sweep manoeuvre at a vehicle velocity of 150 km/h is shown in figure 13. The gyroscopic moments effect the dynamic gain with a damping impact. The gyroscopic yaw torque is the primary reason for the damping. The elastokinematic influence of the gyroscopic torque causes a gain of the yaw rate due to change of toe angles. An increase of the toe torque stiffness of the rear axle, or rather a decrease of the front axle, would cause a higher yaw damping, with the assumption, that all other entities are not affected.

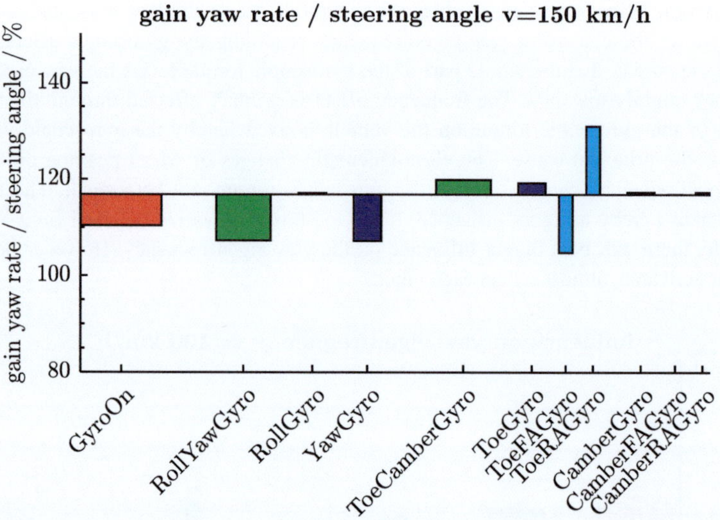

Figure 13: Dynamic gain of the yaw rate relatively to the steering angle

The impact on the steady-state roll behaviour is shown in figure 14. The roll angle is stationary increased by approximately 0.1 °/g through the gyroscopic torque. The origin for this is only the roll body torque, which is evoked by the toe rates of all four wheels.

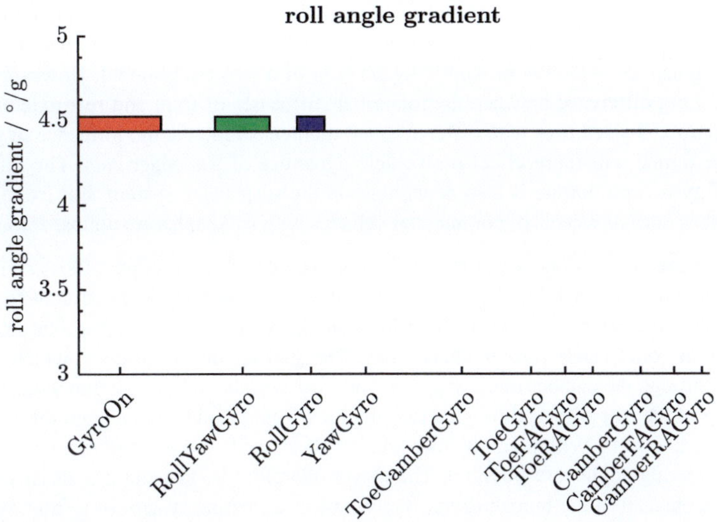

Figure 14: Roll angle gradient

Summarizing the consideration on full vehicle level, the absolute influence of the gyroscopic torque on the vehicle dynamics of this examined vehicle is small. During the examination of the influence of the gyroscopic torque on the vehicle body, it was noticed, that the yaw impact is bigger than the roll impact. The examination regarding the influence on the elastokinematic behaviour exhibits, that the toe angle change, evoked by the camber rates, is higher than the camber angle change, caused through the toe rates. The balance of the rotational axle stiffness as well as the mass moment of inertia of the rotating wheels of front and rear axle are extracted as relevant manipulated variables to influence the vehicle dynamics regarding gyroscopic effects.

7 Outlook and conclusion

The presented research exposes the quantities, which influence the gyroscopic torque, whereby for one thing torque is transferred to the vehicle body and for another thing wheel positions are affected due to axle compliance and the occurring torque.

The call of the design department for bigger wheel diameters first of all causes a proportional decreasing wheel speed. However an increase of the distance of mass relatively to the centre of gravity raises squarely the mass moment of inertia. As a consequence, the bigger wheel dimension leads to a higher mass moment of inertia and thus to higher gyroscopic moments. A balanced stiffness and mass moment of inertia distribution between the axles leads to a small influence of the gyroscopic torque on the vehicle dynamics, which is proved with the help of a two track model. However, if there is a big difference between the rotational stiffnesses of front and rear axle or if there are mixed-sized tires with different mass moment of inertia, the influence of gyroscopic torque will more affect the vehicle dynamics of passenger cars. The influence of gyroscopic torque is also depending on the suspension system. For example the camber rates of wheels of commercial vehicles with solid axles are almost zero.

There are also interactions of active suspension systems and gyroscopic effects. Within an integrated roll stabilization the roll rate of the vehicle body and consequently a huge part of the camber rate could be influenced. Also the usage of an rear axle steering system could cause further interactions. The gyroscopic toe torques, which are evoked through the camber rates, apply on both rear wheels in the same direction. The coupling of the rear wheels through a steering rack would lead to a equilateral interaction of both wheels. Within compliance and backlash inside the steering system undesirable steering angles are possible. The reason therefore is, that there is usually no feedback channel to the human driver in contrast to a steering system of a front axle. Furthermore innovative drive systems could affect the mass moment of inertia of the rotating masses. A wheel hub motor would considerably change the mass moment of inertia. To reduce the possible unwanted gyroscopic effects, an inverse rotation direction of the rotor towards the initial wheel motion using a transmission is conceivable.

References

[1] BILL, KARLHEINZ:
Einführung in die Motorradtechnik
Lecture manuscript, Berlin: FHTW-Berlin, 2000

[2] BISCHOF, GÜNTER ET AL.:
Investigation of a passenger car's dynamic response due to a flywheel-based kinetic energy recovery system
Vol. 52, No. 2, 201-217, Vehicle System Dynamics, 2014

[3] COSSALTER, VITTORE:
Motorcycle Dynamics
2nd English edition, Padova : Lulu.com, 2006

[4] DIN ISO 8855:2013-11:
Straßenfahrzeuge – Fahrzeugdynamik und Fahrverhalten – Begriffe (ISO 8855:2011)

[5] HEIßING, BERND; ERSOY, METIN; GIES, STEFAN (ED.):
Fahrwerkhandbuch
4th edition, Wiesbaden : Springer Fachmedien, 2013

[6] ISO 7401:2011-04:
Road vehicles – Lateral transient response test methods – Open-loop test methods

[7] PACEJKA, HANS B.:
Tire and Vehicle Dynamics
3rd edition, Rotterdam : Elsevier Ltd., 2012

[8] REISNER, MAXIMILIAN:
Abbildung eines Elastokinematikmodells einer Hinterachse einschließlich flexibler Strukturen
Master thesis, Ingolstadt : TH Ingolstadt, 2015

[9] STOFFREGEN, JÜRGEN:
Motorradtechnik: Grundlagen und Konzepte von Motor, Antrieb und Fahrwerk
8th edition, Olching : Vieweg+Teubner Verlag, 2012

[10] THEISSEN, PASCAL:
Unsteady Vehicle Aerodynamics in Gusty Crosswind
Dissertation, Munich : TU Munich, 2012

[11] ZENG, XIAO-HUI ET AL.:
The effect of wheel set gyroscopic action on the hunting stability of high-speed trains
Vol. 55, No. 6, 924-944, Vehicle System Dynamics, 2017

Carbon wheels conquer the market – benefit, state-of-the-art, foresight

Dr. Jens Werner, ThyssenKrupp Carbon Components GmbH

This manuscript is not available according to publishing restriction. Thank you for your understanding.

P.E. Pfeffer (Ed.), *8th International Munich Chassis Symposium 2017*, Proceedings, DOI 10.1007/978-3-658-18459-9_57

Correction to: Steering System Models – An Efficient Approach for Parameter Identification and Steering System Optimization

J. Lin, T. Kloos, P. E. Pfeffer

Correction to:
Chapter 36 in: J. Lin, T. Kloos, P. E. Pfeffer, *Steering System Models – An Efficient Approach for Parameter Identification and Steering System Optimization*
DOI 10.1007/978-3-658-18459-9_36

The author J. Lin was missed in this chapter and it has been corrected.

The updated online version of the original chapter can be found at DOI 10.1007/978-3-658-18459-9_36.

P.E. Pfeffer (Ed.), *8th International Munich Chassis Symposium 2017,* Proceedings, DOI 10.1007/978-3-658-18459-9_58

MIX
Papier aus verantwortungsvollen Quellen
Paper from responsible sources
FSC® C105338

If you have any concerns about our products,
you can contact us on
ProductSafety@springernature.com

In case Publisher is established outside the EU,
the EU authorized representative is:
Springer Nature Customer Service Center GmbH
Europaplatz 3, 69115 Heidelberg, Germany

Printed by Libri Plureos GmbH
in Hamburg, Germany